PRACTISING
HUMAN
GEOGRAPHY

PRACTISING HUMAN GEOGRAPHY

Paul Cloke
Ian Cook
Philip Crang
Mark Goodwin
Joe Painter
Chris Philo

Los Angeles | London | New Delhi
Singapore | Washington DC

First Published 2004

Reprinted 2009, 2012

SAGE Publications Ltd
1 Oliver's Yard
55 City Road
London EC1Y 1SP

SAGE Publications Inc.
2455 Teller Road
Thousand Oaks, California 91320

SAGE Publications India Pvt Ltd
B 1/I 1 Mohan Cooperative Industrial Area
Mathura Road
New Delhi 110 044

SAGE Publications Asia-Pacific Pte Ltd
3 Church Street
#10-04 Samsung Hub
Singapore 049483

British Library Cataloguing in Publication data

A catalogue record for this book is available
from the British Library

ISBN 978-0-7619-7325-6
ISBN 978-0-7619-7300-3(pbk)

Library of Congress Control Number 2003108066

Typeset by C&M Digitals (P) Ltd., Chennai, India
Printed and bound by CPI Group (UK) Ltd, Croydon, CR0 4YY
Printed on paper from sustainable resources

FSC
www.fsc.org
MIX
Paper from
responsible sources
FSC® C013604

Summary of Contents

Contents

Preface

In the last few years, there has been an exciting growth of interest in questions about what we do as human geographers and how we do it. Reflecting the general shift within the social sciences towards a reflexive notion of knowledge, geographers have begun to question the constitution of the discipline – what we know, how we know it and what difference this makes both to the type of research we do and who participates in it with us either as colleagues or research subjects ... An intrinsic part of these debates has been a greater self-consciousness about research methods. (McDowell, 1992a: 399–400)

This book is inspired by these observations from Linda McDowell, about whom we will say more in Chapter 1. While written over a decade ago, they can still be mobilized to frame the concerns of the present book, recognizing as they do a body of writing that was starting to accumulate from the early 1990s as an explicit commentary upon the research methods deployed by human geographers. Over the intervening years 'a greater self-consciousness about research methods' has therefore become more commonplace in the literature, leading to a small industry of textbooks and essays, some very good, on this theme. Even so, there is arguably still much more to be done in this respect, not least to make accessible to a wider audience many of the complex issues bound up in the very acts of 'doing' or, as we like to term it, the *practising* of human geography.

Our purpose in the present book is hence is to take seriously the many tasks entailed in conducting research on given processes and problems, certain types of societies and spaces, and nameable peoples and places. We wish to ask about what is involved in such research, how it happens 'in the field', whether a village under an African sky, a housing estate drenched by Glasgow drizzle or the seeming safety of a local planning department or dusty historical archive. We will on occasion query what precisely is meant by this so-called 'field', but more significantly we will ask about what exactly it is that we do in field locations: what sources are we after, what methods are we using in the process,

and how exactly do we manage to extract 'raw data' ready to be taken away for subsequent detailed interpretation? What kind of interactions occur here, particularly with the people from whom we are often trying to obtain information, whether they be the gatekeepers of sites and documents that we wish to access or, more significantly, the people whose lives in specific spaces and places we are hoping to research (our 'research subjects'). And what goes on once we do get the data back to our office, library or front room: how do we endeavour to 'make sense' of these materials, to bring a measure of order to them, to begin manipulating them to describe and to explain the situations under study, to arrive at the cherished goal of understanding? Moreover, if we get this far, what is then at stake as we try to write through our findings, to offer our interpretations, and as we produce accounts which purport to represent peoples and places more or less different from our own?

Can all this really be as simple as many earlier human geographers tended to assume? Can it all be taken for granted, subsumed under headings such as 'intuition' or the enacting of the 'scientific method', and is it genuinely free from any relationship to the researcher's own values and beliefs, ethics and politics? We would want to answer the latter questions with a definitive *no* and, as the book unfolds, to indicate why we suppose this to be the case by striving to provide answers to many of the previous questions just raised. In line with McDowell's observations, therefore, we are convinced that human geographers do need to become still more self-conscious, more reflexive and more willing to 'reflect back upon' all aspects of their research practices.

We should underline the *specific* contribution that we are aiming to make in this book, then, in that it is designed to provide human geography researchers – from the undergraduate to the more seasoned academic – with an introduction to the many and varied considerations integral to the practising of human geography. We are not reviewing all the near-infinity of possible data sources open to human geographers, since such an attempt has already been made (Goddard, 1983) and there are a number of specialist texts dealing with particular sources such as censuses, inventories and published surveys that may be of use to the human geographer (see various papers in the CATMOG and HGRG series and also the contributions in Flowerdew and Martin, 1997). Neither are we offering a complete 'cookbook' of methods, going systematically through a range of methods in turn and outlining how to do them, although there will be moments in what follows where we dwell on specific methods available for both (as we will say) *constructing geographical data* and *constructing geographical interpretations*. There are already many such 'how-to-do' manuals in the general social science literature, and there is also something of this character in several human geography texts.[1] We would definitely see such texts as complementary to our own, but having a different feel in their focused attention on the nuts and bolts of specific methods through which human geographers both gather and process data.

It might be noted that numerous older geography textbooks lead school children and undergraduate students through hands-on methods of field survey, land surveying and mapping (e.g. Dickinson, 1963), but we must admit to regarding such an interest in what has been termed 'practical geography' (Bygott, 1934) or 'mathematical geography' (Jameson and Ormsby, 1934) as beyond our purview. This is less the case with more recent contributions to the use of statistical, mathematical and GIS procedures in geography, which we do touch upon in Chapters 7 and 8, but we do not assess these in technical detail because such treatments are provided elsewhere by specialist quantitative and GIS geographers (for references, see Chapters 7 and 8). We would also have to acknowledge, perhaps controversially, that we do not think that such developments are *central* to contemporary human

geography. They undoubtedly generate useful 'tools' to be deployed on occasion, and we certainly applaud the notion of combining quantitative and qualitative methods (see also Hodge, 1995; Philo et al., 1998), but we do not see how what are basically technical exercises can be taken as more than a small part of the larger whole which is the practising of human geography. Our own preferences, and maybe prejudices, are of course hinted at by such a statement (and see below). What we should also underline at this point is our rejection of the oft-made assumption that utilising qualitative methods entails a loss of rigour in the research process, a forsaking of the objective, analytical and replicable attributes supposedly integral to deploying quantitative methods within the strictures of the conventional 'scientific method' (see Chapter 9). We resist the accusation that the route to qualitative methods amounts to a 'softening' of human geography, where 'softening' is understood as weakening and making things easier (e.g. Openshaw, 1998). We regard such views as flawed because quantitative work (and science more generally) is just as shot through with human frailties and social conditioning as is qualitative work, a claim increasingly borne out by social studies of science and technology (e.g. Demeritt, 1996), and also because it is possible – as we hope to demonstrate – for qualitative human geography to be practised in a manner combining its own version of intellectual rigour with a responsibility to the realities (not merely the academic's inventions) of the field beyond the academy.

Our intention is also not to provide a compendium of 'stories from the field', relating the experiences of particular human geographers as they have sought to operationalize substantive research projects, although we do recognize the great value of such personalized accounts.[2] We therefore make some use of such experiential materials at various points in what follows. Furthermore, our intention is not to rehearse the complex arguments about either abstract moral stances or 'ethical philosophies' which might be brought to bear in the discipline (see Sayer and Storper, 1997; Proctor, 1998; Smith, D., 1994; 1997; 1998), although we are attuned to more specific issues rooted in the ethics and politics of research practice (e.g. Mitchell and Draper, 1983a; see also Mitchell and Draper, 1983b). This is especially true with respect to claims about the 'positionality' of the researcher, ones which talk about researchers needing to reflect self-critically on the power relations running between them and their research subjects, and we introduce such claims in Chapter 1 and then throughout many of the following chapters.

Neither is our immediate aim to show how practical dimensions of research link with more conceptual orientations, the latter being identified by such daunting terms as 'positivism', 'Marxism', 'humanism', 'feminism', 'postcolonialism', 'poststructuralism', 'postmodernism' and the like. A handful of works do prioritize this linkage, notably Hoggart et al. (2002) in their attempt to tease out the role of epistemological differences – meaning variations in the concepts adopted by different researchers in their attempt to arrive at reliable knowledge about the world – as key influences on what can be achieved in the researching of human geography. Something similar certainly is offered on occasion in the book (giving a link back to a prior text by Cloke et al., 1991; see also Robinson, 1998; Winchester, 2000; Shurmer-Smith, 2002a), and we would insist that some familiarity with the discipline's recent theoretical concerns *is* essential to the self-critical practice of research being urged here. None the less, we will seek to introduce concepts in as gentle a fashion as possible, trying not to assume too much prior knowledge on the part of the reader, and we should make clear that the book arises from six different authors who are themselves differently persuaded (or dissuaded) by the merits of different conceptual orientations. If there is a commonality between us in this regard, it lies in a conviction that the overall conceptual landscape of contemporary human geography is a healthy one, and that there are now available all manner of exciting concepts to guide researchers in their practical enquiries. Such concepts must remain as 'guides', since integral to the whole ethos of this book is the suggestion that researchers should always be reflecting self-critically on every component of their research, concepts included. If given concepts do not appear to 'perform' well in helping the researcher to get to grips with the particularities of an issue, situation, space, place or whatever, then they should be revised or abandoned. (We do realize, though, that recognizing whether or not a concept 'performs' well in this respect is not quite as simple as such a remark might imply.)

From the above listing of what the book is *not*, we hope that an impression may be emerging of what it actually *is*. It is, then, a book that – drawing upon experiences of research, contemplating the ethics and politics of research, and insisting on seeing research practices in the context of wider

conceptual orientations – does aim to *stand back* a little from the hurly-burly of getting actual research projects planned and executed. It wants to inject a pause in the battling with logistics, officials, respondents, tape-recorders, statistical tests, software programs and the like, even though such everyday things at the 'coal face' of research will be mentioned repeatedly. It wants instead to ponder arguably deeper questions about the why, how, what, when and where of the research process, probing more fully into precisely what it is that researchers are searching for and aiming to work upon (this 'stuff' called data), and then discussing the differing strategies for 'making sense' of these data (for forcing them to 'make sense') and for converting them into written-through accounts, findings and conclusions for various audiences (from the dissertation marker to the academic conference participant). It is indeed to explore at some length the dynamics of practising human geography and, in so doing, to offer something distinctive which is neither a treatise on theory in the discipline nor a 'how to do it' manual for disciplinary practitioners, but rather a theoretically informed reflection on the many different twists and turns unavoidably present in the everyday 'doing of it'. Moreover, in part exploring these dynamics should help to inform us, and our readers, when critically reading the written-through research of others.

The chapters that follow take different cuts at this goal, and they differ in how they do this according to the specific concerns of the chapter, the contents of the relevant pre-existing literature, and the particular competencies, interests and theoretical predispositions of the author(s) who have had prime responsibility for each chapter (although we are not going to tell you who did what!). We willingly acknowledge that there is some unevenness between the chapters, some overlaps, doubtless some omissions, a few variations in emphasis, even a few differences of opinion, but such inconsistency is also very much part of both the 'real' human geography (of the world) under study and the 'real' human geographers (from the academy) who struggle to study it.

A 'map' of the book

Following this Preface, Chapter 1 sets the scene for the rest of the book by sensitizing readers to the different ways in which human geography can be practised, first by contrasting the extremes of practice by two semi-fictional human geographers ('Carl' and 'Linda') and, secondly, by sketching out a thumbnail history of changing practices within the 'doing' of (human) geography. While not wishing to imply that older practices were entirely mistaken and have nothing still to teach us, we are critical of the extent to which earlier geographers tended to regard the research process as relatively unproblematic, as either the 'natural' way of proceeding for individuals gifted with geographical insight or the 'logical' way of proceeding if obeying the basic rules of conventional science. In the course of the chapter we introduce a number of key terms which are utilized throughout the book – terms such as 'research', 'field', 'data', 'methodology' and the like – and we also introduce several key themes to do with the heightened attention which human geographers are now showing (rightly, in our view) to the complexities involved in both the 'field' and the 'work' elements of 'fieldwork'.

The heart of the book is divided into two parts, and we propose a division between two fundamentally different sets of practices relating to the treatment of *human geographical data*. By this phrase we simply mean data (in the plural) which have been, are now and could in future be used by human geographers, and it does not necessarily mean data which are obviously related to space, place, environment and landscape (the staple big concerns of academic geography: see Hubbard et al., 2002; Holloway et al., 2003), although in the vast majority of cases such geographical references will be involved somewhere (see Chapter 7). It should be noted too that we will often speak of 'geographical data' rather than of 'human geographical data', but this is merely to avoid the somewhat cumbersome appearance of the latter term, and throughout the book we will almost always mean specifically data pertaining to human or social dimensions of the world. Our discussion will be of little immediate relevance to physical geographers, except in so far that there are overlaps in Chapter 1 with the history of changing practices within the 'doing' of physical geography. We should additionally emphasize that at every turn in the book our concern is for processes of *construction*: accenting that data, how we come by them, and all the many procedures which we then operate upon them, from the most basic of sorting to the most complex of *re*presentations, are all in one way or another constructions found, created and enacted by 'us' (human geography researchers) as people

living and working within specific economic, political, social and cultural contexts. There is nothing natural here; nothing straightforwardly pre-given, preordained or untouched by human emotions, identities, relations and struggles.

Part I is entitled 'Constructing geographical data', and here we tackle two very different varieties of data: first, those data which are 'preconstructed' by other agencies (governments, companies, journalists, poets and many more: see Chapters 2–4) and from which human geographers can then extract materials relevant to their own projects; and, secondly, those data which are 'self-constructed' through the active field-based research of human geographers themselves (when using methods such as questionnairing, interviewing, observing and participating: see Chapters 5 and 6). In the first instance, the focus is very much on how these sources are constructed, on the many different contexts, influences and forces embroiled in the putting together of such sources, which can be both purportedly 'factual' (as in parliamentary reports or news footage) and seemingly 'fictional' (as in novels or films). (We are certainly aware that there is no clear separation between paying attention to how sources are constructed and then trying to interpret what they are telling us, underlining that the boundary between our Parts I and II is not a hard and fast one.) In the second instance, the focus is much more on the roles played by researchers when in the field, and begins to ask about the methods which can be utilized to gain a window on the lives of research subjects, particularly by talking or spending time with them. Quite specific questions hence arise about such methods, about how practically to put them into operation, but so too do a host of questions about the relations which inevitably run between researchers and the researched, thus prompting careful reflection on matters of power, trust, responsibility and ethics (in short, on the micro-politics of engaged research). In addition, we appreciate that even at this stage of research human geographers in the field will begin to write about that field, noting down preliminary findings, their personal experiences and their thoughts on how the research is going and on their interactions with research subjects. In Chapter 6 we briefly discuss this often unremarked feature of the initial research process. It is in the course of such jottings that the seeds of more developed interpretations start to emerge, and it is also at this stage that the 'textualization' of our research – the conversion of it into written forms for wider audiences – begins to occur.

Part II is entitled 'Constructing geographical interpretations', so as to stress that what we are talking about now is indeed still very much a creative act of construction, of making something, and is assuredly not some magical process whereby interpretations drop fully formed out of data. What we are terming 'interpretation' covers various strategies through which human geographers endeavour to 'make sense' of their data, the so-called 'raw data' which they have collected or generated in one way or another, and thereby to provide accounts, arrive at findings and posit conclusions. Somewhat hesitantly, we distinguish between five interpretative strategies that could be conceived of as being complementary, but some of which in practice often end up being positioned as antagonistic by researchers with particular investments in claiming one strategy as fundamentally superior to another. The 'wars' between such strategies is an underlying theme of these chapters, although in large measure we feel such wars to be unhelpful and even an unnecessary distraction from what ought to be the higher goals of arriving at good interpretations. The first of our strategies, which we simply term 'sifting and sorting' (Chapter 7), cannot avoid being present in any study, even if it is rarely given explicit consideration, and it entails the basic activities through which a measure of order is imposed on raw data by the identifying of relevant entities in, to use a shorthand, 'lists and boxes'. The second strategy, which we term 'enumerating' (Chapter 8), is inevitably a follow-on from sifting and sorting, and it entails the whole panorama of numerical methods which are commonly utilized to measure properties of distributions and to detect patterns within data sets. In this chapter we deal with techniques of numerical analysis which have until recently been taken as the core of geographical interpretation, but which we reckon warrant less special consideration than has usually been the case. Although many will not agree, we regard enumeration as essentially a descriptive activity, describing quantitative data sets, their differences from one another and possible relationships, in a manner that requires further steps to be taken in translating back from the formal vocabularies of statistics, mathematics and computing into the ordinary languages familiar to most readers. *Only* when such translation occurs can it be said that the research has shifted from description to something that we, and many if not all other philosophers of science, would accept as something more clearly explanatory.

Our third strategy is therefore what we term 'explaining' (Chapter 9), where we consider what has been the prevailing objective of so much human geography raised in a 'scientific' vein, whether the science be that of a Newton, a Freud or a Marx, wherein successful interpretation (and here the term 'analysis' is often heard) involves being able to answer 'why' questions by specifying the causal processes combining to generate particular human geographical phenomena. Our fourth strategy, which we term 'understanding' (Chapter 10), points to what has recently become a more popular approach with human geographers inspired by the humanities and cultural studies, wherein successful interpretation involves being able to elucidate the meanings that situated human beings hold in regard to their own lives and that inform their actions within their own worldly places. In addition, we appreciate the range of debates which have recently raged over the so-called 'crisis of representation', the deep worries about what exactly happens when academics start to write or to lecture about peoples and places other than their own, and in Chapter 11 we review some of the conventions, rhetorics and other considerations arising as human geographers seek to represent their interpretations to different audiences. Just as all studies cannot avoid containing a moment of sifting and sorting, so all studies, unless never written through in any form, cannot but include and usually culminate in acts of representation. We can regard representation as an interpretative strategy in its own right, if being itself fractured by many different assumptions about the relations running between 'word and world', but in other respects we prefer to regard it as a moment that indeed crops up as an adjunct to the various practices outlined in all of the earlier chapters (including those in Part I of our book).

Finally, we bring things to a close in a chapter (Chapter 12) picking out a particular strain of themes which have been present, albeit not all that expressly tackled, within the preceding chapters. Building upon comments in Chapter 1 about the values of the researcher, notably those with an obvious political edge, this chapter explores the politics of practising human geography, reviewing in particular the thorny questions which surface once researchers begin to reflect upon the entangled politics influencing their decisions about topics to study, data sources to consult and field methods to deploy, and then interpretative strategies to bring to bear on the data derived. Here we pay attention to the politics of the research process itself, notably with respect to the often highly uneven power relations traversing the social realm, the academy and everyday lives that cannot but contextualize this process, energizing but also sometimes constraining a researcher as he or she initiates and seeks to pursue his or her preferred practices of data construction and interpretation. While not suggesting that all researchers should nail their politics clearly to the mast – many of us may not be so certain about our politics and may prefer to allow them to change according to context – we are in no doubt that practising human geographers should offer at least some self-critical reflection on their own, if we can put it this way, 'political investment' in the projects which they conduct.

Paul Cloke
Ian Cook
Phil Crang
Mark Goodwin
Joe Painter
Chris Philo
All over the place, 2003

Notes

1 In the general social science literature see, for example, Blunt et al. (2003), Burgess (1984) and May (1997). For human geography texts see, in particular, Lee (1992), Rogers et al. (1992), Cook and Crang (1995), Flowerdew and Martin (1997), Lindsay (1997), Robinson (1998), Hay (2000), Kitchin and Tate (2000), Limb and Dwyer (2001) and Hoggart et al. (2002).

2 See, for example, Eyles (1988a); see also several pieces in Eyles and Smith (1988), Nast (1994), Farrow et al. (1997), Flowerdew and Martin (1997) and Limb and Dwyer (2001).

Acknowledgements

We have accumulated a mass of debts over the (far too) many years that this book has been in gestation and preparation, and we cannot begin to acknowledge all or even many of them. What we will do, though, is to say a massive thank you to Robert Rojek at Sage for the patience of Job (and then some) in waiting for the disorganized rabble that is the authorial team of this book to get its act together. More particularly, we want to thank David Kershaw for his great contribution to the copy-editing process: his efforts have certainly helped to bring some more discipline to proceedings, and the book has been much improved as a result. Finally, we would like to thank all those at Sage who have been involved in the final production of the book for their hard work in getting the final product to look as attractive (and coherent!) as (we hope that) it does.

Every effort was made to obtain permission for Figures 1.1 and 1.3 and for Box 1.6.

The following illustrations were used with the kind permission of:
Allyn & Bacon (Boxes 5.5, 8.5)
Andrew Sayer (Table 5.1, Figures 9.2 and 9.3)
Association of American Geographers (Figures 9.6 and 9.7)
Blackwell Publishers (Box 7.3 and Figure 7.3)
Cambridge University Library (Figure 4.3)
Cambridge University Press (Figures 7.1 and 7.2)
Continuum International Publishing (Box 5.2)
Meghan Cope (Box 10.1)
The Countryside Agency (Table 8.1)
Paul du Gay (Figure 4.1)
Elsevier Science (Figures 9.8, 9.9)
Hodder Headline plc (Figures 8.3, 8.4, Box 8.9, Figure 9.1)
John Wiley & Sons (Box 8.7)
Mansell Collection (Figure 3.6)
Linda McDowell (Figure 1.2)
The National Gallery (Figure 4.5)
Nelson Thornes (Box 7.2)
Open University Press (Box 5.1)
Pearson Education Ltd (Box 8.6, Figure 8.2)
Philips Maps (Figure 1.4)
Pion Ltd (Box 8.3)
Syracuse University Press (Figure 5.1)
Taylor & Francis Books Ltd (Figure 4.2, Table 5.2, Figure 5.3, Boxes 5.3, 5.4, 5.7, 8.8)
Verso (Figure 3.7)
W.W. Norton & Co. (Box 8.4)

1

The Changing Practices of Human Geography: An Introduction

Practising human geography?

Let us begin by imagining two different human geographers, one we will call Carl and the other Linda. Both these are human geographers who believe that at least part of what human geography entails is the actual 'practising' of human geography: the practical 'doing' of it in the sense of leaving the office, the library and the lecture hall for the far less cosy 'real world' beyond and, in seeking to encounter this world in all its complexity, to find out new things about the many peoples and places found there, to make sense of what may be going on in the lives of these peoples and places and, subsequently, to develop ways of representing their findings back to other audiences who may not have enjoyed the same first-hand experience. Both of them are enthralled, albeit sometimes also a little daunted, by everything that is involved in this practical activity. Both of them are convinced there is an important purpose in such activity, both because it enriches their own accounts and because it can produce new 'knowledge' which will be eye-opening, thought-provoking and perhaps useful to other people and agencies (whether these be other academics, students, policy-makers or the wider public). For both of them, too, this practical activity

is something they usually find enjoyable, fun even, and both of them would wish to communicate this importance and enjoyment of practising human geography to others. Yet Carl and Linda go about things in rather different ways, and it is instructive at the outset of our book to consider something of these differences.

For Carl, the approach is one which does very much involve packing his bags, leaving his home, locking the office door and heading out into the 'wilds' of regions probably at some distance from where he normally lives and works. In so doing he tries, for the most part, to forget about all the aspects of his life and work tied up with the home and the office: to forget about his social and institutional status as a respected member of the community and senior academic, to forget about his relationships with family, friends and colleagues, to forget about the books, reports and newspapers which he has been reading, and to forget about the concerns, troubles, opinions, politics, beliefs and the like which usually nag at him on a day-to-day basis. In addition, he is determined to leave with an open mind, with as few expectations as possible, and even with no specified questions to ask other than some highly generalized notions about what ought to interest geographers on their travels. Instead, his ambition is to become immersed in a whole new collection of peoples and places, and to spend time simply wandering around, gazing upon and participating in the scenes of unfamiliar environments and landscapes. He might occasionally be a little more proactive in chatting to people, perhaps farmers in the fields as he passes, and sometimes he might even count and measure things (counting up the numbers of houses in a settlement or fields of terraced cultivation, measuring the lengths of streets or the dimensions of fields). From this engagement, as Carl might himself say, the regions visited begin to 'get into his bones': he starts to develop a sense of what the peoples and places concerned 'are all about', a feel which is very much intuitive about how everything here 'fits together' (notably about how the aspects of the natural world shape the rhythms of its cultural counterpart or overlay), and an understanding of how the local environments work and of why the local landscapes end up looking like they do. The impression is almost of a 'magical' translation whereby, for Carl, meaningful geographical knowledge about these regions is conjured from simply being in the places concerned, formulated by him as the receptive human geographer from activities which are often no more active than a stroll, the drawing of a sketch, the taking of a photograph and the pencilling of a few notes. And the magical translation then continues, perhaps on return to his office, when Carl begins to convert his thoughts into written texts for the edification and education of others, and through which his particular feel for the given peoples and places is laid out either quite factually or more evocatively. Taken as whole, this is Carl's practising of human geography.

For Linda, the approach is arguably rather more complicated. She is much less certain about being able to manage a clean break from her everyday world as anchored in her home, her office and her own social roles and responsibilities, nor from her prior academic reading, and nor from the accumulated baggage of assumptions, motivations, commitments and formalized intellectual ideas which swirl around in her head. Moreover, her *research* practice, her fieldwork, may not take her physically all that far away from the home or her office: she might end up researching peoples and places that are almost literally just next door, or at least located in the estates, shopping centres, business premises and so on, in a nearby city. The separation of everyday life from the *field*, the regions under study, which Carl can achieve, is not possible for Linda: indeed, it is also a separation about which she might be critical. And, whereas Carl aims to go into the field as a kind of 'blank sheet', Linda's approach depends on having a much more defined research agenda in advance, not one that entirely prefigures her findings but one that will incorporate definite research questions based around a number of key issues (perhaps connected to prior theoretical reading). Like Carl, though, she does wish to become deeply involved with particular peoples and particular places (which might be very specific sites such as 'the City', London's financial centre and its component buildings, rather than the much larger regions visited by Carl). She does want to get to know the goings-on in these micro-worlds, to become acquainted with many of the individuals found in these worlds and to try as hard as possible to tease out the actions, experiences and self-understandings of these individuals in the course of her research. The implication is that what she does *is* very much 'hard graft' research, since she has to be extremely proactive in deploying specific research tools – perhaps questionnaires, but more likely a mixture of documentary work, interviewing and participant observation (all of which will be covered in our book) – so as to generate a wealth of *data* which will enable her to arrive at specific interpretations pertaining to the issues (or, to put in another way, at clear answers to her initial research questions). There is perhaps less the magical quality of Carl's approach, therefore, in that the labour allowing Linda to complete her research is much more evident and probably rather more bothersome, wearisome and even upsetting. The labour also continues to be apparent at the writing–up stage in that Linda reckons it vital to include sections explicitly on the *methodology* of the research, including notes on its pitfalls as well as its advantages, alongside debating at various points the extent to which someone like her – given just who she is, her social being and academic status – can ever genuinely find out about, let alone arrive at legitimate conclusions regarding, the issues, peoples and places under study. Taken as a whole, this is Linda's practising of human geography: it differs enormously from Carl's.

You should notice that several terms in the last paragraph are italicized, and Boxes 1.1–1.4 define and expand upon the meanings of these terms. They are crucial to the book, and

you should ensure that you understand them before proceeding. They are also crucial to our introduction, which will now continue by making Carl and Linda more real. We have talked about them so far as fictional characters through which we could illustrate different approaches to the practising of human geography, but we should also admit to having in mind two real human geographers, one past and one living, who are Carl Sauer and Linda McDowell. Carl Sauer (see Figure 1.1) was a geographer based for virtually all his career in the Berkeley Department in California, and his chief interests lay in the 'cultural history' of long-term inter-relationships between what he termed the 'natural landscape' and the 'cultural landscape', and in teasing out distinctive patternings of human culture as revealed in the mosaic of different material landscapes produced by different human activities (agricultural practices, settlement planning, religious propensities).[1] For the most part, Sauer disliked statements about both theory and methodology (although see Sauer, 1956), and he tended to regard the practising of human geography (and indeed of geography more

Box 1.1: Research

This term describes the overall process of investigation which is undertaken on particular objects, issues, problems and so on. To talk of someone conducting *research* in human geography is to say he or she is 'practising' or 'doing' his or her discipline, but it also carries with it the more specific sense of a sustained 'course of critical investigation' (POD, 1969: 703) designed to answer specific research questions through the deployment of appropriate methods. The ambition is to generate findings which can be evaluated to provide conclusions, and usually for the whole exercise to be reported to interested audiences both verbally and in writing. It contains, too, the suggestion that the exercise will be conducted in a manner critical of its own objectives, achievements and limitations, although we will argue that, by and large, human geographers have been insufficiently self-critical in this respect. The term 'research' is now very widely used in the discipline (e.g. Eyles, 1988a), and its relative absence from earlier geographical writing suggests that geographers prior to c. the 1950s and 1960s were less attuned to the notion of producing geographical knowledge through premeditated and structured procedures.

Box 1.2: Field

This deceptively simple term – the *field* – normally refers to the particular location where research is undertaken, which could be a named region, settlement, neigh-bourhood or even a building, although it can also reference what is sometimes called the 'expanded field' (as accessed in a few studies) comprising many different locations spread across the world (see also Driver, 2000a; Powell, 2002). We would include here, too, the libraries and archives wherein some researchers consult documentary sources, which means that we are also prepared to speak of historical geographers researching 'in the field'. In addition, we suggest that the field should be taken to include not only the material attributes of a location, its topography, buildings, transport links and the like, but also the people occupying and utilizing these locations (who will often be the research subjects of a project). As such, the human geographer's field is not only a 'physical assignation', but it is also a thoroughly 'social terrain' (Nast, 1994: 56–7), and some feminist geographers (e.g. Katz, 1994) have extended this reasoning to insist that a clean break should *not* be seen between the sites of active research and the other sites within the researcher's world (a claim elaborated at the close of this chapter). This being said, we do wish to retain some notion of the field as where research is practi-cally undertaken, but we fully agree that *fieldwork* must now be regarded as much more than just a matter of logistics. Instead, fieldwork should be thought of as encom-passing the whole range of human encounters occurring within the uneven social ter-rain of the field, in which case it is marked as much by social 'work' as by the practicalities of getting there, setting up and travelling around.

Box 1.3: Data

'Data are the materials from which academic work is built. As such they are ubiquitous. From passenger counts on transport systems to the constructs used in the most abstract discussion, data always have a place' (Lindsay, 1997: 21). *Data* (in the plural) hence comprise numberless 'bits' of information which can be distilled from the world around us and, in this book, we tend to think of data, or perhaps 'raw data', as this chaos of information which we come by in our research projects (whether from the physical locations before us, the words and pictures of documentary sources, the state-ments made in interviews and recorded in transcripts, the observations and anecdotes penned in field diaries, or whatever). As we will argue, a process of *construction* necessarily occurs as these data are extracted from the field through active research, ready for a further process of *interpretation* designed to 'make sense' of these data (to substitute their 'rawness' with a more finished quality). Various kinds of distinction are made between different types of data (see also Chapter 7), the most common of which is that between *primary* and *secondary data*. The former is usually taken as data gen-erated by the researcher, while the latter is usually taken as data generated by another person or agency, but we restate this particular distinction in terms of *self-constructed* and *preconstructed data* (see also the Preface and below). For us, therefore, primary data should be taken to include everything which forms a 'primary' input from the field into a researcher's project (i.e. anything which he or she has not him- or herself yet interpreted). These data can include highly developed claims made in a govern-ment report or well-thought-out opinions expressed by an interviewee, in effect inter-pretations provided by others, but they remain primary data for us because the researcher has not yet begun to interpret them. We do not really operate with a notion of secondary data, therefore, except in so far as we might reserve this term for the interpretations of primary data contained in the scholarly writings of other academics.

Box 1.4: Methodology

'In the narrowest sense, [methodology is] the study or description of the methods or procedures used in some activity. The word is normally used in a wider sense to include a general investigation of the aims, concepts and principles of reasoning of some discipline' (Sloman, 1988: 525). On the one hand, then, there are the specific *methods* which a discipline such as human geography deploys in both the construction and the interpretation phases of research (including such specific techniques as measuring, interviewing, statistical testing and coding). On the other hand, there is the *methodology* of a discipline such as human geography that entails the broader reflections and debates concerning the overall 'principles of reasoning' which specify both how questions are to be posed (linking into the concepts of the discipline) and answers are to be determined (pertaining to how specific methods can be mobilized to provide findings which can meaningfully relate back to prior concepts). For some writers (including geographers: e.g. Schaefer, 1953; Harvey, 1969) there is little distinction between methodological discussion and what we might term 'philosophizing' about the basic spirit and purpose of disciplinary endeavour, but we prefer to regard methodology in the sense just noted, and hence as a *standing back* from the details of specific methods in order to see how they might 'fit together' and do the job required of them. In this sense, our book is most definitely a treatise on methodology.

generally) as something fairly obvious, coming 'naturally' to those who happened to be gifted in this respect. Linda McDowell (see Figure 1.2) is a geographer presently based at University College London, and her chief interests lie in the insights that feminist geography can bring to studies of 'gender divisions of labour' as these both influence the spatial structure of the city and enter into the day-to-day gendered routines of paid employment, in the latter connection paying specific attention to senior women employed in the London-based financial sector.[2] While McDowell has not written extensively about methodology, she has contributed significantly to the debates currently arising in this connection (see 1988; 1992a; 1999: ch. 9), and it is apparent that for her the practising of human geography is something necessitating considerable 'blood, sweat and tears'.

Our reasons for now fleshing out the human geographers who are 'Carl' Sauer and 'Linda' McDowell are various and, at one level, simply emerge from a wish to emphasize

that human geography is always produced by individual, flesh-and-blood nameable people whom you can see and perhaps meet. They could be you! But at another level, the differences between 'Carl' Sauer and 'Linda' McDowell are highly relevant to the broader arguments which we are developing in this introductory chapter. Indeed, in what follows we take Sauer and McDowell as exemplars of two very different ways of practising human geography which 'map' on to, respectively, older and newer versions of human geographical endeavour that can be identified within the history of the discipline. We must be circumspect about such a mapping: a Sauer-esque approach is still very much with us today, partly in the continuing works of regional synthesis and description carried out by many who regard this as the highest expression of the 'geographer's art' (Hart, 1982; Meinig, 1983; Lewis, 1985); while a McDowell-esque approach does have its historical antecedents in the use of certain clearly defined methods, such as questionnaires and interviews, long

5

From photograph by K. J. Pelzer, September, 1935.

Figure 1.1 Carl Ortwin Sauer

Source: From Leighley (1963: frontispiece)

before the current eruption of interest in putting such methods at the heart of human geographical research (see below). Yet, we believe that there *is* still some truth in the proposed mapping, and that a profound change *has* occurred in how human geographers envisage and proceed with their practising of academic research: a change which *can* be indexed by contrasting the likes of Sauer and McDowell. By the same token, we wish to resist the impression that older approaches are bad whereas newer approaches are good, an impression readily conveyed by 'presentist' accounts which project a narrative of things steadily improving, progressing even, from a worse state before to a better state now. This means that we still find there is much of value in an older Sauer-esque orientation, in the

Figure 1.2 Linda McDowell

Source: Courtesy of Linda McDowell

ideal of suspending one's everyday and academic concerns in the process of becoming immersed in the worlds of very different peoples and places, and in no sense are we seeking to encourage an 'armchair geography' unaffected by the wonderment, hunches and ideas which strike the human geographer in the field. Yet it would be wrong to deny that we are more persuaded by McDowell than we are by Sauer, and that the basic purpose of our book is very much inspired by the likes of McDowell – complete with her insistence on the labour, messiness and myriad implications of actual research practices, all of which must be carefully planned, monitored, evaluated and perhaps openly reported – than it is by the more intuitive, magical, 'just let it happen' stance adopted by the likes of Sauer.

A thumbnail history of practising human geography

Leading from the above, and to frame what follows in our book, we now want to chart

something of the history of changes in the practising of human geography. It is only recently that serious attention has been paid to 'aspects of disciplinary practice that tend to be portrayed as mundane or localised, but that represent the very routines of *what we do*' (Lorimer and Spedding, 2002: 227, emphasis in original). Various authors are now claiming that we fail to appreciate much about our discipline without recognizing that 'geographical knowledge [is] constituted through a range of embodied practices – practices of travelling, dwelling, seeing, collecting, recording and narrating' (Driver, 2000a: 267). They further worry that many of our 'knowledge-producing activities', old and new, remain largely absent from how we represent our research, suggesting that 'our products of knowledge (our texts and even our emphases in conversations of recollection) could do more to make available this tension of the present tense of the world' (Dewsbury and Naylor, 2002: 254): meaning precisely the fraughtness of our actual practices as *we do them*. It is in the spirit of trying to make more visible such practices, and in so doing to assess them critically, that we now turn to our thumbnail history.

The history relayed here is not intended to be a comprehensive one, particularly since more historiographic research is required to clarify the details of how human geography (and also geography more generally) has been practised by different practitioners during different periods and in different places. (And note that active research is required to find out about this history, even if it be research whose 'field' is the archive and whose 'data' chiefly comprise the yellowing pages of writings, maps and diagrams produced by past adventurers, explorers and academic geographers: see Boxes 1.2 and 1.3.) Our history should also be read in conjunction with other works more specifically concerned with the history of geographical inquiry (Cloke et al., 1991;

Livingstone, 1992; Peet, 1998). The history that we tell will be somewhat arbitrarily separated into three different, roughly chronological, phases: we focus chiefly on the first of these, for which Sauer is an exemplar in his preference for immersed observation; and then on the third of these, for which McDowell is an exemplar in her preference for what we will term reflexive practice based as much on listening as on looking. Reference will also be made to a 'middle' phase in which the practising of human geography did begin to be problematized, rather than regarded as intuitively given, and here we will mention the rise of a 'survey' impulse which ended up being hitched to a particular (and we will argue narrowly) scientific orientation. For each phase, we will outline the basic details of what the geographers involved were doing and arguing, before switching to offer some more evaluative comments about pluses and minuses that we perceive in their practices.

'Being there' and 'an eye for country'

Probably the most longstanding tradition within the practising of (human) geography, albeit one rarely considered all that explicitly, has been one which makes a virtue out of the geographer being personally present in a given place and thereby able to observe it directly through his or her own eyes. There are two interlocking dimensions to this tradition: the travelling to places within which the geographer can become immersed, surrounded by the sights, smells and other sensations of the places involved; and then the actual act of observation, the gazing upon these places and their many components, the peoples included.

Taking the first dimension of 'being there', few would dispute that the very origins of something called 'geography' lie in the earliest travels which people from particular localities began to make to visit other places and peoples further away, and in how such people subsequently returned to convey their 'geographical' discoveries of these distant places and people to their own kinsfolk. H.F. Tozer (1897) duly suggests that the origins of 'ancient geography' must be found here, and he stresses the impetus for particular societies – notably Ancient Greece – to 'trace the increase of the knowledge which they possessed of various countries – of their outline and surface, their mountains and rivers, their products and commodities' (Tozer, 1897: 1–2). Although it is unlikely that the ancient geographers such as Strabo would have reported entirely on the basis of what they 'saw taking place before their eyes' (Tozer, 1897: 2), they probably aimed to witness as much as possible and then to base the rest of their work on the first-hand observations of other travellers. Indeed, it is probably not too fanciful to propose that a fairly direct lineage can be traced from these earliest geographers, many of whom must have been intrepid adventurers, through to the vaguely 'heroic' figure – almost a kind of 'Indiana Jones' character constantly journeying to distant lands – which may still be associated with the role of the geographer in the popular imagination.

Even in more academic circles such a notion is not entirely absent, most notably in the powerful motif of the geographer as 'explorer-scientist' which many (especially Stoddart, 1986; 1987) see as capturing the essence of academic geography's origins and continuing purpose. Leading from the 'Age of Reconnaissance' (c. 1400–1800) when voyages of discovery were attended by a gradual recovery of the lost navigational skills of the ancients, an academic 'geographical science' or 'scientific geography' began to take shape (Kimble, 1938; Bowen, 1981;

Livingstone, 1992: chs 2 and 3). By the eighteenth and nineteenth centuries, European explorers such as Captain Cook were regularly taking scientists who talked of 'geography' on their excursions, while 'geographers' such as Humboldt were themselves mounting remarkable expeditions to the likes of Middle and South America. Through the endeavours of such individuals, specifically their attempts at accurate scientific description, measurement and specimen collection, the field-based production of geographical knowledge became more systematized, rigorous and the herald of a formally instituted academic discipline (taught in universities and boasting its own societies and journals: see Bowen, 1981; Capel, 1981; Stoddart, 1986: chs 2, 7–10; Livingstone, 1992: chs 4–7). Furthermore, organizations such as Britain's RGS (Royal Geographical Society) began to provide detailed guidance to explorer-geographers, offering more than just 'hints to travellers' in specifying procedures of description, measurement and mapping which would enable reliable geographical findings to be procured from their sojourns overseas (Driver, 1998; 2000b). Many controversies attached to this phase in geography's history, however, and considerable debate surrounded the extent to which geographical knowledge derived from the explorations could be trusted. Arguments duly raged both then and more recently over issues such as the value of writings by 'lady-travellers' (Domosh, 1991a; 1991b; Stoddart, 1991) and as to how to regard the bellicose activities of explorers such as Stanley who appeared to be more agents of empire (and European conquest) than exponents of geographical science (Driver, 1991; 1992; Godlewska and Smith, 1994). None the less, the undisputed core of this growing body of knowledge which was increasingly identified as academic geography remained the simple fact of 'being

there', of being present in the places, often far-flung, under examination.

Such a notion has continued to be central to academic geography, and to give one instance it is interesting to read Robert Platt's 1930s espousal of a 'field approach to regions' which wilfully set its face against those in North American geography who were then proposing some system to the practising of geography (see below). Sparked by a strong feeling that one should '[g]o to the field when the opportunity arises without worrying over lack of preparation' (Platt, 1935: 170), he recounted an expedition with students to the regions between James Bay and Lake Ontario in Canada which yielded impressionistic senses of these regions rather than guaranteed accurate findings. He thereby produced a species of regional geography organized as a narrative of the journey, reporting on what had been encountered en route as a window on phenomena such as forestry and trading patterns, and in so doing he offered an almost anecdotal evocation clearly spurred by personal experience of the sites and sights encountered:

There were no signs of human occupance nor animals of respectable size. The air was bright and warm, and the scene pleasant except for one item which spoiled an otherwise agreeable environment: swarms of insects from which we had no means of escape, a few mosquitos and innumerable vicious flies. (1935: 153–4)

In this context it is appropriate to return to Sauer, since he too evidently supposed that 'being there' was essential for the good geographer, a claim that surfaced in an early piece when declaring that 'less trustworthy' are sources 'which have not been scrutinised geographically at first hand' (1924: 20). Here Sauer's favouring of field-based study, one predicated on being immersed in the peoples and places under study, was loudly exhorted, and precisely the same sentiments resurfaced over 30 years later when he stressed the need to be 'intimate' with regions being researched 'in the course of walking, seeing and exchange of observation' (1956: 296).

We will revisit the point about observation shortly, but for the moment it is revealing to add that Sauer echoed Platt in proposing a more informal engagement with places free from too many of the trappings of formal regional 'surveying' (see below):

> To some, such see-what-you-can-find field-work is irritating and disorderly since one may not know beforehand all that one will find. The more energy goes into recording predetermined categories the less likelihood is there of exploration. I like to think of any young field group as on a journey of discovery, not as a surveying party. (1956: 296)

The ideal, he added, was to be in the field achieving 'a peripatetic form of Socratic dialogue about qualities of and in the landscape' (Sauer, 1956: 296). It is also intriguing that, while noting the tradition whereby the geographer 'goes forth alone to far and strange places to become a participant observer of an unknown land and life' (1956: 296), he insisted as well on 'the dignity of study in the superficially familiar scene' (1924: 32) much closer to home. This line of reasoning, which found a wonderfully British inflection in the stress on student studies of 'field geography' and 'local geography' (see Box 1.5), has since led to the emphasis in many undergraduate geography courses on running field classes and field days within an immediate region or country (while overseas field trips may be seen as inheriting the more 'global' aspects of claims made by Sauer and his like). Another result has been papers considering the different forms of locomotion that geographers might employ when on active fieldwork, as in Salter's (1969) neglected note about the value of 'the bicycle as a field aid'!

Box 1.5: 'Field geography' and 'local geography'

In 1945 Charlotte Simpson published a paper entitled 'A venture in field geography', summarizing ten years of 'local geography' fieldwork undertaken by school children and undergraduate students in one particular Gloucestershire village. She stressed the role of 'observational work', based on a field walk taking in a 'viewpoint commanding a ... larger area' (1945: 35), and she outlined her sense of the discipline as 'an intensely practical subject, dealing with realities which can be experienced at first hand' (1945: 43). This paper indexed a whole tradition of running locally based fieldwork for younger geographers, and the mid-century British geographical literature is awash with notes and guides regarding fieldwork in schools.[3] The establishment in 1943 of something called the Field Studies Council (Jensen, 1946) was important here in promoting the ability of 'reading a landscape' (Morgan, 1967: 145), initially publishing a series of countryside *Field Study Books* (Ennion, 1949–52) and from 1959 sponsoring a specialist journal called *Field Studies* (wherein geographers have often published papers). While someone like Coleman was seemingly obsessed by the need to make small children take long walks in the countryside, other writers had a clear sense akin to that of a Sauer or a Wooldridge about why such activity keyed into the core concerns of the discipline: 'the landscape is our subject matter, so we must look at it first hand as well as through the media of books, films and maps ... The need is simple and should not be expressed in quasi-philosophical terms' (E.M.Y., 1967: 228).

The second dimension mentioned above, to do with observation, is obviously tightly linked to the theme of 'being there'. It, too, has certainly been a feature of geographical inquiry down the ages, given that the whole stress on the witnessing of distant lands which became codified in the RGS's 'how to observe' field manuals (Driver, 1998; 2000b) hinges upon the expectation that the individuals involved – whether lay folk, professional voyagers or academic geographers – will be able to see, to look, to gaze upon the peoples and places visited. Most of the more methodological remarks which can be found in the earlier literature of academic geography hence concentrate on the observation issue, and it is telling to recall Platt's simple statement that, once in the field near James Bay, 'we opened our eyes and looked around' (1935: 153). Sauer is again a sure litmus for the prevailing wisdom: 'Geographic knowledge rests upon disciplined observation and it is a body of inferences drawn from classified and properly correlated observations … We are concerned here simply with the relevance of the observations and the manner in which they are made' (1924: 19).

We will shortly review Sauer's reference to both classification and 'properly correlated observations', but at this point let us move to similar statements in his 1950s paper pivoting around the remark that the 'morphologic eye' allows the geographer to evince 'a spontaneous and critical attention to form and pattern' (1956: 290):

> The geographic bent rests on seeing and thinking about what is in the landscape, what has technically been called the content of the earth's surface. By this we do not limit ourselves to what is visually conspicuous, but we do try to register both on detail and composition of scene, finding in it questions, confirmations, items or elements that are new and such as are missing. (1956: 289)

> Underlying what I am trying to say is the conviction that geography is first of all knowledge gained by observation, that one orders by reflection and reinspection the things one has been looking at, and that from what one has experienced by intimate sight comes comparison and synthesis. (1956: 295–6)

Sauer also described the propensity for geographers 'to start by observing the near scenes' (1956: 296), before making the above-mentioned comment about going forth 'to become a participant observer of an unknown land and life' (1956: 296). More recently, and quoting one of the passages above from Sauer, C.L. Salter and P. Meserve have advocated geographers compiling 'life lists' of their accumulated field visits and the like, concluding in the process that:

> To be a real geographer, one must observe. There is great power in observation. For a geographer, there are few skills more important in intellectual growth than the development of the ability to 'see what's there'. The act sounds so very innocent, yet being able to discern patterns on the horizons, what components make up the whole, significance in details, and whole from its parts, represents a critical geographic skill. (1991: 522–3)

We will also return to Salter and Meserve in a moment.

The British literature is full of similar claims advocating the centrality of observation to the geographer's craft, notably in the writing of Sidney Wooldridge, where he celebrates what he describes as 'an eye for country', which should be encouraged in geographers from an early age:

> I submit that the object of field teaching, at least in the elementary stage, is to develop 'an eye for country' – ie. to build up the power to read a piece of country. This is distinct from, though plainly not unrelated to, 'map-reading'. The fundamental principle is that the ground not the map is the

primary document, in the sense in which historians use that term. From this first principle I pass to a second, that the essence of training in geographical fieldwork is the comparison of the ground with the map, recognising that the latter, at its best, is a very partial and imperfect picture of the ground, leaving it as our chief stimulus to observe the wide range of phenomena which the map ignores or at which it barely hints. (1955: 78–9)

A class of young geographers, taken to a viewpoint in the field, should not be made to pore over a map 'instead of concentrating on the work of looking and seeing' (Wooldridge, 1955: 79), and the unequivocal message was that 'eye and mind must … be trained by fieldwork of laboratory intensity' (1955: 82). In arguing this way, Wooldridge also insisted that it was vital for refined powers of observation to be inculcated in young geographers through fieldwork in the 'little lands' close to home, and that the transmission of core geographical skills 'lies in the development of the laboratory spirit and the careful, indeed minute, study of limited areas' (1955: 80). Such beliefs clearly urged the value of 'being there', and provided an even more forceful assertion than did Sauer of the need for observation-based fieldwork in the geographer's immediate locality. These were dominant themes in mid-century British geography, informing the 'field' and 'local geography' initiatives which emerged in schools (see Box 1.5), and they also featured in the efforts of something called the 'Le Play Society', alongside its initially student offshoot called the 'Geographical Field Group', which sought to encourage British professional geographers in the conduct of rigorous fieldwork (Beaver, 1962; Wheeler, 1967). One ambition of the latter society was to get geographers out of libraries, to curtail the practice of many which involved little more than synthesizing facts about regions from second-hand library sources, and to foster in them an imitation of

geologists and botanists in achieving 'the highest qualities of observation and faithful recording in the field' (Beaver, 1962: 226). Moreover, much was made of the role of observing landscapes in the field by an individual called G.E. Hutchings, who wished to blend into geography the skills of the 'field naturalist', and also sought to provide some rigour to geographers in the oft-promoted but rarely discussed art of landscape drawing (Hutchings, 1949; 1960; 1962: see Box 1.6).

Having laid out something of this highly pervasive emphasis on 'being there' and its associated 'eye for country', we should now acknowledge that we see many problems with such a stance on the practising of human geography (our criticisms would not necessarily apply to such a stance on the practising of physical geography). This being said, we should emphasize again that in no way would we wish to deny the importance of spending time in the field, immersed in the worlds of particular peoples and places, and neither would we want to underplay the importance of careful observation. Indeed, at various points in the book we will have much to say about such matters, albeit expanding on them in ways which Sauer, Wooldridge and the like probably would not recognize. There are significant criticisms to be made, however, in part to anticipate the alternative proposals of a more recent turn in the practising of human geography.

Thinking first about 'being there' and, while not wishing to commit us to staying in our armchairs, there is perhaps a certain arrogance in the assumptions of many older geographers about their supposed right to be able to travel widely, to visit wherever they wanted and to do their geographical research wherever they alighted. Such an arrogance also arises when the likes of Stoddart (1986) contemptuously dismiss the likes of Wooldridge for suggesting that much fieldwork should

Box 1.6: Field studies and landscape drawing

G.E. Hutchings (1962: 1) once declared that he sought 'to relieve the bookishness of education with practice in observation and exploring out of doors', thereby explaining his preference for combining geography and natural science through the medium of boots-on field study. He emphasized the importance of *landscape* as 'something that has to be viewed, whether scientifically or aesthetically', and insisted that 'it is very necessary for the geographer to acquire by training in the field what Prof. Wooldridge calls the "eye for country"' (1949: 34). And again, he stated that geography 'is a kind of learning arising in the first place from curiosity about the visible and tangible world, and requiring a capacity for looking beyond the superficial appearance of things' (1962: 2). Revealingly, given Hutchings's clear belief in geography as an observational pursuit, he published a book on landscape drawing which was directed particularly at the needs of geographers, in the course of which he gave technical hints about how to produce a sketch which 'is an honest picture of a piece of country, drawn with close attention to the form of its parts and the appearance of the various objects in it according to the effects of light and perspective' (1960: 1).

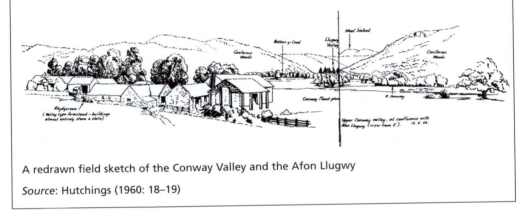

A redrawn field sketch of the Conway Valley and the Afon Llugwy

Source: Hutchings (1960: 18–19)

take place near to home, asserting instead that the wider world should be the geographer's province. For many geographers the belief that the world is *their* 'oyster' has never been questioned, and the possibility that large portions of it are really *somebody else's* world is not one that is often addressed. We are not so much talking here about the complication given by national borders or legal 'ownership' of land, although these can both be pertinent in some human geographical research, but, rather, we are talking about how the field – the specific places to be visited, including the human settlements, homes, workplaces and sites of recreation – is somewhere that we should perhaps be more hesitant to enter than we have often been in the past. These are places where other people do live, striving to scrape a living and to make a life, and these people may pursue all sorts of activities which they would want to keep private from the prying intrusions of outsiders (and a host of considerations duly arise about the preservation of cultural difference, the guarding of spiritual and religious mores, the keeping secret of illegalities and so on). In addition, the places involved might be ones which hold deep meanings for those people who occupy and make use of them, and the presence there of outsiders, particularly intrusive ones taking photographs and writing notes, could be greatly resented.

Figure 1.3 White 'explorer-geographers' being carried across a river by black 'native' bearers

Source: From Brice and Bain (no date, *c.* 1918: 34)

Objections on these counts to the geographer as intrusive alien are increasingly coming from development geographers, persuaded in part by the criticism that geographers working in Africa, Asia, South America and elsewhere effectively reproduce the same structural relationship with 'native' peoples as had arisen in the expeditions of the colonial explorer-geographers from earlier centuries: a relationship in which power, influence and assumptions of superiority lie with the white geographers appropriating knowledge, labour and skills from the peoples of colour in these places (e.g. Sidaway, 1992; Madge, 1993; Powell, 2002). The relationship in question is neatly illustrated in Figure 1.3, which shows a white explorer-geographer, perhaps Stanley, being carried across a river on the shoulders of black bearers. We guess that nothing like this happens in the research of today's development geographers, but is the presence of (say) Anglo-American researchers in the places of their black research subjects so completely free of all the inequalities and embedded assumptions which are coded into this illustration?

The seeming innocence of just 'being there' can also be questioned in situations where geographers are researching closer to home, since the activity of strolling into (say) a Cornish fishing village or an Alpine skiing resort to commence the work of immersed observation is surely one that many people in these places – whether local villagers or people on holiday – might regard as an unwanted imposition. Moreover, while some human geographers have now started to study marginal groupings such as ethnic minorities, children and the elderly, 'Gypsies' and other travellers and so on, it is certainly not obvious that the researcher arriving in the places of such groupings is a good thing for them. It does comprise an intrusion and an imposition, one that may be deeply disturbing to the individuals and families involved, and one which could have dire consequences if researers made public certain information about their precise locations, movements and place-related activities (a concern constantly expressed by David Sibley in his research on travellers: 1981a; 1985). We realize that human geographers will want to continue doing engaged work that requires them to be present in the situations of other peoples and places, and we fully support this important aspect of research, but we note too that – following the examples of critical development geographers, McDowell, Sibley and many others – the apparent rightness of such research practice can no longer be straightforwardly assumed. Rather, the picture must become one of researchers negotiating *access* to peoples and their places, both by formally liaising with the peoples concerned and by thinking much more carefully than hitherto about the politics of 'being there' as bound up with the differing origins, backgrounds, attributes and social standings of the human geographer relative to these peoples and places. We will return to such access issues again in this chapter, and then again later in the book.

Turning now to the issue of 'an eye for country', it should be explained that there is now a sustained critique of the pervasive 'occularcentricism' of much conventional geographical inquiry. Acknowledging this critique forces a thorough-going reappraisal of the obsessive advocacy of observation which figures in many of the statements quoted above. There are various strands of this critique, but all of them converge on what is entailed in geographers setting themselves up as privileged observers able to gaze upon – and, more dubiously, to gaze *down* upon – peoples and places laid out before their eyes like so many exhibition entries. Several historians of the discipline have begun to examine the observational technologies which geographers have deployed, highlighting the extent to which visual images of landscapes are themselves not so much innocent factual records as fabricated or 'staged' representations. David Livingstone (1992: 130–3) assesses the 'artistic vision' which served to compose many of the observations taken by eighteenth-century explorer-geographers, discussing the tensions which existed for both scientific illustrators on the voyages and engravers back in Europe when trying to balance the need for a faithful (empirical) rendition with the prevailing aesthetic tastes of the age:

> Banks always felt a tension between the call of taste and that of pictorial reproduction, and so his painters did devote some of their energies to romantic topics like grottoes, exotic rituals and so on because these suited the then fashionable rococo style. Moreover, even when accurate depictions of native peoples … were provided, it just was very hard to bring an objective account of them before the British public. Engravers *would* dress up the original paintings to keep them in line with their own philosophical predilections. (1992: 131, emphasis in original)

Tackling the role of photography in the 'imaginative geography' of the British Empire, and as linked to the production of geographical knowledge by nineteenth-century British explorer-geographers (many of whom were associated with the RGS), James Ryan (1997: 17) exposes the limits to the Victorian (and still prevailing) assumption that photography comprises 'a mechanical means of allowing nature to copy herself with total accuracy and intricate exactitude'. Alternatively, Ryan finds a suite of cultural constructions running throughout the photographic observations of 'distant places' throughout this period, teasing out the 'symbolic codes' which structured the composing and the framing of the images, and also hinting at the effects of these images on their audiences:

> Through various rhetorical and pictorial devices, from ideas of the picturesque to schemes of scientific classification, and different visual themes, from landscape to 'racial types', photographers represented the imaginative geographies of Empire. Indeed, as a practice, photography did more than merely familiarise Victorians with foreign views: it enabled them symbolically to travel through, explore and even possess those spaces. (1997: 214)

The reference here to 'possessing' spaces through observation will be recalled in a moment, and it also suggests a key claim that might be pursued in critical accounts of what is entailed in the much more recent use by geographers of technologies such as remote sensing (with its self-evident links to military and commercial uses of such technologies).

A second but related dimension of the critique centres on the more metaphorical sense in which many geographers have configured the world as an 'exhibition' for them to wander around, as it were, gazing upon the exhibits (the diverse collections of peoples and places there displayed) and making judgements about them. Indeed, Derek Gregory (1994; see also Mitchell, 1989) borrows the phrase 'world-as-exhibition' when tracing this tendency through different phases and approaches to geographical study, relating it as well to the 'cartographic' impulse which has led many geographers to conceive of themselves flying over landscapes of nature and society laid out below them. Revealingly, some writers have even drawn connections here to the strangely detached sensation which arises when looking down from an aeroplane, and (more worryingly) from the basket of a World War One balloon or the cockpit of a World War Two bomber (Bayliss-Smith and Owens, 1990). Gillian Rose (1993b; 1995), meanwhile, has developed a powerful argument that this version of an academic gaze reflects a distinctively 'masculinist' way of looking at the world, one predicated on an assumed mastery which allows the viewer to see into all corners of the world – all of which are reckoned to be available and amenable to the gaze, transparent to the piercing intellectual eye – and one which also carries with it an inherent desire to possess, to subdue, the phenomena under the gaze. Rose's argument is difficult, hinging on a combining of psychoanalytic ideas with a historical account of how male intellectuals have effectively constructed science (geography included) in their own (presumed) self-image. More simply, though, she outlines the masculinist propensities of fieldwork: both the heroic encounter of rugged individuals with a challenging field which is coded into the 'being there' approach, notably of someone like Stoddart (1986), and the associated figure of the male researcher observing, describing,

measuring and thereby capturing this field for himself (see also Sparke, 1996; Powell, 2002: esp. 263).

Central to the objections raised by Gregory, Rose and others to the prominence of observation is the suggestion that the faculty of sight should not be accorded such a master status in geographical work, whether in actual practices or in how we conceptualize the wider projects of the discipline. One implication is that other senses through which the world is knowable by us, notably hearing and more particularly still the practice of listening closely to what people say, should be brought more fully into our practice as *human* geographers (see also Rodaway, 1994).[4] A second implication is that we should resist the too glib deployment of terms saturated with assumptions about observation and sight in our thinking about the discipline, and in the process to resist a general orientation which conceives of the discipline as trying to make transparently visible all facets of the subject-matters under study. This is not the occasion to expand further upon such lines of criticism, nor upon their implications, although we hope that readers will be able to appreciate how the third approach to practising human geography described below does take them on board.

Surveys and scientific detachment

It should not be thought that the practising of human geography prior to recent years has only been about immersed observation, however, and it is actually the case that efforts to provide a more systematic basis for geographical research – one going beyond intuitive fieldwork to develop a definite technique of 'survey' – *did* figure in the history of the discipline earlier in the century. Sauer himself was instrumental in starting this ball rolling with his 1924 paper (see also Jones and Sauer, 1915) which was entitled 'The survey method in geography and its objectives', in the course of which he urged geographers to develop regimented and replicable methods of geographical inquiry. The similarities between this paper and his later writings notwithstanding, there are also key differences which reflect the enthusiasm of the younger Sauer for developing systematic principles of areally based 'geographic survey' incorporating not only *qualitative* materials but also *quantitative* information, the latter being derived from both 'statistical tables of state and national agencies' (Sauer, 1924: 20) and 'local statistical archives' (1924: 30). (See Box 1.7 for a preliminary note on the distinction between qualitative and quantiative data: this distinction, and its limitations, will be explored further in later chapters, especially Chapter 8.) It is perhaps surprising to hear the younger Sauer's own words in this regard:

Box 1.7: Qualitative and quantitative

This distinction has become rather sedimented in the thinking of many geographers, perhaps to the point where it becomes unhelpful and overloaded with misunderstanding and prejudice (see Philo et al., 1998; see also Demeritt and Dyer, 2002). *Qualitative data* are data that reveal the 'qualities' of certain phenomena, events and aspects of the world under study, chiefly through the medium of verbal descriptions which try to convey in words what are the characteristics of those data. These can be the words of the researcher, describing a given people and place in his or her field diary, or they can be the words found in a planning document, a historical report, an interview transcript or whatever (in which case the words are in effect themselves the qualitative data). Sometimes these data can be visual, as in the appearance of a landscape observed in the field (see Box 1.6), or as in paintings, photographs, videos and films. *Quantitative data* are data that express the 'quantities' of those phenomena, events and aspects of the world amenable to being counted, measured and thereby given numerical values, and the suggestion is that things which are so amenable will tend to be ones which are immediately tangible, distinguishable and hence readily counted (1, 2, 3, ...) or measured (an area of 200 m^2; a population of 20 000 people live there; a per capita earning of £100 000). It should be underlined that such counts and measurements are still only descriptions of the things concerned, albeit descriptions which are arguably more accurate and certain than are qualitative attributions (chiefly because they allow a common standard of comparing different items of data, and also the possibility of repeating this form of describing data: i.e. other researchers would count the same number of things or measure the same areas, population levels, per capita incomes, etc.). The use of quantitative data is hence commonly reckoned to be more *objective* (allowing researchers to deal with data in an accurate, certain and therefore unbiased manner), while qualitative data are commonly reckoned to be more *subjective* (leaving researchers prone to injecting too much of their own 'biases' in their dealings with data). As should be evident from much of this chapter, and of the rest of the book, we do not agree with such a conclusion because it forgets about the countless other issues which militate against the possibility of complete objectivity (which means that being quantitative is no convincing guarantee of objectivity).

The purpose ... is not to make fieldwork mechanical, but to increase its precision. The choice of things to be observed must remain a matter of individual judgement as to the significant relationship between area and population. Out of such field measurements will come the ... ideal of statistical coefficients. From them the geographer will determine ultimately the extent to which the theory of mathematical correlation is to be introduced into geography. (1924: 31)

It is telling that Sauer linked this version of field survey to the possibility of a more statistically minded geography, one which by the 1960s was regarded as the province of a fully scientific discipline, and we will return to this linkage shortly. The thrust of his reasoning here was echoed and extended a year later in D.H. Davis's (1926: esp. 102–3) rejection of 'superficial observations' when calling instead for geographers to evolve a 'mechanical quality' to their 'system of recording essential data accurately', one suitable for 'establishing correlations', which would then lead 'geography ... to be entitled to rank as a science'. Similar discussions of survey as a scientific methodology for geographers can be found elsewhere in the early- to mid-century literature, and it was these discussions, with their thinly veiled criticism of those who favoured a more impressionistic field style, that prompted both Platt's (1935) reactions and certain reservations from the older Sauer (1956), as already mentioned.

More practically, several papers (e.g. Jones, 1931; 1934) appeared in the North American literature which began to itemize the kinds of things which needed to be recorded in a comprehensive field-based geographical survey, the forms and functions of land uses to be mapped, as well as specifying the specific survey methods which might be employed to create this record (field walks and drive-bys, complete with their counting and mapping of phenomena, along with collecting statistical data from 'local depositories'). A review of 'field techniques' available for use by geographers in

their surveys of areal units was provided by C.M. Davis (1954), and a feel for the ground covered by this remarkably thorough early statement of survey methodology can be gained from these claims in the paper's opening paragraph:

There are four sources of factual information: 1) documents such as maps, ground photographs, statistics and written materials; 2) air photographs; 3) direct observation; and 4) interviews with informants. And there are four ways of analysing factual information for the purpose of identifying and measuring areal, functional or causal relations, each requiring the use of symbols: 1) analysis by expository methods, using word symbols; 2) analysis by statistical methods, using mathematical symbols; 3) analysis by cartographic methods, using map symbols; and 4) analysis by photo-interpretation methods, using photo-interpretation keys. (1954: 497)

Davis began to suggest a distinction between the data to be collected (*constructed* in our terms: see below) and the procedures through which those data can be analysed (*interpreted* in our terms), and he also indicated that the data collected could be qualitative or quantitative, with implications for the sorts of analytical techniques to be deployed on these data from the field. Equivalent practical statements also appeared in the British literature, many of which effectively hovered between the celebration of an unsystematized 'being there' stance and providing systematic guidance about what should be found out in locally based field surveys (see Box 1.5). The emerging 'field studies' movement which hooked into academic geography in various ways (e.g. Hutchings, 1962; Morgan, 1967; Yates and Robertson, 1968) articulated a vision not far from the survey orientation of some North American geographers, and the Geographical Field Group (descended from the Le Play Society) was expressly committed to ensuring that '[o]bservation, direct inquiry and documentation, including statistical material, all contribute to the data-collecting process' (Edwards, 1970: 314; and note that this group conducted a 'series of "regional survey" type excursions' described in Wheeler, 1967: 188). In an edited collection on the geography of Greater London, A.E. Smailes described 'urban survey' as the detailed recording of information about town sites from field-based observation ('reconnaissance survey') gleaned from 'traversing the streets' (1964: 221). Additionally, in a manoeuvre paralleling the survey cataloguing recommended somewhat earlier in North America by Wellington Jones (1931; 1934), Smailes proposed a specific 'urban survey notation' which produced a pseudo-quantitative form of data logging ready, presumably, for more sophisticated statistical analysis (see Figure 1.4). Whatever the precise details of the survey systems developed by these scholars, however, what we would immediately emphasize is their list-like, box-filling, counting and mapping ambitions: ones reflecting the primary ambition of the geographers concerned to accumulate data through which they could characterize the areas and sub-areas under study.

It is true that there were some qualitative elements here, as in the significance occasionally placed on talking to field 'informants' (see Chapter 5), but the basic trajectory of the survey approach was none the less towards a self-proclaimed scientific orientation. The prime ambition was to conduct systematic surveys which would produce comprehensive and reliable quantitative data representative of areal units (whether these be regions as large say, the Paris Basin or as small as, say, Glasgow's West End). There was also the beginning of suggestions about being able to conduct statistical analyses on these quantitative data, perhaps by using standard statistical tests to establish the strength of correlations between different sets of data (i.e. to show that certain areas are marked by high values

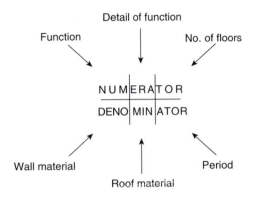

Figure 1.4 A.E. Smailes's pseudo-scientific 'urban survey' notation

Source: From Smailes (1964: Figure 41, 204)

on different variables, say income levels, occupancy rates and car ownership). We have already noted D.H. Davis's (1926) explicit linking of such surveys to a version of geography which could claim the name of 'science', while James Anderson (1961) played up the role of survey work in the context of land-use classifications as enhanced by 'statistical probability sampling procedures' and the use of computers. In such a vision survey work would contribute to providing the data on spatial patterns, chiefly data which might then be deployed in the process known as 'regionalization', the supposedly scientific delimitation of regions or areas fundamentally different from one another, and then in a process of classifying different types of regions according to certain distinctive clusters of attributes (e.g. Philbrick, 1957; Grigg, 1965; 1967; see also Chapter 7).

There is a complicated story to tell in this regard, but at bottom a continuity can be detected from the quantitative impulse in this survey work and the rise of geography as spatial science from the 1950s onwards (see Livingstone, 1992: chs 8 and 9; see also Barnes, 2001a). Spatial science, as is well known, entailed a fusion of quantitative techniques with a form of locational analysis aiming to explicate the basic 'spatial laws' governing the organization of phenomena (human behaviour and productions included) across the earth's surface.[5] In Chapter 8 we say much more about enumeration in geography, but for the moment it is sufficient to emphasize the extent to which quantification became the favoured way of going about things in geography as spatial science, linked into a particular model of how statistical tests (and then more formalized mathematical modelling) could all aid in the explanation of revealed spatial patterns (and see also Chapter 9).

All manner of claims were made for the superior merits of spatial science, complete with its quantitative sophistication, and at the nub of such claims was the assertion that a properly 'scientific' approach to research was one which overcame the potential 'bias' of researchers in ensuring the completely detached (and hence certain, accurate and trustworthy) cast of the research undertaken. In particular, the use of numerical values as measures of quantity, distance, position and the like was reckoned to provide an *objective* representation of what was actually happening in the 'real world', in contradistinction to the much less reliable data obtained through the *subjective* understandings integral to both the intuitive stance of a Sauer or the conversational elements of some survey work. (See Box 1.7 for a summary of the tangled debates about objectivity and subjectivity.) Such were the assumed advantages of a spatial-scientific practising of human geography, one which grew out of the above-mentioned survey tradition, but which came to embrace a much wider set of procedural, technical and explanatory goals. As a coda, and anticipating some of our arguments in Chapters 7 and 8, this version of human geography continues today, notably in the development and application of geographical information systems (GIS) and various forms of geocomputation.

Arguably, there is a level of sophistication about these more recent approaches to quantitative geography that was absent in the early days of spatial science.

Having laid out something of this scientific and survey approach to practising human geography, we should acknowledge that here too we see many drawbacks with what was being proposed. Many of these drawbacks were bound up with the overall philosophical difficulties attaching to spatial science, particularly as have been rehearsed through exposing the somewhat narrow 'positivist' philosophical assumptions which can be said – certainly in retrospect (Gregory, D., 1978a; Hill, 1981; Barnes, 2001a; 2001b) – to have framed this scientific turn within the discipline. At various points in our book we will engage with these problems, demonstrating the ways in which they arguably hamper the practising of human geography, although – as with the 'being there' and 'eye for country' stances – we are not denying that some aspects of the scientific and quantitative turns still have much to offer in the doing of human geography (and in a similar vein, see Sayer, 1984; Philo et al., 1998). None the less, we are concerned at the extent to which methodological treatises in human geography continue to be dominated by an exposition of spatial-scientific techniques (e.g. Lindsay, 1997; Robinson, 1998): a reduction of methodology to matters of technique.[6] Similarly, we are concerned about a rather narrow sense of what 'geographical enumeration' can entail which fails to look much beyond standard parametric tests. Such considerations remain pertinent to the practising of human geography, to be sure, but what concerns us is the lack of serious engagement with more conceptual questions about the limitations of what a self-professed scientific and quantitative human geography can and cannot achieve. Such questions are raised throughout this book, even if not always being conveniently labelled as such (to reiterate: Chapters 7–9 all debate these questions in one way or another).

What we will specifically underline now, since it is so relevant to the third phase in the history being recounted, is that we are deeply suspicious of the claims about the detachment of the researcher which are celebrated in the literatures of scientific, quantitative and survey-based human geography. We are perturbed by the determined erasure of the 'I', the researching individual or group, which serves (in our minds) only to occlude the realities of the active research progress through which flesh-and-blood geographers such as 'Carl' and 'Linda' actually get their feet, hearts and minds muddied in the places and people under study. Spatial science, along with both its antecedents and its derivatives, thus closes off the possibility of debating the practising of human geography in the fashion of this book. As just remarked, spatial scientists tend to reduce methodology to technique, being bothered about the correct running of an appropriate statistical test but less about anything entailed in the deriving of the data on which the test is conducted (unless relevant to deciding on which particular test is suitable), nor about anything following conceptually, politically, ethically or otherwise from choosing to tackle the data statistically rather than in some other way. There is a further and possibly simpler objection to raise in relation to the appearance of spatial science, moreover, in that it evidently led many human geographers to lose interest in field-based *primary* data, given that they rapidly became far more interested in the enticing array of statistical-mathematical techniques available (and being refined) for analysing *secondary* data (see Box 1.3). As Robert Rundstrom and Martin Kenzer neatly put it:

> Although quantitative human geographers were primarily concerned with abstract theory development [specifying the spatial laws of location theory], many of the early spatial analysis papers ... were based on fieldwork. The pattern changed by the middle of the 1970s. Continuing progress in spatial analysis was marked by theoretical developments relying on pre-existing data. Primary data became superfluous. Ackerman (1965) already considered fieldwork a mere chore, only occasionally necessary to validate the analytic, theoretical work of spatial science. James and Mather (1977) noted that some human geographers questioned whether fieldwork was still a necessary part of the discipline. (1989: 296)

Instead of going out into the field to collect data, many spatial-scientific human geographers started to spend the bulk of their time sitting in their offices and laboratories, punching in data found in library sources (e.g. Census surveys), reworking their own older data or even inventing data sets, as a prelude to the *real* work (for them) of using computer facilities to effect statistical-mathematical interrogations, simulations and model-building. To put it another way, 'economic geography [and human geography more widely] moved from a field-based, craft form of inquiry to a desk-bound technical one in which places were often analysed from afar' (Barnes, 2001b: 553). The practising dimension of their inquiries therefore collapsed into the techniques, reinforcing our previous argument, with a loss of concern for nuances of data, the composition of the field or the overall research process. In the terms of this book (see the Preface), this meant a loss of interest in the *construction* of data in preference for focusing, albeit very narrowly, on the *interpretation* of (quantitative) data. It is an exaggeration, but perhaps not too great a one, to state that this version of human geographical inquiry ceased to practise human geography except in the most minimal of senses. Neither a Sauer nor a McDowell would find much here to satisfy them, and the same is probably true of many human geographers today who continue to use numerical data and sophisticated statistical-mathematical procedures but always with an alertness to the origins, meanings and limits of the numbers and their manipulation (e.g. Dorling, 1998).

'Being reflexive' and 'listening to voices'

From about 1970 onwards, many human geographers, unhappy about both older approaches to the discipline and the spatial-scientific version, began to seek for new possibilities. There were numerous bases to their quarrel with how human geography was being practised at the time, centring chiefly on the limited conception of how human beings entered into the 'making' of their own worlds, but also on the almost complete absence of what might be termed a 'political' vision of why research was undertaken in the first place (who was it supposed to benefit and why?). While somewhat oversimplifying the picture, it can be argued that these twin objections to previous approaches, and most especially to spatial science, fed into two rather different alternative varieties of human geography – to be referred to here respectively as 'humanistic geography' and 'radical geography' (see Cloke et al., 1991; Peet, 1998) – which both demanded new ways of practising human geography. Indeed, their emergence and subsequent elaboration, particularly when mixed in with the insights from 'feminist geography' from the mid-1980s onwards, have effectively called forward a sensibility almost wholly unheard of before in the discipline. In short, this sensibility can be described as 'being reflexive', which means that human geographers are now called upon to reflect much more explicitly upon their *own* research endeavours than hitherto, giving careful consideration to precisely what it is that they are doing in their own projects: the conceptual, practical, political and ethical implications arising for these projects, for themselves, for

the people and places under study, and perhaps even for society more generally.

We would argue that, while not often given this credit, humanistic geography was decisive in prompting the developments leading to the new sensibility just mentioned. Humanistic geography was an umbrella term which arose in the 1970s (esp. Entrikin, 1976; Tuan, 1976a; Ley and Samuels, 1978) to denote a range of perspectives highly critical of how most human geographers, but most obviously spatial scientists, tended to conceptualize human beings. Extending an earlier 'behavioural' turn in the discipline and drawing upon various so-called 'philosophies of meaning' (Ley, 1981a; for summary details, see Cloke et al., 1991: ch. 3), the humanistic geographers complained bitterly about the 'pallid' view of human beings present in existing scholarship (Ley, 1980): one that portrayed human beings as little more than mere objects or at best robots with no interior sense of themselves, no intentions, no hopes or fears and no creative role to play in shaping their surroundings. Instead, so they insisted, the discipline needed to be dramatically reforged around a very different conception of humanity, a vision which recognized humans in all their flawed ambiguities as experiencing, perceiving, feeling, thinking and acting beings. Such a vision sought to enlarge the 'space' for human beings within the discipline, to grant them a measure of dignity, to 'people' human geography; in fact, to foster a new emphasis on the *human* part of human geography. The intellectual terrain here was uneven, but one over-riding outcome leading from this expanded conception of the human being was the need to find ways of accessing the human qualities, the sheer humanness, now reckoned to be central to disciplinary concerns. Spurred by a changed appreciation of what is important in the world under study – people and their inner lives, rather than spatial patterns and supposed spatial laws – the humanistic geographers had to consider fresh research practices, novel stances before their subject-matters and unfamiliar methods for getting close to people and their everyday apprehensions, understandings, routines and activities. It meant starting to use methods which provided some structure to the tasks of meeting with people, perhaps interacting with them on an everyday basis, perhaps talking with them in depth and certainly 'listening to their voices'. It meant rediscovering the questionnairing and interviewing techniques of the earlier survey tradition but, more significantly, it meant bringing into human geographical research the 'ethnographic' practices of in-depth interviewing, participant observation and the excavation of meaning which were much more the province of other academics such as anthropologists and sociologists. It meant returning to a measure of immersed observation in the vein of Sauer, but it also meant a much more sustained encounter with the peoples in the places visited. It required the thoroughly involved people-centred fieldwork which led the likes of John Eyles (1985), Michael Godkin (1980) and Graham Rowles (1978a; 1978b; 1980) to spend days, weeks and months in the company of individuals, witnessing the grain of their lived worlds, discussing with them the meanings attached to places, environments and landscapes comprising the spatial contexts of these small worlds.

A key figure in all this was undoubtedly David Ley, whose famous exploration of inner-city Philadelphia, including an attempt to recover the existential meanings of place and 'turf' held by black street gangs, blended the ideas of a humanistic geographer, the interests of a social geographer and the practices of an 'urban ethnographer' (Ley, 1974). Following in part the example of ethnographers associated with the Chicago School of urban sociology (Jackson, 1985), but growing as well from his response to 'the relentless barrage

of everyday pressures in the inner city' (Ley, 1988: 132), Ley created his own distinctive version of what he later came to term 'interpret(at)ive social research':

> With limited experience to fall back on, aside from the intuition gained from a British field tradition [presumably Wooldridge's 'eye for country'] and knowledge derived from a number of ethnographies, devising a method was in part a matter of learning on the job. The principal method was participant observation. The period from January to July was set out as the length of continuous residence in the neighbourhood ... It is essential to establish a systematic procedure for recording field data in any ethnographic research, and my practice was to write up field notes each evening, ranging in size from a paragraph to (occasionally) 1000 words [see also Chapter 6]. These notes were records of impressions, events and conversations, sometimes reconstructed from brief phrases or sentences scribbled down during the course of the day. (1988: 130)

The 'unstructured, everyday encounters' which generated this data were then supplemented with more formal face-to-face 'questionnaire interviews', some taping of public meetings and some reading of 'agency data and documents from police, planning and school board authorities', as well as extensive field observation on phenomena such as 'graffiti, vandalised cars or abandoned properties' (Ley, 1988: 130–1). The result was an eclectic mix of data sources and methods of both data collection (*construction*) and analysis (*interpretation*), containing both quantitative and qualitative moments, and it broke new ground in creating a model for practising human geography which has since been widely emulated. The majority of later researchers would probably not call themselves humanistic geographers, preferring instead labels such as social or cultural geographer (see Jackson and Smith, 1984, or the various studies reported in Jackson, 1989; Anderson and Gale, 1992), but the basic procedures deployed by them did arrive in the discipline with the humanistic experiments of someone like Ley (and note that participant observation was explicitly claimed as a prime method for humanistic geography by Smith, S.J., 1981; 1984; see also Jackson, 1983). It might be added that the version of inquiry being progressively refined in this vein has since also been termed 'interpretative geography' (Eyles, 1988b) or 'interpretative human geography' (Smith, D.M., 1988b), and several fine examples of practice have been collected together in books edited by John Eyles and David Smith (1988) and more recently by Melanie Limb and Claire Dwyer (2001).

Ley has remarked that '[e]thical issues are far more conspicuous in ethnographic research because of the close relationship between the researcher and the community' (1988: 132), thereby indicating that the new forms of research initiated by humanistic geography have forced into the open questions about the researcher's personal involvement with a project, people and place. What are the 'biases' of the researcher? How do these influence how he or she conducts the research, how he or she represents the peoples and places studied in a write-up, and the informal 'contract' which he or she strikes up with a community about what is done, said and finally given back? All these questions start to concern the researcher in a manner which had never been the case for previous generations of geographers, and certainly not for spatial scientists, and the 'ethics and values' of doing geography thus become a subject for debate as never before (Mitchell and Draper, 1983a; 1983b). On an operational level, an extreme instance is reported in Rowles's (1978b: esp. 179) study of the geographical experiences of elderly people, and entailed him sitting at the deathbed of one of his aged respondents, Stan. Here, as the only 'friend' whom Stan had left in the world, the only person remaining who cared enough to sit in that hospital room, Rowles found himself urging Stan not to die because the research was still incomplete. But he hated himself for thinking in this way: was this all that it had been about, getting the research done, and how hollow, how intrusive but meaningless had been his 'befriending' of Stan for the purposes of academic research? The ethics – the turmoil, stress and guilt – attendant upon such a moment were light-years away from anything experienced by, say, spatial scientists working on impersonal numerical data sets at the computer terminal. This is not to suggest that there are any simple guidelines for how researchers should pick their way through the ethical minefield which can confront them in research of this character. As Ley (1988: 133) acknowledges, '[t]here are no set pieces in answering those questions, and indeed answers will vary according to the circumstances of the community', but what can be insisted is that in specific studies 'the questions must be asked and answered in good faith'. This is not to demand that researchers always put down in writing their thoughts, worries and responses on this count, but it is to propose that they should be able to offer some relevant reflections if ever challenged to do so.

The researcher's presence as an 'I', a creative and reflexive figure in the research process who is not erased as a non-issue (as might a Sauer) or cloaked behind a veil of claimed objectivity (as might a spatial scientist), is therefore as much a part of this approach to human geography as are theories, data, methods and so on. On the still deeper level of the researcher's underlying interests, convictions and motivations, furthermore, the appearance of humanistic geography was also crucial in foregrounding such values in a fashion rarely if ever seen before in the discipline. To put things another way, humanistic geography prompted attention not just to the subjectivity of the researched but also to the subjectivity of the researcher. Writing in a determinedly scientific manifesto for human geography, Ronald Abler et al. (1971: 24) had declared that the scientific 'way of life' should be 'total', and that it should be completely divorced from how scientists might 'let their hair down emotionally and theologically … during their off hours'. The latter aspects of their lives, of who they are and of what they feel or believe, should hence be roped off from their efforts as professional human geographers.

But writing only three years later, Anne Buttimer, a humanistic geographer who was also then in religious orders, advocated something entirely different when insisting that such a compartmentalization of the human geographer's personal and professional faces is mischievous because it just cannot happen, since personal values can *never* be systematically erased from the framing, conduct and write-up of research. And for Buttimer such an erasure should never be attempted anyway, being unnecessarily restrictive because it denies many of the well-springs of genuine human concern and creativity, and also depriving us of crucial grounds for sensible ethical judgement. Science alone cannot provide those grounds, so she argued, and the dangers of a science without ethics now become increasingly obvious. Her alternative proposal was quite clear:

> Each reader [each human geographer] should endeavour to explore the values which guide/influence his [or her] mode of being in the world, for it is the contention of this paper that one's geography cannot be considered a separate domain of one's life but is influenced by many personal, cultural and political 'values' surrounding that work. (1974: 5)

Buttimer duly reflected upon many of the values which shaped her own geography, pointing out that they were 'strongly influenced by Christian, and especially existential thought' (1974: 5), and she thereby countered Abler et al.'s (1971: 24) pronouncement that 'God … is not permitted' as a 'concept'

informing human geographical research. She supposed that all human geographers would entertain different, idiosyncratic assemblages of values, making generalizations difficult, but she also recognized the importance of supra-individual intellectual, 'cultural and political' values which can themselves become the focus of careful 'sociological' scrutiny (1974: Part II). While many have disagreed with the specifics of her arguments here (e.g. see the four commentaries appended to Buttimer, 1974), our view is that her insistence on human geographers being reflexive about the diverse values shaping their own work is one which continues to resonate loudly with more recent efforts at practising human geography.

Alongside humanistic geography, a self-professed radical geography emerged during the 1970s (Peet, 1977; 1978), anchored initially in the pages of *Antipode: A Radical Journal of Geography*, and subsequently diffusing to become a wide-ranging critical window on social and spatial inequalities of many different shades. Taking as its focus environmental and social problems with a clear geographical expression, from the devastation of rainforests to poverty, deprivation and disadvantage, a radical-geographical perspective arose which sought to expose the systematic structuring of injustice which leads to a world fragmented into spaces of plenty (occupied by 'the haves') and spaces of deficit (occupied by the 'have-nots'). Starting with approaches which did little more than document, table and map this polarity at various scales from the international to the intra-urban, radical geographers gradually evolved a conceptual basis for explaining these inequalities which included (sometimes contradictory) inputs from welfare economics, anarchist theory and different strands of Marxism.[7] Some of the research undertaken in this vein retained a survey feel, albeit utilizing survey techniques to expose inequalities in phenomena such as income levels, housing conditions and ill-health indicators (this was particularly true of something called 'welfare geography': Smith, D.M., 1977; 1979; 1988b). Indeed, much of the research continued in a fashion not wholly different from the more scientific and quantitative cast of previous work, even if putting the data and techniques to a radical use critical of the social status quo, and even if fuelled by a commitment to radical (even revolutionary) social change wholly absent from the more 'establishment', often policy-orientated studies of previous generations, notably of spatial scientists. We will return to examine this commitment presently.

It may be claimed, then, that radical geography did not usher in as dramatic a change in the routine practices of human geographers as did humanistic geography. Its immediate methodological implications were not so great, even if conceptually and politically it was probably more unnerving to established modes of inquiry. None the less, mention might be made of the 'advocacy geography' experiment associated with William Bunge's attempt to shift the orbit of professional geographical research out of the university campus – together with the geographers themselves, and also their students – and into the streets of the inner city, chiefly the black inner city of Detroit, where the aims of studies should be directed by the articulated needs of poor inner-city residents themselves (see Colenutt, 1971; Horvarth, 1971; Bunge, 1971; 1975; Merrifield, 1995). Rather than offering yet another calibration of a 'central place model', for instance, advocacy geographers should be uncovering the geographies of slum processes, traffic accidents affecting children, diseases of babies and the like, and acting as advocates able to demonstrate the contours of problems to city authorities who might be sufficiently convinced by academic evidence to respond positively. Failing that, the geographers involved should be themselves involved in grassroots projects like building a children's playground. This species of radical geography thus urged an action-based research, predicated on full involvement in a research activity designed to achieve highly practical ends: a policy-orientated research from below, on behalf of those who might with justification be referred to as 'the oppressed'.

Intriguingly, such research did have things in common with humanistic geography in that it depended upon a sustained participation in the lives and struggles of certain inner-city communities – Amaral and Wisner (1970) spoke of 'participant immersion' instead of a less involved 'participant observation' – and also because it forced researchers to deal with concrete ethical issues rooted in their responsibilities towards the relatively powerless people whom they were supposed to be serving. David Campbell expressly reflected upon 'role relationships in advocacy geography', underlining the virtues of a thoroughly democratic practice resistant to the 'elitism' common in most other work by human geographers, and striving instead to empower the research subjects who should 'become problematisers of their [own] situations and ... active creators of their [own] environment' (1974: 103). Moreover, and echoing still further the ethical charge of humanistic geography:

> Constant self-criticism and re-evaluation in an attitude of humility and respect for others is ... a vital and healthy component of advocacy activity ... 'Humanising social change' [to borrow a phrase from Harvey] is dependent ... upon the ability of advocates and academics to create humanising relationships with those whom they work ... Radical science must be based upon a personal commitment to genuine communication with others in an attitude of mutual respect. (Campbell, 1974: 104–5)

The radical, politicized overtones of these remarks must have been anathema to the more 'conservative' geographers of the era, but they are ones with which many human geographers today would have great sympathy, and the notion of democratic, empowering and respectful research practice will certainly reappear at various points in the chapters that follow.

More generally, the rise of radical geography, particularly in a guise which turned to Marxist critiques of the inequalities integral to a globalizing capitalist economic order, carried with it explicit commitments to a coherent political programme: one which oscillated between a reformist line, supposing that the existing state of society can be improved through the standard democratic process, and a revolutionary call for complete social transformation (at its crudest a call for the workers to seize control of the 'means of production' from the capitalists). David Harvey (1973) led the way when self-consciously shifting from a basically reformist line, associated with a welfare position, to an assertively revolutionary line convinced that the only way to create true 'social justice in the city' would require an overturning of capitalist forms of urbanism. This latter way forward would also necessitate an input from, if not necessarily a Leninist intellectual vanguard, then certainly a corpus of Marxist academics, geographers included, prepared to undertake the theoretical and practical work of planning revolutionary change. Radical geographers ever since have been wrestling with this tension between reformist and revolutionary ambitions, as is clear from recent debates played out in the pages of the journal *Society and Space* (Blomley, 1994; Chouinard, 1994; Tickell, 1995), and a further feature of debate has been the seeming gulf between radical theorizing in the academy and radical activism on the streets (see also Routledge, 1996; Farrow et al., 1997; Kitchin and Hubbard, 1999).

The principal point for us here, though, is that – just as Buttimer insisted on humanistic geographers incorporating explicit reflections on their basic values – radical geographers have often entertained some self-interrogation about political values, objectives and involvements. Perhaps the most rigorous formulations in this respect have emerged from Jürgen Habermas (esp. 1972), a famous German Marxist intellectual, who has proposed the refining of an explicitly *critical* science or theory fitted to achieve 'the realisation of a[n] … *emancipatory* interest' (Gregory, D., 1994: 107, emphasis in original) which would free all peoples of the world from the yoke of (capitalist) oppression. Habermas's vision explains how all varieties of intellectual labour are determined by 'cognitive interests' which turn their practices of knowledge production to particular ends, usually 'technical' or 'practical' ones functional to the maintenance of the social status quo (and an extension of his argument would include all varieties of human geography, including both spatial science and humanistic geography, as essentially 'reactionary' in this sense: see Gregory, D., 1978a). Following from such a recognition, however, the argument is that it should then be possible to frame a new version of intellectual endeavour predicated on an emancipatory cognitive interest which would be at once critical (of an inherently unjust world) and *self*-critical (constantly evaluating the extent to which the academic's own practices are emancipatory in both overall design and specific interventions). While rarely presented in such obviously Habermasian terms, except in Gregory's (1978a) statements about 'committed explanation in geography', it is arguably the case that this notion of being simultaneously critical and self-critical has energized the efforts of most radical geographers over the last two decades or so. We will pick up on the arguments about the politics of human geography, returning to some of the materials just outlined, in our final chapter (Chapter 12).

It would be possible to say more here about the burgeoning twists and turns in human geography which have, more recently, built upon the twin pillars of humanistic and radical geography to forge further dimensions for practising human geography. But for the sake of brevity, and yet to cover what have been pivotal new claims relevant to our practising theme, it will suffice now to mention certain aspects of the interlocking contributions made by both 'feminist geography' and 'postcolonial geography'. Feminist geography initially arose to provide an explicit examination of the specific spatial experiences, constraints and worlds of women, that 'other half' of humanity almost never considered by previous generations of male geographers (Tivers, 1978), and it quickly developed as a more fundamental critique of how unequal gender relations shape countless sociospatial structures endemic to a diversity of 'patriarchal' human societies past and present (McDowell, 1983; Foord and Gregson, 1986; WGSG, 1984; 1997).

In the process questions of how to do feminist geography inevitably came to the fore, particularly in the matter of thinking about how research could be carried out which would enable the voices of women to be heard, notably when recounting their experiences of an everyday male superiority, harassment and even abuse accepted by many of them as sadly 'natural'. The task also became one of finding methods which would be sufficiently sensitive to tease out often very subtle dimensions to how women's perception and use of space differs from that of men, whether in terms of a phenomenon like the 'gender division of urban space' (McDowell, 1983) or something like women's fear of public spaces such as parks and subways (Valentine, 1989). It has been argued both within (esp. Rose, 1993b) and beyond the discipline that conventional models of academic inquiry, with their scientific and quantitative emphases, display an inherent 'masculinism' which militates against the kind of grounded research which is probably essential in this

context. The debates are tricky, but we can allow McDowell to be our guide in a passage which also signals the character of the specific methods that feminist academics, geographers included, have tended to favour in their own research:

> Certain feminists ... not only reject the quantitative, 'scientific' approach to research, but argue that it is specifically a patriarchal model as it denies the significance of women's experience of oppression, classifies their concerns as private rather than shared, and embodies the values of traditional views of women's and men's expected positions in society. They have argued that feminist research should recognise and challenge the everyday experiences of women. In order to excavate women's experiences, feminist methods should value subjectivity, personal involvement, the qualitative and unquantifiable, complexity and uniqueness, and an awareness of the context within which the specific [issue] under investigation takes place. (1988: 166–7)

The onus shifts towards being a highly personalized research encounter, in which the most qualitative of methods such as in-depth interviewing and the taking of 'oral histories' of 'life stories' (see Chapter 5) are pursued in a manner which necessarily entails a sustained exchange – potentially one dealing with emotional materials – wherein the personalities of both researcher and researched cannot be arbitrarily suspended. For both parties involved, such an exchange is potentially draining, as well as being fraught with the dangers attendant upon the release of emotions, often resentments and angers, which will usually do far more to illuminate the realities of a given issue than could any other data source.[8] We may be stressing the more extreme end of such feminist research here, and we acknowledge (and hope) that research encounters will not all be of this intensity, but we do wish to underscore just how much the researcher as a whole person – as an embodied individual with his or her own worries and frailties[9] – enters into the feminist research frame. We have therefore travelled a very long way from the mostly fact-finding questions asked of, say, local farmers by a Sauer-esque geographer strolling one evening through a pleasant valley.

Feminist geographers do not only deploy such intensive methods, of course, as McDowell (1988) makes clear and as Hodge et al. (1995) also insist when assessing the possible use of quantitative techniques by feminists conducting geographical research. Yet we will stick with this picture of intense inter-subjective research encounters – ones demanding an intimate meeting of two or more subjectivities: that of the researcher and those of the researched (see also Cook and Crang, 1995; see Chapter 10) – since such a picture is helpful in clarifying an additional set of claims. And what will also be useful in this respect is briefly to acknowledge the influence of postcolonial geography.[10] If feminist geography confronts the axis of gender, problematizing its constitution as well as its effects, postcolonial geography confronts the axis of 'race', problematizing the inequalities between white people and people of colour which feature today in so many different situations under study by geographers (from the relations between 'developed' and 'less developed' countries to the circumstances of racial minorities in predominantly white Western cities). Given their acute sensitivity to axes of social difference, it is feminist and postcolonial geographers who have done most to reflect upon the problematic power relations which can arise in the research encounter, most starkly when men are researching their 'other' (women) or when white people are researching their 'other' (people of colour), but also in many other ways when differences of class, education, sexuality, age, (dis)ability and so on potentially drive a wedge between the world of the researcher and that of the researched. There are many thorny considerations here, but we

would suggest that the consensus emerging from recent texts such as Jackson (1993), Nast et al. (1994) and Cook and Crang (1995) is one insisting upon a 'reflexive notion of knowledge' (McDowell, 1992a: 399), the crux of which necessitates researchers reflecting critically upon their own 'position(ality)' – their own backgrounds, attributes and values, as bound up with their own personal geographies (the sites, localities and networks of their own biographies) – in relation to the 'position(alitie)s' of those peoples and places under study. Such a stance on the doing of qualitative human geography underlines much of the recent Limb and Dwyer (2001) collection, where four chapters explicitly debate matters of 'positionality'.[11] We try to visualize this emerging model of intersecting positions in Figure 1.5, the implication being that the researcher should aim to clarify his or her own position in a wider societal hierarchy of power, status and influence, thereby ascertaining the different sorts of relationships – complete with the many differing roles, responsibilities and possible limitations to what can and should be 'exposed' about the researched – which may surface in a given research project.

From such a model it becomes apparent why it is impossible for a geographer like McDowell to leave behind her personal world in the same fashion as can a Sauer: this is not only because she has personal duties which she cannot forsake as easily as can most male academics, but it is also because she firmly believes that it is wrong in research terms to do so, since who she is (all the baggage of her own position) is so very pertinent to what she can achieve in her research. It shapes her gaining of access to particular research situations rather than others; it shapes her ability (and willingness) to build 'research alliances' of empathy, trust and dialogue between her and the people whom she researches (see also Pile, 1991); it shapes what findings she can obtain, the ways in which she will interpret these findings, and her sense of what is and is not

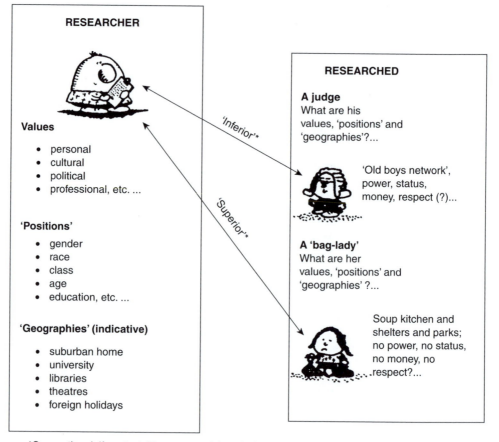

RESEARCHER

Values

- personal
- cultural
- political
- professional, etc. ...

'Positions'

- gender
- race
- class
- age
- education, etc. ...

'Geographies' (indicative)

- suburban home
- university
- libraries
- theatres
- foreign holidays

'*Inferior*'*

'*Superior*'*

RESEARCHED

A judge
What are his
values, 'positions' and
'geographies'?...

'Old boys network',
power, status,
money, respect (?)...

A 'bag-lady'
What are her
values, 'positions' and
'geographies' ?...

Soup kitchen and
shelters and parks;
no power, no status,
no money, no
.respect?...

*Conventional, if contestable, senses of the relative positions of the individuals concerned in a status hierarchy

Figure 1.5   Our visualization of the encounter between the differing 'position(alitie)s' of researcher and researched (the 'research subjects')

appropriate to reveal in final write-ups of projects undertaken. More particularly, it means that everything which she does in this regard cannot help but be influenced by her feminist experiences, values and politics, but she is reflexively aware of these feminist influences on her research and is self-critical about what they enable to be seen and what they might also occlude. She is thereby arguably *more* objective about the determinants of her research practices than are the likes of a Sauer or a spatial scientist.[12] While recognizing drawbacks with a visualization such as that provided in Figure 1.5, we do regard it as one which usefully pulls together

many of the themes which first surfaced in both humanistic and radical geography, but which have now been recast most effectively in the light of both feminist and postcolonial geography. This visualization is also one which readers might find useful to revisit at various points in the chapters which follow.

We should acknowledge that some human geographers may be unhappy about our narrative above, particularly given that we attach priority to 'being reflexive' and 'listening to voices' as the key recent developments in the practising of human geography, and in so doing put less store by more technical innovations such as GIS or computational approaches. None the less, and as should be apparent from what we said above, we do feel that the latter innovations – while undoubtedly of great utility in certain projects – are less significant as contributions to a genuinely *human* geography than is the emergence of a self-critical reflexivity which begins 'to question ... what we know, how we know it and what difference this makes both to the type of research that we do and who participates in it with us as either colleagues or research subjects' (McDowell, 1992a: 399–400). This being said, we appreciate that there are still criticisms to be levelled at a reflexive human geography which claims to be good at listening to the voices of others, and which thereby sets itself up as (striving to be) both ethically sound and politically empowering in relation to (less privileged) peoples and places under study. In particular, Clive Barnett (1997) has suggested that there may be problems with the notion of 'giving voice' to others, in that there are many others in the world for whom silence may actually be a preferred, even more empowering, strategy. Similarly, Gillian Rose (1997a) has suggested that there are problems with the impression of 'transparent reflexivity' which is conveyed by the debate about position(ality) – the assumption that researchers can somehow lay bare the many dimensions which comprise their position(ality) – because, as a psychoanalytic perspective indicates, many of the impulses, desires and passions which feed through into our academic work are ultimately lodged in realms of the unconscious inaccessible to conscious reflection. Additionally, Rose criticizes notions of empowerment which operate with a 'map' of power such as that implied in Figure 1.5, given that it hints at the possibility of researchers being able to redistribute stores of power from position to position (from the powerful to the powerless, from themselves to

the people under study). As she rightly points out, the notion of power here is perhaps too simplistic, in that power arguably operates more relationally than both the map and claims about redistribution imply, as indeed has been claimed in various recent texts on the messy geographies of power (e.g. Hannah, 1997; Sharp et al., 2000). The arguments by Barnett and Rose are very much set within the horizons of thinking previously rehearsed in this subsection, however, and they comprise a gloss (albeit an important gloss) on recent debates about practising rather than the critical demolitions with which we concluded the two previous subsections of the chapter.

Conclusion

In this chapter we have introduced many of the themes relevant to the practising of human geography, initially by contrasting two extreme examples of how different human geographers ('Carl' and 'Linda') go about the research process, and then by providing a more sustained review of changing ways in which human geography has been practised over the years. While wishing to suggest that there are still things of value to take from the first two approaches assessed here, in that their respective attributes of immersed observation and systematic rigour do contain much of merit, we are more firmly persuaded by the claims integral to the 'being reflexive' and 'listening to voices' orientation. Indeed, this latter orientation is now highlighting all manner of complications with the practising of human geography, countless issues which were either ignored in the past or were not called into play because different (arguably simpler) research questions were being asked, and in the course of this chapter we have sought gradually to draw out these complications for closer inspection. They are all ones which feature at various points in the book.

To conclude this chapter, though, we will provide a summary listing of the major themes growing out of the above narrative. In terms of the *construction of data*, we have shown how past generations of human geographers have tended to regard the construction of data as a fairly unproblematic matter, something that 'simply happens' in the field or occurs as packets of statistics are sent to you in the post. Instead, it is now argued that much more attention really does need to be paid to precisely how these data are come by. Although this has not really been a theme above, consideration must be given to the composition of *preconstructed data*, as derived from sources other than the researcher's own primary research (see Chapters 2–4). Rather more has been said here about *self-constructed data*, those that have been pieced together through the researcher's own endeavours (see Chapters 5 and 6), and hence about the precise methods which need to be deployed. Rather than simply 'being there', having an intuitive 'eye for country', conducting list-like surveys or seeking out suitable large-scale numerical data sets, it has become vital to ponder more carefully than hitherto the researcher's methods. In particular, the necessity for formalizing and extending qualitative methods has become increasingly evident with the heightened concern for what people under study think, feel and do in their everyday lives. Questionnaires and, more especially, interviews and participant observation have thus become popular, with humanistic and feminist geographers being at the forefront of this sustained qualitative turn. Moreover, and particularly as these geographers have made plain, the researcher's own underlying values and ethical views cannot be discounted as an influence on how data are constructed, notably in the context of the uneven power relations running between (the positions of) the researcher and the researched. The many implications of this research relationship can never again be regarded as unimportant.

In terms of the *interpretation of data*, we have shown how past generations of human geographers have tended to regard the interpretation of data as equally unproblematic, something that involves gifted intuition or batteries of quantitative analysis. Instead, it is argued that a broader span of attention is now required to the overall interpretation of data, demanding much more than just the learning and refinement of new statistical-mathematical procedures. No longer is it assumed self-evident how researchers move from data to conclusions, with the whole terrain of interpretation becoming something requiring consideration, and different possibilities for interpretation needing to be explicitly weighed up by researchers (see Chapters 7–11). Moreover, and particularly as the humanistic, radical and feminist geographers have made plain, the researcher's own underlying values and political commitments cannot be discounted as an influence on how data are interpreted, and the clear message from such geographers is that we should be fully aware of – and prepared to reflect explicitly on – how the whole cast of our research is shaped by such values and commitments (and see also our Chapter 12).

Notes

1 Some of Sauer's key writings are collected in Leighley (1963).
2 Key writings by McDowell include (1983), (1989), (1999); and, for the women in the City work, see McDowell and Court (1994; McDowell, 1997).
3 See, for example, French (1940), Coleman (1954), Dilke (1965), Wheeler and Harding (1967), Archer and Dalton (1968), Yates and Robertson (1968) and Coleman and Lukehurst (1974).
4 A related issue is that the unthinking occularcentricism of the discipline is insensitive to geographers who are visually impaired, and it is rare to come across a proposal such as Kingman's (1969) regarding 'field study for the non-sighted'. More generally, those who celebrate the being there of fieldwork tend

to assume the able-bodied status of all geographers, another aspect of neglecting the inaccessibility of many fieldwork sites to many individuals who are differently abled: see Hall et al. (2002).

5 Prime statements at the time about this variety of geography included Bunge (1962), Burton (1963), Harvey (1969), Abler et al. (!971) and Amedeo and Golledge (1975).

6 In passing, it is worth repeating Barnes's (2001b: 552) point about the extent to which the early spatial scientists were fixated on their new 'machines', IBM mainframe computers and the like, and on the numerical analyses of large quantitative data sets now made possible by such technology.

7 Key texts included Peet (1978), Smith, D.M. (1977; 1979), Harvey (1973; 1982); for summary details, see Cloke et al. (1991: ch. 2).

8 See also recent claims about taking seriously the emotional registers of the researcher: Widdowfield (2000), Anderson and Smith (2001).

9 See, for example, Pile (1991), Nast (1998), Parr (1998a), Dewsbury and Naylor (2002: esp. 256–7).

10 See Crush (1994), Radcliffe (1994), Jacobs (1996) and Nash (2003).

11 Butler; Ley and Mountz; Mohammad; Skelton.

12 See also the claims in Philip (1998) drawing in part on Wright's (1947) notion of 'objective subjectivity'; see, too, Box 1.7.

Geographical data

Human geographers are almost always using data (see Box 1.3) in their research. We may sometimes conduct highly abstract theoretical meditations on, for instance, the nature of social space, the character of human places or the deeper philosophy behind how we struggle to know what is happening in the world of people. In these cases the active engagement with data may be quite limited, but it is unlikely to be entirely absent. Alternatively, we may occasionally conduct more synthetic work that seeks to pull together, perhaps to find new ways of understanding, the findings from previously completed research projects by geographers and other academics. In these cases there is an active engagement with data, even if these data have been collected and processed beforehand by scholars other than the person doing the synthesizing, but it remains imperative that the synthesizer is aware of how the relevant data have been come by and worked upon in these prior research projects.

In probably the majority of our studies as human geographers, however, we *are* deeply preoccupied with data from first to last. Whether the undergraduate student undertaking a dissertation or the established professor conducting funded research, we are largely dependent upon data for the successful completion of our research projects: we need to acquire data, to appreciate both their possibilities and their limits (what they can and cannot tell us) and to be able to deal with data, often in a variety of ways, so that we can arrive at findings, conclusions and further speculations. And yet, despite the recent raft of textbooks on matters of methodology that have appeared in the geographical literature, referenced in our Preface, there is still little systematic reflection on the data we use, nor on the means by which we endeavour to get data to reveal things to us. The two parts of this book that follow, therefore, strive to tackle both these issues in turn, beginning with five chapters exploring what is involved in the different sorts of data which we routinely bring to the table of research. Our key claim spread across these five chapters is a deceptively simple one: that all these data, whatever their source, are *constructed* in one way or another. They do not magically or spontaneously arise in the world just waiting for the geographer to come along; rather, they are 'made' by somebody for given reasons and in specific ways.

Preconstructed geographical data

Most human geographers have to put a lot of effort themselves into 'getting hold' of data with which to work. For a few of us this is a seemingly quite straightforward process. Perhaps we photocopy a few pages from the British Census, or download these from relevant websites, or perhaps we buy some data from another organization, say profiles held by a company on its customers and their home neighbourhoods, which arrive as either hard copy or on a computer disk ready to be processed. For some of us there is more effort involved, since it entails visiting a national or local archive to hunt out materials, perhaps a Parliamentary Select Committee report, perhaps an inventory of people resident in a nineteenth-century institution, the minutes of city council meetings, someone's diary or letters. For others of us it might require trawling through back issues of magazines or copies of old television programmes, which might necessitate buying these from some source or travelling to a specialist archive. Sometimes we will be lucky, easily locating and being able to purchase, borrow or access the relevant data. On other occasions we will not be so lucky and may experience considerable difficulty in tracing data appropriate to the research questions being asked. Alternatively, we may find ourselves unable to access the data once they are located because they are confidential to an organization or individual, or because they are simply too expensive to buy, or because they are housed in an archive that is closed, private or too far away. It is not unknown for research projects to hit a dead-end because it is simply impossible to track down relevant data or to get a sight of relevant data sources if they do exist.

The crucial point to notice here is that in all the instances listed above the data may have the appearance of 'simply sitting there' waiting for us to locate and to access them, but such an appearance must not disguise the fact that all the data sources involved have, indeed, been constructed. These data sources have all been 'made' by somebody for reasons which need to be ascertained and in particular formats and using particular methods which need to be uncovered. In our vocabulary, these data have been *preconstructed*, which means they have not been constructed or 'made' by the researcher but by somebody else at a prior moment in time (whether long ago or last week). These data have not been constructed by the researcher and have been put together for entirely other reasons than the purposes of a researcher carrying out a research project. A census is put together because a national government wants to know things about its subject population, their age and sex composition, their occupations and the like, chiefly because such information helps it to formulate socioeconomic policies, and it is only as a by-product that researchers can start to ask their own questions of the census data so gathered and tabulated. Minutes of committee meetings are taken because it is important that members of the committee have a record of proceedings, of decisions taken and arguments won or lost, and perhaps, too, so that the committee can be accountable to non-members. A diary is kept by an individual or letters are written for a multitude of very personal reasons, while a film or a documentary is made for reasons of entertainment and instruction.

If human geographers are going to utilize these data sources, to extract things from them for interpretation (see Part II), it is vital that we are as familiar as possible with the subtleties of these sources: that we do know the why, the how and also the when and the where of their construction, since it is only on the basis of such knowledge that we can sensibly assess what such sources can and cannot tell us. It is only by having a clear understanding of these data that we can decide the extent to which we 'trust' them to be revealing an accurate picture of the 'reality' beyond (e.g. the population structure, a committee's deliberations, a diary-writer's experiences), or can appreciate the basis for effects, messages and the like – we might say 'biases' – more or less wittingly installed into particular kinds of data (e.g. the political content of a report or the emotional charge of a documentary, both of which might themselves become the focus of a researcher's attention). In the first three chapters that follow, we strive to lay bare many of the issues bound up in the making of three different if overlapping varieties of preconstructed data, suggesting along the way numerous things the human geographer must keep in mind when drawing upon such data. First, in Chapter 2, we look at 'official sources' (e.g. a government report); secondly, in Chapter 3, we look at 'unofficial sources' (e.g. a newspaper article); and, thirdly, in Chapter 4, we look at 'imaginative sources' (e.g. a novel). In each case, we focus on the data involved, the documents that are part of what some writers now term a wider 'thing world' of documents, devices and artefacts of all kinds, asking about how these data are constructed and made available for us as human geographers to labour upon in our own research.

Self-constructed geographical data

For many human geographers, however, the data required are arguably rather more difficult to come by than it is for those who utilize preconstructed sources. For many of us, then, there are no pre-existing data sources available with which we can answer our research questions. We may want to know things about the characteristics, activities, worldviews and the like of people living in a given locality, or we may want to gain detailed information about the thoughts and actions of a very particular group of people (residents of a gentrifying neighbourhood, for instance, or the senior executives of a multinational company). There may be some extant data sources which we could consult in these respects – censuses, opinion poll surveys, petition leaflets, company reports and so on – but it is unlikely that such sources will contain precisely the information that is required, for the very reason that these sources have not been compiled with the specific research aims of the human geographer in mind. Sometimes we have no choice but to make do with these sources, notably if we are historical geographers dealing with eras beyond living memory whose only access to the people, places and events of the past is through documentary remains in dusty archives.

Yet for many of us studying more contemporary subject-matters, we do have the option of trying to generate appropriate data for ourselves. We have the opportunity to 'go out into the field' to conduct what is normally called fieldwork – although note our wish, as stated in Chapter 1 (see Box 1.2), to keep an expanded rather than a narrowed conception of both the field and fieldwork – in the course of which we seek to generate data through our own active engagement with, or questioning of, all manner of relevant people going about their business in the everyday world (whether they be a planner, a salesperson, a musician, a refugee or whatever). There are numerous methods deployed by human geographers in this attempt to generate data through their own endeavours, and some of these are perhaps all too familiar, such as the questionnaire survey, while others of them are probably not so familiar (certainly to undergraduate students), such as using focus groups or conducting participant observation. All these methods are relatively well known to social scientists of different persuasions, including historians and anthropologists, and there are established theories about their strengths, guidelines about the practicalities of their conduct and debates about the ethics wrapped up in their operation. Human geographers must be aware of these theories, guidelines, ideas and the like, and should also contribute to reflections on the methods involved (see Box 1.4) on the basis of their own special insights about, for instance, what it means to be trying to generate data about ordinary people occupying specific spaces or possessing feelings about particular places.

In this regard, therefore, we as human geographers strive to make our own data, to produce our own sources of data: in a word, we are now talking about data that are not preconstructed but *self-constructed* by us through our active fieldwork. If we conduct a questionnaire survey, it is we who are deciding upon the questions to ask, the sample of people to be asked, the procedures for administering the questionnaire, the tactics for dealing with respondents and so on. Any data arrived at in the process are indelibly shaped by the decisions we have taken, and the reliability of these data, however exactly we might conceive of what is or is not 'reliable', is deeply dependent upon the manner in which we have gone about the questionnaire exercise from the first moment of question formulation to the last moment of questionnaire delivery in the street, on the doorstep or over the telephone. If we conduct interviews, it is similarly we who are responsible for every aspect of how schedules of interviews are drawn up, who is interviewed and how interviews run in the flesh, and the data produced are unavoidably a measure of how successful or otherwise we have been in topic selection, interviewee choice and interview management. If we deploy other methods, such as different versions of participant observation, the data that are embodied by entries in our field diaries cannot but reflect our own abilities and inabilities in playing a role in field settings, our confidence or inhibitions when it comes to directing interpersonal interactions and the quality or otherwise of our observational skills, among other things.

It is essential that we do think very carefully about these methods, that we stand back from the logistics of undertaking them to consider the broader picture of exactly how they work, how what we do affects their execution and how relations with people under study are fostered, maintained or perhaps threatened by their utilization, to name but a few salient considerations. In so doing, both the possibilities and the limitations of data constructed by these methods should become apparent, and what is appropriate and what is inappropriate to demand of such data should also become much clearer to us. In Chapters 5 and 6 we strive to lay bare many of the issues bound up in the making of several different if overlapping varieties of self-constructed data, suggesting along the way numerous things that the human geographer must keep in mind when generating such data. First, in Chapter 5, we look at different methods for what we refer to quite simply as 'talking to people', examining questionnaires, interviews and focus groups; and, secondly, in Chapter 6, we look at doing ethnographies', tackling those methods that return in some respects but certainly not others to the 'being there' tradition of fieldwork examined in Chapter 1. In both cases, we focus on the practices involved as we encounter a 'world of people', stressing that such practices are indelibly social, marked by the complexities of negotiating social relations both directly with our research subjects and more broadly in the social settings under study. Moreover, because of these social dimensions, it is inevitably the case that things may not always run smoothly and that our best efforts to structure research encounters will sometimes go awry – people will not turn up or be around, will answer our questions 'oddly', will be hostile to us or will simply introduce a host of wholly unexpected issues – and hence leave us realizing that the self-constructing of data has to be flexible, open to disasters and prepared to change in the process. Self-constructed data are hence unlikely to be neatly self-*controlled* data, precisely because the methods involved are social practices rather than akin to laboratory techniques.

2   *Official Sources*

The centralization of knowledge requires *facts* – and the legitimization of some facts, and the methods used to collect them, against other facts – to justify features and forms of *policy*. (Corrigan and Sayer, 1985: 124, emphasis in original)

Introduction

One of the central arguments of this book is that in all human geographical research it is very important for the researcher to consider the processes through which data sources are constructed. In some cases the researcher is directly involved in the processes of data construction. In others he or she is using data that have been produced by others. In this and the next two chapters, we consider sources of geographical data that are not produced by the researcher him or herself but constructed by others before the research takes place. Such sources may have been produced privately by individuals, social groups, voluntary organizations or firms (these types of data are considered in Chapters 3 and 4). Alternatively they may have been produced by government agencies or public authorities; that is, by the state. Such 'official data' are the subject of this chapter. We are treating 'official information' separately from other types because it is produced in distinctive ways for particular purposes and these need to be understood when using the information in research.

Official information is of enormous importance for research in human geography. The population geographer's migration flows, the economic geographer's local labour markets, the social geographer's crime patterns and the political geographer's election analyses (among many others) all usually depend, at least in part, on official information. Historical geographers often rely heavily, though by no means exclusively, on official records of past events (for examples, see Baker et al., 1970). Given the importance of historical sources to their work, they have often also been particularly sensitive to the impact of the circumstances under which those sources were produced, an issue that is of particular significance to the arguments of this chapter. Census data are of great value to both historical and contemporary geographers and it is interesting that geographers are now at the forefront of technical and intellectual developments in census design and data collection (Martin, 2000). Official voting data provide the raw material for numerous studies of electoral geography.[1] Official documents are also textual, as well as statistical, and geographers make good use of these too. For example, Moon and Brown (2000) examine the 'spatialization' of policy discourse through an analysis of health policy documents in Britain.

In addition to the sheer number of studies which use it, official information is important because of its particular authority. When we read official statistics or government reports, we often assume they are reliable and accurate simply *because* they are official. Furthermore, some of this authority rubs off on academic research based on such data, making it seem sound, well founded and valid. There are certainly some good grounds for these assumptions. First, governments (especially in industrialized countries) have very large resources at their disposal. These resources (staff, finance and so on) are usually much larger than those to which academic researchers have access. Governments can thus draw larger samples, undertake more thorough analyses, use longer timescales and cover wider areas than other researchers and organizations. Secondly, governments frequently give themselves legal powers to support their research activities. For example, in most countries, members of the public are obliged by law to complete a census return on a regular basis, often every ten years. In many cases information may be a useful by-product of legislation primarily enacted for other purposes. For instance, the collection of taxes may also generate information about employment and income, while vehicle licensing leads to information about car-ownership patterns. Thirdly, in the modern world, governments are interested and involved in much of the everyday life of the population. Through the breadth of its activities the government knows something about what we are doing: whether we are shopping, working, earning, marrying, having children, travelling, suffering illness, claiming welfare benefit or finally dying. Of course we are not always identifiable as individuals within this information, but our activities contribute to it, none the less.

Yet despite the resources, legal powers and breadth of involvement of governments there are also good grounds for supposing that official information is not as reliable as is frequently assumed and that its special authority

is undeserved. Governments are not neutral referees overlooking society but players actively involved in the game. Like other organizations in society, they have particular objectives in obtaining, processing and presenting information and particular interests at stake in its content. Official information should therefore be treated with the same healthy scepticism which most good researchers bring to the study of *un*official sources.

Throughout this book we will be stressing that no source of research material may be taken at face value, even, or perhaps especially, where it carries the cachet associated with 'official' information. In all cases, data used by researchers in human geography are constructed in specific cultural, political and economic contexts which influence their character and content. If we are to construct valid interpretations of data, therefore, we need to understand, and to take account of, those contexts. Thus, in this chapter we consider how, why and in which contexts governments construct the data they do and the consequent implications for their use by human geographers engaged in research.

Types of official information

Governments produce information in a variety of forms. A simple typology might show that information can be textual, graphical and cartographical, aural or numerical in form. A government report is a textual source, official photographs are graphical, government radio broadcasts are aural and government statistics are numerical. These differences are important and they affect how the data can be interpreted. In this chapter, however, we are focusing on data construction rather than data interpretation. As far as construction is concerned, three broad categories of data can be distinguished.

First, governments are large bureaucracies and produce information as part of the bureaucratic process. A minister may commission a report before reaching a decision. Parliament produces bills and Acts in the process of legislating. Local councils publish the agendas and minutes of their meetings. Official inquiries absorb written and oral evidence and produce reports. The courts produce written judgments. In all these cases, documents and information are the by-products of the process of governing and of the operation of large bureaucratic organizations.

Secondly, governments monitor the societies they govern. As we shall see shortly, monitoring and surveillance are very important in shaping the nature of the modern state. They also produce huge amounts of information. Taxation records, population records, censuses, health records, financial statistics and a whole host of other sources of information are generated simply because the state and the government need or want to keep a check on what is going on.

Thirdly, governments communicate. From public health announcements to political propaganda and from communiqués at international summits to the contents of the school curriculum, governments address their citizens (and the wider world) in a variety of ways. Such communication may involve social engineering of a mild kind, such as trying to persuade parents to immunize their children, or it may involve giving direct orders prohibiting certain activities. It may involve the relatively straightforward provision of information (about welfare benefits or health care, for example) or it may be sophisticated political publicity intended to change people's minds.

These three categories may overlap. For example, official statistics are often produced through the bureaucratic process of governing, and may then be used in public information campaigns.

Information and state formation

Record-keeping and the origins of states

All organized human societies, past and present (including tribal societies based on kinship relations), have some form of government or collective rule. Following the rise of settled agriculture, early urban centres and divisions of labour by about 5000 years ago, government increasingly became institutionalized in *states* of various forms (Mann, 1986). According to the sociologist Anthony Giddens, the origins of states are closely related to the invention of writing. Giddens (1985: 41–9) challenges the idea that writing is straightforwardly a representation of speech. He argues instead that writing initially took the form of record-keeping:

> Many linguists have regarded writing as no more than an extension of speech, the transcription of utterances to transcriptions on stone, paper or other material substances that can be marked. But neither the first origins of writing in ancient civilisations nor a philosophical characterisation of language bears out such a view. Writing did not originate as an isomorphic representation of speech, but as a mode of administrative notation, used to keep records or tallies. (1985: 41)

There are examples of non-modern states which apparently used no form of writing. In the Andean Mountains of what is now South America, for example:

> About A.D. 1400–30, one 'tribal' grouping and chiefdom, the Inca, conquered the rest. By 1475, the Inca had used massive corvée labor gangs to build cities, roads, and large-scale irrigation projects. They had created a centralized theocratic state with their own chief as god. They had taken land into state ownership and had put economic, political, and military administration into the hands of the Inca nobility. They had either devised or extended the *quipu* system whereby

bundles of knotted strings could convey messages around the empire. This was not exactly 'literacy' ... Yet it was as advanced a form of administrative communication as any found in early empires. (Mann, 1986: 122)

What seems crucial, therefore, is not writing per se but the resources and ability to *keep records*. The significance of this development should not be underestimated since, without the capacity for record-keeping, an organization would lack a key means of what Giddens terms 'administrative power' (1981: 94–5; 1985: 19). Indeed it is doubtful whether it could be an organization at all. The exercise of administrative power based on the collection and organization of information is thus a defining feature of all states. It is both an outcome of state activity and constitutive of state power: states generate information through their functions but also require information in order to undertake those functions in the first place. Generating information, therefore, is not an optional extra for state organizations but seems to be central to the very possibility of states existing at all.

Official information and the development of the modern state

While record-keeping is a necessary feature of non-modern states like the Inca Empire, it is in modern states that it reaches its most developed form. Drawing on the work of the classical sociologist Max Weber, Christopher Dandeker argues that 'bureaucratic decisions and calculations depend on *knowledge of the files,* that is on a mastery of the information stored centrally in the organization' (1990: 9, emphasis in original). For Weber, the institutions of the state are exactly such archetypal bureaucratic organizations. The construction and processing of very large quantities of information are defining features of government in the modern world. The amount of

information routinely collected by state organizations is vast. The UK government's annual summary of official statistics (the *Annual Abstract*) contains over 300 separate tables of data dealing with topics as diverse as population, iron and steel production, government debt and academic research. The US government's *Statistical Abstract of the United States* runs to over 1000 pages. By contrast, in medieval England the famous Domesday survey of 1085 contained just 11 questions relating to the ownership of property (Darby, 1977: 4–5). Furthermore, medieval governments undertook such data collection very infrequently: no more often than once every several decades. By contrast, modern governments are engaged in an almost continuous process of monitoring the various indicators which are the basis of 'official statistics'.

Furthermore, *statistical* information forms only a part (albeit an important one) of the total information produced by the state. Policy decisions depend on 'knowledge of the files' but the contents of the files can take many forms. Records concerning individuals, for example, may include reports, photographs, transcripts and forms. The governing process itself involves the production of large quantities of textual and graphical, as well as numerical and statistical, information. For example, governments often establish commissions of inquiry relating to particular areas of policy. The reports of such bodies are often invaluable sources of research data. Other kinds of records include financial accounts and registers of various kinds. An example of a geographical study that makes use of this kind of source is Driver's account of the geography of the workhouse system in Victorian England and Wales, which draws on the unpublished Registers of Authorised Workhouse Expenditure (1989; 1993: 73–94).

The 'outputs' of the policy-making process can also be a rich source of data for research. Modern states produce copious quantities of planning and policy documents which both present government policy and enable its implementation. They also contain very particular representation of the social and economic world (the objects of governance, as it were). Governments have also significantly expanded the quantity and range of information they publish under the umbrella of government publicity or public information. In this case states are addressing their populations directly, and examples include the educational curriculums, health advice, edicts compelling people to do certain things and to refrain from doing others and so on. Whichever the form in which state information is produced, one thing is clear: the rate of its production has increased dramatically over time and continues to do so.

How should we account for the extensive growth in government activity in this area? One important reason concerns technology. Today governments have the technological capacity to conduct data-gathering exercises on a scale which their medieval predecessors could scarcely have imagined. Printing, transport, telecommunications and storage facilities are all involved. Yet while this is clearly a necessary condition of the growth, it is not, in itself, a sufficient one. It does not tell us why governments might *want* to gather information on this scale. To answer this question we need to consider the ways in which states themselves change over time and how those changes are related to changes in civil society.

All official information is gathered for specific government purposes. However, those purposes vary through time and across space. The worries and concerns of the Norman monarchy in medieval England were not the same as those of the drafters of the Constitution of the USA or of the present-day European Commission. A good example of this historical development is provided by the English (subsequently the British) state. According to Philip Corrigan and Derek Sayer in their book *The Great Arch* (1985), the English state underwent a succession of phases of formation. Each phase involved what they term a process of 'cultural revolution' (Corrigan and Sayer, 1985: 1–2): a transformation in the ways in which the agents and agencies of the state made sense of and represented civil society. Part and parcel of this process was a growth and extension of the generation and use by the English state of official information of various sorts.

The purpose of the Domesday survey, for example, has been the subject of academic debate, but whatever its purpose it is clear that official information in medieval Europe was collected, when it was collected at all, for a limited number of reasons. By modern standards, medieval states had a small number of functions centred on the security and finance of the royal household, the defence of the realm and the regulation of feudal ownership rights. Beyond these, there was relatively little contact between the state and its subjects. Most of what are regarded as state functions today in the fields of education, health care, support for economic activity and even the preservation of law and order were carried out, in so far as they took place at all, by other agencies, particularly the Church and the feudal manor.

The state did need information in some spheres of activity, notably taxation, but feudal ties of rights and obligations and religious conviction ensured that for the most part government was carried on and social order was maintained without the state needing to know in detail about most of the activities of its population.

In the 1530s, during the so-called Tudor revolution in government, Thomas Cromwell, an official at the Court of Henry VIII, was responsible for a significant increase in the information-gathering activities of the English state (Corrigan and Sayer, 1985: 48). Much of this was aimed at countering threats to the

King, but it included making compulsory, from 1538, the keeping of parish registers of baptisms, marriages and burials. According to Corrigan and Sayer these registers:

> provoked widespread resentment, and people feared their use for taxation purposes. Cromwell's own justification, responding to such fears, is interesting. The registers were being instituted, he said, 'for the avoiding of sundry strifes, processes and contentions rising upon age, lineal descents, title of inheritance, legitimation of bastardy, and for knoledge of whether any person is our subject or no'; 'and also' he added, 'for sundry other causes' (Elton, 1972: 259–60). Concern with property and concern with rule are equally witnessed here. Registers are not merely a technical device, they materialize new sorts of claims of a state over its subjects. The Elizabethan Poor Law, for example, was to rely for its administration on the kind of detailed knowledge of individuals the registers supplied. (1985: 50)

Here then was a definite extension of the functions of government accompanied *and enabled* by an extension in the information-gathering activities of the state. Tudor England also saw the systematic mapping of the country with the county maps of Christopher Saxton. According to Derek Gregory, 'these paper landscapes represented the administrative apparatus of the state with an apparently indelible authority, revealing the steady encroachment of surveillance into still more spheres of social life' (1989c: 376). Such piecemeal encroachment continued during the seventeenth and eighteenth centuries (Brewer, 1989), but it was not until the nineteenth century that the information-gathering activities of the state took on anything like their modern form in terms of both quantity and quality.

This development both reflected and partly influenced the changing relationship between state institutions and the society they governed. In this, the twin processes of urbanization and industrialization were central. In Britain, the agricultural 'improvements' of the

country during the seventeenth and eighteenth centuries had created a large class of landless people. These people, separated from the traditional religious, cultural and economic ties of rural life, and with no means of support other than the sale of their labour power, made their way to urban areas. There, new, factory-based systems of production were hungry for the labour that they could provide and the industrial towns and cities quickly expanded. In 1801 (the year of the first Census of Population in Britain) 33.8% of the population of England and Wales lived in urban centres. By 1901 the urban population had expanded eightfold and its proportion had increased to 78% (Carter, 1990: 403). This startling transformation in the geography of the country caused *and made visible* social problems on a scale never previously imagined. Working-class districts were regarded as overcrowded and the people who lived in them uneducated and at risk of disease and crime. In addition they were uprooted from established ties of kinship and community, which had often served to mask, if not to prevent or solve, similar problems in rural areas. In the new industrial areas, however, there was no disguising the suffering and poverty which existed and which was exacerbated by exploitative conditions of employment in the factory system.

It was not long before the mass of the urban poor became a focus for crusading social investigators and thence for government policy. Seduced by its own rhetoric of progress, bourgeois society in Victorian Britain was hardly about to blame industrialization itself for the plight of the poor. Instead it preferred to locate the causes of social distress in the moral failings of individual sufferers themselves and the overcrowding and dirt of their immediate environments. According to Felix Driver:

> The association of moral and medical concerns was a central feature of nineteenth-century social science. The question of

density was a special focus of interest. There were many attempts to relate population density to crime and other indices of 'demoralization', at various ecological scales. The most common foci were the overcrowded, unsupervised and disorganized spaces of the nineteenth-century city ... What seems to have most concerned middle-class commentators was their own lack of control over and within such areas; indeed, their obsession with hidden recesses, narrow turnings, dark alleys and shadowy corners was quite overwhelming. The literature on the rookeries of London, for example, was predicted [sic] on the assumption that they were located beyond the public gaze, *outside the ambit of official surveillance.* (1988: 280–1, emphasis added)

A mixture of fear, loathing and social concern of the middle class and the government for the mass of the urban poor prompted a desire to know precisely what were the social conditions of the population, to enable both reform and social control. The rapid growth in the collection and interpretation of social statistics in nineteenth-century Britain can thus be related directly to the urban–industrial transformation. Similar processes took place in other countries which experienced similar industrial development, albeit sometimes with different emphases. In France, where the development was related particularly to concerns about public health, 'a professionalization and regularization of population statistics, linked to a range of reformist projects, occurred during the first decades of the nineteenth century' (Rabinow, 1989: 60). In the USA the focus of concern was frequently immigration and the formation of ethnically segregated 'ghettos' (Ward, 1978). What unites all these transitions is the use made of statistical data to try to understand new, radically different and much larger social worlds: 'without the combination of statistical theory ... and arrangements for the collection of statistical data, ... the society that was emerging out of the industrial revolution was literally unknowable' (Williams, 1979: 171).

These desires to make knowable the unknown both to 'improve' and to control it led to an upsurge in the collection of statistical information. In Britain, the government was accompanied, and frequently assisted, in this task by interested members of the Victorian middle class. The establishment of 'statistical societies' in Manchester (1833) and London (1834) (and later in other provincial cities) laid the foundation of a long national tradition of empirical social research (Cullen, 1975). The studies undertaken by these societies and individuals were the forerunners of non-governmental research, which is the subject of the next chapter. They culminated in the 1890s with the publication of Charles Booth's huge survey of *Life and Labour of the People in London*.

For the government, statistics were compiled with increasing frequency throughout the nineteenth century. The Statistical Department of the Board of Trade was founded in 1832. The Registration Act 1836 led to the establishment in 1837 of the General Register Office for England and Wales to record births, deaths and marriages and to undertake the decennial Census of Population. A General Register Office for Scotland was set up in 1855.

The close links between the government and the growth of statistical information are expressed in the very word 'statistics' itself. Originally the word was simply an adjective associated with the noun 'state' (*statist* + *ic*). According to one nineteenth-century definition, statistics is 'that department of political science which is concerned in collecting and arranging facts illustrative of the conditions and resources of a State' (Corrigan and Sayer, 1985: 134).

The purpose of this information gathering, particularly in the first half of the nineteenth century, was to enable the development of state policy. Policies of moral and environmental improvement were introduced on the back of statistical inquiries. The workhouse system, the development of public health measures, medical activities and education were all central to the government of Victorian Britain and all depended on information. Furthermore, all were intended to treat social ills by extending state influence, regulation and control into all parts of civil society.

Effective public policy was not the only consequence of the growth in state information gathering. The development of statistics also affected the object of study (i.e. society) itself. This is a difficult notion to grasp at first as we tend to assume that statistics are first and foremost a simple description of social 'facts'. However, the production of official statistics is not the neutral, technical and scientific exercise it appears to be. As we have suggested, official data do not provide complete or transparent pictures of social reality. Rather they are influenced and conditioned by the interests at stake in their production. The demands of government policy or the moral concerns of the middle classes are reflected in the statistics that are constructed. At the very least they require that certain topics of inquiry are selected as relevant over others. At worst there may be active manipulation of figures to provide justifications for particular government activities. But we can go further than this. Not only are statistics sometimes an inaccurate picture of 'reality' but, by being collected at all, they also change the nature of that 'reality'. Catherine Hakim (1980) has analysed the census reports accompanying the British Census since it began in 1801, and her findings help to show what we mean.

As she points out, the census has been an important source of information on occupations. Yet the particular choice of occupation categories included in the survey has played a part in making some occupations respectable and others illegitimate. Estimates suggest that 20% of unmarried women in London in 1817 were involved in prostitution (Hakim, 1980: 563). Yet the category of 'prostitute' is not listed in the occupational classification of the British Census. This refusal to give formal recognition to the economic activity of a substantial proportion of the city's population had the effect of making prostitution appear less socially acceptable, as well as calling into question the accuracy of the data.

In the early years of the census the full separation of paid employment from home life, and the resulting partial exclusion of women from the labour market, had yet to occur. The economic unit was considered to be the family as a whole, rather than an individual male breadwinner. By the middle of the nineteenth century, however, the census had begun to ask for information on the occupation of individuals and this led to numerous women being classified as economically inactive, despite continuing to be engaged in domestic labour in the home. The change in the form in which these figures were collected partly reflected, *but also partly produced*, these developments in the labour market. By categorizing women as 'inactive' the official census contributed to the exclusion of women from the labour market and to their confinement in the home.

The power of such official 'discourses' and 'representations' to affect as well as reflect society is recognized by Corrigan and Sayer:

> Whilst we should reject states' claims as description – they are, precisely, claims – we need equally to recognize them as material empowerments, as emphases which cartographize and condition the relations they help organize. Here, as elsewhere, to adapt an old sociological maxim, what is defined as real (which is not to say the definitions are uncontested) is real in its consequences. (1985: 142)

Commenting on Hakim's research, Corrigan and Sayer point out that the British Census is above all an inquiry into who is a subject of the Crown, and is thus itself centrally implicated in 'the formation of social identities' (p. 132). Giddens makes the point forcefully:

> it might well be accepted that, given certain reservations about the manner of their collection, official statistics are an invaluable source of data for social research. But they are not just 'about' an independently given universe of social objects and events, *they are in part constitutive of it*. The administrative power generated by the nation-state could not exist without the information base that is the means of its reflexive self-regulation. (1985: 180, emphasis original)

Recent writings by human geographers and other social scientists have drawn on Michel Foucault's (1991) concept of 'governmentality' to capture the relationship between state power, information (or knowledge) and the constitution of the objects of governance.[2] 'Governmentality' refers to the ways in which knowledge, information and understanding operate as tools (or 'technologies') of government. For Foucault, knowledge, including the kinds of knowledge produced by official information, is not a neutral, transparent window on to a pre-existing social world, in which social problems are somehow simply revealed by objective fact-finding on the part of state organizations. On the contrary, Foucault's approach implies that the social problems that form the objects of government policy in some sense come into being through the production of official knowledge of them. Take the example of 'regional economic development' which has been of longstanding concern to both economic geographers and governments. Official economic statistics are used to identify regions with economic problems, such as high unemployment or an overdependence on a narrow range of (perhaps declining) industrial sectors. In order to do this the state must first define the boundaries of each economic region (or 'regional economy'). Yet as recent work by geographers has shown, the very idea of coherent regional economies is at best complex and problematic and, at worst, downright misleading (Allen et al., 1998). Regional territories are certainly not natural or self-evident containers for economic processes. In fact, the definition of regions is itself as much a contestable political process as the development of regional economic development policy. From the perspective of the literature on governmentality, 'problem regions' as objects of state policy-making exist only because of the state's acquisition and mobilization of particular kinds of economic knowledge and information (Painter, 2002). This is not to deny the existence of high levels of poverty or other kinds of economic distress in particular places, but it does call into question the way in which economic problems come to be framed as such and, in this example, how they come to be framed as 'regional problems' requiring 'regional' solutions.

The contemporary informational state

One of the important insights of the Italian political theorist, Antonio Gramsci, concerned the relationship between force and consent in the maintenance of political power. For Gramsci, in the modern world power is not solely a question of control over armed force but also involves the active co-operation of the population. Both the creation and management of this 'consent' by the state and its use of 'coercion' (police, prisons, etc.) require information to be gathered on the population. As Giddens puts it, 'surveillance as the mobilizing of administrative power – through the storage and control of information – is the primary means of the concentration of authoritative resources involved in the formation of the nation-state' (1985: 181).

During the twentieth century information collection, storage and use by states have become increasingly sophisticated and differentiated. After the Second World War, in particular, the role of states has expanded considerably, with a concomitant increase in their informational activities. First, in the decades since the 1940s (at least until recently), most developed or industrialized states have been *welfare* states. This has involved governments in the widespread provision of public services in the fields of social security (e.g. old-age pensions), health care and education. While these expenditures have seen a degree of retrenchment in the 1980s and 1990s, they remain substantially above the level of state provision typical of the nineteenth and early twentieth centuries. Planning and providing these services require substantial amounts of information relating to the geographical distribution, age and health of the population.

Secondly, over the same period, state claims to legitimacy have shifted in primary focus from 'high' politics (national security and the maintenance of state borders) to 'low' politics (trade, economic growth, social welfare). This has involved a relative decrease in the collection of 'intelligence' information (i.e. spying) and diplomacy and a relative increase in the collection of information relating to economic performance, industry and employment.

Thirdly, information is now held not only on populations or geographical areas as a whole but also to a very great extent on members of the population as individuals. In contemporary Britain every adult is assumed to be registered with the National Health Service, the national insurance system and the electoral registration officer. Most will also be recorded by the Inland Revenue and many by the Driver and Vehicle Licensing Authority and by the Passport Office. Other specialist agencies such as the social services and educational organizations have files on particular indviduals. The police maintain the Police National Computer which records not only the details of convicted criminals but also details of police contacts with other members of the population, including many who may be completely innocent of any wrongdoing. This type of information reaches its most developed form in so-called totalitarian countries, in which large parts of the state apparatus are given over to the surveillance and monitoring of individuals. As Giddens points out, however, the difference between 'totalitarian' states and 'democratic' ones is a difference of degree, rather than kind. According to Giddens, all modern states have totalitarian *tendencies* (1985: 310). Indeed, since totalitarianism depends upon surveillance, a key precondition of its emergence exists by definition in all modern states.

Information collected at the level of the individual is undoubtedly highly significant in terms of states' social control activities, especially at the more 'coercive' end of the consent–coercion spectrum. However it is not a significant source of data for geographical research. The reason for this is, of course, its confidentiality. Information on individuals is confidential to the state, for two main reasons: first, for the protection of the individual concerned and, secondly, for the protection of the interests and activities of the state. Thus information relating to crimes and criminals, in particular, is kept secret in order to maintain its power. If suspected criminals were aware of the information held about them, it is argued, they would be better able to evade the efforts of the state to police them.

New technology has added a further twist to debates about the secrecy of state information. The use of microelectronic information technology has vastly increased the capacity of both the state and other organizations to store, sort, process, duplicate and analyse information, including information about individuals. This has led to concern among civil rights organizations about the potential uses to which such readily available large quantities of data might be put. As a result of such pressure, some countries now have statutory protection for those about whom records are kept. In the USA the Freedom of Information Act 1966 guarantees (within limits) the rights of individual citizens to have access to large parts of the state's information archive. In Britain the provisions of the Data Protection Act are much weaker and relate only to access to records on oneself, and until recently only where these are held on computer. The Act was intended primarily to allow individuals to correct errors of fact in records relating to them. There are many exclusions, including, notably, police records. Historically there has been no general freedom of information in Britain, maintaining the reputation of the British state as one of the most secretive in the Western world (Michael, 1982).

Through the highly restrictive operation of the Official Secrets Act, most information held by the British government is confidential unless expressly released to the public. Notwithstanding some recent amendments to remove some of the more absurd examples of Whitehall secrecy, there is still much information gathered by the British state which is reserved for the confidential use of government alone. This confidentiality extends well beyond information on particular individuals to cover government-commissioned reports on a whole variety of subjects of potential interest to academic researchers. Again, in the USA there is considerably more access through the Freedom of Information Act. In some cases this results in 'secret' British information being publicly available in the USA. This picture may change in the future, as the British government has introduced a Freedom of Information Act. However, the legislation has been heavily criticized by civil rights organizations as likely to restrict further, rather than to open up, access to government information.

For the most part, therefore, the official information most used by academic researchers is that which is actively made public by the state, usually through publication through a government publishing house, such as Her Majesty's Stationery Office in Britain (now mostly privatized under the trading name the Stationery Office Limited) or the US Government Printing Office, or through a multilateral agency such as the European Commission or the United Nations. The remainder of this chapter considers this subset of government information which is most widely available, even though it may represent only a small proportion of the information in the state's possession.

Government information organizations

Most contemporary governments publish large amounts of statistical and other information relating to their populations and economies and social systems. This information is made available as a public service to researchers, journalists and other interested individuals through the publication of statistical reports on a regular (e.g. annual or monthly) basis. Such reports are normally available in large public libraries, in most university libraries and can be purchased from government publishing houses. Typical examples include the British government's *Annual Abstract of Statistics* and the US government's yearly *Statistical Abstract of the United States*.

These materials are very useful to human geographers undertaking research. A wide range of topics is covered and the figures are usually detailed and can be broken down into smaller geographical areas and/or time periods. Frequently even more detail, and the possibility of cross-tabulation, is available from the government department concerned. Increasingly this material is being, or will be, made available on various electronic media, such as computer-readable magnetic tapes or disks. In the future more and more of these data will be available for use in computer mapping and geographical information systems, allowing rapid cartographical representations to be constructed from official statistics. In many countries the state has its own cartographic service (in Britain, the Ordnance Survey), reflecting the important role of territory in the formation and security of modern states.

The data contained in official statistics are often population (rather than sample) data. This makes them, at least in principle, more reliable than similarly constructed sample data as there will be no sampling errors. Sampling errors arise because the characteristics of a sample of the population are unlikely to be precisely the same as those of the whole population. However, there may be errors associated with survey design and implementation. Where population data are available, descriptive statistics for the population can be derived, and it is not necessary to infer population characteristics from a sample. Where samples are used in official statistics, they are likely to be large (or at least larger than in most academic research) and this will tend to reduce the significance of sampling errors. In many cases government agencies have been at the forefront of the development of social survey techniques, and the reliabilty of the data they produce should thus, again in principle, be good. (See Chapter 5 for further information on the use of samples and surveys, and Chapter 8 on the interpretation of numerical data.)

For all these reasons (availability, coverage, detail and reliability) official statistics are very useful in human geographical research. However, as has been discussed above, official statistics are not first and foremost produced as a public service. The principal aim of their construction is to inform government policy. Furthermore, as we saw in the work of Hakim and Giddens, official statistics influence the society they represent by defining and categorizing society in particular ways rather than others. At the end of this chapter we will consider some of the technical questions which arise from these theoretical insights and look at some specific examples to illustrate the problems and possibilities of using official statistics in geographical research. Before doing so, however, we will briefly discuss the changing role of government statistical agencies, again using the British state as an example.

In Britain, the modern data-construction activities of the state date from 1801, the year of the first national Census of Population. Since then, a Census of Population has been taken every ten years, with the exception of 1941. In the twentieth century this basic survey was joined by an ever-increasing output of official information. The first four censuses (in 1801, 1811, 1821 and 1831) were undertaken in England and Wales by the Overseers of the Poor under the guidance of John Rickman, the Clerk to the House of Commons. A General Register Office (GRO) was established in England and Wales in 1837 and in Scotland in 1855. The Registrar General in each case was responsible not only for the registration of births, marriages and deaths but also for undertaking the census (Mills, 1987). Initially the GRO's brief was limited to these two main functions. Social survey research was in its infancy and the usual means, apart from the census, for gathering information for the government were inquiries carried out by royal commissions on the basis of subjective evidence from public officials around the country (Burton and Carlen, 1979; Whitehead, 1987). In the late nineteenth and early twentieth century the investigations of Charles Booth and Beatrice and Sidney Webb established the usefulness of social surveys, but it was not until the Second World War that the population census was joined by regular *sample* surveys carried out on behalf of the government.

The wartime government created the Government Social Survey to 'gather quantitative information on, among other things, the state of public morale in wartime conditions' (Whitehead, 1987: 45). After the war the survey organization, part of the Central Office of Information (formerly the Ministry of Information), was retained and in 1957 it was given the job of undertaking a regular Family Expenditure Survey, which continues to the present day. From 1967 to 1970 it was made an independent department, before being merged with the GRO to form the Office of Population Censuses and Surveys (OPCS). The OPCS was responsible for the census, social surveys, health statistics from the National Health Service and population registration, and came under the responsibility of the Secretary of State for Health.

The OPCS was complemented by a separate Central Statistical Office (CSO). Like the Government Social Survey it started life in 1941 to provide better statistics for the management of the wartime economy. It remained part of the Cabinet Office until 1989 when it became a separate government department responsible to the Chancellor of the Exchequer. Its main responsibility lay in the collection and publication of trade, financial and business statistics, and the national accounts. It also housed the government Statistical Service and published a wide range of official statistics on behalf of other government departments.

In 1996, the OPCS and CSO merged to form a new 'Office for National Statistics' (ONS), an independent government department responsible to the Chancellor of the Exchequer. The government claimed that the new organization would 'meet a perceived need for greater coherence and compatibility in government statistics, for improved presentation and for easier public access' (ONS website). Further changes took place following the election of the Labour government of Tony Blair in 1997. A new organization, National Statistics, was launched in June 2000, headed by the country's first National Statistician. A Statistics Commission was also established to oversee the quality and integrity of official statistical data.

In many cases the statistics produced by the government reflect information which is collected in the course of other activities. The unemployment figures, for example, are a product of the operation of the unemployment benefits system. Sometimes statistics will be collected specifically to inform new policy initiatives. In many cases today this work is contracted out to external, non-governmental researchers. These contractors may be dedicated survey and research organizations, such as the market research companies, or they may be academics working in universities. For academics there are substantial attractions to this type of work. It represents a source of research funding at a time when other funds are difficult to come by and it appears to provide a way for academic researchers to influence the policy-making process. Very often, however, there are unforeseen difficulties. There are frequently restrictions placed on the publication of results, and often the information required by the government takes a different form from that suggested by the academic judgement of the researcher. It is also doubtful how far government policy is changed as a result of such research. Frequently, broad policies have been agreed and research is conducted either to justify the policy after the event or to prepare for its implementation. Where results contradict assumptions of government policy-makers it is likely that the research will be suppressed rather than that the policy will be changed.

Understanding the construction of official information

Government statistical organizations are perhaps the most organized and focused examples of the information and knowledge production activities of modern states. As we have already noted, however, governments also produce huge quantities of non-numerical information. In both cases, however, similar motivation is at work. The process of government both requires and produces information. At the same time, the ways in which information is constructed and presented help to shape the 'policy landscape' in particular ways; some policy approaches are made possible and legitimated, others are blocked off and discredited. In this sense no information is 'neutral' and this is certainly true of government information, whether statistical or textual. Using official data sources in geographical research thus effectively requires careful attention to the process of data construction and means that the researcher should ask some hard questions about the information he or she is proposing to use:

- Why was the information constructed?
- To which government policies does it relate?
- Have policy concerns influenced which data were constructed and how?
- In what ways?

The question 'What policy concerns or political ideas motivated the construction of the information?' is important even where the answer seems obvious. Take inflation for example. We take it for granted that information about price increases will be published on a regular basis. But why is this important? One could argue that collecting retail price information at all sets 'inflation' up as a policy problem and that it leads to particular policy responses. Some economists, for example, reject the idea that a goal of low inflation should be as absolutely central to government policy as it has been in recent years, but the high profile of the inflation statistics tends to reinforce the idea that inflation is a very bad

thing. Until recently in Britain there was not seen to be a need to compile information on a national basis about the educational attainment of school children while they were still at school. Today, testing of pupils on a national basis at 7, 11 and 14 provides information allowing 'league tables' of school performance to be constructed. Leaving aside the very important issue of whether the information is a good measure of school performance, there are profound implications from the fact the information is constructed at all. There is a close relationship here between the compilation of the information on a national basis and the former Conservative government's political aim of increasing competition among schools for the best pupils and scarce resources. Whether one agrees with this aim or not, the data-construction activity here cannot be thought of as a neutral activity separate from the political controversy. Rather it is part of the political process.

These questions about motivation and the relationship between information and policies can be asked of data sources of all types (statistical, textual and graphical). In addition, there are some further questions that need to be considered in the case of textual data:

- Which 'voices' are present in the text? Who is 'speaking' – a supposedly disinterested expert? A professional politician? An interested organization or individual? Perhaps several voices are present, as may be the case in a report of public inquiry, for example.
- By contrast, whose voices are absent? Are there 'silences' in the text representing issues or points of view that have been neglected or deliberately left out?
- What mechanisms led to the production of the report? How was the text put together – verbatim transcription? Summaries of longer drafts? Single or joint authorship? And so on.

- What rhetorical devices and figures of speech are used to convey the message? What metaphors are used? What effect do these have on the content of the document?

There is also a set of questions that relate specifically to statistical data:

- Which categories have been used and why?
- What would be the effect of using different categories?
- What and who are included and excluded from the count and why?
- What would be the effect of including other variables or groups?
- Which sampling and survey techniques have been used?
- What are the likely errors and biases associated with these techniques?
- What corrections and adjustments have been made to the results to allow for errors, and what effects have these had?

None of these questions (with the partial exception of those concerning the mathematics of sampling error) can be answered in a purely technical way. They are matters of judgement and debate. Much will depend on the extent to which information is available on the way data have been constructed and on the political and theoretical stance taken by the researcher (see Chapter 12).[3]

We will now consider these factors in more detail in relation to specific examples.

Domesday Book

On the face of it, it might be thought that Domesday Book would be of interest only to those studying the historical geography of early medieval England. While this may be true of the evidence it supplies of the economic and political landscape of the time, it also provides a fascinating illustration of some of the more general issues that arise in using

official sources of data. Many of the problems and difficulties of using Domesday Book as a research source are also present in using contemporary official sources, albeit in more muted ways. Domesday Book, therefore, gives us a way of highlighting in acute form many of the factors that need to be considered in using *any* official data.

As every British schoolchild knows, Saxon England was invaded in 1066 by William, Duke of Normandy, who established a much stronger system of government in place of the relatively weak and divided Saxon rule. As part of this process, a fully fledged feudal system was put in place, under which all the land belonged to the sovereign who divided it among the nobles as the chief tenants. The chief tenants could divide up their lands and sublet them to other tenants, who could sublet again and so on all the way down the social and political hierarchy. Each feudal tenancy involved sets of mutual rights and obligations. The lord provided protection to his tenants (vassals) and they in turn provided military service and paid certain dues. After his military victory, William abolished most of the existing Saxon earldoms and parcelled out the land afresh, much of it to his Norman supporters. The smallest unit of feudal organization was the manor which was the seat of a local feudal lord to whom the local population owed allegiance. Manors possessed their own land (known as the demesne) and also sublet land to the lowest tier of tenants.

To understand the construction of the document, is necessary to appreciate the administrative geography of Norman England. The major territorial divisions of the country were the shires (from Old English *scir*), also known as counties (from Old French *conté*). Counties were divided in turn into 'hundreds' so-called because they corresponded nominally to an area containing one hundred units of taxation known as 'hides'. In those parts of the country that had been heavily affected by the Danish invasions the Scandinavian terms *wapentake* (instead of 'hundred') and *carucate* (instead of 'hide') were used. The word 'hide' derives from the Old English word for household and refers to a unit of land large enough to support a family and its dependants. Hundreds were subdivided into *vills* (feudal townships). However, the geography of the vills did not correspond directly to that of the feudal manors because a manor could possess land in more than one vill.

The historical geographer, H.C. Darby, made the interpretation of Domesday Book his life's work and he was well aware of the complexities involved in the data–collection process. The precise circumstances that led to the survey that produced Domesday Book are obscure. However, the *Anglo-Saxon Chronicle* records that, in 1085, following discussion at a council at Gloucester, 'King William sent his men into each shire to enquire in great detail about all its resources and who held them' (Darby, 1977: 3) and that all the surveys were subsequently brought together. There is very little in the way of direct evidence about how the surveys were undertaken. However, it is possible to make some deductions about the process of data construction from Domesday Book itself and, especially, from a number of subsidiary documents associated with it (Darby, 1977: 3–9).

Some evidence of the operation of the information-collection process is provided by a document called *Inquisitio Eliensis*, which is a survey of the estates of the abbey of Ely. According to Darby:

> The *Inquisitio* tell us that the king's commissioners heard the evidence 'on the oath of the sheriff of the shire, and of the barons and of their Frenchmen, and of the whole hundred – priests, reeves and six villeins from each vill'. There was a separate jury for each hundred, consisting of eight men, whose function 'was apparently to approve and check the information variously assembled' ... half the jurors were English and the other half Norman ... We cannot say whether the commissioners themselves attended every hundred court ... or whether ... they merely held one session in the county town. A number of entries make it clear that they sometimes heard conflicting evidence ... It has been suggested 'that fiscal documents already in existence were drawn upon to help with the compilation of a partly feudal and partly fiscal enquiry'. The making of the inquest may have been a far more complicated process than was at one time thought. (1977: 5)

While it is impossible to be certain that this procedure was followed in all the local inquiries that produced Domesday Book, the overall picture is fairly clear and suggests that commissioners appointed by the king were sent out around the country to conduct the survey, and that they did so, not through a house-to-house inquiry as might happen today, but by gathering together 'juries' of local representatives including a cross-section of the population and questioning them about the local situation.

The *Inquisitio Eliensis* also lists 11 questions to be asked in each hundred (see Box 2.1). As Darby (1977: 5) notes, 'whether these were the "official instructions" for all counties, we cannot say, but they, or at any rate a similar set of questions, must also have been asked elsewhere'.

This 'fieldwork' seems to have been conducted on a county-by-county and hundred-by-hundred basis and provided the initial survey returns. It appears that these initial returns were then summarized in documents that grouped together several counties, still organizing the information on an area-by-area basis. From these summaries the final Domesday Book proper was prepared. In the final version some of the detailed information was edited out but, more significantly, the areal organization of information was replaced by a structure that mirrored the feudal system. That is, for each county the data were:

Box 2.1 The Domesday questions

1   What is the name of the manor?
2   Who held it in the time of King Edward?
3   Who holds it now?
4   How many hides are there?
5   How many teams, in demesne and among the men?
6   How many villeins? How many cottars? How many slaves? How many freemen? How many sokemen?
7   How much wood? How much meadow? How much pasture? How many mills? How many fisheries?
8   How much as been added or taken away?
9   How much was the whole worth? How much is it worth now?
10 How much had or has each freeman or each sokeman? All this is to be given in triplicate; that is in the time of King Edward, when King William gave it, and at the present time.
11 And whether more can be had than is had?

Explanation of terms not defined elsewhere:

'In the time of King Edward' refers to the reign of Edward the Confessor (1042–66). 'Teams' means ploughteams made up of (usually eight) oxen. 'Villeins' were tenants tied to the land on which they worked. 'Cottars' were tenants who occupied a cottage in return for labour. 'Freemen' (unlike 'villeins') were allowed to leave the land on which they worked. 'Sokemen' were a class of personally free peasants attached to a lord rather than to the land.

rearranged under the headings of the main landholders, beginning with the king himself and continuing with the ecclesiastical lords, the bishops followed by the abbots, then with the great lay lords, and finally with the lesser landholders in descending order ... It follows that if two or more lords held land in a village, the different sets of information must be assembled from their respective folios in order to obtain a picture of the village as a whole. (Darby, 1977: 8)

There are two documents that are thought to be surviving regional summaries. These are the so-called Exeter Domesday Book covering the south west of England and the survey of the eastern counties of Norfolk, Suffolk and Essex. The information in the Exeter summary was edited and incorporated into the main Domesday Book, but that for the eastern counties, was, for some reason, never

incorporated. Domesday Book today, therefore, consists of two volumes, the Great Domesday (also known as the Exchequer Domesday) that was compiled in Winchester at the king's Treasury and the Little Domesday that covers Norfolk, Suffolk and Essex and which represents an earlier stage in the data-construction process.

Darby (1977: 8) provides an example of typical entry from the Great Domesday Book in which all the information relating to the village of Buckden is to be found in one entry, because all the land was held by one landlord, the Bishop of Lincoln:

In Buckden the bishop of Lincoln had 20 hides that paid geld [i.e. tax]. Land for 20 ploughteams. There, now on the demesne 5 ploughteams, and 37 villeins and 20 bordars having 14 ploughteams. There, a church and a prieset and one mill yielding 30s (a year), and 84 acres of meadow. Wood for pannage one league long and one broad. In the time of King Edward (i.e. in 1066) it was worth £20 (a year), and now £16 10s.

From the hundreds of entries like the one for Buckden, historical geographers such as Darby have been able to construct detailed accounts of the geography of Norman England (Darby, 1950–1; 1951; 1977). In this process of interpretation, all the questions raised above about the construction of data are relevant.

In the case of Domesday Book, the motivation for the gathering of the information is a matter of some debate. Traditionally it was viewed as a taxation assessment (Round, 1895; Maitland, 1897). More recently it has been suggested that it provides a record of the patterns of feudal ownership and of the King's relations with his senior feudal tenants (Galbraith, 1961). Its role could also have been primarily symbolic, serving to secure the Conqueror's authority over his new subjects (Clanchy, 1979: 18). To some extent, the view that is taken of the purpose of the document will affect the interpretation. If it is treated as a fiscal document, for example, issues of accuracy become very important. On the other hand, if its role was largely symbolic, then whether the Bishop of Lincoln's holding in Buckden was precisely 20 hides or some other figure arguably becomes less important than the insight provided into the relationship between Church and state by the allocation of lands to bishops in the first place. The questions about the character of the information gathered are also pertinent. Livestock, for example, are largely ignored by the survey, while the information on industry and towns is very limited. While this reflects the rural and agricultural basis of the Norman economy, it also influences the interpretations that can be made of the source.

The UK unemployment figures

Unemployment has been one of the most important social and political problems since the mid-1970s. Politicians of all parties recognize this, although proposed solutions vary radically. For those on the free-market right unemployment is said to be caused either by the unwillingness of individuals to work or by the effects of state intervention in the labour market producing disincentives to do so. The traditional left-wing view has been that unemployment is a product of unfettered free-market capitalism and that state regulation or economic intervention is required to respond to market-generated economic crises and to more general problems of market failure. Whichever view is taken, it is clear that measuring the level of unemployment accurately provides an important indication of economic success and, by implication, of the success of social and economic policies of governments. However, this raises the spectre of manipulation; if the measure of unemployment falls, governments can claim success for their policies even if unemployment itself has remained stable or increased. According to many commentators, this was precisely what happened in Britain during the 1980s and 1990s.

The geography of unemployment and employment has also long been of interest to economic geographers. In order accurately to interpret the geographical distribution of employment changes and trends and to map and analyse the pattern of employment opportunity and labour market activity, geographers use official statistics. Accuracy, and the process of data construction, is therefore not just an issue for assessing policy success; it also has profound implications for academic analyses, including those on which policies are partly based. Writing at the end of a long period of government by the Conservative

Party (1979–97) during which the integrity of official British unemployment data had frequently been a matter of political debate, Levitas (1996) provided a detailed review of the measurement of unemployment in the UK. This section is based on her analysis.

At the heart of the problem is the question of definition. Prior to 1982, the official measure of unemployment was based on a count of those registered as seeking work. Those who were claiming welfare benefits were required to register, but others could also register if they wished, so that the count included those who wanted to find work but who were for one reason or another ineligible for benefits. In 1982 the official measure was changed so that it included only those claiming benefits. This led to a reduction in the measure of unemployment of about 200 000 people (out of a total of over 3 million). As Levitas points out (1996: 48), this change was highly significant because it turned the unemployment count into an administrative measure. This meant that changes in the rules relating to benefit entitlement would have an immediate effect on the unemployment measure which was not produced by any real change in the economy. By the time of Levitas's study, 30 further changes had been made to the unemployment count. This meant that the official headline figure for unemployment in the UK in 1996 was based on a very different definition from that used in 1979. This, according to Levitas, had three consequences:

First, there is the effect upon unemployed people of a benefit system using increasingly stringent criteria for the receipt of declining amounts of benefit. Secondly, the count cannot be treated as a continuous series, as its coverage becomes narrower over time. Thirdly, it cannot be regarded as a measure of unemployment. (1996: 48)

The changes included the following:

- 1986: introduction of more stringent availability for work criteria, especially affecting women with child-care responsibilities (107 000 claims for benefit disallowed under this rule in 1987).
- 1988: removal of benefit entitlement from 16 and 17-year-olds (reduction in the count: over 100 000).
- 1989: introduction of requirement that claimants must be able to prove they are 'actively seeking work'.
- 1996: introduction of Jobseekers Allowance (JSA) in place of unemployment benefit involving means testing of benefit claims after only six months, instead of a year.

All these changes affected the unemployment count regardless of the real state of the economy. Whatever their merits in terms of the operation of the social security system, they produced entirely administrative reductions in the measure of unemployment. For many commentators, therefore, the official count became largely useless as a measure of unemployment. Even the retiring head of the government statistical service asserted in 1995 that 'nobody believes' the claimant count (Levitas, 1996: 62).

However, as Levitas pointed out, the problems with the claimant count do not mean that official statistics could not be used to provide better measures of unemployment. She reviewed a number of alternative approaches, including that provided by the Labour Force Survey (LFS). The LFS is a sample survey in which 60 000 households are interviewed about their labour market activity.

Although it produces a better measure of unemployment than the claimant count, and one that can be improved further by academic researchers undertaking secondary analysis of the data, the LFS does have some drawbacks of its own. Because it is a sample survey rather than a count it is subject to sampling errors. There are also problems with the sample frame. Because it is based on addresses, it excludes those living in hostels and with no fixed abode (two groups whose numbers have increased with the rise in homelessness and who are disproportionately unemployed). Finally, where individuals are absent at the time of the interview, proxy responses are recorded by asking other members of the household. The accuracy of some of this information may be rather questionable, and it may also be subject to gender bias. For example, a woman seeking work who spends her time undertaking domestic chores might describe herself as unemployed but be described by her male partner as a 'housewife' (Levitas, 1996: 53–4). 'Housewife' is a particularly gendered term and was itself popularized, in part at least, by official policy after the Second World War when governments were faced with large numbers of men returning from military service, and wanted to encourage women to return to domestic labour and give up the paid employment many had undertaken during the war. Despite these drawbacks, the LFS does provide significantly more reliable data on unemployment rates, has the twin advantages of following international definitions (allowing for international comparison) and allowing detailed analysis to examine relationships among other social variables, such as gender, ethnicity and age.

While the LFS serves much better than the claimant count for measuring unemployment according to international definitions, and in a more consistent way, there are some wider issues of definition that call into question conventional measures of unemployment altogether. The population is conventionally categorized not into two groups of employed and unemployed but into three: employed, unemployed and 'economically inactive'. The existence and definition of this third group raise further problems for the measurement of unemployment and as a consequence for the definition of unemployment as a political problem. The LFS definition of unemployment is explained by Levitas as follows:

To be counted as unemployed in the LFS, respondents must have:

(a) done no paid work in the week in question;
(b) wish to work;
(c) be available for work within the next fortnight;
(d) have made some effort to seek work in the past four weeks, or be waiting to start a job already obtained. (1996: 55)

Take the example of a middle-aged man who has been out of work for two years and who believes there are no jobs available for him. It is not unlikely that he will have made no effort to seek work in the past four weeks but, at the same time, be available for work and wish to work. On the LFS criteria he will be recorded as economically inactive, rather than unemployed. Similarly, many married women are defined as economically inactive, rather than unemployed, because, while they may wish to undertake paid work, they are not actively seeking employment because they are occupied with domestic responsibilities. Anomalies such as these have led to employment and unemployment falling simultaneously in some parts of the country at some times. If the population is static but employment is falling, this is hardly a picture of economic dynamism and success, and yet changes in 'economic activity' rates might allow a government to claim as much. At bottom the problem is precisely the one that forms the basis of the arguments in this chapter, namely, that statistics are a social construction. As Levitas (1996: 60) puts it: 'The line between "unemployment" and "non-employment" or "economic inactivity" is imposed upon a reality that is far more complex and fluid.'

Far from reflecting reality in a neutral and objective way, official statistics depend upon particular understandings of reality, and these understandings affect (often in profound ways) the picture that is painted of social and economic life by official sources of data. In a further shift in the official statistical view of unemployment since Levitas's analysis was published, the New Labour government elected in 1997 decided to adopt the Labour Force Survey definition as the source of Britain's official unemployment figures.

Conclusion

In this chapter we have deliberately not presented detailed case studies of geographers' use of state information as to do so would be to focus more attention on the *interpretation* of official sources than on their *construction*. By considering the issue of construction in detail we have tried to demystify official sources – to knock them off their pedestal, as it were – and to show that just because information comes with an official seal of approval does not mean that researchers can take it simply at face value. In fact, as we have suggested, official sources can be just as partial and subject to errors and evasions as any other kind of data. Indeed, when it comes to political influence, official sources are probably more affected than others. First, official information is always collected for a governmental (and hence political) purpose. Secondly, partly because of the cachet of the 'official' label, official sources can appear to be more neutral,

reliable, comprehensive and authoritative than other kinds so that their political character can be disguised. For this reason, paying attention to the process through which they are constructed is particularly important, as this should highlight the purposes for which they were produced and allow the researcher to take these into account in their interpretation. None of this should lead to an outright dismissal of official data, though. For many geographers they will continue to be an invaluable source, partly because of their comprehensiveness and quality. Handled carefully, official information can be expected to fuel the practising of human geography for a long time to come.

Notes

1 Prominent examples include Johnston (1979), Taylor and Johnston (1979), Johnston et al. (1988), Shelley et al. (1996) and Johnston and Pattie (2000).
2 Murdoch and Ward (1997), Blake (1999), Dean (1999), MacKinnon (2000) and Raco (2003).
3 Those interested in considering this latter point in more detail should also see the companion volume to this book, *Approaching Human Geography* – Cloke et al. (1991).

3   *Non-official Sources*

Introduction

The state, in all its guises, is not the only agency to provide the human geographer with rich sources of preconstructed data. Indeed, in tandem with the extensive growth in government record-keeping, which we detailed in the previous chapter, there has been a dramatic increase in the extent and scope of 'non-official' data. The inexorable rise in information technology and electronic communication (including the Internet and World Wide Web), the ease of desktop publishing and printing, and the continued expansion of the technical and bureaucratic nature of capitalist society have all combined to accelerate this process. The trend towards increased bureaucracy, administration and record-keeping in the private sphere was first noted by Weber (1978) at the beginning of the last century. Since then, of course we have witnessed a veritable 'information explosion' as companies, news agencies, research organizations, political parties and voluntary groups, to name but a few, have all struggled with the collection, storage and communication of an ever-increasing tide of information. In this chapter we will be concerned with those aspects of this information which purport to tell the truth, through understanding the construction of supposedly 'factual' data. In contrast, the next chapter will explore the construction of self-proclaimed

imaginative data through an examination of literature, music and the visual arts.

As we will maintain over the next two chapters, this distinction between 'factual' and 'imaginative' sources should not be held too rigidly. Both types of data are socially constructed, and both are produced for specific purposes in specific social contexts. Moreover, all attempts to describe the world in a factual manner involve particular ways of imagining it, and all such accounts will contain some element of subjectivity. Indeed, any construction of factual or documentary knowledge is based on a particular view of what is real or what constitutes the truth. However, because we are concentrating on the ways in which these sources are constructed, and because the processes behind the production of 'imaginative' and 'factual' sources do differ (to varying degrees), it is useful to consider the two types separately.

In this chapter we will examine the production of supposedly 'factual' data by the multifarious agencies and institutions which lay outside the official realm of the state. In doing so we will be concentrating on those sources which are of greatest use to the human geographer. These vary enormously. The archetypal source, and the one that we are all most familiar with in our academic inquiry, is the text printed or handwritten on paper, usually produced in the form of books, journals and magazines. But the potential field of 'factual' documentary sources is much broader than this. The historical geographer, for instance, might make use of sources carved in stone; those working with geographical information systems will use electronic and magnetic sources; the urban geographer interested in housing change may well use display boards in estate agents' windows. It is also the case that the text of a factual source does not have to be written. Promotional or campaigning videos, documentary radio programmes, television news broadcasts and audio tapes of oral histories are all valuable sources of factual information about social phenomena, but none of them are written documents.

In short, the possible scope of non-official sources is extremely wide and can range from the most specialized company report costing thousands of pounds to a torn and tattered notice in a shop window. Useful sources are also produced by individuals just as much as by organizations and institutions. A geographer looking at a person's 'sense of place', for instance, might use diaries and letters containing snippets of information about the place where the individual lives. The message here is that potential researchers need not feel bound by written sources, or by those which are overtly geographical in nature, or by those which appear formal and thereby authoritative. Indeed, we would stress that the human geographer can find a rich and rewarding range of material in a whole variety of non-official sources – published and unpublished, public or private, written or spoken, wittingly or unwittingly provided. Later in the chapter we will explore some of this potential material by examining the production, and use, of selected examples in more depth. But before we do so we will outline some general advantages in using non-official sources and then consider the main types of such data you might use.

Non-official sources in geographical research

There are a number of reasons why geographers might wish to turn to non-official 'factual' data as a source of research material. We have already hinted at the breadth and scope of the sources available and by looking across these we can identify five potential advantages in using such material:

1 Non-official sources can open up social worlds which are inaccessible and relatively closed. Davis (1990) has used a variety of newspaper reports, police documents and verbatim records of televised interviews to look at the military-style policing of contemporary Los Angeles, with particular reference to the so-called 'war on drugs'. In this instance both the police and the drug gang members represent relatively closed worlds, but Davis is able to construct a vibrant picture of current events from his use of contemporaneous news material. From this he is then able to construct vivid accounts of the changing social and cultural geographies of urban Los Angeles. In a similar manner, Jackson (1988a) has used unpublished archival material from voluntary and community groups, school reports and newspapers to examine the experience of neighbourhood change in Chicago.

2 Documentary sources of 'factual' data offer a major chance for the geographer to conduct historical and longitudinal studies. Thus Massey (1984), in her pioneering work on the geography of economic change, uses company records and reports to sketch out the historical performance of particular industrial sectors over the last century, and Martin et al. (1993) use OECD sources to chart the decline in trade union membership over the last

30 years, before drawing on current membership records to assess the geography of contemporary unionization.

3   In a similar manner, just as they can be used to compare events across time, non-official 'factual' sources be used to analyse cross-cultural processes. Because similar records tend to be kept by similar organizations across national boundaries, they offer a reasonably safe way of comparing like with like (although see Dickens et al., 1985, for a comment on the pitfalls of comparative research, in their study of housing provision in Britain and Sweden). Moreover, some agencies and institutions are international in character, and they provide comparative preconstructed data as part of their own research. The World Bank, for instance, publishes a yearly report of development indicators, and the International Labour Organization produces regular comparative material on issues concerning employment change. In a similar way, the World Health Organization produces international comparative data on health issues. All these will be of potential use to the human geographer.

4   Another reason why 'factual' data from non-official sources are popular is because they are potentially very cheap and often easy to access. The sources are often available in reference libraries, and they are often collected together in a usable form. One of the advantages of using any form of preconstructed material is the very fact that it is already constructed – someone else has spent the time and money to collect the research data and put them together. A collection of these sources may well have to be drawn together or used selectively in a manner which suits the individual researcher, but the bulk of research expenditure has usually been undertaken by someone else. Allied to

this is the fact that the researcher can draw on a potentially vast sample size, without being limited too much by considerations of cost and time. To return to an earlier example, by using the membership records of just eight trade unions, Martin et al. (1993) were able to document the geography of some 3 million individual union members.

5   Finally, the researcher can use documentary material for ideas as well as for data. An examination, for instance, of the data contained in the reports of the Anti-slavery Campaign will raise a number of issues concerning child labour, the production techniques of multinational companies, gender relations within the Third World, the consumption habits of those in the West, the international debt problem and so on. Researchers should not feel they can consult these materials only in search of ready-made data but, instead, non-official sources can offer a rich vein of ideas and stimuli for possible areas of study.

As we have already noted, the range of potential non-official data sources for the human geographer is vast. Some are intended specifically for research purposes, but most are not. Some are meant for public consumption in one way or another, but others are intended to be private. Some are the product of international organizations and agencies, while others are produced by individuals. All, however, are the end result of deliberate social construction, and all purport to tell the truth about social events. Because we are stressing that such sources can be understood only within the context of their construction, we will briefly outline the major data sources available to the researcher, according to the type and intention of production. We will then look in more depth at a smaller selection of these, at their detailed construction and at the potential pitfalls in their use by geographers.

Reference material

Some source material is designed specifically for reference purposes. This ranges from large-scale electronic databases to individual research papers. Often such material is collected together and published in the form of directories, manuals, almanacs, guides, registers, yearbooks and calenders (Scott, 1990: 156). Economic and historical geographers have long made use of commercial directories, for instance, which list industrial and commercial activities on an area-by-area basis. Other commercial manuals are sector based and can offer technical information regarding a particular industry or financial information regarding investment and shareholding. Social and urban geographers regularly use reference material on housing costs supplied by building societies to monitor the changing nature of housing markets (see Hamnett, 1998).

Academic books, journals and articles are a major source of reference material which can get overlooked simply because they are so familiar to researchers. In fact, a potential problem here is the sheer diversity of academic material, but bibliographic databases, abstracts and digests can be used as tools for literature searches. Some primary academic material used in the writing of books and papers is also available to the researcher. The Economic and Social Research Council helps to fund two data archives at the University of Essex, one concerned with qualitative data and one concerned with computer-readable quantitative data. The former, QUALIDATA, aims to locate, assess and document qualitative data and arrange for their deposit in suitable public archives; to disseminate information about such data; and to encourage the reuse of these data. A major function of the centre is to maintain an information database about the extent and availability of qualitative research material from a wide range of social science disciplines in general, whether deposited in public repositories or remaining with the researcher. The centre has surveyed a huge range of ESRC-funded qualitative social research projects dating back to 1970 and is cataloguing the data from classic postwar sociological studies as well as monitoring current ESRC and other funded projects. The Qualidata catalogue (Qualicat) is available via the Internet (www.essex.ac.uk/Qualidata), from which researchers can search and obtain descriptions of qualitative research material, its location and accessibility. Quantitative data are available from the Data Archive, originally established in 1967 and now holding some 4000 data sets. While many of these sets are based on official data, others are not and use researchers' own data in addition to those from a range of other organizations. The material can also be searched and accessed via the Internet (www.data-archive.ac.uk). A quick look at these two archives will give some idea of the range and scope of 'factual' material available to the researcher.

Research reports

Recent years have seen an explosion in semi-autonomous research centres, sometimes linked to universities but more usually funded through charitable or private sources. These produce extensive series of research reports, covering areas of interest to the geographer such as housing, urban policy, economic development, environmental change, planning and health.

Other organizations and institutions produce research reports which, although not aimed primarily at the academic researcher, contain a range of useful materials. Many of these reports are now produced for lobbying purposes, and the rise of official state information which we considered in the previous chapter has been paralleled by the rise of oppositional and lobbying material. Pressure groups such as Shelter, the Low Pay Unit, Labour Research or the Child Poverty Action Group produce regular publications and reports on housing, employment, trade union and poverty issues, respectively. An interesting collaboration between the Child Poverty Action Group and the Social and Cultural Geography Research Group of the Royal Geographical Society–Institute of British Geographers resulted in a publication on the social geography of poverty in the UK. This contains several examples of geographers using research reports by unofficial agencies to build up a picture of spatial inequalities in housing, health, employment, educational achievement and income, and the volume itself can also be used as a rich documentary source (see Philo, 1995). Aside from pressure groups, organizations such as the Economist produce regular briefings on financial issues, and 'think-tanks' such as the Institute for Public Policy Research and specialized research institutes, such as the Joseph Rowntree Foundation, the Policy Studies Institute or the Tavistock Institute, all have extensive publication lists. Political parties also have research departments which publish their findings, as do trade unions and employer organizations such as the CBI and the Institute of Directors.

Reports, memos and minutes

Commercial and non-commercial institutions also produce a wealth of internal material in the form of reports, memos, messages and minutes which, while not based on primary research itself, can provide a rich source of research material. Large-scale companies are constantly producing reports and strategies, and any large institution will normally produce at least annual institutional plans and forecasts. Sadler (1997), for instance, has used company reports from a variety of car manufacturers and suppliers to study the changing geography of the European automobile industry. Although most of this material will not be aimed at the public market it is often possible for the researcher to negotiate individual access, especially if constraints on the use of the material are accepted. Through such access one can glean realms of information on employment patterns, labour processes and productivity, investment decisions, marketing strategies, production targets, management techniques and so on.

Documentary media

A whole range of purportedly factual material is produced by the news media. Broadcast news, documentary programmes, photographs, newspaper and press reports and specialist magazines between them provide many hours and column inches of 'factual' material. Indeed, there are now cable TV stations, radio channels and satellite broadcasting networks solely devoted to the production of news. Much of the written material is available on microfiche or via the Internet, and CD-ROMs are now able to hold several years' worth of newspaper reports on one disk. These can be searched via keywords so that ten years of press reports can now be 'filleted' by the researcher in a matter of hours, if not minutes. Geographers regularly make use of such material in their work. Cresswell (2001), in his work on the US tramp, has drawn on newspaper reports, magazine articles and documentary photography in order to examine how tramps were both perceived and treated by the media. Interestingly, his work shows how the news media, far from simply documenting the 'truth' about tramps, actually played a critical role in constructing our knowledge about tramps, and was instumental in the making up of the tramp as a figure on the social and geographical margins of society. In a similar vein, Beauregard (1993), in his book on postwar cities in the USA, shows how those who were reporting and reflecting on the state of urban America for newspapers and magazines were central to the construction of a discourse which promoted the idea of urban decline – and which reshaped both the public understanding of, and an explanation for, the fate of the postwar US city.

Publicity and promotional material

This is a source produced for external consumption, but not primarily for research purposes. Commercial organizations, and increasingly their non-commercial counterparts, will market themselves and their products via a host of promotional and publicity devices. These can again be a good source of information for the geographer. Goss (1999), for instance, shows how the marketing and selling of products in the contemporary shopping mall are based on the consumption of particular spatial and temporal archetypes, each depicting a particular geography. In this case, examining publicity and promotional devices allows us to see how the retail environment partly operates through the specific consumption of an idealized geography – based around archetypes such as the marketplace, the festival setting, nature, heritage, childhood and primitiveness. On a larger scale, Molotch (1996), in a chapter entitled 'L.A. as design product', has explored how the 'local aesthetics' of art, design and advertisement have contributed to the changing economy and urban form of Los Angeles. Indeed, he claims these aesthetics have influenced modes of expression and economic production on a global scale.

Personal documents

These are usually not intended for research purposes, although politicians and media personalities are now producing personal documents with one eye on the Sunday newspapers and the other on their bank accounts. In the main, though, by personal documents we mean private material such as letters, diaries, household accounts, address books and personal memos – although the range can be extended to cover oral recordings of life histories, photographs and family portraits, and personal possessions and belongings. These are a potentially fascinating source of material on community change, gender relations, consumption habits, employment trajectories, local politics, generational change and so on. This type of material can be especially helpful in historical research where it can be used to piece together a portrait of a time and a place which have long passed. Blunt (1994a) and Bell (1993), for instance, use the letters and diaries of Victorian women travellers to investigate issues of gender and exploration, and gender and health, respectively, in colonial Africa.

Maps

These have provided a key source of geographical material which, in the past, has gone largely uninterrogated. But like other factual sources, maps help to construct, as well as represent, our view of the world and as such they are useful to the researcher in a variety of ways. They are singled out here as a particular type of factual document precisely because they have been so privileged as a source of information within the discipline of geography. Geographers have tended to construct and use maps rather uncritically – they have been seen as relatively straightforward representations of landscape and of other geographical 'facts'. Recently, however, authors such as Harley (1988) and Wood (1993) have explored how maps act as particular representations of the world, serving some interests over others. Thus they can be read as statements about the world rather than as simple reflections of it. In this sense they are the same as other 'factual' sources, and to appreciate fully their use as a research source requires an understanding of the issues behind their construction. We will move on to explore this more fully later in this chapter, where the use of maps provides one of our set of case studies. First, though, before examining different types of non-official sources, we will chart some of the more generic issues in their use which are common to all types of factual data.

Critical issues in the use of non-official data sources

Scott (1990: 19–35) sets out a number of concerns which are common to the use of all types of documents. He groups them under the four areas of authenticity, credibility, representativeness and meaning. We will keep to these headings as they have become established in accounts of documentary data. It should be noted, however, that Scott's work is mainly centred on written textual sources, but his framework can be drawn on to encompass other types of data.

The question of a document's authenticity is concerned with the two related issues of soundness and authorship. The former refers to whether the document in question is complete or reliable in terms of its content if it is a copy or damaged in any way. The latter refers to reliability of authorship. With regard to soundness, a source may have words, sections or pages missing, or have contents which have been distorted through copying, whether manually or mechanically. Visual text may be unfocused or unclear in other ways. Audio text may be difficult to hear clearly or may have been edited since the original production. In each instance the researcher should attempt to establish the extent of the problem and decide whether it is sufficient to invalidate the source's use. It is not enough merely to assume that everything is in its place and that the source is sound and complete.

What is also common is that a range of records covering a variety of documents will be incomplete in some way. A particular year may be missing from a historical run of records or the contribution of one person to a meeting or discussion may be missing or distorted. Documents are often lost or misplaced when institutions and agencies close, move or amalgamate. Some are kept while others are not, and the source will be incomplete in some way or another. A set of diaries may have a month missing or a year. A collection of letters may have those from one particular correspondent destroyed. Memos and minutes will often have sections deliberately blanked out. Again it will be up to the researcher to decide whether these missing pieces are vital or necessary to the research in question, and thus whether the source is sound enough to be used for his or her particular purpose.

In many instances the veracity of the authorship of a source will not be an issue. In some instances, though, geographers will need to establish the authorship of the document or record. Sometimes, those signed in the name of one person have been compiled and written by another. Letters or papers which appear in the name of politicians, an organization's director or a company manager may well have been authored by one or more of those responsible for handling their correspondence. Again, this need not necessarily matter and may produce interesting research questions in its own right, but the full interpretation of any text cannot be separated from issues of its production. If the researcher is aware of the organizational structure or political circumstances in which the document was written, this will inform us not only about its production but also about its content and meaning.

The notion of credibility involves assessing how distorted the contents of a document are likely to be. As Scott points out:

> All accounts of social events are of course 'distorted', as there is always an element of selective accentuation in the attempt to describe social reality. The question of credibility concerns the extent to which an observer is sincere in the choice of a point of view and in the attempt to record an accurate account from that chosen standpoint. (1990: 22)

Accepting that there will be an inevitable bias in any document, then, we are left to examine the issues of sincerity and accuracy. The former refers to how much the author of the document believes what he or she recorded; the latter to how accurate the account is likely to be. There are many reasons as to why we may doubt the sincerity of a source. One way of assessing this is to ask what interest, material or otherwise, the author has in its content. Politicians will have an obvious interest in putting a particular gloss on to a document – whether for self-promotion and aggrandizement or for purposes of party propaganda. In an era where senior civil servants admit to being 'economical with the truth' when giving written evidence to a court, the ethical standards of those producing political sources make insincerity more rather than less likely. The same can be said of the financial inducements offered by much of our press for 'scooped' news stories and, in any event, the circulation battles between titles will inevitably affect the ways in which news is reported. There are now well-documented cases of 'factual' news and documentary stories, in newspapers and on television, simply being invented. Business records may also be written deliberately to deceive where, for instance, tax avoidance is an aim.

Where the author acts with the best of sincerity, the accuracy of the source can still be questioned. Although the accuracy of any 'factual' source can never be precisely judged, we can examine those circumstances which might have led to inaccuracies. Most notably this involves assessing the conditions under which the source was produced and, in particular, the proximity of the compiler to the events described. Was the account first hand, whether made from contemporaneous notes or those made later, or was it produced from secondary sources, or using memory alone? If the latter, how long after the event was the account produced? An interesting exchange of letters took place in the *Independent* newspaper in August 1993 concerning the accuracy or otherwise of the political diaries of Tony Benn, a Labour Cabinet minister during the 1960s and 1970s. When their accuracy was questioned, Benn's biographer wrote of times spent in Benn's archive with the diary editor, who constantly checked back to contemporary documents, including the minutes and agendas on which Benn took his notes. He also stated that comments made when the hardback version was initially published were collated and checked and, if necessary, the paperback text was altered (*Independent*, 31 August 1993: 25). This may add credibility to the paperback version but raises interesting questions of research methodology for the contemporary political geographer. The same event would in this case be described differently (by the same author) depending on which edition he or she consulted.

There will also be an inevitable sampling bias in many of the sources we consult, whatever their sincerity and accuracy. The bias will be towards the literate middle classes or those with access to documentary production facilities. Only recently, for instance, through history workshop schemes, have working-class people started to produce and document their own versions of their own histories on any kind of scale. Rose (1990) has used some of these oral histories in her account of community politics in Poplar, and deliberately draws a contrast with accounts given in the 'official' histories of working-class politics, written mainly by male middle-class political activists to suit party political purposes. In the same manner as material from telephone polls will never represent those without a telephone, or that produced by a house-to-house survey based on the electoral register will never reflect the views of the homeless, each 'factual' source used by the researcher will be biased towards those who have the means to produce it. This brings us on to the issue of representativeness more generally. We may be forced to use certain data because these are quite literally the only ones available to us, whether for reasons of selective survival, denial of access, lack of knowledge or pressures of time and resources. The fact that the source is perhaps not representative of its type is again not necessarily a problem, as long as the researcher is aware of this and adjusts the claims that can be made, and inferences that can be drawn, accordingly.

The fourth issue identified by Scott in the use of factual sources concerns their meaning. In some cases this will be quite literal – our understanding may be prevented by the use of technical terminology, acronyms or shorthand notes of one kind or another, as well as by the use of an unfamiliar language. In other instances we will be confronted with the problem of interpretative understanding. The essential idea here is that the text of any document will hold different layers of meaning, which go beyond the words themselves. For now we will simply state that an understanding of the practices, styles and definitions employed by those producing the source will help us understand its meaning. (We return to explore the different ways of addressing these interpretative understandings in Chapter 10.) It is important that we investigate these different understandings because very few sources are actually intended solely for research. Most sources are generated – whether by a company, a voluntary group or an individual – for reasons wholly unconnected with the research needs of the geographer. It is therefore important to acknowledge the conditions under which the sources were produced and the purposes for which they were intended. As Hammersley and Atkinson put it (1983: 137):

> rather than being viewed as a (more or less biased) source of data...documents...should be treated as social products: they must be examined, not simply used as a resource. To treat them as a resource and not a topic is to...treat as a reflection or document of the world phenomena that are actually produced by it.

We should therefore seek to establish the purpose behind the production of the source rather than just accept the record as somehow 'given'. Factual sources are not impartial and autonomous accounts of particular events and processes. We have already shown that this is the case with regard to state documents.

We now turn to examine some of the conditions of production of non-official data sources.

CASE STUDY 3.1: THE WRITTEN TEXTS OF NEWS MEDIA

Newspaper articles and stories form a major part of our lives, alongside audio and visual news media. They tell us what is going on in the world; and they tell us about important changes taking place in different cities, regions and nations; and they explain to us why such changes are occurring and their likely implications. In short, our geographical imaginations of places, people-groups and individuals are influenced significantly by 'the news', which can form a very important source of information on which to base our geographical understanding of the world around us.

The news is often presented as a neutral account in which facts are collected, analysed and reported in an objective fashion, without bias, and in unambiguous and undistorted language. Take, for example, the following account by Andrew Neil (the then editor of *the Sunday Times*) of his newspaper's stance on the 1984–5 miners' strike:

> From the start, the *Sunday Times* took a firm editorial line: for the sake of liberal democracy, economic recovery and the rolling back of Union power, and for the sake of the sensible voices in the Labour Party and the TUC, Scargill and his forces had to be defeated, and would be. It was a position from which we never wavered throughout this long, brutal dispute. Our views, however, were kept to where they belong in a quality newspaper: the editorial column. For us the miners strike was above all a massive reporting and analysing task to give our readers an impartial and well-informed picture of what was really happening. (cited in Fowler, 1991: 1–2)

The desired code for interpreting text in newspapers, then, is to search the editorial column for *views* and to find unbiased, factual *news* in the remainder. However, in order to make critical use of newspapers as sources of information, it is necessary to deconstruct this code, realizing that language in the news is used to suggest ideas and beliefs as well as 'facts'. In order so to do, it is necessary further to recognize how the news is socially constructed (see Thompson, 1997).

In short, all news is reported from some particular angle (see Hall et al., 1980; Franklin and Murphy, 1990). This case study deals with the language of news as expressed in written text, but the principles of socially constructed news also apply to visual and verbal presentations. The structure of the written news medium encodes particular significances which stem from the positions within society of the publishing organizations concerned. We therefore have to understand both how and why the news is made. The 'how' involves a *selection* of what is newsworthy, and thus a series of judgements in which a partial view of the world is expressed. Different newspapers will have different partialities – in Britain, the *Mirror* is more likely than *The Times* to report on a Lottery win, but *The Times* is more likely than the *Mirror* to cover famine in Rwanda.

CASE STUDY 3.1 *(Continued)*

Moreover, the selection of news is likely to be accompanied by a *transformation* of material, such that the same story will be given differential treatment by different newspapers. Particular transformations can fuel specific geographical myths. For example, Burgess (1985) shows how tabloid coverage of the urban riots in England in 1981 helped to create a 'myth of the inner city', constructed through very selective representations of the physical environment, along with sometimes extreme commentaries on the nature of black and white working-class cultures. The myth was, in Burgess's view, created deliberately for ideological reasons in order to distance the rioters (and the reasons why they were rioting) from 'normal' society. Hence the events of the riots, and the social and political challenges which they represented, could be packaged up in a mythological inner city, which was conveyed as discrete, hostile and alien so as not to allow them to disturb existing political, economic and social orders.

The danger for human geographers seeking to interpret such partialities is that bias will be recognized in all but our own favourite outlet. It is important, therefore, to recognize why the news is made in a partial manner. Here, the goals of maximizing readership and advertising, and therefore revenue and profit, are obvious considerations. In addition, however, there are deeper-seated goals relating to the strategic support given to governments or political parties which are most likely to favour the other commercial interests of newspaper magnates such as Murdoch, Packer and Thompson. In that newspaper publication is an industry and a business, with a particular position in the nation's and world's economic and social affairs, it is to be anticipated that the output of newspapers will be at least partially determined by that position. This is particularly important given the concentration of newspaper ownership in recent years. In 1965 there were 11 companies owning 19 newspapers in Britain. Some 30 years later there were 7 companies owning 21 titles.

In some ways these partialities are straightforward. For example, the press is often seen to be preoccupied with money, fame and royalty, each of which serves to symbolize hierarchy and privilege (Fowler, 1991). To interpret the world in such a way as to naturalize hierarchy and privilege will serve the interests of capitalist concerns such as newspaper empires. Other partialities are perhaps less obvious, although equally important. Box 3.1 presents some conclusions from an interesting study by Reah (1998) about how newspapers interpret social groups. From these examples it is clear that the language of the news is used not only to stereotype particular social groups – as being other to male, white heterosexual orthodoxy – but also to discredit, belittle and misrepresent them for the purposes of reinforcing dominant cultural norms and the power vested therein. Thus geographies of gender, race and sexuality, for example, will of necessity be most interested in the ways in which news media present particularly partial views in these respects (Figure 3.1 shows a montage of the original articles).

Box 3.1 Reah's (1998) conclusions on how newspapers portray particular social groups

Women

A random selection of stories suggests that certain stereotypes are operating in the press in relation to women. There is a tendency to depict them as existing primarily in relation to their families – their children, their husbands or partners rather than as individuals in their own right. Younger women are frequently described in relation to their physical appearance. Women are often depicted as weaker – they are victims, they are on the receiving end of action rather than the performers of it. If this selection is representative, then the newspaper-reading public receives a series of images depicting women according to a very limiting stereotype that values them in only a narrow set of roles. (1998: 69)

Ethnic groups

In this example Aboriginals and white Australians:

The Aboriginals are described as being 'treacherous and brutal'. The qualities 'gentle' and 'trusting' are mentioned as being falsely attributed to them by others. The descriptions contained in the naming compare the Aboriginals to 'children', and 'Ancient Brits who painted themselves blue'. These descriptions portray either the concept of the savage – a sub-human species; or the noble savage – a sub-human species with a child-like innocence. Both of these representations are caricatures that are rooted in early colonialism.

The white Australians, on the other hand, are 'tough' and 'bloody-minded'. They are named as 'individuals', a description that contrasts strongly with the portrayal of the Aboriginals.

Therefore, closely linked to naming are the attributes and qualities that groups or individual members of groups are credited (or discredited) with when they are discussed in newspaper texts. Newspapers are artefacts of the dominant cultural norm...The attitudes of the dominant culture tend to be reflected in the language of news stories, particularly those about minority power groups. (1998: 62)

Gay homosexuality

This concept of homosexual behaviour as abnormal is carried through the story via the word choice. A love poem is 'slushy'. They are said to have 'committed', a verb normally associated with criminal activity, 'a lewd sex act' together. This word choice carries strong connotations of illegal and abnormal behaviour. There is also an element of belittlement in the use of idioms such as 'caught with their trousers down', 'caught with his pants down', that have connotations of a stage farce. These, combined with words such as 'slushy' and 'toyboy', refuse to recognise the possibility of serious emotional involvement in such a relationship. (1998: 73)

Express chief's wife arrested on suspicion of stealing Fergie book

THE fierce battle for circulation between mid-market tabloid newspapers took a fresh twist yesterday when police confirmed they had arrested the wife of a senior executive at Express Newspapers on suspicion of theft.

Anita Monk, married to the Express's deputy editor Ian Monk, was detained at an hotel at Heathrow airport in possession of a pre-publication copy of a damning biography of the Duchess of York.

A second copy of the book, Fergie: Her Secret Life, which the rival Daily Mail is serialising after paying a reported £170,000 to the publisher, was later found at her home.

The duchess tried to have the book banned but had to give up after she was ordered to lodge £500,000 before her legal action could go ahead. When she failed, a fierce battle broke out over it between, the Express and the Daily Mail. The Mail ran extracts of book, which included claims that the

duchess would have killed herself but for her daughters, as 'the book she tried to ban'.

But on the first day of the Mail's serialisation, November 2, the Express also ran a story with many of the same revelations, which, the Express told its readers, had been published by a US magazine.

The previous day Mrs Monk had been arrested at Heathrow.

A police spokesman said: 'A woman aged 52 was arrested on suspicion of theft and handling stolen goods.

'Officers seized from the woman a copy of a book yet to be published and another copy was seized at her home. It was Fergie: Her Secret Story. The woman was taken to Uxbridge police station and bailed to return on 18 November pending further inquiries.'

A spokeswoman for Express Newspapers said: 'Ian said he was not available to comment on this.'

Guardian

A STUNNING pal of comedy star John Cleese miraculously escaped death yesterday when her plane careered through an airfield fence at 100mph and was rammed by a van.

The £1million Lear jet carrying 27-year-old Lisa Hogan broke in two as it shot across a packed main road.

Terrified drivers braked and swerved, fearing a fireball.

But as a waterfall of fuel poured from the plane's shattered wing tanks, 6ft 2in Lisa cooly climbed from a smashed window.

A stunned witness to the crash on the A40 at Northolt, West London, said: 'We found her sitting on a wall. All she wanted was a cigarette.'

The Irish-born beauty, who stars with Cleese in his new film Fierce Creatures, was treated in hospital for a cut leg.

Sun

I lay wide awake as docs cut me open

Mum June's op hell

A MUM who had a heart attack during major surgery claimed yesterday that she was WIDE AWAKE when doctors operated.

June Blacker said she suffered agonising pain from the scalpel because the anaesthetic hadn't worked.

Now June, whose heart stopped for **THREE MINUTES** is suing health chiefs over her horror ordeal. 'It was every patient's worst nightmare', she said.

The 43-year-old catering worker, who was in hospital to be sterilised, couldn't raise the alarm because she had been given muscle-relaxing drugs

Mirror

GLENDA TO FIGHT ON THE BEACHES

LABOUR will be fighting for votes on the beaches of Spain tomorrow.

Shadow Transport Minister Glenda Jackson will hand out sticks of rock to holidaymakers in Benidorm and Alicante.

Mirror

Figure 3.1   Newspaper articles and the geography of gender, race and sexuality
Source: Reah (1998)

Women wanted for building site work

And it's not just so we can whistle at 'em, say

BRITAIN'S most sexist industry has been told to recruit women – or risk a manpower crisis.

Construction chiefs are being urged to shed their macho image to make room for female brickies.

And wolf-whistling at pretty girls could be a thing of the past among tomorrow's female scaffolders.

Many firms already report a shortage of skilled electricians and joiners. Now there are fears that a drop in the birth rate 20 years ago means that were will not be enough young men to fill the jobs.

Industry bosses will be told to make construction more attractive to women by providing all-girl training courses, flexi-time and proper childcare facilities.

The move could lead to creches on building sites once dominated by hairy, beer-swilling men with tattoos. Sexy pin-ups would almost certainly be banned to make women feel more 'comfortable' in the workplace.

And women's support groups to rival the old boy's network could be introduced.

The shake-up will be formally recommended by the Government-sponsored Construction Industry Board next month.

A CIB survey has found that schoolchildren think the industry is 'dirty, dangerous and demoralised'.

A CIB source warned: 'If the industry fails to respond to the report's recommendations to recruit more women it will face increasing skill shortages and problems with recruitment'.

'Demographic changes and the increasing success of women in both education and employment will mean that employers will face a shrinking number of applications from young men'.

'We are not calling for women to be hired to be politically correct'.

'It is a question of productivity and necessity in an industry increasingly hit by skill shortages'.

'The industry should realise it is ignoring half the country's workforce by not recruiting women. But why shouldn't women work as bricklayers or electricians?'

The construction industry has an unenviable reputation for sexism. CIB figures show it has the lowest proportion of women employees of any mainline British industry.

Women make up 46 per cent of Britain's working population but only 10 per cent of the construction industry – and most of them are secretaries or clerks.

Not only are fewer women recruited but their drop-out rate is higher than the men's – possibly because of the macho lifestyle.

The battered huts of building sites are notorious for their explicit pin-ups and the crude jokes of their occupants.

The CIB wants 'mentoring and support' schemes for women to stop them feeling isolated.

Daily Express

Figure 3.1 *Continued*

THE SUN SAYS

Salute from the Poms

DOWN UNDER they are having the biggest and longest birthday party in history.

It is exactly 200 years since the first British fleet, with its mixed cargo of sheep stealers, pickpockets and naughty ladies, dropped anchor in Botany Bay.

However, there are skeletons at the feast.

'What is there to celebrate?' demand a thousand placards.

The Aussies are being asked to tear out their hearts over the plight of the poor old Abos.

They are asked to believe that, before the white man stole their land, Australia was a paradise inhabited by gentle, trusting, children of nature living on the fat of the land.

In fact, the Aboriginals were treacherous and brutal.

They had acquired none of the skills or the arts of civilisation.

They were nomads who in 40,000 years left no permanent settlements.

The history of mankind is made up of migrations. Australia no more belongs to them than England does to the Ancient Brits who painted themselves blue.

Left alone, the Abos would have wiped themselves out.

Certainly, they have suffered from the crimes (and the diseases) of the Europeans.

Tamed

That is inevitable when there is a collision between peoples at different stages of development.

The Aussies tamed a continent.

They have built a race of tough, bloody-minded individualists.

We have had out differences – over politics and more serious things like cricket – but when the chips were down they have always proved true friends.

Not bad at all for a bunch of ex-cons!

We hope that they enjoy their long, long shinding and crack a few tubes for the Poms.

Sun

Figure 3.1 *Continued*

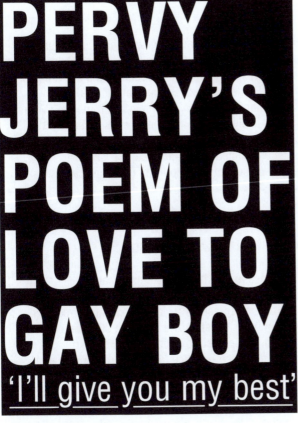

PERVY JERRY'S POEM OF LOVE TO GAY BOY

'I'll give you my best'

TORY MP Jerry Hayes sent his gay toyboy a Valentine's card containing a slushy poem which pledged: 'I promise to love you in good times and in bad.'

Hayes, 43, signed the card he gave to teenager Paul Stone with a single kiss.

The poem, by Dorothy R Colgan, is called My Promise Of Love To You and contains the line: 'I promise to give you the best of myself.'

It adds: 'I fell in love with you for the qualities, abilities and outlook on life that you have.'

Hayes sent the card to Stone in 1992. It was in an envelope franked in the House of Commons. The lad was then 18 – at a time when the legal age of consent for homosexuals was still 21.

Their 16-month affair took place when the MP, who is married with two children, was on an IRA hit list.

Security chiefs found the list during a raid in Ulster.

Threat

Hayes' name appeared together with other prominent Tories and senior members of the Armed Forces.

The fling with former Young Conservative Paul could have left Hayes vulnerable to blackmail attempts by the Provos. The MP was targeted by the IRA after he became a Parliamentary aide to Northern Ireland minister Robert

Eight police marksmen guarded his family home in Wenden's Ambo, Essex, round the clock for three months.

His garage doors were bomb-proofed. And behind them armed cops secretly set up a control room watching all approaches to the property.

An ex-cop who was responsible for the MP's security said yesterday: 'We took the threat very seriously indeed.'

Paul told The Sun's sister paper the News of the World how Hayes got him a House of Commons Security Pass by claiming he was his unpaid researcher.

He said the MP often chatted to him about Ulster. Paul added: 'When I joked to him that I could blow up the House of Commons he just ignored me.'

Paul, now 24, said he first met Hayes at a gay rights group meeting.

Hours later they committed a lewd sex act in a park, it was claimed.

Paul, who was studying for his A-levels at the time, said Hayes also wined and dined him at expensive restaurants.

Magic

Hayes, vice chairman of the Commons AIDS committee, also sent Paul intimate letters on Commons notepaper.

One read: 'I love you. I miss the magic of your hugs and watching you when you are asleep.'

Another declared: 'Just remember, 1 We love each other **VERY** much. 2 We miss each other **VERY** much. 3 We have fun together and can help each other.'

Paul had access to virtually every area of Westminster.

He said: 'Jerry liked having me around for when the mood took him, which was quite in lot. I was being introduced to people like Michael Portillo and Virginia Bottomley.'

He said Hayes always professed to be devoted to his wife Alison, 40, and their children aged seven and nine.

Hayes' political career has been coloured by a string of game-for-a-laugh stunts on TV.

He has dressed up as a bumble bee, sky-dived in a chicken costume – and was even whipped in kinky bondage gear on a late-night show

Figure 3.1 *Continued*

CASE STUDY 3.1 *(Continued)*

Human geographers have been taking a sharper interest in news media as a source of geographical information and imagination over recent years (see Burgess and Gold, 1986). Particularly effective studies include the investigation of the production and consumption of environmental meanings in the media (Burgess, 1990; Burgess et al., 1991); the examination of newspapers as sites of regional definition and regional promotion (Myers-Jones and Brooker-Gross, 1994); and the analysis of newspaper stories to highlight different discourses on the Falklands/Malvinas conflicts (Dodds, 1993). Each of these studies demonstrates how the socially constructed text of newspapers can be deconstructed in order to provide nuanced accounts of important geographical issues.

CASE STUDY 3.2: THE VISUAL TEXTS OF TOURIST ADVERTISING

Visual images are an integral part of the means by which we communicate in contemporary society. These images can be still or moving (see Case Study 3.4), but in either form the visual portrayal of information represents a vital and culturally relevant source of data for human geographers. In this case study we focus on still images, of which there is a wide range – for example, those used to supplement words in newspapers or magazines; those used to advertise products on billboards or in leaflets; those used to portray places on postcards; and those used to convey the activities of a particular business, government department or voluntary agency on the front cover of a report. In these and other cases, photographic and other forms of image seem to present us with reality – a mirror image which shows the 'truth' about people and places. After all, we are told, the camera never lies. However, just as with written text, visual media images are like a language, and human geographers need to be aware of how that language has been constructed, both technically and socially, before they can interpret the meanings locked therein (Gamson et al., 1992). Visual texts then make a form of reality rather than simply mirroring reality. They are carefully constructed and edited to present a particular view of the subject they portray.

The technical construction of visual images consists of a number of formal characteristics and conventions (Wells, 1997). The different elements of the image are formally organized; like grammar and syntax in a sentence, elements are made meaningful by their organization. Such organization will involve the relationship between words and pictures, any narrative styling – for example, when a number of sub-images are to be read in a particular order so as to produce an unfolding story like a comic book or a Peanuts cartoon – and the precise framing of any photographic visual text. Photographs will be composed so as to include some things and to exclude others, a process which inevitably results in partialities because of the choices involved. Within the content, some characteristics of places of the people in those places will be located in the foreground or the background. Horizons will be set, lengths of field will be selected and focal points will emerge. Thus the 'reality' of places and people will be restricted to the inclusive, exclusive, focus/non-focus and foreground/background decisions of the image-maker.

CASE STUDY 3.2 *(Continued)*

Many of these technical decisions are organized around recognizable cultural conventions of how to 'read' visual texts (see Panofsky, 1970). They make sense because we understand how they work. For example, the close-up view of a place or person draws us in to study the emotional or place-related state of the subject in detail. By contrast the landscape image offers a convention for viewing a wide-angle depiction of the scale and grandeur of a wider area, an idea extended first by aerial photography and more lately by satellite images.

The context of visual images is similarly constructed. Carefully manufactured visuals will often seek to represent multiple meanings, some of which will be at the surface and others more hidden. The image-maker may employ myths or ideologies (Gold, 1994) so as to convey a scene from a particular viewpoint. Iconographic symbols representing a wider whole are important here. Thus images of the 'Wild West' of America will convey literal meanings about the movement west and the conversion of wilderness into cities and farmland, but they will also suggest symbolic meanings about the freedom and adventure inherent in the making of America. They are far less likely to convey the exploitative nature of such processes with regard to American Indians or the profit motives involved.

Just as the written texts described in Case Study 3.1 were shown to reflect biases against 'other' social groups, so visual texts can be equally discriminatory. An example of the ways in which visual images can convey gendered myths and ideologies can be found in Brandth's (1995) analysis of advertisements for tractors (Figure 3.2). She argues that these advertisements serve to inform male farmers of what positive characteristics of masculinity they should be striving for, and thus that they participate in the social shaping of masculinity by reinforcing it and reshaping it (see Box 3.2).

Box 3.2 Brandth's (1995) analysis of tractor advertisements

In presenting the products they want to sell, the advertisers connect the tractor to images of masculinity both in text and pictures. What they anticipate will help sell the tractor is seen from the abundant use of words like horse**power**, **supreme** engine**power**, hydraulic **power** and pull**power**, **great** capacity, position **control**, reactions **control**, driving **control**. Keywords are power, precision, control.

The tractor is pictured large and strong. Big, heavy, greasy, dirty and noisy machines are symbols of a type of masculinity which often is communicated through the male body, its muscles and strength, and what is 'big, hard and powerful' (Lie, 1992). The analogy to male potency is quite clear, and visualized by the phallus, the exhaust pipe standing up in front. In an illustration from one of the ads, the machine is not pictured out in the field, as we first perceive it to be. The tractor is placed upon a landscape consisting of hills, valleys, forests and mountains...It is placed on top of nature, 'virgin land', and photographed from below – an angle which makes anything else, including the potential buyer, seem/feel small in comparison unless, of course, the buyer fancies himself as being on top of the machine (and that is probably the intention). (1995: 126, emphasis in original)

Vi sier at John Deere 1950 har 62 Hk.
Tyske tester sier noe annet!

 Tror du motoreffekt ved maks. turtall forteller alt om en traktor? I så fall tar du feil. Moderne motorkonstruksjon gjør det nå mulig å lage motorer med svært stor seigdragingsevne. Dette fører til at traktoren er sterkest ved turtall litt lavere enn det maksimale.

EKSTRA KREFTER NÅR DU TRENGER DET MEST.
Vi har lenge snakket om våre "konstant-kraft" motorer. Nye tyske OECD-tester viser at John Deere 1950 ikke bare har konstant effekt i det øvre turtallsområdet. Den blir faktisk enda sterkere. Øker belastningen, synker turtallet, og en vanlig motor får lyst til å gi seg. John Deere-motoren derimot henter fram ekstra krefter fra "kjelleren".
Ved standard kraftuttakshastighet (2075 omdr./min) er motoren fortsatt på sitt sterkeste. I den tyske testen ble effekten målt til hele 68,8 HK!

PRØV DEN PÅ DITT EGET JORDE.
Vi tror ikke ord og bilder alene klarer å overbevise deg om de suverene motoregenskapene til John Deere 1950. Derfor vil vi gjerne at du prøver den hjemme på ditt eget jorde. Vær bare forberedt på at traktoren vinner seg ved nærmere

bekjentskap. Ta kontakt med ditt nærmeste Felleskjøp og avtal tid for hjemlån av en demonstrasjonstraktor.

VINN EN TUR FOR TO TIL TYSKLAND!
Hvis du tar deg tid til å prøvekjøre John Deere 1950 og i tillegg fyller ut et enkelt spørreskjema, er du med i trekningen av to Tysklandsturer for to personer. Vinnerne får se John Deere fabrikkene i Mannheim samt oppleve tyske slott og vingods i Rhindalen. Alle som prøvekjører får en hyggelig overraskelse i tillegg. Ikke dårlig bare for å prøve en traktor! Ta turen innom ditt Felleskjøp snarest!

(OECD - Organisasjonen for økonomisk og kommersiell utvikling)

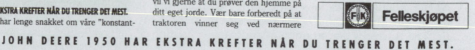

JOHN DEERE 1950 HAR EKSTRA KREFTER NÅR DU TRENGER DET MEST.

Figure 3.2 John Deere tractor advertisement

Source: Brandth (1995: 129)

CASE STUDY 3.2 *(Continued)*

The use of visual images as sources of geographical text and data is on the increase. Some of the best-known examples are found in accounts of visual texts in geographies of tourist places. One of the problems inherent in this kind of work is the very subjectivity of analysis of images (see Selwyn, 1996). Put bluntly, what one geographer interprets from an image may well be different from that of another. Nevertheless, in studies of tourism it is now recognized that the visuality of brochures has a significant role in the cultural practice of tourism, and so image subjectivity becomes a necessary part of understanding the anticipatory expectations of tourists (see Cloke and Perkins, 1998).

A good example here is Goss's (1993) study of the textural strategies employed in magazine advertisements to see Hawaii as a tourist destination for mainland Americans. He identifies five aspects of the imagined geographies of Hawaii presented in these images, which differentiate Hawaii both from the US mainland and from other tourist destinations.

Paradise

The advertisements consistently reference paradisal icons such as beaches, palms, waterfalls, tropical gardens and exotic flowers. They also valorize the unspoiled, unhurried way of life of the natives and a sense of their unrepressed sexuality. He suggests that some advertisements implicitly offer the good prospects of sexual liaisons with natives (who 'wore little more than a smile' – Figure 3.3), and that others emphasize the renewal of passion for tourist couples by depicting them in intimate or preintimate settings (Figure 3.4).

Marginality

The advertisements construct Hawaii as a place on the geographical, temporal and social frontier of North America, offering an exotic experience within the security of the familiar. The location of Hawaii is mystified and presented as halfway between East and West, and the life of Hawaii is often represented nostalgically as a precapitalist village economy, where the past meets the present.

Liminality

Visual images from the advertisements suggest a liminoid tourism which provides a break from the everyday life of the modern world. Tourists will discover not only the authentic world of the native but they will also (re)discover their authentic self.

Feminization

Goss sees the advertisements as constructing the Hawaiian people and environment as female, and constructing tourism as an act of masculine possession. The visual texts are dominated by female characters (Figure 3.5), and he suggests that the viewers of the visual texts are likely to be men, whereas the readers of the verbal texts are likely to be women.

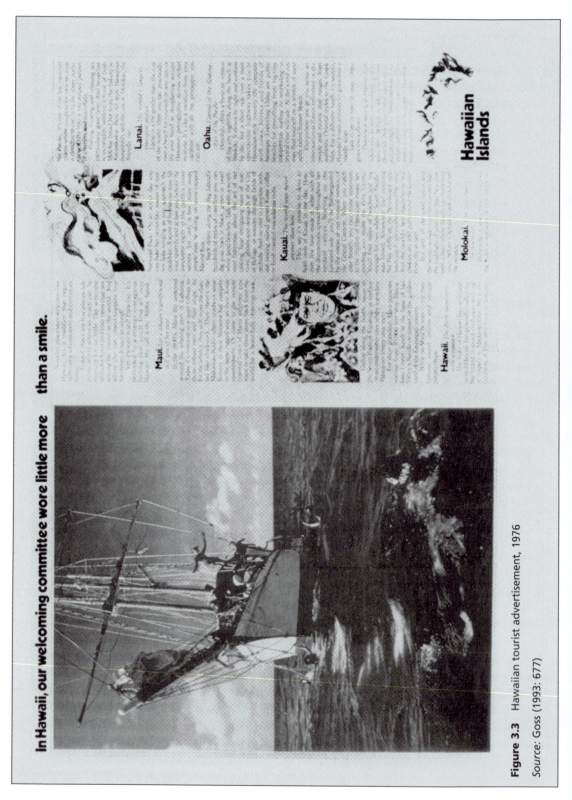

Figure 3.3 Hawaiian tourist advertisement, 1976

Source: Goss (1993: 677)

Figure 3.4 Hawaiian tourist advertisement, 1985

Source: Goss (1993: 670)

Aloha

The word 'aloha' is used in most of the advertisements and seems to guarantee the hospitality of Hawaii (Figure 3.3). It defines the relationship between the tourist and workers in the tourist industry and, in Goss's view, systematically exploits and commercializes the feelings and emotions of the Hawaiian people for the profit of the tourist industry.

Figure 3.5   Hawaiian tourist advertisement, 1983

Source: Goss (1993: 669)

In this example Goss demonstrates how visual images are able to reconstruct the geography of a place – Hawaii – replacing it by a series of signifiers which represent difference and symbolize the 'other'. Similar analyses of visual texts in many other areas of human geography will add beneficial perspective to how we imagine the worlds in which we live and the people who live either alongside us or in far-off exotic places.

CASE STUDY 3.3: THE VISUAL TEXTS OF MAPS

'The map is not the terrain', the skinny black man said.

'Oh, yes it is', Valerie said. With her right hand she tapped the map on the attaché case on her lap, while waving with her left at the hilly green unpopulated countryside bucketing by: '*This* map is *that* terrain.'

'It is a quote', the skinny black man said, steering almost around a pothole. 'It means, there are differences between reality and the descriptions of reality.'

'Nevertheless', Valerie said, holding on amid bumps, 'we should have turned left back there.'

'What your map does not show', the skinny black man told her, 'is that the floods in December washed away a part of the road. I see the floods didn't affect your map.' (Donald Westlake, *High Adventure*, cited in Wood, 1993: 26–7)

The brochures discussed in Case Study 3.2 have a straightforward and usually recognizable purpose. Although they will use all kinds of visual devices in support of their tasks, tourist brochures are inescapably designed to promote or sell particular places as tourism destinations. Maps, on the other hand, will often take on a more authoritative characteristic. In general, they are recognized, and used, as sources of information, something to be read for the purpose of seeing what is in a place, where things are or how to get there. Maps have been central to the study of geography, even to the extent of forming half of such unfortunate defining epithets as 'maps and chops'.

Yet, as Harley (1988) has shown, maps are very seldom recognized as socially constructed forms of knowledge, which through the decision of map-makers with regard to ideological intent, the use of cartographic devices and the choices of what to include and exclude should not be read as straightforward 'thin' descriptions. Rather, maps should be scrutinized in terms of the conditions in which they are made and the social constructions which they employ and embody. Only then do we begin to realize how appropriate a medium they are for manipulation by powerful elements in society.

Harley offers three sets of insights on the ideological power of maps. First, he sees maps as a kind of language (Mitchell, 1986), thereby opening up ideas of translating and interpreting the 'language' of maps, as well as ideas of writing for particular readerships, the conditions of authorship, the potential for secrecy and censorship and so on. Secondly, maps employ iconology which imports deeper-level symbolic meanings as well as superficial, literal meanings. It is at this iconological level which power is most effectively conveyed. Thirdly, cartography is a form of knowledge. As Harley (1989: 279) points out: 'Whether a map is produced under the banner of cartographic science – as most official maps have been – or whether it is an overt propaganda exercise, it cannot escape involvement in the processes by which power is deployed.'

Maps are almost always produced for a strategic purpose. For example, most official mapping by the state is connected to defence and military purposes and is often constructed from satellite imaging techniques that were specifically developed for the military for surveillance objectives. It is inevitable, then, that maps as a form of

CASE STUDY 3.3 *(Continued)*

knowledge will carry with them at least some of the characteristics from the sociopolitical arenas in which they have been constructed.

Human geographers continue to be fascinated by maps and analogous visualizations as sources of both information and representation (Dorling and Fairbairn, 1997). However, an essential corollary of this fascination is to understand how maps in the past have constructed their own history at different scales. For example, at the scale of *empire* maps have served as weapons of imperialism, with land often being claimed on paper before it was occupied by armies. Surveyors served alongside soldiers, and maps served to naturalize the exploitation of colonial civilizations (Driver, 1992). Maps also served to legitimize the politics of empire, as Figure 3.6 illustrates in terms of the portrayal, both graphically and symbolically, of the British Empire in the late nineteenth century. Maps have also served as weapons of nationalism and of national security. Thus the post-Soviet era has involved, both metaphorically and literally, the rewriting of the map of Eastern Europe. Similarly the Bosnian and Albanian crises of the 1990s have focused on the perceived mismatch between legitimate ethnic groupings and the national boundaries of states as conveyed by maps.

One interesting perspective on the national boundaries conveyed by maps is to be found in Bunge's (1988) *Nuclear War Atlas*, in which he graphically illustrates the need to interpret maps in terms of more than one-dimensional boundary lines:

> If the United States rolled a nuclear missile across the border from Minnesota into Ontario, Canadian customs, among others, would have a sovereign fit. If the same bomb were hoisted up 10 feet, Canada would be equally upset. If it were lifted 100 miles, it would not even be noticed. How high (in feet) is Canadian Sovereignty? (p. 71)

The flat map of boundaries would suggest national sovereignty in this case, but it would obscure the power of the US military to roam across 'international' skies in order to carry out its purposes. Equally, the need for such power can also be illustrated by the notion of how flat-map boundaries are transgressed. To quote from Bunge again: 'To state the problem in the exciting militaristic outlook of the 1990's, "The Russians are not coming. They are already here". They are not "ninety miles off our shores" in Cuba or sitting in missile silos in the Soviet Union. They are up in the sky' (1988: 83–5). Boundary transgression can be used to create fear, and therefore legitimize defence, as well as to hide notional strategic power from the purview of traditional mapping.

Alongside these scales of empire and nation, maps have also been used in more localized situations – for example, to convey the power of property rights. Thus in Britain, the much-loved Ordnance Survey maps of (among other areas) the great countryside outdoors are an essential waymarking guide to desirable footpaths and routeways through scenic areas. However, they have embedded within them the notion of 'right of way' through privately owned land, and therefore might also be considered as means of legitimizing and upholding the power of private landowners over clearly demarcated

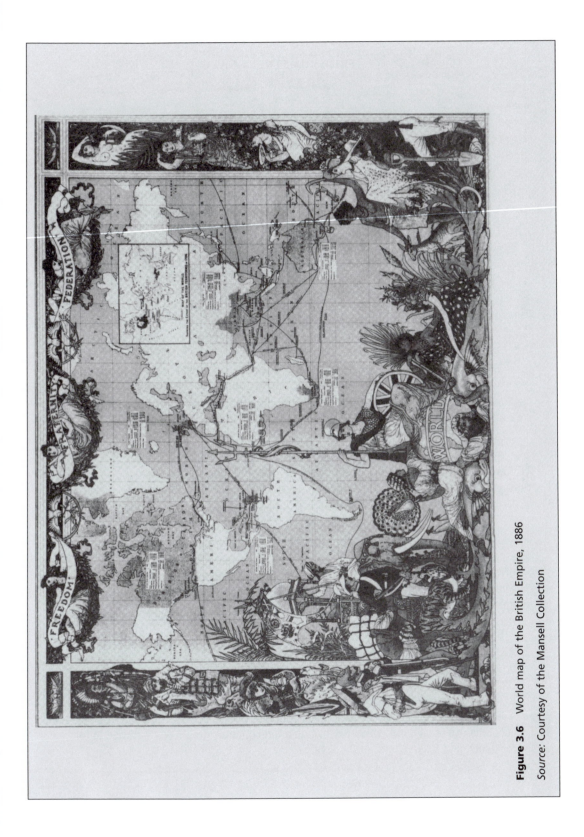

Figure 3.6 World map of the British Empire, 1886

Source: Courtesy of the Mansell Collection

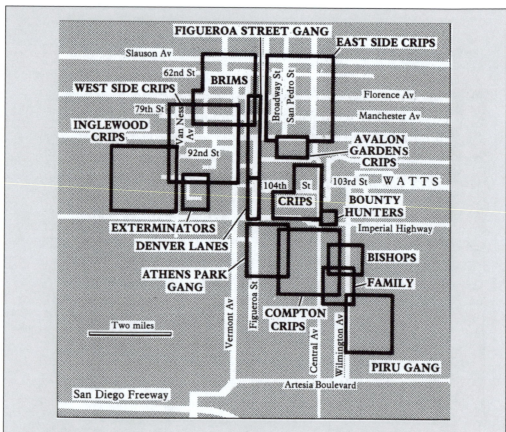

Figure 3.7   Gang territories, Los Angeles, 1972

Source: Davis (1990: 321)

private scapes. Leisure maps are also increasingly an agent of directing the 'gaze' of the tourist (Urry, 1990), pointing out 'viewpoints' and thereby organizing both the destinations and the perspectives from which landscapes are seen or gazed upon by many visitors (Wilson, 1992).

Local maps can also be used to convey alternative constructions of power. A 'conventional' street map of Los Angeles might convey key lines of communication, housing and business areas, civic buildings and so on. However, Figure 3.7 is Mike Davis's map from his account of Los Angeles in *City of Quartz* (1990). His perspective is to map out the territories of power in south central Los Angeles' sociologically distinct gang culture – a mapping which is usually unseen and indeed even hidden in official accounts of the city, but which represents a fundamental street-level knowledge for residents, and researchers, of the area.

CASE STUDY 3.3 *(Continued)*

Maps, then, require very careful interpretation from human geographers in order to understand the 'thick' descriptions as well as the more obvious 'thin' descriptions being conveyed. Some maps will carry deliberate distortions for political or strategic purposes. Others will be influenced much more subtly by the values of the map-producing agency and society. Some distortions will, therefore, be subliminal (for example in the use of geometry). Others will occur because a particular map is 'silent' on particular characters of what is being represented. Maps exert an influence through omission as well as through inclusion. Careful interpretation is particularly required because the social history of maps is predominantly one of power not protest, and even in the current age of more communication, map-making continues to be dominated by powerful groups in society.

CASE STUDY 3.4: THE MOVING VISUAL TEXTS OF 'NATURE' DOCUMENTARIES

Moving visual texts are an essential facet of the media age. Morley and Robins (1995) talk of the existence of global media in contemporary society and suggest that, in these terms at least, national frontiers are a thing of the past. On the one hand, the onset of global media raises optimism for bringing the world closer together, with one-world citizenship responsibilities, interconnecting cultures and international democracy. On the other hand, global media are part of private giant economic interests pursuing traditional capitalist objectives of profit-making. The struggle for power and profit between such concerns of Time-Warner, Sony and News International is therefore being waged on this global scale.

So it is that television programmes and films have now crossed many boundaries and have achieved world awareness of their plot, characters and merchandise. These imaginative moving visual images are dealt with in the next chapter, but here we recognize that such images are also used for documentary purposes. News programming and documentary output from the global media companies are screened worldwide, allowing us to experience the world through our television sets, and increasingly also our personal computers. These filmic texts are crucial to any contemporary theoretical concern for space (Cloke, 1997). We are much more likely to know about spaces and places and the people who inhabit them through moving visual text than either through direct experience or by reading written text. As Aitken (1994) suggests: 'I'd rather watch the movie than read the book.' Indeed, it is not just that filmic texts represent the real world to us, it is that such representations become our real world. It might even be suggested that the real world is increasingly being produced to match the representation.

Once again, we want to stress that documentary visual images are technically and socially constructed rather than being mirror images of 'real' life. It follows that in order

CASE STUDY 3.4 *(Continued)*

to understand the messages and meanings conveyed therein, we need to deconstruct the codes and devices being employed by the image-makers. Many of these devices are similar to those discussed in Case Study 3.2 in the context of still, visual images. The formation and conventions of the image, its use of myth and ideology, its capacity to convey discriminating impressions of particular social groups, its use of iconography and signification, are all important in the understanding of filmic images. There are, however, other conventions which are specific to the interpretation of moving visual images.

Filmic texts employ a number of techniques which audiences will have come to accept, subconsciously, as normal (Aitken, 1994). For example, we have learnt to suspend our disbelief and accept the impression of reality given by the rapid succession of static serial images in motion over time. We also find it unproblematic to consume filmic images which transpose unreal timescales, through the use of flashbacks, action replays and slow motion. Camera techniques of close-up, panning, tracking and zooming are also entirely acceptable signifiers of scale, intensity and interconnections between places and people. Documentary and news techniques which pose a foregrounded reporter against a background which represents an issue or place are viewed as an uncritical signification of the proximity of the production of the film to the topic itself. The background thus somehow signifies the newsworthiness of the place or issue. Documentary or news interviews present an 'in-touchness' with the other. These are just some examples of the filmic conventions which we take for granted in our viewing of filmic representations of the 'real world'.

Geographical research on film (Kennedy and Lukinbeal, 1997), however, also has to contend with a further series of interpretative claims borrowed from film criticism (see hooks, 1997). First, filmic images often constitute voyeurism. Much voyeurism is associated with the sexuality embedded within imaginative filmic texts. Yet documentary films can also be constituted around the voyeuristic impulse. News and documentary programming often deals with death – human massacres, famine and warfare, as well as powerful animals (lions, sharks and so on) devouring their prey – and injury and illness. Attention is also often focused on problematic life experiences such as homelessness, drug addiction, squalor, hunger, orphaned children, mad cows and so on. In these kinds of ways, filmic text organizes the gaze of the watcher as that of momentary tourist, gaping at the pain and distress of other people and animals.

Secondly, filmic images will often construct 'otherness' as the direction of news or documentary establishes through pictures on dialogue what is the mainstream self and what is the marginalized other. Filmic text is therefore used to exoticize 'foreigners' or indigenous populations, to regulate what the viewer will regard as 'normal' practice (and therefore what is dysfunctional) and to regularize particular cultural or political biases. Even in news and documentary formats, hidden ideological biases can pervade the production, and therefore consumption, of filmic text.

Thirdly, filmic texts can distort particular spaces, places and the activities of people within them. There is often a sense of accepting the authenticity of places, people and events in news and documentary coverage. However, the very presence of cameras and

CASE STUDY 3.4 *(Continued)*

reporters may itself ignite a smouldering situation, with the actions and place meanings caught on film only being 'real' because the film was there ready to catch them. Equally, there are many different stories and images which can be conveyed about a particular place, and news and documentary coverage will often choose a dramatic portrayal, even if different voices are incorporated within the narrative.

There are still relatively few examples of research which brings human geography interests to bear on 'factual' filmic texts.[1] One interesting example of this genre of interpretative research is Wilson's (1992) discussion of Disney nature films, in which he notes a number of key conventions of filmic narrative. For example, the stories are always told starting at the beginning, and ending at a new beginning. Thus in *The Lion King* (admittedly not a documentary!) the picture starts with the birth of Simba, and ends with the birth of Simba's first male cub. Here, according to Wilson, is a transparent allegory of 'progress', symbolizing the forward-looking notion of modernity, with better standards of living in the next circle of life.

Wilson also captures the changes in the locational authenticity of Disney's nature programmes. Early films were short on location and often in the West. However, by the 1950s and 1960s the West had become tainted both environmentally and politically as the virgin landscapes of previous films were transformed by resource extraction industries and the industrialization of timber production and water power. These landscapes thus became too laden with contradictions for the shooting of further nature films, and so the Disney wildlife features were relocated, either into the Burbank Studio lots, using trained animals, or into more remote locations, often in deepest Africa. Such relocations changed the focus of 'wildlife' and 'wildlife locations' in many ways. More recently, Disney's wildlife programmes have switched to highly technological shots of the private lives of animals, insects and plants, using trick photography, extremely intrusive camera techniques and 'trained' creatures in non-natural conditions. These changes have often been mirrored by other nature programmes.

Wilson's analysis demonstrates how filmic interpretation of wildlife constituted the addressing of nature as a social realm. Animals were inserted into human stories and usually treated as social beings. In so doing Disney precursed what McKibben (1989) has termed 'the end of nature'. Nature as filmed became a cultural construction on to which human stories were told and projected. It is this kind of story-telling through the filmic images of news and documentary programming which provides an exciting and very significant source of information for human geography researchers.

Conclusion

What each of these sources have in common is that they purport to tell, or represent, the truth as accurately as possible. Yet what they also have in common, as Harley says of maps (1988: 278), is that 'in the selectivity of their content and in their signs and styles of representation, [they] are a way of conceiving, articulating, and structuring the human world which is biased towards, promoted by and exerts influence upon particular sets of social relations'. An understanding of this selectivity and bias will enable the researcher to appreciate the issues involved in the use of these sources. In the next chapter we will examine the construction and use of imaginative sources of data, which we would expect to promote partial and particular viewpoints. However, the non-official sources we have considered here also do this, but under the banner of faithfully reporting the truth. As we maintained at the outset of the chapter, imaginative and factual sources do have a lot in common. This issue will be explored further in the next chapter.

Note

1 But see Burgess and Gold (1986), Burgess (1987), Rose (1994) and Benton (1995).

4   *Imaginative Sources*

Introduction

As we have shown in Chapters 2 and 3, in recent years geographers have adopted an increasingly critical approach to conventional sources of data such as official statistics and other 'factual' documentary sources. These sources are no longer seen as providing direct and unmediated access to geographical reality but as social constructions produced for particular purposes in specific social contexts. At the same time there has been a growing acceptance among geographers of the legitimacy of using sources of information which are overtly products of human imagination and whose primary original purpose is not to make factual statements about the world but, instead, to entertain, provoke, inspire or move the reader, listener or viewer; in short, to engage the emotions and, indeed, the imagination. Such 'imaginative sources' are the focus of this chapter, but before considering their construction in detail there are a couple of ambiguities to address.

In some ways, the distinction between 'factual' and 'imaginative' sources is more apparent than real. Geographers are not alone in coming to realize that all attempts to describe, explain and understand the world involve particular ways of imagining it, and that there is no way of presenting a purely factual account devoid of imaginative content. Nevertheless, because we want to emphasize the ways in which sources are *produced*, we find it helpful to deal with 'imaginative' and 'factual' works separately because the conventions governing their construction are (at least to some extent) distinct from one another.

Another ambiguity in this chapter concerns the concept of 'data source'. Frequently, geographers are interested in imaginative works not as sources of data which will provide information about something else but as research topics in their own right. For example, one geographer might read the novels of Charles Dickens in order to learn about the geography of nineteenth-century London, while another might use the same books in a study of the rise of the novel as a literary form and its relationship to nineteenth-century industrialization. In keeping with the structure of the book, the main emphasis in this chapter will be on the use of imaginative works as sources for research, rather than objects of research, not least because in the latter case the sources for the research may well include the statistical, documentary and interview materials discussed in other chapters. However, since studying imaginative works as objects of research often involves an investigation of the conventions and conditions of their production, we cannot draw a rigid distinction between data sources and research topics. Thus, in writing this chapter we have often found it useful and necessary to consider geographical studies which treat imaginative works as research objects as well as research data.

Geographers are turning to imaginative works with increasing frequency. There are many reasons for this and, to some extent, they depend on the particular nature of the work involved and the issues under investigation. At a more general level, however, we may identify five claims about imaginative works which are commonly used by geographers to justify their use:

1. The production of imaginative works ('artistic creation') is a significant social activity, that has a geography to it and should therefore be one of the activities geographers study.
2. Artistic works embody imaginative and imaginary geographies. Geographers bring a sensitivity to issues of space, place, landscape and environment to their analyses and can therefore add an additional perspective to the interpretations of art historians, literary critics, film theorists and so on.
3. The products of creative people provide a window on to the human condition and thereby illuminate issues of interest to geographers, such as human feelings for place.
4. Imaginative works are themselves the products of particular geographical contexts and thus help geographers to understand those contexts: Renaissance drama tells us something about Elizabethan England and Hollywood movies give us insights into contemporary America.
5. The form and content of imaginative works are related to wider social and geographical relations and processes; they both reflect and affect these wider social relations.

As will become clear, these claims are by no means unproblematic but they provide a useful starting point for thinking about the relationships between imaginative sources and geographical research.

Understanding the construction of imaginative sources

Just as in the cases of official and non-official 'factual' sources, it is particularly important to understand the conditions under which imaginative works are produced when using them as research sources in human geography. These conditions are of two related sorts. First, there are the changing conventions and regimes in which creative people work and through which the meanings of their products are conveyed to and understood by the reader, listener or viewer. These include such things as the technique of perspective drawing in Western art (which was developed in the Renaissance and which governed the construction of most painting and drawing for 500 years), the conventions of narrative construction in literature or the use of particular styles of music in film soundtracks to convey emotions or to support the action. Secondly, there are the social contexts within which imaginative works are produced. These might include the organization of the publishing industry, the relations of patronage between a painter and a client, or the wider social relations of class, gender and ethnic difference that influence (unconsciously or otherwise) both the form and the content of imaginative works.

There are two main reasons why it is important to understand the processes through which imaginative sources were produced when seeking to use them in geographical research. First, we would argue that it is important to 'demystify' the process of artistic creation. Writers, artists, poets, architects should not, we feel, be seen as special beings endowed with powers of insight and revelation denied to others. This is not to suggest that emotions and what is often called 'spirituality' are unimportant in imaginative works. Rather it implies that emotional response and spiritual feeling are not the unique preserve of specially creative people but are available to all. By understanding artistic creation as a social process which takes place in an economic, political and cultural context, we hope to show that while imaginative works do possess certain distinctive qualities, these arise from particular cultural norms, conventions and practices rather than from some mysterious capacity of their producers.

Secondly, we would argue that imaginative works cannot be adequately interpreted without an understanding of their construction. For one thing, in some cases interpretation of a source and analysis of its construction blur into a single process. Thus the meaning of a painting frequently depends on the conventions and context of its production. However, even where interpretation can be separated from an understanding of its construction, the conditions of production still make a difference. Thus one can interpret a film like *Blade Runner* for its symbolic content without a detailed knowledge of the Hollywood film industry, but such a knowledge would expand the range of questions one could ask. Why, for example, are action films like *Blade Runner* the mainstay of so many companies' production schedules? Why was the later 'Director's Cut' version of the film seen as 'more authentic' than the initial release?

It is easy to stress the importance in principle of understanding the conditions of production of imaginative sources but rather more difficult to describe how it should be done given the apparently huge differences between forms like poetry, architecture and virtual-reality simulations. One way would be to undertake an in-depth study of each type. This would involve studying in turn the television industry, the media, the publishing industry, the conventions adopted by painters, poets, architects and film-makers and so on. However, not only would each study require a chapter, or even a whole book, to itself but such an approach also neglects what is common to all imaginative productions. At first sight is seems difficult to see what the common threads might be, but it is possible to outline an approach to the construction of imaginative sources that can be applied to all or most of the forms we have discussed above. There are a number of ways of doing this (for an overview see du Gay, 1997, and Hall, 1997). Our approach, set out below, draws mainly on the work of Raymond Williams (1981). Williams's writings have provided much inspiration for subsequent accounts of the production of culture and for cultural studies more generally. Williams's approach came to be known as 'cultural materialism' because of the emphasis he placed on the role of material interests and material processes in the production of culture. Although the 'cultural turn' in geography marked something of a shift away from avowedly materialist approaches towards a focus on textuality, discourse and rhetoric, we want to emphasize the continuing relevance of cultural materialism for geographical research. For one thing we consider the distinction between the discursive and the material to be overdrawn. As Williams shows in his work, even something as apparently wholly 'discursive' as speaking or singing entails all kinds of material transformations involving the body (lungs, vocal chords, tongue and teeth) and the environment (air and the acoustic properties of the performance space). Intriguingly, these kinds of concerns have recently come into the heart of geographical theory with the growth of interest in the radically materialist actor-network theory of writers such as Bruno Latour and Michel Serres (Serres and Latour, 1995; Bingham and Thrift, 1999; Whatmore, 2002).

Moreover, despite the passage of time since Williams's ideas about the production of culture were first published in the 1960s and 1970s, they continue to exert a substantial influence on cultural studies and cultural geography (Milner, 2002).

The process of cultural production

Conventional cultural and literary theory accords a central place to the role of the author or creator as the prime or even only source of a work's meaning. Some more recent writers have proclaimed the 'death of the author' and have asserted that the meanings of texts or cultural representation escape entirely from their authors' control. In our view authorial or creative intention has some role in the production of culture but, and it is a big but, it is rarely, if ever, the most significant aspect of the production process. At the same time the input and impact of an author on his or her work are not limited to conscious intention. Conscious intentions are structured and influenced by unconscious motivations and desires and constrained by cultural conventions and material and discursive contexts.

All cultural and artistic works are necessarily surrounded by material processes and practices which influence, constrain and enable the development of particular cultural phenomena in specific ways in different times and places. For example the institutions of the artist's patron in the Renaissance period affected the nature of the art produced, just as the art market does today. The different means of cultural production (from the human voice and body to the television camera and the compact disc) influence what is produced. In addition to these material conditions, non-material (or 'discursive') practices also play an important part. For example, the 'genre' of the soap opera or the detective novel helps to set the conventions for new soap operas or thrillers, while variations in symbolic relations are crucial in differentiating Greek tragedy from Renaissance drama from the realist novel (Williams, 1981). All these aspects (material and discursive) of the production process are also related in complex ways to the wider society of which they are a part. Thus revolutions and changes in dominant artistic forms often coincide, and are intertwined, with major transformations in social, political and economic relations. This is not to say that the content of the cultural change is mechanically determined by social change, but neither is it wholly independent.

There are seven sets of relations that, together, explain what, we are calling the construction of imaginative sources of data (Williams, 1981): cultural *institutions*, cultural *formations*, the *means of cultural production*, cultural *identifications*, cultural *forms*, cultural *reproduction* and cultural *organization*.

INSTITUTIONS

Virtually all artists work in some relationship to social institutions of various kinds (institutions comprising artists themselves – for example, artistic 'movements' such as the Pre-Raphaelite Brotherhood considered in the next section on 'formations'). These institution vary historically and geographically. The dominant institutions in medieval England were very different from those in twentieth-century America. Nevertheless it is possible to boil down the enormous variety of institutions that have existed to four main types (Williams, 1981). These are the instituted artist, patronage, markets and post-market institutions.

The instituted artist was particularly a feature of early societies and involved the artist (usually a poet) taking a key part in the central social organization. The archetype is probably the Celtic bard. Despite apparent similarities with patronage systems, this situation is distinct because patronage relations depend on a deliberate decision to recognize or patronize a particular artist. In the bardic case, by contrast, 'the social position of this kind of social producer was instituted as such, and as an integral part of general social organization' (Williams, 1981: 38).

Patronage in some cases evolved from the case of the instituted artist. In Wales, for example, there was a gradual transition in which the role of bard became increasingly detached from the court and began to be attached to individual noble households, or sometimes to be peripatetic between households, seeking support, shelter and commissions (Williams, 1981: 39). In many societies, however, there was no institution of 'court poet' and the more normal form was full patronage. For many centuries and in many countries artists have been retained by individual or institutional patrons and have produced their work to order on a commission basis. This means, of course, that the content of these works has been determined not solely, or even mainly, by the conscious intentions of the author but by the instructions and specifications of the patron. A further distinct form of patronage involves protection, support and recognition for the artist but little direct commissioning. Here the artist may have considerable control, within limits, over content and form. Williams identifies the theatrical companies of Elizabethan England as examples. Sponsorship represents another type of patronage, which sees the survival of some form of patronage relation into the age of the market and the general commodification of cultural products. In eighteenth-century England this often took the form of a subscription list in which a number of sponsors would support the work of an artist at an early stage before it had found general success in the market. This form continues as commercial sponsorship of the arts where it seeks a commercial return through the use of advertising or financial investment. These forms also influence the character of art works. For example, it is frequently asserted by artists that large companies prefer to sponsor 'safe', 'mainstream' and 'conventional' work, as opposed to innovative and avant-garde material. Finally, patronage can also come from the state through the public expenditure of taxation. In these cases there is continuity with earlier forms but also the scope for extending and developing the arts as a matter of public policy. This can also lead to some works being favoured over others for political reasons. For example, in Britain in the 1980s the Conservative government was motivated by a right-wing political agenda to restrict the public funding of the cultural activities of gay and lesbian groups by limiting the range of things on which local authorities could spend money.

In all relations of patronage, the patron (whether an individual or an organization) has considerable power, and it is this which unites these very varied social forms. The patron can withhold or provide support at will, and such support is withheld whenever the activities of the cultural producer are unacceptable to the patron. For the human geographer, therefore, using artistic sources as research data requires that close attention be paid to the nature of any relations of patronage involved in the production of the work.

Wherever and whenever cultural products are produced for sale and exchange, they are commodities and the artist becomes a commodity producer (Williams, 1981: 44). This institutional relationship is qualitatively different from that of patronage. The form and content of the work are determined not by the whims of the patron but, at least in part, also by the perception of actual or projected market demand. However, this can take many forms. Production may be artisanal. In this case the artist is producing under his or her own control and offering the work for direct sale on the immediate market. A contemporary example would be a fashion designer with his or her own shop. There are also two forms of post-artisanal production. The first, in which the artist sells to a distributive intermediary, is akin to some modern commercial art galleries in which the gallery owner buys from the artist. The second, in which the artist sells to a productive intermediary, is akin to the modern book publisher in which the publisher not only buys the artistic work (such as a novel) but also produces (by printing and binding) the finished book. In both cases the artist's relation with the market becomes indirect.

A third type of market relation is the market professional. This phase sees the artist gain specific professional rights, such as those associated with copyright and royalties, over the work. This is associated with a new type of intermediary, the literary agent. In this phase the artist does not have direct relations with individual buyers, or completely indirect relations with the market through intermediaries, but has 'relations with the market as a whole' (Williams, 1981: 48).

The fourth and so far final phase of the marketization of cultural production is the corporate phase, and this is particularly associated with the development of new media such as film and television. Here the creative artist acts not as an individual author but as part of a production process undertaken by an organization, such as a magazine publisher, a television station or a record company. The cultural product is the outcome of the activities and interactions of dozens of individuals, contracted and employed by the corporation. The input of the 'artist' (pop star, film director, etc.) is now just one component in the production process which depends to a large extent on the labour and creativity of others (technicians, designers, actors, musicians, editors and producers).

All the 'phases' of marketization (artisanal, post-artisanal, professional and corporate) can and do coexist in time and space, and market relations can and do coexist with other institutional arrangements, such as patronage.

In interpreting such sources in geographical research, the impact of different forms of market relations on the nature of these works must be analysed and understood. The bottom line (literally) for market production is that it must, by definition, make a profit, at least in the medium term. This means that the many artistic forms that are not 'profitable' could not exist in modern societies without other forms of support.

These other forms of support are provided by a final group of institutions: 'post-market' institutions. These in turn can be divided into the modern patronal, the 'intermediate' and the governmental (Williams, 1981). The first is common and involves foundations, subscribers and private individuals supporting works financially in a continuation of the traditional forms of patronage discussed above. The intermediate category accounts for organizations which are not private in the patronal sense but not fully part of the state either. The Arts Council and the BBC are examples of organizations which are publicly funded but which 'direct their own production' (Williams, 1981: 55). Finally, governments produce art themselves. This situation is most common in state socialist countries such as China. Here, artists may be direct employees of the state and, to a greater or lesser extent, their output may be controlled by state officials.

FORMATIONS

The second group of relations comprises the grouping of cultural producers themselves into formations such as schools, movements and tendencies. The social institution of the Celtic bards has already been mentioned, and the bards acted together to sustain and to police the rules of their art. This was an early example of a formation. In the Middle Ages, the craft guilds were influential and in the English mystery plays each guild took responsibility for a particular aspect of the production. The Italian Renaissance gave birth to the *academy*, which marked an increasing distinction of arts from crafts and a growing concern with the education and training of artists. The academies, in turn, gave rise to exhibitions. The rise of the market professional documented above was associated with the development of professional societies, particularly societies of writers, who came together to negotiate with publishers and governments over issues of copyright and so on.

More diffuse and less institutionalized than all these are the artistic 'movements', 'schools' and '-isms' which are so much a feature of the modern (post-eighteenth-century) cultural landscape. These can be grouped according to both their external and internal relations.

Internally, artistic movements can be based on formal membership, a collective public manifestation or on a conscious association and group identification (Williams, 1981: 68). The first of these may well have a formal constitution and forms of internal authority and is exemplified by the German Order of St Luke. The second, exemplified by the English Pre-Raphaelite Brotherhood, involves an exhibition, a publishing imprint, a periodical (e.g. the Pre-Raphaelites' *The Germ*) or a manifesto. The third category, such as the French *Nabis,* has no formal structure or public manifestation but, none the less, involves group meetings and a shared identity.

External relations are also of three types. These are the specializing formations, the alternative formations and the oppositional formations. Specializing formations work to sustain and promote work in a particular medium or branch of the arts or in a particular style. Alternative formations aim to provide facilities for the production and circulation of certain kinds of work for which there is no outlet elsewhere. Oppositional formations (like the surrealists, the dadaists or the

situationists) raise this alternative activity to active opposition to the existing institutions and practices. Clearly, different forms of internal relation can go together with different external ones. Oppositional groups might be highly internally structured or fairly diffuse.

For human geographers using artistic data sources the question of formations is important because the character of particular works, and their meanings, are partly determined by their formational heritage. Moreover, as was the case for institutions, formations themselves have a geography: they vary in their form and purpose from place to place and from time to time, and they are themselves internally heterogeneous (Williams, 1981: 85–6).

MEANS OF PRODUCTION

The means of cultural production can be divided into those that inhere in the human body and those that involve the use and transformation of non-human materials. The use of the inherent capabilities of the human body allows, principally, dance, song and speech as forms of artistic expression. These may be traditional forms or may involve long periods of professional training. In either case the physiology of the human body places limits on what can be danced, sung or spoken. It should not be supposed, however, that these forms of production have been surpassed by technology; rather, both coexist today (Williams, 1981).

In the use of non-human means of production the social relations involved are much more complex. External objects may be combined with the human body, as in the use of masks and costume in dance. Instruments of performance may be developed, notably musical instruments of various types. Then there are the 'selection, transformation and production of separable objects' (Williams, 1981: 90) such as the use of clay or metal in sculpture and pigment in painting, and systems of signification separate from human gesture or speech, notably writing. Finally, there is the development of 'complex amplificatory, extending and technical systems' (Williams, 1981: 90) from printing, through photography and recording to computer graphics and 'virtual reality'. The first three of these five are essentially extensions of the resources inherent in the human body, at least as far as the social relations involved are concerned. The fourth and fifth, however, involve qualitatively new social relations of production.

IDENTIFICATIONS

The further set of social relations concerns the ways in which particular works and cultural activities are identified and categorized socially. Put more simply, what is art? For Williams there can be no eternal or universal answer to this question. Defining and categorizing cultural products is itself a social process, and conventional universal definitions of 'art' fall down because they fail to recognize the implications of this insight.

For example, it might be argued that 'art' is consciously performed or exhibited for an audience yet this may leave out works like cave paintings, commonly thought of as an early form of art but situated in dark and inaccessible areas which are hardly the first choice for a site selected deliberately with a view to exhibition. Perhaps art is distinguished by the special skill with which it is executed. But a wide variety of other activities involve high levels of skill and are not usually regarded as 'art'. Finally, 'art' might be defined as possessing 'aesthetic' qualities such as beauty, harmony and so on. However, this definition fails too, as many things have aesthetic appeal without being counted as 'art'.

All these problems suggest that the process of defining art and categorizing activities and objects into 'art' and 'not art' is a social process and there are no universal definitions. Williams's account emphasizes changes in categorization through time, but he rather neglects an issue of central interest to geographical research – namely, the ways in which the process varies over space.

Williams argues that 'art' must be seen as a social form and that society identifies certain activities and works as art through systems of signals. Locating something in an art gallery, for example, identifies it as a work of art. Drama is conventionally signalled by location in a theatre. On television, where imaginative works are joined by factual programmes, news broadcasts and so on, 'art' is signalled by conscious titles and introductions which clarify for the viewer that the programme is drama rather than documentary, for example.

There are also systems of signals internal to imaginative works. Specific devices and techniques (such as the soliloquy in drama or the sonnet in poetry) provide cues to the audience or reader through which to identify the activity. These devices themselves can be understood as social practices and analysed in terms of the social relations they embody. Thus in drama the use of the soliloquy (an individual performer talking to him or herself) depends on (and perhaps contributes to producing) a particular conception of 'selfhood' as, for example, self-conscious individuality, which means in turn that the device would have been simply unthinkable in social contexts where different conception of the self were found. In this sense it is significant that the soliloquy was strongly developed in Renaissance drama in early modern Europe.

FORMS

At the start of the chapter we outlined different types of imaginative works used by geographers. There are, however, further differences within each type. Epics are different from detective stories, religious icons are different from landscape paintings and the silent comedies of the early years of cinema are distinct from modern adventure movies. Williams calls these different subcategories 'forms' and he uses the example of drama to illustrate his argument because of the long historical development of drama as an activity and the variety of forms through which it has passed.

The earliest historical example of drama is classical Greek tragedy. This form emerged from earlier predramatic forms such as the dithyramb or choric hymn. There was then a series of innovations which together produced the emergence of drama as a cultural phenomenon. These included dialogue between a single figure and the chorus, then dialogue between two figures, and then the addition of a third actor. Over time, more actors were added and the chorus gradually became more marginal. At this stage drama was fully distinctive and would have been recognized by modern audiences *as* drama. Similar processes of development could be traced for other cultural activities – the emergence of formal dance from folk dance or of individual paintings, for example.

There were distinct limits to the Greek dramatic form, however, which mean that while it was clearly drama it was different from modern drama. Thus the modern convention of having different roles spoken by different actors was not part of the Greek form, and it also had its own distinctive modes of production combining singing, speaking and recitative, which mark it out from modern forms defined by spoken dialogue.

Another example is English Renaissance drama (the theatre of Shakespeare and his contemporaries) which, while still 'drama', is quite distinct from the Greek case:

> Meanwhile, within the different social order of Renaissance England, quite other formal innovations were made. The 'dramatic', by the late sixteenth century, was a highly specific combination of acted dialogue between individuals and developed spectacle. Making its way in popular rather than primarily aristocratic theatres, it drew heavily on those arts of visual representation (in both acting and scene) which had been central in a popular pre-literate culture. Acts of violence, for example, were now directly staged, rather than narrated or reported. Drama as visible action, without words, was available in the simple form of the dumb show or in the highly developed forms of staged processions, battles or visions … Music and song were also used, but with rare exceptions not integrally, but as isolated elements of performance. (Williams, 1981: 154–5)

What really distinguished the new drama, however, was a new form of dramatic speech that is particularly interesting to the social researcher because of its diversity. For the first time the speech of all social groups was represented, from the most educated to the most 'vulgar'. However, this endured for only a relatively brief period before an increasing social exclusiveness of the theatre as a cultural experience was linked to a corresponding reduction in the linguistic range of English drama. As we have stressed before, the form as well as the content of imaginative works can't be separated from the social context in which they take place.

Williams used drama to illustrate his argument partly because it was his own specialism and partly because of the long historical record, which allows the changes to be documented. The same principles could, though, be applied to other imaginative works, such as painting, the cinema or the novel. The central point remains the same: that new forms and developments from one form to another can't be understood in abstraction from the historical (and we would add geographical) setting in which they occurred. Where imaginative works are concerned, 'what' and 'how' are intimately related to 'when' and 'where'.

REPRODUCTION

'Reproduction' includes both the ways in which cultural practices are transmitted from generation to generation and the ways in which particular cultural forms are reproduced.

The issue of reproduction is important because it is related to social and cultural change. As societies change and develop, the character, content, role and status of cultural activities change too. Two aspects to reproduction may be identified. First, there is the reproduction of cultural practices through society. Here education and tradition are of central importance but need to be treated with care. In both cases a process of selection occurs. In an education system particular cultural practices are valued over others and are included formally or informally in the educational curriculum. A 'tradition' can be thought of as 'the process of reproduction in action; ... a selection and reselection of those significant received and recovered elements of the past which represent not a necessary but a *desired* continuity ... [This] "desire" is not abstract but is effectively defined by existing general social relations' (Williams, 1981: 187, emphasis in original). Traditions are struggled over, and what gets to count as tradition (such as the 'canon' of great authors in English literature) depends on the relationships between social groups.

This is not to say, however, that culture is simply the direct expression of more fundamental social processes. The 'closeness' between cultural practice and the conditions of cultural practice varies widely. In some cases, such as television production, the connection is a close one; in others, such as sculpture and poetry, there is a significant distance between the reproduction of the cultural activity and wider social relations.

The second aspect of reproduction is internal reproduction, or reproduction *within* cultural forms. The concept of form may be refined by considering two different levels. 'Modes' include the dramatic mode, the lyrical mode, the narrative mode and so on. Within each mode traditions are established and practitioners learn from them and thus reproduce the mode through time. Within modes there are 'genres', such as 'tragedy', 'comedy', 'epic' or 'romance'. Williams argues that modes tend to survive major changes in the social order while genres either die out or are radically redefined.

For the geographer interested in imaginative works, issues of reproduction are important because they enable particular works to be located in relation to traditions, modes and genres which in turn relate to changes in social relations. For example, as we noted above, landscape painting as a genre was closely connected in the eighteenth century to the power of the landed gentry and their desire to display their wealth and to maintain the social relations on which it was based.

ORGANIZATION

Finally, there is 'organization'. Organization has already been considered in part in the discussion of institutions and formations, of forms and kinds of art, and of the processes of cultural production and reproduction. However it is possible to go further than this and to consider the organization of culture in more general terms. Here Williams outlines an approach to culture which many geographers using imaginative works have adopted and adapted. He argues that culture is organized as a *signifying system*, a definition which he contrasts with both the too specific notion of culture as 'works of art' *and* the too general notion of 'culture as a whole way of life'.

Individual systems, such as the economic system, the political system and the kinship and family system, each has its own signifying system and these are part of a wider 'general signifying system'. While economics, politics and so on cannot be reduced to processes of signification alone 'it would also be wrong to suppose that we can ever usefully discuss a social system without including, as a central part of its practice, its signifying systems, on which, as a system, it fundamentally depends' (Williams, 1981: 207).

For example, a national currency is both a crucial factor in economic activity and, simultaneously, a signifying system which conveys meanings such as economic value and political control over a delimited territory. (If you doubt the latter, think of the dispute over the European single currency; clearly for some a national currency like the pound carries meanings of national identity and political authority as well as economic value.)

A signifying system is one in which meanings are produced, transmitted and interpreted in a cycle of 'encoding', 'decoding' and 'recoding'. This can be thought of as a 'circuit of culture' (Johnson, 1986) involving the production of meanings, which are then coded and embodied in texts of various sorts. Texts are then read ('decoded' and interpreted) and thereby affect lived culture which in turn influences the production of further meanings, new texts and so on. Both production and reading are conditioned by the social relations in which they take place (Figure 4.1).

This 'circuit of culture' is part of everyday life but it is also significant in geographical research. Thus geographers interpret texts and, as we have been suggesting throughout this chapter, in order to do this effectively they need to understand the process of the production of texts as well. For a geographer considering an imaginative source, therefore, it is important to consider not only the material and institutional conditions of production and the conventions and forms which govern the creation of artistic work, but also the signifying system (the cultural circuit) in which the work is embedded and from which it acquires meaning. Geography matters here because signifying systems vary through time and across space. The field of meaning in which eighteenth-century landscape

paintings were produced was very different from that which gave rise to twentieth-century novels.

Two caveats need to be entered, however. First, the circuit of culture involves changes to meaning. The embodiment of meaning in texts detaches it from its original context and allows it to be read in another, and the meanings decoded by the reader will differ from those encoded in the production of the text. This would only be a 'problem' if the purpose of interpretation was to recover the original intentions of the author. However, as we have shown above, this cannot be the purpose of interpretation, in part because authorial intention influences the meaning of texts in only a limited way. The transmission of meanings in signifying systems is always mediated (hence 'media') and thereby changed. The circuit of capital is thus a dynamic and evolving process, not a wheel spinning in perpetual equilibrium.

Secondly, there is no guarantee that two analysts will both decode a text in the same way. For example, the film *Blade Runner* was interpreted by David Harvey (1989a) as conveying the phenomenon of time–space compression and the domination of corporate capital, and by Doreen Massey (1991a) as conveying the inequalities of patriarchal gender relations. Again, this would be a 'problem' only if it was assumed that all texts have a 'true' meaning which it is the job of the analyst to uncover. In fact, texts usually convey multiple and often conflicting meanings, reflecting the complexities of the signifying systems in which they are produced, and it is perhaps therefore no surprise if different readers and viewers derive conflicting interpretations as a result. On the other hand, it would be a mistake to assume that 'anything goes' where interpretation is concerned. Just because a text will bear many interpretations does not mean it will bear any interpretation you like!

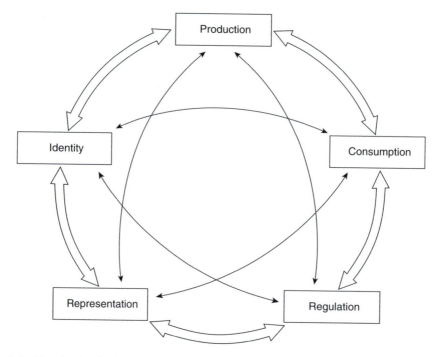

Figure 4.1   The circuit of culture

Source: From du Gay (1997: vi)

Imaginative sources in geographical research

We have relied heavily on Williams's approach because the emphasis he places on the process and context of cultural production is in keeping with our concern in this part of the book with the ways in which sources of data are generated, either directly by the researcher or by other actors whether institutional or individual. In the case studies at the end of this chapter we look at the way this kind of approach can be adapted to help to understand the genesis and nature of the some of the kinds of imaginative sources used in geographical research. First, though, let us take a look at the range of sources that geographers might use.

The range is large and growing and includes literature, music, the performing arts, the visual arts, film, photography, architecture and electronic media. Some of these, notably film, still photography and electronic media, are also used for 'factual' documentary purposes, although the boundary between documentary and imaginative uses is often blurred (sometimes deliberately, as in drama-documentary films). In general, however, where the primary impulse behind the creation of the work is imaginative or artistic it is covered by this chapter, and where it is documentary or factual it is covered by Chapter 3. Let us now consider each type in turn.

Literature

Literature (in which we include novels, poetry and plays) is probably the commonest type of imaginative source used in human geography, and geographers have been engaging with literary works for longer than most other forms. During the so-called 'quantitative revolution' in human geography in the 1950s and 1960s, the only sources of data which were considered valid by many human geographers were 'factual', and particularly numerical, ones. In the 1970s the positivist bases of the 'new' human geography were criticized from a number of positions. Foremost among these was humanistic geography, which was also one of the first strands in contemporary human geography to use literary and artistic sources of data.[1]

In the 1970s and early 1980s, humanistic geographers began to turn to literature (especially novels) as a source of insights into human experience, most notably the experience of place (see Figure 4.2).[2] In part this work was motivated by the belief that novelists and poets possess special skills of evocation and an aesthetic sensibility that allow them to develop such insights and convey them to the reader. The implication is that literature provides an alternative, richer 'truth' about the nature of place and landscape that is not available from more conventional data sources:

> The truth of fiction is a truth beyond mere facts. Fictive reality may transcend or contain more truth than the physical or everyday reality. And herein lies the paradox of literature. Although different in essence, and therefore a poor documentary source for material on places, people or organisations, literature yet possesses a peculiar superiority over the reporting of the social scientist ... Literary truth has a universality: it evokes a response in Everyman's breast while apparently concerned with the particular ... [It] is a truth that is more humanly significant. (Pocock, 1981b: 11)

While not doubting the skills of writers or the evocative nature of their work, other geographers argued that literature should not be seen as a mysterious, transcendent or otherworldly activity, but should be understood and analysed as a social practice in its own right, embedded in networks of economic, political and cultural relations (Thrift, 1983a; Silk, 1984; Daniels, 1985). This perspective implies that before literature can be used as a source of evidence in geographical research

we need to consider the geography of its production and consumption. In other words, we need to understand the contexts in which it is written and read and the impact these contexts have on the meanings which are conveyed. In line with the overall framework of this book, this is the approach which we will adopt.

The study of literature from a geographical perspective continued during the 1980s[3] and has since begun to move beyond the novel to consider other literary forms such as folk tales (Kong and Goh, 1995). More recent work can to some extent be distinguished from earlier studies in three ways. First, it places more weight on the social context of the production and consumption of literary works and on the literary conventions used by writers. Secondly, it is increasingly linked to a wider resurgence in cultural geography, which has seen a growing emphasis in all kinds of geographical research on issues of textuality, meaning and representation. Thirdly, in line with an ongoing reconfiguration of the social sciences more generally, contemporary work is marked by its interdisciplinary character. Today the relationship between geography and literature is studied by sociologists, cultural theorists and literary critics, as well as geographers, who have themselves drawn more and more on the insights of other disciplines such as literary theory and cultural studies. This more diverse approach is evident in the collection edited by Peter Preston and Paul Simpson-Houseley (1994) entitled *Writing the City*. Finally, while geographers have become more sophisticated in their treatment of literature, literary theorists and literary critics have also begun to explore geographical approaches to their work (Page and Preston, 1993).

Travel writing

In one sense travel writing is just another type of literature along with novels, poems and so

Humanistic Geography and Literature

Edited by
Douglas C.D. Pocock

Figure 4.2   Cover of *Humanistic Geography and Literature*

Source: Pocock (1981a)

on. However, we are treating it as a separate category for two reasons. First, travel writing has made a particular contribution to the geographical literature over a long period of time. Initially, travellers' accounts of their journeys added directly to that literature and were seen by geographers and others as valid contributions to geographical knowledge. As geography became dominated by the view of knowledge common in nineteenth-century natural science, travel writing was increasingly seen as 'unscientific' and dismissed by the geographical establishment (Domosh, 1991a). More recently there has been a resurgence of interest in travel writing among geographers and others (e.g. Blunt, 1994b; Gregory, 1995). Contemporary geographers do not simply take travellers' accounts at face value as earlier generations might have done, but nor do they dismiss them on the grounds that they are 'unscientific' (not least because scientific writing is itself no longer thought to be a straightforward and transparent activity). Rather, travel writing, such as that of Mary Kingsley (Figure 4.3), is interpreted for the insights it can provide into the ways in which the world was understood by the writer and the society in which he or she lived, and the ways in which those understandings in turn affected that society and its relationship with the places described in the writing.

Secondly, we distinguish travel writing from other literary forms because of the distinctive context in which it is written. Unlike novels which are intended as fiction, travel writing involves the blurring of genres. Travel writing is part reportage, part documentary, part factual description and part literary embellishment. The precise mix depends on the author and the stylistic conventions of the time and also varies across a range of styles from guidebooks to much more literary accounts. As we have suggested above, we wish to emphasize the role played by the conditions of production and consumption of the

works in informing their interpretation. Therefore the fact that travel writing involves conditions of production distinct from those of more strictly literary forms means that we should treat it as a separate category while recognizing the links that do exist between the two.

Music and the performing arts

By contrast with their now well-established study of literature, geographers have only recently begun to express a professional interest in music in any sustained way.[4] Even less attention has been paid to dance and other performing arts, despite the integral role of space and spatiality in those activities. However, the current strong interest being shown by cultural geographers and others in the human body suggests there is much scope here for future research.

Given the extraordinary importance of music in most Western and non-Western cultures, the lack of attention paid to it by geographers may seem strange. This may have been due in part to the emphasis in traditional cultural geography on *material* culture and to the apparent difficulty of defining a distinctively geographical perspective on music. Nowadays, however, geographers work with less constraining definitions of both 'culture' and 'geography' and there is in any case a growing recognition of the intellectual value of interdisciplinary work which refuses to be constrained by the traditional narrow academic division of labour.

As with the more recent work on literature and travel writing, and in keeping with the perspective adopted in this book, contemporary research undertaken by geographers using musical sources does emphasize the importance of the context in which (and the conventions through which) it is produced and consumed. As Leyshon et al.

Figure 4.3   Mary Kingsley in 1897

Source: From the Royal Commonwealth Society Collections, Cambridge University Library

(1995: 424) put it: 'key issues of geography and music [are]: the nature of soundscapes, definitions of music and cultural value, the geographies of different musical genres, the place of music in local national and global cultures.'

Moreover, the role of geography is central: 'Space and place are ... not simply the sites where or about which music happens to be made, or over which music has diffused, but rather different spatialities are suggested as being formative of the sounding and resounding of music (1995: 424–5). Leyshon et al. go on to outline how an understanding of geographical context and of the role that space and place play in constituting music and musical activity might illuminate studies of both the Western classical tradition and the popular music industry (1995: 425–31).

Painting and the visual arts

The resurgence of interest in cultural geography has also led to much greater attention being paid to the visual arts as sources of geographical insight (e.g. Bonnett, 1992; Daniels, 1993). Within this trend, painting has provided most of the inspiration. Because of its obviously geographical theme, many geographers have become particularly interested in landscape painting, though in principle many different types of subject-matter could be considered. The interpretation of landscape has been one of the most enduring themes in geography, and many geographers are interested equally in landscapes on the ground and in the representation of landscapes in paintings and other media. Recently, there has been a growing realization in geography that physical landscapes (trees, fields, parks, gardens, buildings and so on) are frequently as much products of the human imagination as paintings, and that the techniques involved in interpreting paintings (especially for their symbolic content) can be applied to great effect in the interpretation of the physical landscape too (Cosgrove and Daniels, 1988). This process of describing and interpreting visual imagery for its symbolic meanings is known as *iconography*. While painting is the most widely used medium among the visual arts for geographical research, other forms have started to generate some interest. David Matless and George Revill (1995), for example, have examined the 'land sculpture' of Andy Goldsworthy.

As with music and literature, our view is that a full understanding of a visual source requires a knowledge of the conditions of its production and consumption, including the conventions through which its meanings are conveyed. Landscape painting in eighteenth-century England, for example, was not simply (or even mainly) about the innocent and realistic depiction of the countryside. On the contrary, landscape artists were frequently commissioned by major country landowners to represent visually the power and wealth of their estates. Such depictions then in turn helped to shore up that wealth and power by making them appear to be part of the 'natural' order of things (Daniels, 1989; see also Rose, 1993b: 91–3).

Photography

Although geographers often approach photography in a similar fashion to the other visual arts, it is worth distinguishing as a separate category because of the distinctive conditions in which photographs are produced and consumed. 'Art' photography, such as the landscapes of Ansel Adams, are often very 'painterly' of course, but photography also includes popular 'snapshot' forms as well as a variety of approaches which depend for their effect on the qualities of the technology itself, such as the mosaics produced by artist David Hockney.

Landscape, field and aerial photography has always played an important, if mundane, role in conventional geographical research and cartography. Although these uses of the medium may seem straightforward and descriptive they are by no means a merely technical matter. The use of aerial photography in map-making, for example, implies a particular view of the role and status of maps. Although it appears to reveal the whole landscape it has the effect of excluding from the map everything which is invisible to the camera. In landscape and field photography, apparently neutral and truthful representations depend on consciously or subconsciously selected viewpoints, framing and composition which in turn convey particular sets of meanings and imply particular values. Even the common appeal to photography as a guarantor of truth ('the camera never lies') privileges the visual over other means of perceiving the world.

In this chapter, however, we are not mainly concerned with the geographer's own use of photography in geographical research but with the production of photographs by others which are then used as a source in geography. Cultural geographers have begun to examine photography for the meanings and symbols it conveys and for its social content. In doing so, the context within which photographs are made is of central importance. For example, Phil Kinsman (1995) has shown how the landscape photography of Ingrid Pollard cannot be understood without reference to the photographer's identity as a black woman or to the role of the countryside and ideas about 'race' in the construction of British national identity. In David Harvey's (1989a) discussion of the photographs of Cindy Sherman, he argues that the unsettling approach to appearance and identity which they reveal is closely bound up with a contemporary shift from modernity to postmodernity, which in turn needs to be interpreted in the context of the restructuring of capitalism. Doreen Massey (1991a) and Ros Deutsche (1991) also argue that Sherman's photographs can't be understood separately from their cultural context but, in sharp contrast to Harvey, they argue that the gender relations in which their production and consumption are entwined are just as important as their relationship to capitalism.

Cinema and television

Related both to photography and to the performing arts, cinema and television are nevertheless distinct because of their particular conventions, production technologies and organization. They are also widely used for documentary work, and this form has received some attention from geographers. Gillian Rose, for example, has studied the use of film by community groups in east London (1994).

Cinema and television are hugely important in daily life and in the economy and are thus potentially of considerable interest to geographers. Although a central element in the discipline of cultural studies, fictional work in these media has only recently begun to be studied by cultural geographers. This is despite the importance of space and distance in the construction of films and of representations of place in their content.

This past neglect is now being addressed by geographers working from a variety of perspectives. Gold (1985) looked at the futuristic visions of the city in early cinema. Tomaselli (1988) used Afrikaans cinema to investigate the role of the myths of the trek, the outside and Eden in the production of Afrikaaner national identity. In his analysis of *The Condition of Postmodernity*, David Harvey (1989a) draws on two films, Ridley Scott's *Blade Runner* and David Lynch's *Blue Velvet*, to show how the experience of time–space compression is exhibited in cultural phenomena (see also Deutsche, 1991; Massey, 1991a). Stuart Aitken and Leo Zonn (1993) have discussed the connections between gender relations and the environment in the films of Peter Weir, while Matthew Gandy (1996) examines the films of Werner Herzog to investigate the portrayal of nature in Western thought. The breadth of potential for geographers in studying film is demonstrated by the range of essays published in a collection edited by Aitken and Zonn (1994a).

While professional painters work mainly alone, either for themselves or to a commission, professional cinema and television production is a highly organized activity involving dozens or even hundreds of people, large amounts of capital equipment, and big organizations such as film companies and commercial or public service broadcasters. These contexts form the conditions of production of most films and television programmes and have been analysed from geographical perspectives. Michael Storper (1994), for example, argues that the US film industry has moved from early craft production through the mass-production era of the big studios in the 1940s and 1950s, to the increasingly vertically dis-integrated forms of organization of today, in which particular parts of the production process, from screenwriting to lighting and from editing to catering, are contracted out to specialized companies (see also Scott, 1984; Storper and Christopherson, 1985). These studies are important because, as with other media, the conditions of production need to be understood in constructing interpretations.

As well as the impact of industrial organiza-tion and the material processes of film and tele-vision production, however, it is necessary to consider the cultural and aesthetic context in which movies and programmes are produced. For example, Aitken and Zonn (1994b) note that the conventions governing the portrayal of place in film have changed over the course of cinema history. Larry Ford (1994) shows how the representation of cities in film has developed from the passive backdrops of early cinema, through the shadowy, menacing settings of *film noir*, to contemporary experiments with colour. Furthermore, the geography of film is not only about the ways in which places are shown but also about the space of the medium itself. Far from being a direct portrayal of life, films are carefully constructed through the composition, framing and lighting of particular shots, through techniques such as panning, zooming and jump-cutting and through special effects. The con-struction and manipulation of space are central to many of these aspects of film-making.

Finally, grassroots and amateur traditions of film-making and video production also exist and these too are potential sources for geo-graphical research. Mike Crang, for example, has argued that geographers' growing interest in the use of video by police and the local state in attempting to control urban crime and disorder should be matched by an equal attention to popular and domestic uses of (increasingly accessible) video technology (Crang, 1996). Video is also becoming a direct tool of geographical research. Adopting prac-tices developed in visual anthropology (Turner, 1991; 1992), geographers are begin-ning to make videos as part of the process of research itself, often in collaboration with those being studied and sometimes handing control of the camera to the 'objects' of the research (Byron, 1993; Crang, 1995).

Architecture

Geographers have long been concerned with the built environment, with the planning of towns and cities and with urban morphology. However, it is only recently that they begun to move on from a concern with the structures and functions of whole urban areas or neigh-bourhoods to cultural and political analysis of the buildings which make them up. Lily Kong (1993), for example, has analysed the relation-ships between religious buildings in Singapore and the conceptions of 'sacred space' associated with them. David Harvey (1985) has written a detailed narrative of the construction of the basilica of Sacré-Coeur in Montmartre, Paris (Figure 4.4), revealing the political conflicts and suffering which surrounded its develop-ment. In more contemporary vein he argues (1989a) that the transition from modernism to postmodernism in architecture must be understood in relation to the changing char-acter of the capitalist economy. Derek Gregory (1994) has discussed the work of Walter

Figure 4.4 The basilica of Sacré-Coeur

Source: Harvey (1985: 201)

Benjamin, whose writings on the *passages* (arcades) of Paris provide a powerful evocation of the cultural dynamics of modernity.

Paris again provides the source of another example. Woolf (1988) interprets the Paris Opera House in the context of the opulence and conspicuous consumption of the Parisian elite during the Second Empire, but challenges the conventional view that the flamboyant building accurately symbolized Parisian life at the time. By setting her analysis against the picture of a divided and conflictual urban environment, Woolf shows that the lifestyle represented by the Opera House was limited to a tiny proportion of the population and that the building symbolizes aspirations, desires and political power rather than everyday life.

Mike Davis, in his account of 1980s Los Angeles as the *City of Quartz* (1990), shows how the increasingly fortified architecture of the city reveals new social divisions between the wealthy, security-conscious residents of the privately guarded housing developments and the increasingly marginalized sections of the population whose physical exclusion from 'Fortress LA' mirrors their growing social exclusion from American society.

As with other forms of imaginative expression, architecture is used as a research source by geographers in a number of ways. What these examples suggest, though, is that issues of context are as important here as they are in interpreting other media such as film and literature.

Objects and material culture

Buildings are expressive of the human imagination but they are also functional objects that are used for all kinds of purposes by people who may rarely if ever think about the imaginative meanings bound up in their construction. The same is also true of all kinds of other everyday objects (from clothes to cars) that simultaneously express meaning and serve functional purposes. The interpretation of 'material culture', as the study of such objects is commonly known, has become increasingly important in geography. Work on the geography of consumption, for example, has blurred the distinction between cultural and economic geography to a remarkable extent, recognizing the ways in which functional commodities become loaded with cultural meanings. For example, a study two of us undertook with Mark Thorpe examined the relationship between food products and multicultural identities in Britain (Cook et al., 1999). Other geographers have focused on

fashion and clothing retailing in both the 'designer' and 'charity shop' segments of that market (Crewe and Goodrum, 2000; Gregson et al., 2000), the iconography of banknotes (Gilbert, 1998), lifestyle magazines (Jackson et al., 1999) and household furniture (Leslie and Reimer, 2003).

This focus within geography on material culture or, more prosaically, 'things', has been given impetus by related developments. First there is geographers' increasingly close relationship with anthropologists such as Daniel Miller, whose work on shopping, consumption and material culture has been particularly influential. Secondly, economic geography has developed in new directions, including a focus on commodities and commodity chains. Thirdly, the popularity of actor-network theory in geography has drawn attention in new ways to the transformation of objects as they are produced, circulated and consumed.

Electronic media

With the rapid development of new information and communication technologies based on microelectronics, fibre optics and the storage and transmission of information in digital form, a whole new range of potential sources for geographical research has emerged. Most geographers are now familiar with computers used for data analysis, wordprocessing, cartography and graphics. In addition, more and more researchers rely on local networks of computers for communicating with colleagues and on the Internet to send and receive electronic mail, to search library catalogues, databases and archives, and to view, download and disseminate a wide range of multimedia resources (incorporating some combination of text, graphics, photographs, video and sound) through the World Wide Web.

In addition, electronic media are potentially as much an object of study for geographers as a tool for geographical research. In some cases, electronic technology is used simply to store and transmit sources which have been discussed above, such as literature, photographs and video. Increasingly, however, the distinctive capabilities of electronic media are being exploited by creative artists, writers and designers to generate new forms which cannot be thought of merely as electronic versions of old forms, and which need to be understood on their own terms (Gilbert, 1995). These qualitatively different forms include the following.

MULTIMEDIA PRODUCTIONS

In a multimedia production, a musical work is presented, for example, simultaneously as sound, video of the musicians playing, text relating to the composer, the work or the performance, graphical display of the score, and electronic links to related documents or presentations.

HYPERTEXT

Hypertext involves the creation of electronic links between parts of the text of one document and a part or all of another document. Similar links can be made between photographs, video clips and so on. At its simplest this may involve providing hidden definitions of technical terms or links from, say, the name of a character in a novel to a short biographical note to remind the reader of who the character is. More elaborately, it has been suggested that hypertext potentially involves a new relationship between writer, reader and the text. Text is no longer constrained to the linear form it usually takes on paper but can become layered or multidimensional. There are clear possibilities here for creative writers of all sorts but so far there are relatively few 'hyper-novels' available to the public.

COMPUTER GAMES

Although the market for computer games may be dominated by 'shoot-em-up' productions, there are examples which engage the imaginations of both the programmer and the player in more complex ways, including elaborate role-playing and simulations of social activities, such as *Sim City*.

GRAPHICS AND ANIMATION

Although they appear to mimic conventional drawing and cartoon films, respectively, computer graphics and computer animation often use the qualitatively distinctive features of electronic technology to produce their effects. These capabilities make it much easier, for example, to incorporate real actors with animation in films like *Who Framed Roger Rabbit?* or to provide dramatic special effects like those in *The Mask*. Increasingly, full-length feature films, such as *Bug's Life,* are being produced entirely using computer-generated animation.

INTERACTION

Electronic media provide for new forms of interaction between the 'author' and the 'audience'. At its simplest, readers can provide rapid feedback through electronic mail, but more sophisticated uses will allow readers to assist in scripting plots or directing characters, for example. Increasingly, however, mass access to the Internet is challenging conventional notions of writer and reader. Bulletin boards, discussion groups and self-publishing are all growing as channels of expression and debate and provide, some argue, more democratic and participatory access to 'the media' than ever before.[5]

VIRTUAL REALITY

Virtual-reality (VR) technology combines interaction with special video and sound facilities and attempts to create the illusion that the viewer/user is participating directly in experiences and environments which are created or recreated by a computer or transmitted electronically from elsewhere. While VR is still developing it may offer the potential to engage in radically new ways with imaginative sources. Instead of sitting in a cinema to watch a film, for example, it may be possible to 'walk through' the screen and 'take part in the action'. Here, too, the traditional distinction between reader and author and between the real and the imaginary is blurred or eliminated.

Geographers have begun to study the new electronic media in a variety of ways. From the perspective of political economy, Manuel Castells (1989) has investigated the relationships between cities and the growing role of information and communication technology. Economic geographers have made many studies of the geography of the computing, software and electronics industries.[6] More recently the imaginative aspects and cultural implications of the use of new media have begun to be explored.[7]

Case studies

To conclude the chapter we will consider three case studies drawn from the geographical literature to show how understanding imaginative works in geographical research requires an appreciation of their conditions of production. As we shall show, the cases apply some of the principles we have outlined. However, the framework above cannot be used as a mechanical template when considering imaginative works. Rather, it should be seen as a set of guidelines which need to be applied to particular cases in an appropriate way.

The case studies illustrate the overall argument of this chapter, which is that in order to use imaginative sources in geographical research it is necessary to understand the conditions in which they were produced. In the first case study we can see how an adequate interpretation of the role of landscape metaphors in Elizabethan drama depends on an understanding of the Elizabethan worldview, which provided the discursive context in which the plays were written. Similarly, the second case study demonstrates that interpreting popular music (in this case in Singapore) involves much more than examining the geography in the lyrics. Issues of identity, the structure of the music industry and processes of 'transculturation' all need to be taken into account in considering the context in which such work is produced. In the final case study the gender relations of eighteenth-century England are central to interpreting the landscape painting of Thomas Gainsborough.

CASE STUDY 4.1: LANDSCAPE METAPHORS IN THE WORK OF WILLIAM SHAKESPEARE

The first case study is provided by a paper by Paul Chamberlain published in *The Canadian Geographer* in 1995. It presents an interpretation of the role of metaphor in Shakespeare's plays, with particular reference to the portrayal of the landscape and the human body. Chamberlain is interested in '"the metaphorical vision" or the use of metaphor to understand the human environment relationship' (1995: 309).

Chamberlain identifies five (overlapping) types of metaphor:

1. Dead metaphors such as 'table leg'. In dead metaphors, the figure of speech has become so much a routine part of the language that most of the time it is not seen as metaphorical at all.
2. Metonymy, in which one word is used to stand for another, as in the use of 'bottle' to refer to strong spirits.
3. Personification, in which inanimate objects are given life-like features. Chamberlain's example is the use of the word 'angry' to describe the sea.

CASE STUDY 4.1 *(Continued)*

4. Metaphors in which there is ambiguity between the object of the description (the metaphrand) and the word doing the describing (the metaphier).
5. Extended metaphors, in which an author makes continuing use of a metaphor through the whole or a large part of a work.

An analysis of Shakespeare's work is interesting geographically, according to Chamberlain, because it means that 'we can not only explore an intriguing geographical aspect of the literary text of a highly creative Elizabethan, but can also improve our understanding of the relationship that human beings had with their environment during the Renaissance' (1995: 318–19).

Chamberlain then discusses two sets of metaphors which appear in Shakespeare's plays. First there are those in which the human body is described in terms of the landscape. These are 'landscape:body metaphors'. Secondly there are those in which the landscape is described in terms of the human body: 'body:landscape metaphors'. In each case the landscape includes the celestial landscape of the stars and planets as well as the terrestrial landscape. This may seem odd to modern readers (we might use the term 'skyscape' for the former case) but in the context of Elizabethan literature it is important to stress the unity of the earth and the heavens for reasons which will be apparent later.

Chamberlain shows how Shakespeare uses a range of metaphors in each of these two categories. These include simple uses such as likening a human being to a mountain: 'Mountains are frequently superimposed on a variety of characters such as Luce in *The Comedy of Errors,* whom Dromio of Syracuse describes as being a "mountain of mad flesh"' (1995: 313). Celestial images are also used: 'in *Antony and Cleopatra* when Mark Antony commits suicide by falling on his sword, the guard exclaim[s] that "The star is fall'n"' (1995: 313).

More elaborate metaphors include allusions to bodies as gardens:

Some of the most important metaphors in Shakespeare's work involve the garden. This element of the cultural landscape is likened to the body by Iago in *Othello* when he tells Roderigo: 'Our bodies are our gardens, to the which our wills are gardeners; so that if we will plant nettles or sow lettuce, set hyssop and weed up thyme, supply it with one gender of herbs or distract it with many – either to have it sterile with idleness or manured with industry – why, the power and corrigible authority of this lies in our wills.' (1995: 314)

In *Richard II* Shakespeare uses the idea of a disorderly garden to convey the king's (gardener's) lack of effective governance over the nation (garden). Body:landscape metaphors also abound. In *As You Like,* 'the exiled duke fondly likens the chilly gusts of wind to "counselors" (II.i.10) and later finds "tongues in trees" and "books in running brooks" (II.i.16)' (1995: 317).

Similarly, cities are given human properties and celestial imagery recurs, as in the following speech by Henry in *Henry V*:

I know you all, and will a while uphold.
The unyoked humor of your idleness.
Yet herein will I imitate the sun,
Who doth permit the base contagious clouds

CASE STUDY 4.1 *(Continued)*

To smother up his beauty from the world,
That, when he please again to be himself,
Being wanted, he may be more wond'red at
By breaking through the foul and ugly mists
Of vapours that did seem to strangle him
(I.ii.192–200) (1995: 316)

In using works such as Shakespeare's plays in geographical research, geographers seek to interpret the texts and their meanings (in this example, especially the use of metaphor). However, as Chamberlain shows and as we have been arguing in this chapter, such interpretations, if they are to be adequate, must be informed by an understanding of the conditions of the production of the texts. Chamberlain argues that this involves giving attention to the textual, the extra-textual and the inter-textual (references to other texts).

A full analysis of the conditions of production of Shakespeare's texts would include, among other things, a discussion of the organization of Elizabethan theatre, of the conventions of Renaissance drama, of the political conditions of sixteenth-century England and a host of other 'extra-textual' factors. To inform his interpretation of the landscape/body metaphors, Chamberlain identifies three sets of conditions as of particular importance. These are the technical aspects of metaphor which were outlined above, the role and importance of astrology and alchemy in Renaissance systems of thought, and Elizabethan cosmology based on the doctrine of the 'Great Chain of Being'

The importance of astrology to the knowledge systems of late medieval Europe clarifies the significance of Shakespeare's use of celestial imagery. Elizabethans took it for granted that the movements of the sun, the moon and the planets influenced the affairs of humans, and such ideas are frequently used by Shakespeare including, as we have seen, in his use of metaphor. The idea of the Great Chain of Being in which all creation is seen as a linked hierarchy from God at the top to the humblest inanimate object at the bottom, helps to explain the power and apparent naturalness with which Shakespeare links people, the earth and the heavens in his use of metaphor. In order adequately to understand the geography of Shakespeare, therefore, it is essential also to understand what the literary critic E.M.W. Tillyard called the 'Elizabethan world picture' (1972).

CASE STUDY 4.2: POPULAR MUSIC IN SINGAPORE

Our second case study is drawn from the work of geographer Lily Kong. In a research paper published in the journal *Society and Space* in 1996, she presents an interpretation of the work of the Singaporean pop star, Dick Lee. While separated from Elizabethan England by 400 years and 7000 miles, here, too, we find that relations between texts, contexts and other texts are central to the analysis (Kong, 1996: 279). Indeed Kong's work shows just how fruitless is any attempt to separate rigidly the interpretation of texts from an understanding of their production.

Dick Lee was born in 1956 and was brought up in Singapore surrounded by English culture. It was only when he travelled to England for the first time that his Asian identity became clear to him (Kong, 1996: 278). Lee grew up in the 1960s and 1970s and was

CASE STUDY 4.2 *(Continued)*

heavily influenced by a range of Western popular musicians of the time. Since then he has pursued his own successful recording career and has developed a distinctive musical style which combines local Singaporean and wider Asian influences with Western idioms. Kong elaborates on the three geographies present in the work of Dick Lee.

First, she argues that 'a sense of the local is strong in Lee's music' and 'that this is evident in two ways: the injection of distinctive local contexts, creating a sense of place, and the recovery of personal, communal, and national heritages, in which a combined sense of history and nostalgia is often drawn upon' (1996: 279). Kong shows that Lee's lyrics contain frequent references to local places in Singapore and to particular aspects of Singapore's cultural, economic and political milieu. In addition, Lee draws on his own roots to emphasize the importance of local identity. Thus a local market, 'Beauty World', remembered from his childhood, provides the title of one of his albums and a musical. As Kong (280) puts it, 'for Lee, his sense of place and feeling of nostalgia are all rolled into one and find expression in his music'. She also shows how he incorporates a mixture of traditional musical genres into his output, reflecting the rich mix of Singapore's multiracial society.

Secondly, Kong illustrates the global influences on Lee's music. This produces 'local music with a transnational flavour' through a process of 'transculturation' (1996: 281):

> What are the evidences of such transculturation? Instrumentation, styles, language, costumes, and guest artistes can all be called into play here. In *Life in the Lion City* (1984) Lee mixed traditional instruments with synthesisers, and claimed this to be his first introduction to the search for a new identity via music. This has become a recognisable feature of his music, described as blending pop funk beats with Asian instruments … and a fusion of different idioms … In addition, Lee's music has often been described as 'anglicised' versions of local songs … For example, a Chinese folk song, 'Little White Boat', is given English lyrics; a traditional Malay folk song, 'Rasa Sayang', is infused with a modern rap; and traditional folk tunes ('Alisham', 'Springtime', and 'Cockatoo') and English pop styles are cleverly incorporated. (1996: 282)

Thirdly, Kong describes how Lee has begun to search for a regional Asian identity through his music. This, she says, was for three reasons:

> First, he wanted to change what he saw as the Western image of Asians … Second, he recognised that there was a sense of shared identity crisis. It was no longer a case of 'Who am I?' or 'Where am I going?', but one of 'What role do I play as an Asian?' – a question he felt many modern Asians were asking. Third, he wanted to recover what he says to be the spirit of the 1960s when Asian pop enjoyed a significant degree of popularity. (1996: 284)

These personal motivations are only part of the story, however. As an artiste under contract to the recording company WEA, Lee is bound into economic circuits which stretch beyond the confines of his native Singapore. Noting that Singapore has a population of only 3 million, Kong argues that the fact:

> That Lee was signed up with WEA which, like all record companies, is first and foremost a large business enterprise governed by laws of profit and loss, meant that it was imperative to reach out to as large an audience as possible. Given that the

CASE STUDY 4.2 *(Continued)*

local market is very small, to sign up any local artiste would necessitate marketing and distribution in the regional market. Music that is produced is therefore guided by this need and it is of little surprise that Lee's music addresses a question of Asian import. (1996: 285)

In Williams's terms, therefore, Lee's attempts to stretch the genre of his work to develop an Asian rather than just a Singaporean popular music is matched by institutional factors, such as the structure of the record industry and the size of markets. According to Kong, however, Lee's attempts to link his work to a pan-Asian identity is only partially sucessful in musical terms, in part because it is unclear whether the notion of a pan-Asian identity, bringing together very diverse traditions, makes sense. Nevertheless, even if a unifying identity is elusive, the music can still have wide regional appeal (1996: 287).

Kong shows how these three geographies of Lee's music (the local and Singaporean, the global and Western, and the Asian and regional) can be related to broader processes of international cultural change. She outlines two contrasting interpretations of change. First is the globalization thesis, which posits that cultures are becoming more and more alike, as global networks of communication and transnational media corporations operate to homogenize cultural production. Secondly is the 'music is local' thesis, which suggests that 'music is necessarily local in its production, consumption, and effects' (1996: 275) because music is about giving expression to ethnic, cultural and other local differences so that 'unique local sounds are produced because of the unique local milieux in which they are produced' (1996: 275).

For Kong, however, both these theses are inadequate:

Although there are clearly those who advocate global and local positions, and argument can be made that global and local cultures are in fact relational rather than oppositional … As Friedman (1990, page 311) argues, 'Ethnic and cultural fragmentation and modernist homogenization are not two arguments, two opposing views of what is happening in the world today, but two constitutive trends of global reality'. (1996: 276)

Instead of globalization or localization, what is happening, Kong suggests, is transculturation:

It could be argued that the process at work is actually one of transculturation, a 'two-way process that both dilutes and streamlines culture, but also provides new opportunities for cultural enrichment' (Wallis and Malm 1987: 128). Transculturation is apparent when musicians are influenced dually by their own local cultural traditions and by the transnational standards of the music industry. The result is local music with a transnational flavour or transnational music with a local flavour … Some … have argued that transculturation in fact gives rise to 'third cultures' which draw upon the culture of the parent country but also take into account local cultures and local practices. (1996: 276–7)

To interpret the musical works of Dick Lee, therefore, Kong finds it necessary to consider not only the immediate context of their production in Singapore and in Lee's own life story, and the economic relations of the international music industry, but also the wider processes of cultural change by which Lee, Singaporean culture and the music industry are all affected but which they, in their turn, are affecting as well.

CASE STUDY 4.3: MR AND MRS ANDREWS

Our final case study is a painting that has become so widely cited by human geographers that we feel it has become the one cultural artefact no self-respecting commentary on the practice of human geography can afford to ignore. Thomas Gainsborough (1727–88) was an English portrait and landscape painter. *Mr and Mrs Andrews* was painted around 1750, probably to celebrate the marriage of Robert Andrews and Frances Carter in 1748. The picture (see Figure 4.5) shows the couple in an English country landscape, part of their estate. Mr Andrews carries a gun and a hunting dog is at his feet. It seems he has been out hunting. Mrs Andrews is seated under an oak tree. In front of the couple the sheaves of corn imply that it is autumn. The painting, though unfinished, is brilliantly executed and is testament to Gainsborough's status as one of the great artists of eighteenth-century Europe.

Landscape has long been a central category in geographical analysis and in the last 20 years landscape painting has also become a focus of great interest. By and large, geographers studying landscape painting take it for granted that no painting provides unmediated access to the reality of the landscape depicted, no matter how realistic the style of expression chosen. Landscape painting, as is the case with other art forms such as Shakespearean drama or popular songs, is heavily conditioned by conventions and the process of production. With this in mind, *Mr and Mrs Andrews* has attracted a certain amount of interest from geographers as well as art critics.

In his influential book (1972) and television series, *Ways of Seeing*, the critic John Berger argues that *Mr and Mrs Andrews* exemplifies the relationship between oil painting and property. Mr and Mrs Andrews are not simply a couple taking delight in their natural surroundings. Rather, 'they are landowners and the proprietary attitude towards what surrounds them is visible in their stance and expressions' (Berger, 1972: 107). The pleasure the couple took from the painting was not purely aesthetic but included 'the pleasure of seeing themselves depicted as landowners and this pleasure was enhanced by the ability of oil paint to render their land in all its substantiality' (1972: 108).

The painting adopts realist conventions, in particular the convention of perspective, which 'centres everything on the eye of the beholder ... the visible world is arranged for the spectator as the universe was once thought to be arranged for God' (1972: 16). Moreover, Berger argues that oil painting specifically, more than other arts, provided an particularly appropriate medium for the expression of the emerging social relations of capitalism: 'oil painting did to appearances what capital did to social relations. It reduced everything to the equality of objects. Everything became exchangeable because everything became a commodity' (1972: 87).

Berger's analysis of oil painting and of the *Mr and Mrs Andrews* picture is taken up by the geographer Stephen Daniels. Daniels (1989: 213–14) points out that Berger himself implies that landscape painting cannot be wholly complicit in bourgeois values because the sky, which is central to the landscape genre, 'cannot be turned into a thing or given a quantity' (Berger, 1972: 105). This, for Daniels, is an aspect of what he calls the 'duplicity' of landscape.

Another geographer, Gillian Rose, takes the interpretation of *Mr and Mrs Andrews* further still and, in doing so, is critical of what she sees as Daniels's unwillingness to offer a sufficiently critical analysis of the pleasurable aspects of looking at landscape painting. Rose points out that there are important differences between Mr and Mrs Andrews in the painting:

> They are given different relationships to the land around them. Mr Andrews stands, gun in arm, ready to leave his pose and go shooting; his hunting dog is at his feet already urging him away. Mrs Andrews meanwhile sits impassively, rooted to her seat with its wrought iron branches and tendrils, her upright stance echoing that of the tree directly behind her. If Mr Andrews seems at any moment able to stride off into the vista, Mrs Andrews looks planted to the spot. (1992: 14)

Figure 4.5 Thomas Gainsborough, *Mr and Mrs Andrews*, c. 1750, oil on canvas

Source: From the National Gallery Picture Library

CASE STUDY 4.3 *(Continued)*

For Rose, this distinction between the two figures reflects the dominant association of women and femininity with nature and masculinity with culture. Thus it is Mr Andrews who is in fact the landowner, while Mrs Andrews's function in eighteenth-century England was primarily to provide children, and ideally a male heir, to carry on her husband's dynasty. Rose suggests that Mrs Andrews's seated position under the oak tree and beside the field of crops reflects the assumption that women were closer to nature. Effectively she becomes part of the landscape over which Mr Andrews holds sway.

In addition, Rose suggests, the passive attitude of Mrs Andrews and her status as an implicitly sexual being (because of the association between women and fertility) are connected with the visual pleasure afforded to the spectator. In this view the active gaze of the spectator is associated with masculinity, while the observed landscape is associated with femininity because of both its apparent passivity and its closeness to nature and fertility. Thus the pleasures of looking at landscape are not only dependent upon a particular set of property relations to do with the ownership of land but are also connected with highly unequal gender relations in which 'the gaze' involves a masculine appropriation of a sexualized and naturalized femininity (Rose, 1992: 14–18).

These geographical readings of Gainsborough's painting suggest that, as with the other two case studies, effective interpretation will depend heavily on an understanding of the context in which the work was produced, and also consumed. The qualities of oil paint, the conventions of perspective, property relations in eighteenth-century England and the gendering of nature and of the spectator's gaze all feed into and influence the meaning of the picture.

Conclusion

In this chapter we have considered the geographer's engagement with imaginative sources from a perspective that stresses the importance of the conditions of production of cultural phenomena. An enormous and ever-increasing variety of imaginative sources is available for geographers to analyse and interpret, from classical myths to the latest electronic simulations. These texts and artefacts take many different forms and carry a huge range of meanings, but they all arose through a process of cultural production (du Gay, 1997). By drawing attention to the materiality of culture and to the importance of institutional and other social relations, Williams's approach provides a useful framework (though not the only one) within which to interpret that process of production. Of course the conditions of production also shape the data that are produced by

geographical researchers themselves, and it is to those kinds of sources that we turn in the next two chapters.

Notes

1 For a discussion of humanistic geography, see Cloke et al. (1991).
2 See, for example, Seamon (1976), Tuan (1976b) and Pocock (1979; 1981a; 1981b).
3 For example, Barrell (1982), Lutwack (1984), Mallory and Simpson-Houseley (1987), Sandberg and Marsh (1988), Shortridge (1991), Cresswell (1993) and Daniels and Rycroft (1993).
4 See, for example, Smith, S.J. (1994), Cohen (1995), Hudson (1995), Kong (1995a; 1995b), Leyshon et al. (1995), Revill (1995) and Valentine (1995).
5 For a less optimistic view of the 'information age', see Roszak (1994), Stoll (1995).
6 For example, Hall and Markusen (1985), Scott and Angel (1987), Hall and Preston (1988) and Morgan and Sayer (1988).
7 For example, Benedikt (1991), Hillis (1994), Shields (1996) and Crang et al. (1999).

Talking to People

Introduction

The history of human geography is often conceived of as a series of blockbuster eras in each of which an all-encompassing mix of philosophy, theory and method dominates the landscape of scholarship until the next 'revolution' heralds the next era with its different concepts and toolkits. As we have discussed in Chapter 1, such a view is not only historically erroneous but it also creates misleading assumptions about the heritage of particular methodologies. As a prime example, recent texts on human geography's methods have tended to convey the impression that any formalization of qualitative methods is an entirely recent occurrence, to be understood in term of supposedly 'postquantitative' cultural turns in the subject. Yet it is important to begin this chapter on 'talking to people' with a clear acknowledgement that some considerable consideration was given to more qualitative procedures throughout the twentieth-century history of human geography.

In North America, for example, there is evidence of a tradition of 'talking to' field 'informants' that stretches back to some of Sauer's earliest work. For example, in his 1924 piece, Sauer recognized the value of using 'the tally sheet or questionnaire' to gain and to record information derived from questions to local people about 'the standard of living and the movement of population' (1924: 30).

Around the same time, Whittlesey (1927) discusses the importance of conducting questionnaires alongside observation and the collection of statistical information. He acknowledges that 'systematic questioning is a necessary supplement to observation, because certain facts required for an understanding of the geography of any area are not observable by the most skilled worker' (1927: 75). Whittlesey outlines a research methodology which includes 'conversations with well-posted people' (1927: 77–8), including *officials* such as farm agents and school superintendents, and *individuals* such as clergy or bankers. The assumption here is that those unobservable facts which are necessary for understanding the geography of an area are to be found in the knowledges and attitudes of these key local figures. C.M. Davis (1954: 523–7) went so far as to include several pages of discussion devoted to 'the interviewing of informants', distinguishing between 'interviewing by questionnaire' (when asking specific sets of questions, preferably face to face) and 'interviewing by informal conversation'. In the case of questionnairing, Davis stressed that chiefly factual information would be the main information sought, but added that attempts could be made to discern more attitudinal information, perhaps using the (somewhat dubious) kind of question format shown in Figure 5.1. In the case of what we might regard as true interviewing, he stated that the objective

'How would you feel if a person in any one of the following groups should ask to marry your daughter? Mark a cross in the appropriate column for each group.'

Table I

	Welcome with enthusiasm	Accept willingly	Accept but not happily	Refuse
Physician				
Retail merchant				
Army officer				
Porter				
Farm laborer				
Lawyer				

Figure 5.1 C.M. Davis's formatting of questions

Source: James and Jones (1954: 42)

would be chiefly to 'gather qualitative information, such as the seasonal variations of farm practice, or individual preferences for competing market centres' (1954: 525). Revealingly, he offered this further remark:

> Information obtained from informants is an ancient source of geographical field data. Since early Greek times, if not long before, geographers learned how people lived by asking them questions. It seems probable that more of the information included in current geographical writings is derived from informants than from direct field observation. Yet the techniques used are not often described. Formal descriptions and experiment with the techniques of the interview are found in the literature of anthropology or sociology but not of geography. (1954: 527)

We begin to get the impression, then, of a rich history of geographers asking questions and of geographical information being heavily derived from 'informants'. There are even glimpses of how such information was being gathered. In a paper gloriously titled 'The bicycle as a field aid', Salter (1969) discusses the role that the bicycle plays in 'opening routes of information which might otherwise

be unfound' (p. 361), notably because the bicycle provides an excuse for a 'chat' to local people – it gets the fieldworker into a position where 'informants' can readily be accessed.

In Britain, the evidence of an emergent methodology of 'talking to people' is more sparse, with few explicit acknowledgements of the role of qualitative interviewing until recent times. However, it does seem clear that among the 'field studies' favoured by many British geographers from at least the 1930s onwards there could not but have been some engagement with field 'informants'. So, for example, Beaver's (1962) discussion of the Le Play Society – a field-based organization encouraging practical exercises in geography – outlines typical field meetings, involving academics and lay folk, which ran according to a clear 'scheme of study' in a given locality. For instance, under the heading of *climatology,* there was supposed to be the accessing of long-term statistical data and *verbal inquiries* were to be made about local peculiarities, including oddly enough 'extraordinary meteorological phenomena, e.g. fireballs' (1962: 235, emphasis added). *Social* data, meanwhile, were to include attention to 'social organisation (local government, education, religious organisation, household life), social psychology – customs, folk-lore, art' (1962: 235). It is difficult to know how many of these subject-matters could be explored *without* talking to local people, but nothing is said about how this might be done in practice. Intriguingly, though, a window on what might have been happening arises where Beaver comments on 'those in the world of university geography who stood aloof, who scoffed at middle-aged schoolmarms prying into peasant cottages' (1962: 239).

Around the same time, Hutchings presents an account of 'village studies' and wonders about the extent to which pure observation will be sufficient to understand what is really happening there:

> Students quickly discover that the story of a village, and especially of its social and economic working, is by no means legible in visible signs. Indeed, their observation, far from explaining the place, raises a lot of questions, *the answers to which they can only obtain by talking to the inhabitants* or from local literature. Whatever the quality or quantity of this harvest of knowledge, it is pretty safe to say that not much of it will be geography as we know it. (1962: 10, emphasis added)

There is an interesting sense, or even fear, here that 'talking to people' may end up generating information that was not normally conceived of then (in the early 1960s) as 'proper' geography.

Hence Anglo-American geography, does involve a tradition of talking to informants 'in the field', with both questionnairing and interviewing being part of emerging techniques of regional survey. The impression, though, is that by the 1950s this tradition was being deflected into a kind of a 'regional science' mould, thus leaving behind the Sauerian cultural-geographical roots and perhaps too becoming more interested in the urban regions of modernity than the rural-agricultural

regions of premodernity. Interestingly, the references to 'talking to people' in these early literatures tend to envisage researchers in rural-agricultural backwaters talking to farmers and rural communities. With attention turning away from these 'premodern' regional worlds (see Entrikin, 1981), there seems also to have been a methodological shift from a more qualitative survey approach, in which questionnaires and interviews might figure, to a more quantitative approach preoccupied with statistical indicators of input–output flows through urban (modal) regions. This shift was not uncontested, however. In the context of a discussion of the use of field research in place-name studies, Berleant-Schiller (1991) suggests the importance of information given orally by field informants. She concludes that, at least since the early 1950s, geographers had been failing to develop these methods to their full advantage, not least because of a general retreat from fieldwork in human geography until the coming of spatial science (see Chapter 1). She writes:

> Few American geographers of the last 30 years have imitated the model of field research that Cassidy (1947) established in his *Place-Names of Dane County, Wisconsin,* which incorporated materials from many informants as well as from maps and historical sources. Robert C. West (1954: 67) corresponded with local informants, but his work on the term 'bayou' marks the beginning of the movement away from fieldwork in place name research. Miller (1969: 245) denied vigorously any need for or advantages in field research, dismissing it as 'slow', 'unrewarding' and subject to 'bias and misinformation'. (1991: 92)

Berleant-Schiller is hostile to this view and wishes to oppose such a writing out of court as 'an irreplaceable source of primary data in human geography' (1991: 92). Here she raises a claim which has far greater generality than the specific context of place–name research.

She argues that 'the implications of regarding local residents and native speakers as mis-informed about their own culture are deep, especially since Miller's judgements were not so much part idiosyncratic as part of a larger trend' (1991: 92). The loss of interest in talking to people, then, in Berleant-Schiller's view is not simply a matter of human geography's 'progress' towards spatial/regional science. It marks a significant junction in philosophical and methodological pathways – namely, that the preference for scientific 'objectivity' was constructing negative arguments about the 'bias' and 'misinformation' inherent in methods involving talking to people. Such arguments were even infecting more 'peripheral' fields of cultural-geographical inquiry such as place-name research. Yet even then there were alternative arguments about the 'irreplaceability' of allowing local residents and native speakers to speak for themselves and thus present essentially informed accounts of their own situations.

So, what are the implications of these traditions of talking to people in human geography? We should recognize that a use of more formalized qualitative methods *does* have a rather longer history in geography than we normally realize, and that qualitative procedures prior to the last decade or so *could* involve more than just intuitive 'hanging out'. We should also take seriously the suggestion that many data deployed by human geographers have *always* come via conversational means and that we are not the first generation of geographers to appreciate this nor to urge some explicit reflection upon how such a conversational approach operates and might be improved in practice. This being said, it must be acknowledged that in earlier years a qualitative method such as interviewing was mostly deployed as a supplement to the cataloguing mentality of the surveys discussed above, and so the purpose to which it was put does *differ* considerably from the more

issue-based probing characteristic of more recent studies. Yet even in these rather different contexts important arguments were already being rehearsed about the inherent philosophical advantages of allowing people to present informed qualitative accounts of their own circumstances. These are themes which reverberate throughout this chapter.

The practices of talking to people

We now turn to a discussion of where 'talking to people' fits into contemporary methodological strategies in human geography. Continuing with the themes established in previous chapters, our emphasis here is on the construction of data by researchers themselves, and our assumption is that we are now dealing with situations where preconstructed sources of data will have already been considered as possible bases for practising human geographical research but rejected because they are inappropriate for the research questions which are being pursued. Here, therefore, we look at different ways in which human geographers can ask questions of other people in order to construct information on issues which are important to the researcher and/or the researched. Our aim is to avoid what Gouldner (1967) has called *methodolatry*, in which there is a danger that we become 'compulsively preoccupied with a method of knowing, which is exalted without serious consideration of how successful it is in producing knowledge' (Eyles, 1988a: 3). Rather we will attempt to show how some forms of questioning are best suited to particular research circumstances, and these circumstances will include the political and the ethical as well as the technical.

There is considerable scope for confusion in the way in which we consider question-naires and interviews. People delivering face-to-face questionnaires are often labelled as 'interviewers', and the resulting questionnaire findings can thus easily be regarded as the product of 'interviewing'. Moreover, there are grey areas between the more unstructured end of questionnairing (for example, the use of open-ended questions) and the more structured end of interviewing (for example, the use of precisely worded and structured interview schedules). In this context we want to make clear at the outset what we see as an important distinction between questionnairing and interviewing. Questionnairing is usually part of a wider quantitatively driven strategy of social survey where representative samples of people can be questioned in order to produce numeric measures of behaviour, attitude, attribute and so on. Interviewing is usually a qualitative exercise aimed at teasing out the deeper well-springs of meaning with which attributes, attitudes and behaviour are endowed.

Choosing an 'appropriate' practice for asking questions should therefore be considered as part of a much wider research strategy, fully interconnected both with careful formulation of research questions and with a clear idea of how the constructed data are to be interpreted. One framework for beginning to make these interconnections is the choice of what Harré (1979) has termed 'extensive' and 'intensive' research. Sayer (1984; 1992) has summarized what each of these can mean in terms of typical research strategies (Table 5.1), with extensive work focusing on patterns and regularities found among a large-scale representative group of people using methods of data construction such as formal questionnaires and standardized interviews, and data interpretation typically involving statistical analysis. By contrast, intensive work pursues specific processes with a small number of people using interactive interviews and ethnographies for data construction, and qualitative analysis as the strategy for data interpretation. This distinction between extensive (= questionnaires and statistics) and intensive (= ethnography and qualitative analysis) can tend to be over-rigorous in 'pigeon-holing' particular practices of constructing and interpreting data, a point acknowledged by Sayer (1992) in the second edition of *Method in Social Science* where he adds the following note to his previous account of extensive and intensive research: 'Note that the extensive/intensive distinction is not identical to the more familiar distinction between survey analysis and ethnography. Intensive research need not always use ethnographic methods to establish the nature of causal groups, and surveys need not be devoid of attempts to understand the social construction of meaning' (p. 244). This is a key point. It is, therefore, also important to note that different methods of asking questions *can* incorporate attempts to identify different depths or layers of meaning, and the dangers of 'methodolatry' are as applicable to those researchers favouring ethnography as to those preferring social scientific methods.

In order to move away from these important but sometimes partisan categorizations, we prefer to think of the range of different questionnairing and interviewing methods as reflecting a changing set of relationships between the researcher and the researched. This chapter thereby discusses a range of social situations in research in which particular researcher/researched relationships signal different kinds of social construction of data. To illustrate the range of these relationships, we can briefly consider four types of social situation commonly encountered in human geography research. First, there is a series of relationships which might be regarded as *robotic*. Here the researcher will be making use of other people's surveys, as in the use, for example, of opinion polls. Data in this case are effectively preconstructed – there has been no contact or direct interaction between the

Table 5.1 Intensive and extensive research: a summary

	Intensive	Extensive
Research question	How does a process work in a particular case or small number of cases? What produces a certain change? What did the agents actually do?	What are the regularities, common patterns, distinguishing features of a population? How widely are certain characteristics or processes distributed or represented?
Relations	Substantial relations of connection	Formal relations of similarity
Type of groups studied	Causal groups	Taxonomic groups
Type of account produced	Causal explanation of the production of certain objects or events, though not necessarily representative ones	Descriptive 'representative' generalizations, lacking in explanatory penetration
Typical methods	Study of individual agents in their causal contexts, interactive interviews, ethnography. Qualitative analysis	Large-scale survey of population or representative sample, formal questionnaires, standardized interviews. Statistical analysis
Limitations	Actual concrete patterns and contingent relations are unlikely to be 'representative', 'average' or generalizable. Necessary relations discovered will exist wherever their relata are present, e.g. causal powers of objects are generalizable to other contexts as they are necessary features of these objects	Although representative of a whole population, they are unlikely to be generalizable to other populations at different times and places. Problem of ecological fallacy in making inferences about individuals. Limited explanatory power.
Appropriate tests	Corroboration	Replication

Source: Sayer (1984: 222)

researcher and the researched. The researcher has had no freedom to pose the questions and certainly no opportunity to follow up on the answers given. The nature and form of the data are fixed and cannot be influenced by the researcher's own subjectivity. Significant epistemological issues arise in the interpretation of information which has been preconstructed in this way and, although the

researcher will have some scope to impose personal and positional interests during the use of such data for interpretative analysis, much of the relational power of the research is vested in those who set the questions and co-constructed the answers with respondents.

Secondly, we can envisage *remote* relationships between the researcher and the researched (for example, as occur in the deployment of postal questionnaires). The involvement of the researcher in the direct framing of questions is a significantly powerful process in the construction of data. However, the lack of face-to-face contact with the respondent means that the researcher has no interpretative history to draw on when seeking to make sense of the answers. Issues relating to the inappropriateness of the categories provided for answering questions, the potential ambiguity of the questions themselves, the unanticipated special circumstances of the respondent, the different layers of meaning which are variously revealed or cloaked by answers and so on will be unavailable to the remote researcher whose co-constitutive role is hampered by the lack of direct interaction with the respondent. Naturally, however, such remoteness of the research relationship appeals to researchers who are seeking to objectify their research methods. To pose the same questions in the same way to all survey respondents can be positioned tactically as a manoeuvre which screens out the perceived subjectivity, or 'bias', of the interviewer.

Thirdly, we can consider more *interactive* relationships (for example, those which occur during semi-structured face-to-face encounters). Here, data are co-constructed as interviewer and interviewee work their way through questions which begin as the 'property' of the researcher but which become co-owned and co-shaped in the unfolding interactivity of questioning, answering, listening and conversing. Here, the original scheme of intended data construction can be diverted or even subverted by both the researcher and the researched: the former as he or she follows up what appear to be interesting conversational angles; and the latter as he or she demarcates consciously or subconsciously the boundaries of what he or she will reveal in the interview. Interactive research relationships will differ markedly according to the power relations involved – interaction with a high-flying business executive, for example, will be very different from that with a homeless person or a child. Interactivity, then, is replete with social, political and ethical differences, and the resultant co-construction of data will reflect these issues.

Fourthly, research relationships can be *involved* as in the use of reflective interviews or discussion groups. In these situations, the researcher is yet further immersed in the social situation of conversation, leading to a higher profile and potential impact for the subjectivities of both the researcher and the researched. In these contexts the notion of getting 'true' answers to 'straightforward' questions is increasingly replaced by the potential for dialogue, to sponsor an unfolding performance in which roles and topics are fashioned in the give-and-take of the event. Involved research relationships represent a deliberate departure from ideas about neutrality and observational 'distance' so as to avoid treating people like objects and to find ways of treating people like people. The search for essentialist truth is thus replaced by a desire to see the world through different windows and to hear the world via a polyphony of different voices (Crang, 1992).

In these kinds of ways, the judgement of how appropriate a particular approach is to a defined question will depend on the role of the researcher, as well as of the researched, in constructing data. A number of underlying epistemological issues will be crucial in this judgement. We take the view that all knowledge is both situated and struggled over (Smith, S.J., 2001). The researcher's own *positionality* – his or her subjectivity and positioning – will represent a significant contextualization of his or her role in co-constructing and then interpreting interview data. However, as Rose (1997a) has pointed out, we need to be aware that we cannot ever fully recognize or represent our own positionality. So while it is important to engage with the values and subjectivities of the researcher, because these will be important factors in the co-construction of knowledges in research encounters, it is also important to realize the limits to knowing subjectivity. Equally, the *intersubjective* nature of the research encounter is a necessary reflection of both the researcher as co-constituent of resulting knowledges and the researcher becoming (however briefly) part of the life of the researched. Intersubjective encounters can be designed as a means to understand the lifeworlds, experiences and meanings of the researched (Buttimer, 1976), but the consequences of such encounters can never be fully predicted or known. Finally, the *power relations* of the encounter are crucial. Although there will be times when power is exercised upon the researcher (as with the high-flying business executive mentioned above), it is far more common for power to be exercised by the researcher in his or her investigation of, and interpretation of, the lives of others. The extent to which we as researchers are able to represent other lives is a crucial and tricky epistemological question, given that all such representations will be in some senses fictional in that they go beyond actual utterances of the research encounter (see Bennett and Shurmer-Smith, 2001). The choice of appropriate approaches to 'talking to people' will, therefore, necessitate critical appraisal of these epistemological issues which underpin the relationship between researcher and researched.

Questionnairing

We begin our discussion at that end of the range of interviewing methods in which the researcher is most remote from his or her respondent. Johnston (2000b: 668) defines the questionnaire as an instrument of data construction 'comprising a carefully structured and ordered set of questions designed to obtain the needed information without either ambiguity or bias... Every respondent answers the same questions, asked in the same way and in the same sequence'. As such, questionnaires are an essential tool of social survey analysis, potentially delivering research data which represent particular attributes of the sample. Questionnaires allow researchers to count up differing kinds of responses to questions, particularly where the questions are 'closed' (that is, referring to a fixed range of potential answers). This survey technique also allows the counting up of scorings from questions involving sematic differentials (such as very good, good, satisfactory, bad, very bad). The practices of counting up produce numeric measurements of what people think and how they behave, alongside information about their gender, age, occupation and so on. This information can then be cross-tabulated and used to make quantifiable inferences about the wider sample from which the sample is drawn.

Questionnaire surveys are a familiar fact of life in the developed world. Not only do households have the regular state census to contend with, but people walking down the high street will also often be accosted by surveyors requesting information for political opinion polls or market research, and a knock at the front door can mean an encounter with questionnaire surveys used by anyone from the Jehovah's Witnesses to human geography researchers. This very familiarity has in turn led some people to develop their own responsive strategies, such as a polite but quick refusal to participate – perhaps a reflexive response of 'survey fatigue' to the intrusive nature of surveys in society.

Overfamiliarity with the idea of a questionnaire survey can also pervade the practice of research in human geography. Questionnaire surveys have become part of the standard baggage of human geography researchers, being used in the main to produce quantitative information. Indeed, they have become so standard that there is sometimes a sense in which the thought 'I'll do a questionnaire' often *precedes* the arrival at clearly thought-out research questions, and research can sometimes, therefore, be reduced to a question of 'what can I do with this information from the questionnaire?' (see Bridge, 1992). This is not to suggest that the questionnaire cannot perform a perfectly legitimate role in practising human geography, either as a sole method of constructing data or in tandem with other methods. Indeed there are many recent instances in which combinations of different data constructions are positively advocated. For example, Jayaratne (1993: 109), writing on issues of methodology in feminist research, states:

> I also advocate the use of qualitative data, in conjunction with quantitative data to develop, support and explicate theory. My approach to this issue is political; that is, I believe the appropriate use of both quantitative methods and qualitative methods in the social sciences can help the feminist community in achieving its goals more effectively than the use of either qualitative or quantitative methods alone.

The important guiding principle in considering a questionnaire survey is therefore whether it is the most appropriate means of constructing data with which to address a particular research issue. Such a judgement will in turn necessitate a recognition of both the technical capabilities and restrictions of questionnaires, as well as the potential for variations in data construction because of unspoken assumptions in the questions asked, difficult-to-interpret differences in the answers given and the potentially different outcomes of questionnaires in different social situations.

When is a questionnaire appropriate?

A very substantial literature exists on survey research using questionnaire techniques.[1]

Issues of what form a questionnaire should take, what questions should be included and how the survey should be implemented are the subject of exhaustive commentary. Much less, however, is said about when a questionnaire is or is not appropriate. For some, the questionnaire survey is an integral part of the application of scientific method to social science, and its function is therefore one of permitting enumeration to take place (see Chapter 8). Thus Oppenheim (1992: 101) maintains that 'the detailed specification of measurement aims must be precisely and logically related to the overall plan and objectives. For each issue or topic to be investigated, and for each hypothesis to be explored, a precise operational statement is required about the variables to be measured'.

In cases where questions are used as vehicles for 'scientific' hypothesis-testing, problems of obtaining a statistically representative sample will often be paramount (see Chapter 8). For others, the usefulness of a questionnaire survey will be less tied to a research framework of hypothesis-testing. Here, the emphasis is on the requirement for a highly structured research technique which can be used to construct a body of information consisting of the answers of a large sample of respondents to a consistent series of questions. As such it can form part of Sayer's extensive research strategy (see Table 5.1) although a more flexible mix of structured and less structured information can be constructed during one face-to-face interview, and so intensive information may also be available using these methods. Perhaps Bell's (1987: 58) advice is most appropriate here:

> Ask yourself whether a questionnaire is likely to be a better way of collecting information than interviews or observation, for example. If it is, then you will need to ensure you produce a well-designed questionnaire that will give you the information you need, that will be acceptable to your subjects and

which will give you no problems at the analysis and interpretation stage.

Questionnaires, then, can be useful in a number of situations: where you are interested in constructing data based on the responses of a large number of people; where you have only time-constrained, face-to-face access to your respondents and therefore only limited interaction is possible; where you are unable for one of a number of reasons to gain face-to-face access to your potential respondents and therefore cannot interview them personally; where you require data to feed into a preconceived statistical form of interpretation; where the presentation of your research findings will achieve greater impact (for example as with some government agencies) if they are based on a recognizable (and politically acceptable) questionnaire survey methodology; and so on.

What form can a questionnaire take?

There is a range of potential forms of questionnaire which should be carefully considered before deciding on a survey research strategy, although these boil down to three basic formats:

1. The questionnaire which is posted or handed to the respondent, who completes the survey and then posts or hands it back to the researcher (contemporary use of Internet or email surveys also conform to this format).
2. The questionnaire which is carried out over the telephone, with the researcher asking questions and recording answers.
3. The questionnaire which is carried out face to face, in which the researcher again asks questions and records the answers.

These formats can be mixed, however. Respondents can be asked to read and record answers to particularly sensitive areas of questionnaires during face-to-face surveys; combinations of telephone and face-to-face interviews might be necessitated in circumstances where incomplete telephone number information is available; and respondents might be asked to complete and return a second element of a survey having completed a face-to-face questionnaire – this sometimes happens when a 'diary' of activities and trips is requested for the following week or month.

It is possible to outline some typical pros and cons for these different survey formats (see Table 5.2), although the various areas of advantage and disadvantage are more arguable than this kind of tabulation suggests. To take a glaring example, the response rates from postal questionnaires are often thought to be much lower than those conducted face to face or by telephone. For example, Bridge (1992: 198) notes that 'response rates of 30–40 per cent from a survey of residents in an ordinary neighbourhood is considered good'. Perhaps less obvious but equally important, given our comments at the beginning of this chapter about interviews as sets of relationships between the researcher and the respondent, is the suggestion in the table that the *quality* of answers is problematic in face-to-face surveys. The assumption here is that questionnaires allow researchers to avoid some of the 'distortions' which are seemingly the inevitable product of the face-to-face social interaction involved in less structured interviews. These 'distortions', so-called, represent the fear that in an interview questions can be asked in a 'leading' way, with the likelihood that there will be too many forceful prompts and too frequent misunderstandings of responses. Questionnaires, on the other hand, are sometimes thought to be able to provide consistent, representative and bias-free answers across a large sample.

This question of bias is an interesting one. One approach is to create a 'bias-free' questionnaire by avoiding issues relating to attitudes,

Table 5.2 Advantages and disadvantages of face-to-face, telephone and mail methods of administering questions

	Face to face	Telephone	Mail
Response rates			
General samples	Good	Good	Good
Specialized samples	Good	Good	Good
Representative samples			
Avoidance of refusal bias	Good	Good	Poor
Control over who completes the questionnaire	Good	Satisfactory	Good
Gaining access to the selected person	Satisfactory	Good	Good
Locating the selected person	Satisfactory	Good	Good
Effects on questionnaire design			
Ability to handle:			
Long questionnaires	Good	Satisfactory	Satisfactory
Complex questions	Good	Poor	Satisfactory
Boring questions	Good	Satisfactory	Poor
Item non-response	Good	Good	Satisfactory
Filter questions	Good	Good	Satisfactory
Question sequence control	Good	Good	Poor
Open-ended questions	Good	Good	Poor
Quality of answers			
Minimize social desirability responses	Poor	Satisfactory	Good
Ability to avoid distortion due to:			
Interviewer characteristics	Poor	Satisfactory	Good
Interviewer's opinions	Satisfactory	Satisfactory	Good
Influence of other people	Satisfactory	Good	Poor
Allows opportunities to consult	Satisfactory	Poor	Good
Avoids subversion	Poor	Satisfactory	Good
Implementing the survey			
Ease of finding suitable staff	Poor	Good	Good
Speed	Poor	Good	Satisfactory
Cost	Poor	Satisfactory	Good

Source: de Vaus (1991: 113), adapted from Dillman (1978)

beliefs or prejudices altogether, thereby excluding respondents' biases. Such a strategy may be suitable in some situations where straightforward factual data are required on an extensive scale. Another, and often more fruitful, approach may be to use questionnaires precisely to tease out these abilities, beliefs and prejudices. In many contexts researchers *do* want to know about the 'biases' of their respondents and, by restricting a questionnaire to what seems like objective, factual accounts of human geographical situations, the potential richness of the encounter may be lost. It is

this perspective which often informs the choice between remotely delivered questionnaires and undertaking face-to-face encounters in which the structured questioning of the questionnaire can be mixed with, or even replaced by, less structured interview techniques. Face-to-face encounters with respondents will greatly enhance the researcher's ability to interpret some of the context in which questionnaire answers are set. What is lost in the consistency of questioning necessary for hypothesis-testing may represent a significant gain in the questionnaire as a hybrid interview method which is able to combine extensive information with some intensive insight.

Thus far it has been implicit in our discussion that the questionnaire is carried out between *one* researcher and a series of individual respondents. This can frequently not be the case. For example it is possible to administer a questionnaire to a group of people at the same time, such as a class of school children or people attending a public meeting, although the use of focus groups is often more appropriate in these situations. Sometimes group questionnaires permit visual texts – video, film clips, slides, photographs, etc. – to form the basis of questions, or the questionnaire can be completed subsequent to a particular 'live' debate (for example, between politicians or between local council representatives and local people protesting over a proposed development). Considerable care is required in these contexts because respondents will tend to talk to each other about questions, ask the interviewer questions about how they should answer and even copy each other's answers. Again, the use of focus groups could be a more appropriate choice in such situations. Perhaps a more common form of 'group' questionnaire is when the researcher requires information about a 'household'. Often one member of the household answers on behalf of others, giving a particular view of the lifestyles and opinions of others which would be different if the person concerned were answering. For example, if questions are asked of the 'head of the household' who will often be assumed to be male, the responses are highly likely to involve gender-biased representations of that household. It is better (although much more difficult to organize) to undertake the questionnaire when all members of the household are present to speak for themselves. A similar dilemma (though often impossible to resolve) is where

one individual responds to a questionnaire on behalf of an agency. The difference between individual and 'corporate' attitudes in these cases is often very difficult to discern, and much will depend on the status and role of the individual within the agency. In addition there may be crucial differences between agency information for *external* consumption, which often constitutes a well-rehearsed public relations narrative, and information which reveals some of the (often political) debates going on *inside* the agency.

Constructing the questionnaire

Those people who have experience of using questionnaire methods will almost invariably conclude that it is enormously difficult to construct a good series of questions. We would suggest that the questionnaire survey is by no means the 'easy option' that it is sometimes assumed to be by human geographers making their methodological choices. We would reiterate that the key factor here is to think through very carefully the precise research questions that are to be focused on and, if a questionnaire survey seems appropriate, to apply specific parts of the questionnaire schedule to specific elements of the overall research questions. The idea will be to ask purposeful questions in as unambiguous a manner as possible, with a view to providing useful information of a type suitable for the interpretative strategy you will have already chosen. The literature on questionnaire surveys abounds with illustrations of good practice – indeed, whole books have been written on the subject of how to construct questions (see Foddy, 1993). It will often be useful to supplement a reading of these manuals of good practice with scrutiny of questionnaires that have actually been used by other researchers. For example, Forbes (1988: 112) writes candidly about how he got together a questionnaire of informal economic activity in urban districts (*kampungs*) of Indonesia:

> Socio-economic surveys were the mainstay of the work being done on the informal sector at the time, due in large part to the inadequacy of official data sources when it came to the marginal groups in the informal sector. I found a number of interview schedules that had been used in Indonesia for *kampung* surveys, informal sector surveys and migration surveys, and drafted out a schedule of nearly 50 questions.

A good questionnaire will often be a combination of previously used and innovative questions which are closely focused on the overall aims of the research.

The types of question employed will depend very much on what constructions of data will be suitable for the interpretative strategies to be used. If interpretation is to be in numeric form using descriptive or other statistics (see Chapter 8), questions will normally be closely structured in a 'closed' form. Bell (1987), following Youngman (1986), suggests six types of questions which can be used to construct structured data (see Box 5.1). If, however, the chosen interpretative strategy permits the construction of written text as data, a seventh more open-ended type of question can be used. When using questionnaires for the first time, it is often tempting to deploy a large number of open questions without having thought through how the resultant chunks of texts can be used effectively in providing answers to the big research questions which are the focus of the project. These passages of text will then need to be 'coded' and interpreted (see Chapter 10). Open questions can bear much interpretative fruit, however, and may open up some fascinating lines of intensive inquiry in situations either where extensive research is the main priority (thus necessitating questionnaire rather than less formal interviews) or where an ethnographic approach is not possible for one of a variety of reasons. An example of a

research report where qualitative text (constructed by using open-ended questions) has been used in conjunction with more numeric tabulated findings is the report on 'Rural lifestyles in England' (see Cloke et al., 1994). Here the research was designed in part to replicate a previous survey of rural households in England to investigate changes in housing, employment, income, accessibility, service provision and other key factors of rural life. The methodological constraints of having to use an extended version of a previous survey instrument were partially freed up by opening out spaces for respondents to add commentary on aspects of their lifestyles. The result, though imperfect when measured against the demands of rigorous qualitative research, did at least allow the 'voices' of rural people to be included in the published output.

Having decided on the types of question required, it will be necessary to give some considerable thought to the precise wording of particular questions. It is all too easy for questions to be ambiguous, imprecise or replete with inherent assumptions. Although piloting the questionnaire (see below) will help to expose any difficulties with wording, it is worth a considerable amount of preparatory time in sorting and sifting particular questions. Oppenheim (1992) provides 13 basic rules for the careful framing of questions (Box 5.2). To these we would add the warning to be fastidious in ensuring that the language used in questions does not make unacceptable assumptions about gender, sexuality, ethnicity, age, localism, religious faith and the like.

These rules will not guarantee a successful questionnaire but, given a well-thought-out series of research aims and an awareness of the interpretative manoeuvres to be carried out on the constructed data, the wording of questions will be important. Many students who do give due care and attention to the pitfalls of questionnaire technique become somewhat

Box 5.1 Different types of questions

More structured questions

List: A list of items is offered, any of which may be selected. For example, a question may ask about qualifications and the respondent may have several of the qualifications listed.

Category: The response is one only of a given set of categories. For example, if age categories are provided (20–29, 30–39, etc.), the respondent can fit into only one category.

Ranking: In ranking questions, the respondent is asked to place something in rank order. For example, the respondent might be asked to place qualities or characteristics in order (e.g. punctuality, accuracy, pleasant manner). Number '1' usually indicates the highest priority.

Scale: There are various stages of scaling devices (nominal, ordinal, interval, ratio) which may be used in questionnaires, but they require careful handling.

Grid: A table or grid is provided to record answers to two or more questions at the same time. For example, 'How many years have you taught in the following types of school?' Here there are two dimensions – type of school and number of years.

	1 or 2 years	3 or 4 years	5 or 6 years	More than 6 years
Grammar				
Comprehensive				
Middle				
Independent				
Other (please specify)				

Less structured questions

Verbal or open: The expected response is a word, a phrase or an extended comment. Responses to verbal questions can produce useful information but analysis can present problems. Some form of content analysis is required for verbal material unless the information obtained is to be used for special purposes. For example, you might feel it necessary to give respondents the opportunity to give their own views on the topic being researched – or to raise a grievance. You might wish to use verbal questions as an introduction to a follow-up interview or in pilot interviews where it is important to know which aspects of the topic are of particular importance to the respondents.

Source: Bell (1987: 59–60), following Youngman (1986)

Box 5.2 Oppenheim's rules for question design

Length: Questions should not be too long; they should not contain sentences of more than, say, 20 words.

Avoid double-barrelled questions: 'Do you own a bicycle or a motorbike?' In such instances the respondents are left in a quandary if they want to say 'yes' to one part of the question but 'no' to the other.

Avoid proverbs and other popular sayings, especially when measuring attitudes, for such sayings tend to provoke unthinking agreement. Instead, try to make the respondent think afresh about the issue by putting it in other words.

Avoid double negatives: For example, 'Cremation of the dead should not be allowed', followed by agree/disagree. Respondents who favour cremation have to engage in a double negative (that is, they do *not* agree that cremation should *not* be allowed). Try to put the issue in a positive way.

Don't know and not applicable: Categories are too often left out, both in factual and in non-factual questions. It has been argued that some people give a 'don't know' response in order to avoid thinking or committing themselves, but do we really want to obtain 'forced' responses which are virtually meaningless?

Use simple words, avoid acronyms, abbreviations, jargon and technical terms – or else explain them. When in doubt, use pilot work.

Beware the dangers of alternative usage: This applies to even the simplest words which 'could not possibly be misunderstood'! To some people, the word 'dinner' indicates an evening meal but to others it means a cooked meal, as in 'Usually I give him his dinner at lunch time' [*sic*].

Some words are notorious for their ambiguity and are best avoided or else defined – for example, the words 'you' and 'have' as in 'do you have a car?'. Other notoriously ambiguous words are 'family', 'bought', and 'neighbourhood'.

All closed questions should start their lives as open ones so that the answer categories will be based as much as possible on pilot work. Even then, always consider the inclusion of an 'Other (please specify)' category.

Beware leading questions: 'When did you last borrow a video tape?' This assumes (1) that all respondents have access to a videotape player; (2) that they generally borrow tapes; and (3) that it is not done by someone else on their behalf.

Beware loaded words such as democratic, free, healthy, natural, regular, unfaithful, modern.

Box 5.2   *(Continued)*

Don't overtax the respondents' memories by asking questions such as 'in the past 13 weeks, which of the following branded goods have you purchased?' For recurrent items of behaviour it is risky to go back more than a few days. For more important events such as major acquisitions, periods in hospital or being mugged, the respondent first has to be taken through the preceding few months with the aid of 'anchoring dates' such as public holidays, family birthdays, school holidays and the like, followed by an attempt to locate the event(s) in question in an individual time frame.

Pay due attention to detail such as layout, show cards and prompts (note especially the danger of overlapping categories, as with 'age 25–30, 30–35'), the wording of probes (follow-on questions), routing or skip-to instructions and even question numbering.

Source: Abridged and adopted from Oppenheim (1992: 128–30)

overwhelmed at the prospect of meeting all these various requirements. With care, however, a questionnaire can be constructed and implemented effectively. Figure 5.2 shows a questionnaire which was designed and carried out by a small group of first-year geography students at Lampeter. Their research aims related to the diversification of agricultural enterprise and, although they would not want their work to be represented as any kind of model of best practice, it does indicate that a simple questionnaire can be put together rather nicely so as to construct useful data with which to discuss manageable research issues.

Thus far we have presented only a rather simple account of the kinds of questions that can be asked in a questionnaire. In practice there is a vast array of possible devices which can be used in this context, and there is plenty of scope for innovation in questionnaire work. To illustrate some of these different avenues we will briefly discuss the construction of data on *attitudes* using questionnaire techniques. The instinctive reaction when asking someone about his or her attitude towards a particular event, circumstance or characteristic of his or her lifestyle is simply to ask 'what do you think …'. Indeed this may

be a productive method of constructing open-ended qualitative text on a particular issue, and delivered consistently it may also produce attitude statements capable of comparison. There are, however, other ways of approaching this task. For example, those researchers seeking numerical data for statistical interpretation will view attitude statements as mere raw materials for attitude-scaling techniques. In a uni-dimensional form, attitude scaling often means asking respondents to indicate their feeling about something on a numeric scale, say between 1 and 5, where 1 indicates strong agreement and 5 indicates strong disagreement. Such scales can also be used to indicate a semantic differential. Here particular adjectives – for example, 'peripheral' and 'core' – can be chosen to represent the two extremes of the continuum and respondents are asked to mark their attitude on a scale between the two extremes.[2]

Attitude scales can also, however, be constructed in a multidimensional form. According to Oppenheim (1992: 174, emphasis in original), this requires that attitude scales

should be phrased so that respondents can agree or disagree with the statement. Usually, after the pilot work, a large number

Please note: ALL INFORMATION VOLUNTEERED WILL BE USED IN THE STRICTEST CONFIDENCE.

Instructions: Please tick boxes where relevant. Where you are asked to specify your answer, please use a brief statement or sentence. Sufficient room has been left for the questions that may require slightly longer answers.

1   Name: ....

2   Address: ....
....

3   Size of farm/holding (please tick relevant box)

SMALL ☐   MEDIUM ☐   LARGE ☐

(70–299 acres)   (300–499 acres)   (500–800 acres)

OTHER ☐ (please specify) ....

4   LENGTH OF TENANCY/OWNERSHIP (years or months) ....

5   (Please complete following table)

FAMILY MEMBERS (Living or working at farm)	RELATIONSHIP TO 1	AGE	OCCUPATION
1	....	.....	....
2	....	.....	....
3	....	.....	....
4	....	.....	....
5	....	.....	....
6	....	.....	....
7	....	.....	....
8	....	.....	....

OTHER (please specify) ....

6   TYPE OF FARMING ACTIVITIES (please tick relevant boxes)

DAIRY   ☐   HORTICULTURE ☐
ROUGH PASTURE ☐   CROPS   ☐
  (GRAZING)
STUD FACILITIES ☐
LIVESTOCK   ☐   SHEEP   ☐   PIGS   ☐   COWS ☐
                           POULTRY ☐   HORSES ☐   GOATS ☐

OTHER (please specify) ....

7   (Please complete following table)

EMPLOYEES	DUTIES	FULL/PART TIME or SEASONAL
1	....	....
2	....	....
3	....	....
4	....	....

OTHER (please specify) ....

8   HAVE YOUR FARMING ACTIVITIES CHANGED DURING YOUR TENANCY/OWNERSHIP OF THE FARM? (please tick relevant box)

YES ☐   NO ☐

(Continued)

b   IF YES, HAS THIS INVOLVED DIVERSIFICATION? (please tick relevant box)

   YES ☐   NO ☐

9   IF YES TO QUESTION 8b, PLEASE BRIEFLY DESCRIBE THE FORM(S) OF DIVERSIFICATION
(you may want to use the list below as a reference when answering this question)

ACCOMMODATION AND CATERING   ANSWER: . . . .
(e.g. bed and breakfast)   . . . .
AGRICULTURAL SERVICES   . . . .
(hay, cutting and baling, ...
machine rental, etc.)   . . . .
EQUINE (e.g. pony trekking, ...
stud, livery, etc.)   . . . .
CROP-BASED PRODUCTION AND   . . . .
PROCESSING (e.g. processing   . . . .
of foodstuffs, animal feed)   . . . .
TRADING (e.g. farm shop, ...
retailing, etc.)   . . . .
MANUFACTURING (e.g. craft   . . . .
goods, etc.)   . . . .
FORESTRY (e.g. timber   . . . .
processing, rental of land for   . . . .
plantations, etc.)   . . . .
BREEDING (e.g. rare breeds, ...
livestock, etc.)   . . . .
MISCELLANEOUS (e.g. land-   . . . .
building facilities, landscape   . . . .
services, pet care, vehicle   . . . .
repair, etc.)   . . . .

b   HOW LONG AGO DID YOU DIVERSIFY? (years or months) . . . .

10 DO YOU RENT   and/or LEASE   OUT ANY PART OF YOUR LAND
       YES ☐   NO ☐   (please tick relevant boxes)

b   IF YES (please specify) . . . .

11 PLEASE COULD YOU DESCRIBE, IN YOUR OWN WORDS, WHY YOU DIVERSIFIED?
(you may want to use the list below as a reference when answerinng this question)

To raise or maintain farm or   ANSWER: . . . .
family income   . . . .
To utilize available human   . . . .
resource, and/or hired labour, ...
land, buildings or machinery   . . . .
Personal interest   . . . .
Utilization of spare capital   . . . .
Market opportunity   . . . .
To gain a competitive   . . . .
advantage   . . . .
Tax incentives   . . . .
Introduction of government   . . . .
legislation (e.g. quotas)   . . . .

(Continued)

12 DID YOU RECEIVE ANY (please tick relevant boxes)

	ADVICE	ASSISTANCE	GRANT	RESISTANCE
FROM THE FOLLOWING:				
NATIONAL PARK AUTHORITY	☐	☐	☐	☐
MINISTRY OF AGRICULTURE, FISHERIES AND FOOD	☐	☐	☐	☐
NATIONAL FARMERS UNION	☐	☐	☐	☐
DEVON COUNTY COUNCIL	☐	☐	☐	☐
YOUR DISTRICT COUNCIL	☐	☐	☐	☐
BANKS/FINANCIAL INSTITUTES	☐	☐	☐	☐
COUNTRYSIDE COMMISSION;	☐	☐	☐	☐
FORESTRY COMMISSION	☐	☐	☐	☐
PUBLICATIONS	☐	☐	☐	☐

OTHER (please specify) . . . .

b   WOULD YOU LIKE TO EXPAND ON YOUR ANSWER TO QUESTION 12?
(e.g. the advice, assistance, grant received, the resistance met, etc.)

. . . .
. . . .
. . . .

13 HOW DID YOU FIND OUT ABOUT THE DIFFERENT SCHEMES, OR INFORMATION, OFFERED BY VARIOUS ORGANIZATIONS? (please identify the organizations in your answer)

. . . .
. . . .
. . . .

14 DID YOU APPROACH ANY OF THESE ORGANIZATIOINS (or the ones you have specified)?

YES ☐   NO ☐   (please tick relevant box)

15 IF YES TO QUESTION 14. WHY DID YOU FINALLY APPROACH THESE ORGANIZATIONS?
(please identify the organization in your answer)

. . . .
. . . .
. . . .

16 IN YOUR OWN WORDS, HOW WOULD YOU EXPRESS, OR DESCRIBE, YOUR LEVEL OF ACHIEVEMENT IN DIVERSIFICATION?

. . . .
. . . .
. . . .

17 WHAT FACTORS HAVE HAD AN AFFECT ON YOUR DIVERSIFICATION? (you may want to use the list below as a reference when answering this question)

Changes in the market   ANSWER: . . . .
Lack of market growth   . . . .
Labour difficulties   . . . .
Lack of finance   . . . .

(Continued)

Operating difficulties ....
Inappropriate support ....
from external sources ....

18   HAS THE NATURE OF YOUR PRIMARY FARMING ACTIVITY CHANGED DUE TO DIVERSIFICATION?

YES ☐   NO ☐   (please tick relevant box)

b   IF YES, WOULD YOU LIKE TO EXPAND ON YOUR ANSWER?

....
....
....

19   WOULD YOU LIKE TO ADD ANY FURTHER COMMENT, OR INFORMATION, ON ANY RELEVANT ISSUES, OR FACTORS, YOU BELIEVE WE HAVE, OR HAVE NOT, COVERED IN OUR QUESTIONNAIRE?

....
....
....
....
....
....

THANK YOU FOR YOUR CO-OPERATION IN COMPLETING THIS QUESTIONNAIRE. IF YOU WOULD BE WILLING TO PARTICIPATE IN A SHORT (15–30 MIN.) INTERVIEW AT A LATER DATE, PLEASE LET US KOW BY TICKING THE RELEVANT BOX.

YES ☐   NO ☐   TO INTERVIEW

Figure 5.2   A postal questionnaire to farmers on diversification of agricultural enterprise

Source: Clare Bover, Mike Gadd, Pam Hovell, Karen Inston and Pauline MacKenzie

of attitude statements (probably somewhere between sixty and 100) is assembled to form an *item pool*. These items will be analysed and submitted to a scaling procedure ... If all goes well, the outcome will be a much smaller selection of items (say, between one and two dozen) which together have the properties of an attitude *scale* that can yield a numerical score for each respondent. The analysis will also have thrown light on the topic or dimension which the scale is designed to reassure.

In its simplest multidimensional form, then, attitude scaling will consist of a group of questions for which the respondent will receive a score for his or her answer. Figure 5.3 illustrates a set of questions on the subject of

traditionalism. The score for each answer reflects its association with the particular attitude being measured, and the total score will therefore be used to reflect the respondent's 'position' on the abstract continuum of opinion on this issue. Thus a high score in Figure 5.3 is held to denote a very highly traditionalist attitude, and vice versa. Much more complex statistical manipulations can be undertaken, often using factor analysis to produce a mathematical value for how well each question 'loads on to' the overall mathematical value for the attitude in question.[3]

As with many statistical analyses, these multidimensional scaling techniques take us

		Strongly agree	Agree	Undecided	Disagree	Strongly disagree
a	If you start trying to change things very much you usually make them worse.	[ ] 4	[ ] 3	[ ] 2	[ ] 1	[ ] 0
b	If something grows up over a long period, there will always be much wisdom in it.	[ ]	[ ]	[ ]	[ ]	[ ]
c	It is better to stick by what you have than to try new things you do not really know about.	[ ]	[ ]	[ ]	[ ]	[ ]
d	We must respect the work of our forebears and not think that we know better than they did.	[ ]	[ ]	[ ]	[ ]	[ ]
e	A person does not really have much wisdom until he is well along in years.	[ ]	[ ]	[ ]	[ ]	[ ]

Person			Item			Scale score	Interpretation
	a	b	c	d	e		
1	4 +	4 +	4 +	4 +	4 =	20	Very highly traditionalist
2	4	2	1	0	3 =	10	Moderate
3	2	1	2	1	1 =	7	Moderate
4	0	0	0	0	0 =	0	Very non-traditional

Figure 5.3 Multidimensional attitude scaling: an illustration

Source: de Vaus (1991: 251)

further and further away from the actual source of the attitude – that is, the respondent. In practice it is extremely difficult to connect a series of different questions with *one* specific attitude because, typically, the answers given to questions represent a confused conflation of a number of different attitudes. By the time such an analysis reaches a factor analysis it is difficult to know what the over-riding mathematical factor actually represents. Many researchers, then, have chosen either to stick with simple open-ended questions and with uni-dimensional scales, or to seek out other ways of dealing with questions

about attitudes. These range from questions which invite respondents to complete a sentence in their own words to the use of vignettes and projective techniques. Vignettes tend to be short stories in which particular circumstances are specified within a particular standardized context (Box 5.3 illustrates a typical vignette). Respondents can then be asked simple questions such as 'what should they do?' and 'why?', and attitudes towards these kinds of circumstances could then be interpreted from their answers. Projective techniques rely on a spontaneous interpretation by the respondent of a written, verbal or

Box 5.3 Using a vignette

John and Barbara Dixon are both in their mid-forties and their children now have jobs overseas. Barbara is a research scientist in a medical laboratory and John is an accountant. They own their own home and are comfortably off. [JOHN/BARBARA] (A) is offered a promotion that would mean [A LARGE RISE IN INCOME/ONLY A SLIGHT INCOME RISE BUT A MUCH MORE ENJOYABLE JOB] (B) but would also mean that they would have to sell up and move several hundred kilometres. [BARBARA/JOHN] could also get a job in the new location [BUT IT IS NOWHERE NEAR AS GOOD AS THEIR PREVIOUS/JOB WHICH IS VERY SIMILAR TO THEIR CURRENT JOB] (C). [JOHN/BARBARA] would like to make the move but [JOHN/BARBARA] does not want to give up his/her current job. [The references to John and Barbara in the last sentence would be varied so that the person who is offered the new job wants to move while their partner does not want to move.]

Source: de Vaus (1991: 91)

visual stimulus. This can involve asking for a response to a calculated statement on a particular issue or showing respondents a series of visual images, often on cards which represent human figures in various situations that respondents are asked to describe and interpret in the form of a story. Saarinen (1966) used one such 'thematic appreciation test' in his study of farmers' attitudes to drought hazards in the Great Plains of the USA. Once popular, many of these techniques have become largely disused because of the difficulties experienced by researchers in interpreting the responses given in such a way as to be relevant to the research question they were investigating, although they may well enjoy a revival now that the inherent difficulties of interpreting any form of data are being more widely acknowledged. As with all the procedures discussed in this section, the decisions taken about constructing a questionnaire should be on the basis of logical and practical interconnections between research question, data construction and data interpretation. Employing an innovative technique which is only loosely connected to your research question and which constructs uninterpretable data will not be a fruitful way of practising human geography!

Carrying out the questionnaire

Unless the questionnaire is directed to all members of a particular target group, one early decision to be made in carrying out a questionnaire will be the selection of a particular *sample* of people to interview. Again it will be important here to use a method of sampling which is in tune with the interpretation strategy for the research. For example, some statistical analyses make clear assumptions about the characteristics of the sample from which numerical information has been drawn. In these cases *probability* samples – that is, those where each person in the population has an equal, or at least measurable, chance of being chosen as a potential respondent – will be preferable. On the other hand *non-probability* samples are those where some people have a higher chance (although not a measurable one) of being chosen and may be appropriate in other circumstances. There are four main types of probability sample (Box 5.4) and three main types of non-probability sample. Selection from these different sampling methods will very much depend on the necessity or otherwise for a statistically representative or otherwise indicative sample of interviewees. A 'representative'

Box 5.4 Sampling methods

Probability samples

1 *Random sample*: Each potential interviewee is given a number and then the required sample size is drawn from the list using random number tables. This sampling method is best where good information exists about the total population concerned, and where that population is concentrated geographically.

2 *Systematic sample*: Interviews are drawn at regular intervals from a population. If a 10% sample is required, every 10th person from the list of potential interviewees is selected. This method shares the disadvantages of the random sample but, in addition, exposes the survey to the risk that the list of potential interviewees may contain a certain kind of person at regular intervals. For example, if a membership list of a pressure group consisted of male and female partners listed together, a 50% systematic sample would pick out all male or all female.

3 *Stratified sample*: The list of potential interviewees is first grouped by any characteristic which the researcher wishes to see evenly represented in the sample – for example, age, gender, location, etc. The sample can then be selected either randomly or systematically from within these groups.

4 *Multi-stage cluster sample*: The original list of potential interviewees is sampled randomly in a number of stages, first to select a sample of areas (say, within a city), then to select a sample of smaller areas (perhaps neighbourhoods), then sample streets and households within those neighbourhoods and, finally, sample individuals within those households.

Non-probability samples

1 *Purposive sample*: Cases which are judged to be of particular interest to the researcher.

2 *Quota sample*: where stipulated quotas of cases will be samples *without* random selection procedures. Interviewers will simply select cases which conform to the criteria set (for example, males and females between the age of 50 and 60). By not using random sampling the cases may 'self-select' because they will be the kinds of people who will most readily agree to be interviewed.

3 *Availability sample*: Where anybody who agrees to be interviewed becomes part of the sample. This can happen, for example, where an advertisement is placed in the newspaper to ask whether people are willing to answer questions of a certain type. Those who respond are thereby 'self-selected'.

Source: After de Vaus (1991)

---

sample can be used to make assumptions about the total population from which it is drawn. 'Indicative' samples may give impressions about the total population but are inappropriate for statistical analyses which seek to interpret that total population. The *size* of the sample is also important in achieving a representative group of interviewees. Broadly, the necessary sample size will depend on both the extent to which the population varies and the precise requirements of the survey (see Ebdon, 1985; Williams, 1986).

It will also invariably be necessary to pilot the survey before starting on the main data–gathering stage. Piloting serves many functions: it familiarizes the questioner with the questionnaire design in practice; it provides an opportunity to find out which questions do not work well so that they can be improved or replaced; very importantly, it allows for the removal of extraneous questions which are found not to construct data which are useful for the overall research question(s); and it gives a clear idea about how long the survey takes to complete. Pilot surveys are often treated haphazardly or perfunctorily by researchers of all levels of experience but they are fundamental in allowing major improvements to be made prior to the main process of data construction.

The implementation of questionnaires subdivides sharply according to the form of survey chosen. For a postal questionnaire it is important to remember that the paperwork which is sent is effectively a research encounter by proxy. Just as the researcher will be looking to interpret aspects of the respondent through the medium of his or her replies, so the respondent will be interpreting the researcher through the medium of the survey instrument. Even without meeting, therefore, a socially constructed relationship will be established between the researcher and the researched by using a postal questionnaire. Most researchers will therefore want to represent themselves as efficient, professional, yet careful over important issues of ethics. They will also implicitly be seeking as high a response rate as possible. Therefore a carefully worded letter establishing the position of the researcher, the purpose of the survey, what will be done with the results, the safeguards for confidentiality and the anonymity of the respondent, and the hope for a returned completed questionnaire by a specified date will usually accompany the questionnaire itself, along, of course, with a stamped, self-addressed envelope. Figure 5.4 illustrates the letter from the group of first-year Lampeter students which accompanied the postal questionnaire on farm diversification (Figure 5.2). It may be useful to send a carefully worded reminder to non-respondents after a reasonable length of time.

Much the same information has to be given to potential respondents in telephone

and face-to-face questionnaires. A letter in advance of the call or visit and perhaps even the booking of appointments for calls and visits are increasingly appropriate as a means of respecting people's privacy and time management. Over the telephone, voice and language will feed information in both directions which will embellish the relationship between researcher and respondent, and in face-to-face questionnaires matters of dress code, body language and personality will additionally contribute to this relationship. Our experience is that some researchers will tend to 'hide behind' their survey, giving very little of themselves in establishing a rapport with their respondent, while others tend to present the survey and themselves as distinctly different entities allowing the researcher to comment on the technicalities of the survey when the respondent finds it hard going.

The textbooks on survey methods present a very rigid view of 'doing a standardized questionnaire'. Fowler (1988), for example, although using the term 'interviewer', sets out a strong prospectus for standardization when undertaking face-to-face questionnaire work. He suggests that

> survey researchers would like to assume that differences in answers can be attributed to differences in the respondents themselves – that is, their views and their experiences – rather than to differences in the stimulus to which they are exposed – that is, the question, the context in which it was asked, and the way it was asked. The majority of interviewer training is aimed at teaching trainees to be standardised interviewers who do not affect the answers they obtain. (1988: 109)

There is also an insistence on standardization in a number of different areas: presenting the study; asking the questions; probing for fuller answers; recording the answers; and interpersonal relations. On this latter point, Fowler says:

> Inevitably, an interviewer brings some obvious demographic characteristics into an

Saint David's University College
University of Wales

Professor Peter Beaumont: Department of Geography

24th February 1992

Dear Sir/Madam

We are a group of Human Geography students at St. David's University College, who are currently interested in researching the causes of upland hill farm diversification in the Dartmoor area, as a compulsory unit of our degree course.

We are contacting 40 farmers asking for 10 minutes of your time to fill in a short questionnaire about diversification. We will also be conducting a few short follow-up interviews in April.

We would like to reassure you that your participation in this questionnaire will be treated with the utmost confidentiality. Similarly, results from the study will be treated in the strictest confidence.

We would greatly appreciate a quick response as we have a tight deadline to meet. Therefore to facilitate this, please find enclosed a stamped, self-addressed envelope.

Thank you for your co-operation

Karen Inston

On behalf of:

Pam Howell
Mike Gadd
Clare Borer
Pauline MacKenzie

LAMPETER, DYFED, SA48 7ED, WALES, UK   TELEPHONE: LAMPETER (0570) 422351   FAX: 0570 423423

Figure 5.4   Letter to accompany a postal questionnaire

Source: Karen Inston

interview such as gender, age, and education. However, by emphasising the professional aspects of the interaction and focusing on the task, the personal side of the relationship can be minimised. Interviewers generally are instructed not to tell stories about themselves or express any views or opinions related to the subject matter of the interview. Interviewers are not to communicate any judgements or answers that respondents give. In short, behaviours that communicate the personal, idiosyncratic characteristics of the interviewer are to be avoided. (1988: 110)

Such comments are clearly designed to prompt us about maintaining the reliability of our surveys, and the goal of reliability continues to be an important one in many circumstances. However, in our view, these rules of standardization can place an unreal and an unjustifiable burden of expectation on human geographers who choose to practise their subject through questionnaires. It is well nigh impossible to exclude personal characteristics from the social relationship formed in this context. The researcher carrying out a questionnaire is not a white-coated automaton who can treat all respondents in exactly the same professional manner. The very differences of personality and experience which inspire variations in the geographical imagination in the first place will inevitably surface during the questionnaire. There are some respondents you just don't relate to, or they don't relate to you, and encounters with them are just *not* the same as those with people with whom you establish a much more mutually enjoyable and supportive relationship during the meeting, regardless of your 'professionalism' in recording answers and asking questions. Equally, the 'researcher as robot' model is also flawed in its own terms because the supposed objective neutrality of the researcher will be responded to differently by different people. Being scientifically neutral may appeal to some respondents, giving an impression of professionalism. To others, however, it may appear rude, uncaring and exploitative. Clearly, then, the idea that adopting a position of scientific neutrality to ensure that social situations of research are either asocial or totally determined by one of the participants is doomed to failure. 'Research as robot' is itself a social role.

Equally it is our view that questionnaire researchers should respect the respondent's rights to know about the survey and about what will happen to the results. This *will* sometimes involve researchers in telling stories about themselves and commenting on subject-matter relevant to the issues involved in the survey. When you have sat down in someone's house for an hour or two, accepted his or her refreshments and been extraordinarily nosy about his or her life and lifestyle, it is perhaps only fair to respond to his or her questions about who you are and what you are doing with something other than 'my training manual doesn't allow ...'. There is obviously a balance to be struck here. An on-street questionnaire may be conducted as the textbooks suggest, but it seems to us that in general there are relational and ethical considerations which make it extremely difficult to accede to the demands of doing a standardized questionnaire.

It may be that in coming to this conclusion we are attempting to move the questionnaire survey away from its position as a technique of conventional scientific method. Perhaps if the flow of the research issue (to data construction, to data interpretation) is compelled by a positivistic programme of hypothesis-testing and statistical analysis, the need to eradicate interviewer bias will be viewed as crucial in order to set reliable standards throughout the survey. Our view, however, is that any such form of data construction involves a variable social relationship between the researcher and the respondent. While it is entirely valid to attempt to construct comparable data about a potentially large number of respondents, thereby accepting the strictures of a structured survey format, it is often also valid to listen to and record other relevant information which is offered by the respondent, even if the definition of 'relevant' in this case is dependent on the researcher as an individual rather than as a standardized research instrument.

Interviewing

The transition from questionnaires to interviews is by no means a straightforward one. Many face-to-face questionnaires will contain elements of interviewing, and some of our discussion of questionnaires will be relevant to the carrying out of interviews. However, we want to argue that there are important distinctions between questionnaires and interviews and that these distinctions arise as much from the intent and purpose of the interviewer as from the ways in which questions are asked. Interviewing has been described as 'conversations with a purpose' (Webb and Webb, 1932; see also Burgess, 1984) and, although the conversations vary across a range of structured, semi-structured and unstructured formats, their purpose is to 'give an authentic insight into people's experiences' (Silverman, 1993: 91). If the scientific methods so often associated with the use of questionnaires attempt (often unsuccessfully) to *mirror* the social and geographical worlds of respondents, interviews will employ knowingly interactive research so as to gain access to the *meanings* which subjects attribute to their experiences of these worlds. Human geographers throughout the 1980s and especially the 1990s have increasingly found a match between what interview research can offer and the overall questions and issues which spark their geographical imaginations. At the same time, however, there has been increasing concern over the *rigour* of this research. Baxter and Eyles (1997: 511), for example, argue the need

> to be explicit about the principles for making this work rigorous. Being forthcoming about these criteria will better equip those who do not traditionally work within the qualitative paradigm to judge its approach and findings and, perhaps more importantly, these criteria will be made public for constructive scrutiny and debate.

The criteria to which they refer are framed very much in the context of an accusative culture in geography in which battle-lines are readily drawn up between the qualitative and the quantitative:

> questioning how things are done – an essential component of self-reflection – allows qualitative research to demonstrate the relevance of the single case (credibility) and to move beyond it (transferability) with a degree of certainty (dependability or confirmability). Context, contingency and the specific positioning of subjects (including researcher-as-instrument) are central to qualitative inquiry and are not threatened

> by the application of a general set of criteria for evaluating rigour. These criteria provide reasonable anchor points for a paradigm which is often inappropriately accused of engaging with 'anything goes' science. (1997: 521; see also Bailey et al., 1999)

It is evident from this that the seemingly simple goal of gaining access to meanings and experiences via the use of interviews is by no means an unproblematic task. It is possible to approach interviewing in a rather blasé fashion, somehow assuming that all you have to do is to ask the right questions and the reality of your respondent's feelings, thoughts and experiences will fall into your notebook or tape-recorder like picking ripe apples from a tree. Even with the now extensive literature on how to conduct interviews without contaminating them with 'bias' (see, for example, Fowler and Mangione, 1990) researchers can seem to approach interviews as methods of detecting existing truths from willing respondents. As Holstein and Gubrium (1997) point out, the supposed 'trick' is to formulate good questions, create an atmosphere conducive to rapport and produce undistorted communication between the interviewer and the respondent. They term this the 'vessel-of-answers' approach:

> In the vessel-of-answers approach, the image of the subject is epistemologically passive, not engaged in the production of knowledge. If the interviewing process goes 'by the book' and is non-directional and unbiased, respondents will validly give out what subjects are presumed to merely retain within them – the unadulterated facts and details of experience. Contamination emanates from the interview setting, its participants and their interaction, not the subject, who, under ideal conditions, serves up authentic reports when beckoned to do so. (1997: 117)

We believe that interviewing in this conceptual context usually turns out to lack a critical sensitivity to the interactions and exchanges between the interviewer and the interviewee. The interview thus becomes *less* rather than more rigorous because it assumes that a constant set of 'truths' is available to be garnered from the interviewee regardless of the social and cultural conditions in which the interview is constructed and carried out. Instead, we argue strongly that interviewers are themselves implicated in the construction of meanings with their interviewees. Such intersubjectivity is crucial and unavoidable, and the data which result are essentially collaborative (see Alaasutari, 1995).

This explicit recognition of the intersubjectivity of interviewing means that the 'respondent' is acknowledged as an active subject. He or she mediates and negotiates what is told to the interviewer and he or she engages in a continuous process of assembling and modifying answers to questions. The supposed vessel of objective knowledge is actually an active maker of meaning. Answers on one occasion may well not be replicated on another because the circumstances of producing those answers will differ, and because conveying situated experiential realities is by no means the same process as supposedly testing answers against an immutable bank of 'true' answers within the respondent. These precepts throw a rather different light on the key components of interviewing:

> understanding *how* the meaning-making process unfolds in the interview is as critical as apprehending *what* is substantively asked and conveyed. The *hows* of interviewing, of course, refer to the interactional, narrative procedures of knowledge production, not merely to interview techniques. The *whats* pertain to issues guiding the interview, the content of questions, and the substantive information communicated by the respondent. A dual interest in the *hows* and *whats* of meaning production goes hand in hand with an appreciation of the constitutive awareness of the interview process. (Holstein and Gubrium, 1997: 114, emphasis in original)

Interviews, then, are rather like an 'interpersonal drama with a developing plot' (Pool,

1957: 193). The interview has a script, at least in outline, roles to be played out and a stage on which conversation occurs. However, it has a plot which develops rather than being static. The waxing and waning of the interview will fashion script, roles and stage, sometimes unpredictably. There is plenty of room for improvisation, both by the interviewer and by the subject; each is complicit in the production of a narrative. The result is far from being an objective, representative and technical exercise. Indeed, the strengths of using interviews lie in the very acknowledgement of intersubjectivity which permits a deeper understanding of the whos, hows, wheres and whats of many aspects of human geography research.

When are interviews appropriate?

Practising human geography is often a matter of choosing horses for courses, and the use of interviews is no exception to this. It will be clear from the above that broad surveys of surface patterns involving large numbers of people will be more compatible with questionnaire techniques than with in-depth interviewing. However, human geographers have become increasingly interested in issues which fall between or below analyses of surface-level comparability. Instead there is an emphasis on explaining processes, changing conditions, organization, circumstances and the construction, negotiation and reconstruction of meanings and identities. Here there will be a turn to intensive rather than extensive research strategies (see Table 5.1), and the use of interviews is both legitimate and necessary in order to create the kind of information which lends itself to the types of explanation required (see Chapter 9).

As an example, in a recent study of homelessness in rural areas it was found to be entirely appropriate to undertake a postal questionnaire survey to homelessness officers in local authorities in order to garner a wide spread of comparative information on the scale and perceived significance of homelessness in rural areas of the UK. However, when seeking to understand something of what homelessness is like in rural areas, the informal interview presented a more appropriate way of approaching and talking to homeless people about the feelings and meanings they attach to their experiences of being homeless (Cloke et al., 1999; 2000a; 2000b).

A second context for using interviews is when we as researchers wish to consider ourselves as active and reflexive in the process of constructing information. To some extent it is inevitable that the researcher *is* an active influence throughout the research process, whether or not he or she is seeking to be neutral or objective. However, interviews offer a more deliberate route than this suggests for those who believe that 'people's knowledge, views, understandings, interpretations, experiences and interactions are meaningful properties of the social reality' (Mason, 1996: 39) and who wish to practise these beliefs by interacting with people by talking to them, listening to them and negotiating with them over narrative accounts. It used to be the case that the word 'I' was persona (1 pronoun) non gratis in human geographical writing. However, the recognition of the active subject and the reflexive self in human geography has rendered entirely legitimate an approach which makes explicit the intersubjectivities inherent in interview practices.

A third reason for using interviews lies in the desire of human geographers to 'give voice to' others as an integral part of the research process. Questionnaires by their very nature will tend to aggregate individuals into the categories which constitute extensive patterns. Not only are human geographers now interested in the inclusion of lay discourses in academic discourses (see, for example, Halfacree, 1993) but many are also determined to look through different 'windows' on to the geographical world by undertaking research in which a range of different voices can be heard (see, for example, Philo, 1992). Such research might be aimed at understanding the position(s) adopted by one or more social groups on a particular place, practice or process, or the focus might be on particular individuals in the form of oral histories, life histories or life stories. Interviews offer a methodological strategy by which these various aims can be realized although, once again, care should be exercised in too blasé an assumption of what interviews can offer. Indeed, it is important to reiterate that the 'voices' that are heard are very much subject to the editorial and authorial concerns of the researcher, as well as to the interpersonal drama of the interview itself.

A final set of reasons for choosing to undertake interviews reflects rather more pragmatic circumstances. In some contexts (for example, an undergraduate dissertation) the time and resources available often do not permit the undertaking of an adequately representative questionnaire survey. Here, the pragmatic view could dictate that a more manageable programme of interviews be undertaken, and research aims might have to be altered accordingly. From a position in which the questionnaire was the 'orthodox' research tool, many human geographers have come to realize that in circumstances of limited resource or limited access to certain types of respondent it is more appropriate to undertake a series of interviews rather than a questionnaire survey. Even so, we would reiterate that interviews are neither a soft option – they need to be thoroughly prepared, carefully undertaken and then painstakingly transcribed and interpreted – nor are they a direct replacement of other research tools such as the questionnaire.

Preparing for an Interview

Having established that interviews can be regarded as conversations with a purpose, which represent interpersonal drama with a developing plot, we now turn to more practical issues of how to prepare for and carry out an interview. As a very practical starting point to our discussion of 'doing' an interview we want to emphasize the need for considerable forethought and care about how the interview is to be recorded. If the interviewee is willing for a tape-recorder to be used, taping is by far the best way of recording the proceedings. Researchers need to be aware, however, that the presence of a tape-recorder can sometimes present a psychological barrier to the interviewee, whose responses might be more guarded and 'formal' than if the tape was not running.

It is also important to be well prepared for tape-recording. Prior familiarization with the machine and regular battery checks are invaluable in avoiding the interviewer's nightmare of finding out, after the event, that the interview has not yielded any text to quote from due to the tape-recorder not working properly! Interviewers should also be aware that the decision to tape an interview commits them to long hours of transcription afterwards! Note-taking, on the other hand, will reduce the opportunity to use verbatim quotes from the interview and can distract the interviewer's concentration from the task of listening to the interviewee, thus reducing some of the potential intersubjectivities of the exchange. If you are thinking of note-taking, various forms of shorthand can be deployed to make the task more simple and speedy. These technical details may appear trivial but, in our experience, they matter because they fundamentally determine the character of the data constructed, and shape the purposes to which they can be put.

Aside from these technicalities it is also important from the outset to acknowledge a constructive and critical tension here between methodological rigour and dramaturgical spontaneity. In our collective experience, the unexpected 'chat' with key individuals has often proved to be a most fruitful research moment. Equally, a well-prepared interview can launch off into unanticipated territory with beneficial consequences for the understanding of particular meanings and experiences. In contrast to many textbooks on geographical methods, then, we would suggest that preparation, though important, is not the be-all and end-all of interviewing. Indeed, one of the key skills of interviewing is a sensitivity of listening to what is being said, linked with an innate flexibility to permit and encourage encounters with the unexpected.

For these reasons, the idea of the 'textbook interview' can be misleading and, at worst, can render interviewing immune to the very intersubjectivities which lie at the heart of the process. Nevertheless, methodological rigour does have a significant part to play in the creative and critical tensions of interviewing, and part of that rigour involves thorough preparation where applicable. Box 5.5 outlines Jennifer Mason's excellent guide to planning and preparing for qualitative interviewing. In the context of her research on how families deal with issues of inheritance, she works through the process with very helpful detail and contextualized expectation. From the identification of the *big* research question, a number of smaller *mini questions* are developed, each of which can be thought through in terms of 'what you need to know' and 'what'. Cross-referencing of these smaller research issues helps to keep the big research question in mind. The next 'stage' is to think through how the interview might go – at least how the researcher might want it to go – and thereby to develop a loose sense of format or structure for the interview. This, of course, has to be flexible but it often leads to a series of 'topic' or 'issue' prompts that can be used in

Box 5.5 Planning and preparing for qualitative interviewing

Step 1
List or assemble the 'big' research questions which the study is designed to explore.

Example of one of the 'big' research questions in the 'Inheritance' project
1 How do families handle issues of inheritance?

Step 2
Break down or subdivide the big research questions into 'mini' research questions. The links between the big questions and the subcategories of them – the mini questions – should be clearly expressed, for example by using corresponding numbers or codes, or by laying the two sets of questions out in a chart, or by using cross-referenced index cards. It is possible to establish a perfectly workable manual system, or you can use a computer graphics package and/or database to help you.

Example of mini research questions which are subcategories of the big research question given above

1 (a) *Are negotiations about inheritance treated as part of a wider set of negotiations about support in families? Or is inheritance treated as a totally separate matter?*
    (b) *Do people in any way take into account the possibility of inheritance in for-mulating their own life plans?*
    (c) *Is a clear distinction maintained between 'blood relatives' and 'in-laws' in the process of negotiating inheritance?*

Step 3
For each mini research question, start to develop ideas about how it might be possible to get at the relevant issues in an interview situation. This means converting your big and mini examples of 'what you really want to know' into possible interview topics, and think-ing of some possible questions – in terms of their substance, and the style you might use to ask them. These will not form a rigid 'script' for you to use in the interview, but the process of developing possible topics and questions will get you thinking in ways appropriate to an interview interaction. Again, make sure that the links between this set of questions and the other two (that is the big and mini research questions) are clearly expressed.

Examples of interview topics and questions related to mini research questions

1 (a) *Family inheritance history, and history of other family support – what hap-pened in practice in relation to specific events and instances? How did people decide what was the most appropriate course of action?*
    (b) *Knowledge of the inheritance plans, content of wills etc., of other family members. Have people thought about inheritance at all? Have they made wills? Do people have life plans, for example, do people have a sense of what they will be doing, where living, and so on, in 5 or 10 or 20 years' time? How were these plans arrived at?*
    (c) *Ascertain composition of family and kin group, and what kinds of relationships exist with specific others. Explore whom people count as 'blood kin', whom as 'in-laws' or 'step-relatives' – establish this so that family inheritance history, and specific events and instances, can be contextualized in the sense that we will know the 'kin status' (as conceptualized by the interviewee) of relevant parties. Explore the detail of dis-tributions of assets, and negotiations about them, in relation to kin of different status. Who has legitimate interests? How do people decide whom to include and exclude? Possibly ask directly whether people think about their blood relatives and their in-laws in different ways in relation to inheritance, and other matters.*

Box 5.5   *(Continued)*

Step 4

Cross-reference all the levels, if you have not done so already, so that you know that each big research question has a set of corresponding mini research questions, and each of these has a set of ideas about interview topics and questions. Make sure the cross-referencing works in reverse, so that your interview topics and questions really are going to help you to answer your big research questions.

Step 5

Start to develop some ideas about a loose structure, or format, for interviews. You will want this to be highly flexible and variable, but you should be able to produce some kind of guide to the key issues and types of questions you will want to discuss.

Examples of loose interview structure/format developed for the 'Inheritance' project

In this project we developed a loose interview format, based on key topics and types of questions we were likely to want to ask. With each interviewee we anticipated following up lines of enquiry specific to their circumstances, which we would not be able to anticipate in advance. We therefore wanted maximum flexibility, but also some kind of guide or prompt for the interviewer about the key issues and questions with which the study was concerned. We did not produce a script of questions, but rather a set of index cards to take into each interview. One card contained a flow chart of a possible interview structure, which could be readily modified on the spot. The other cards contained shorthand notes about specific topics and issues for the interviewer's use at relevant points in the interview. These notes were non-sequential, so that they could be drawn upon at any time, in relation to the specific context of the interview in progress. Here are examples of each type of card:

'Loose structure/format of interview' card

Possible main structure	**Specific topics and issues – to be asked in relation to any of the main structure sections (there are cards for each of these sets of questions)**
Introductory explanation ↓	
Brief social/personal characteristics ↓	
Composition of kin group and spouse's kin group ↓	Inheritance history, other responsibilities and relationships, inheritance family and kin group
Family inheritance history ↓	
Specific questions (if not covered elsewhere) ↓	Formal and external factors, including the law
Questions about the law check ↓	Principles and processes of inheritance and family responsibility
Personal characteristics check	Social and personal characteristics (current and over time)

Box 5.5   (Continued)

'Example of specific topics and issues' card
Inheritance history, other responsibilities and relationships, inheritance family and kin group

Experience of inheritance: personal/others – as testator, beneficiary, executor; patterns characteristic of own family; how many generations; experience of legal procedures and services; expected and unexpected; experience of will making; when, why; professional advice; intestacy laws; lifetime transfers.

Inheritance and other aspects of kin relationships/wider patterns of responsibility: family relationships affected by inheritance? conscious of possible inheritance in relationships with relatives?; conflicts – how resolved; life plans and inheritance, e.g. housing, geography, timing; death and how it is dealt with; making formal statements about relationships?; part of ongoing reciprocity and exchange – explicit/implicit?; idea of final settlement?

The inheritance family or kin group: who has legitimate interests?; in-laws/ step-relatives/secondary kin?; excluions and principles of exclusion/inclusion; inheriting via someone else.

Step 6
Work out whether you want to include *any* standardized questions or sections in your interviews. There may be certain questions which you want to ensure that you ask everybody. In the example above the introductory explanation was fairly standardized, as were some of the questions about personal and social characteristics (for example, age, marital status). You might also want to think of some standardized comments and assurances which you will make about confidentiality of data to your interviewees.

Step 7
Cross-check that your format, and any standardized questions or sections, do cover adequately and appropriately your possible topics and questions.

Source: Mason (1996: 44–51)

the interview as a checklist to ensure coverage of the required ground. Such a checklist will often involve important standardized statements at the beginning and at the end of the interview to explain to the subject what will be done with the information which is constructed therein. Issues of confidentiality and further communication will be important here.

Helpful though these guidelines for preparation are, they form only part of the picture. Intersubjective conversations with a purpose involve an active collusion between participants.

A mechanical trudging through of an interview checklist or schedule by the researcher will restrict the possibilities for interpersonal drama, and therefore of plot development. We deal with issues relating to *conducting* interviews below. However, the participation of the research subject will also vary enormously. Part of the preparation for interviews, then, lies in making decisions about whom to interview. Much of the traditional methodological training for human geographers points them towards the idea of dealing with 'representative samples' in their research. Although principally associated with the construction of data which can be analysed statistically, the sense of somehow wanting to talk to people who represent a wider social group has also spread across into qualitative interviewing. Indeed, where interviews are designed to document a series of different properties or discursive principles, the researcher will want to choose to interview across that range of difference, and will therefore use a loose form of 'representative' samples of interviewees. However, such choices are not representative in the statistical sense, and the idea of spreading interviews across axes of difference – be they geographical, social, identity related, practice related or whatever – soon runs into problems if individuals are somehow being asked to represent their 'category'. For example, when interviewing an elderly, middle-class rural woman who is active in local politics or a young gay urban man working in financial services, the multiple and shifting identities involved will usually defy easily categorized explanations relating to, say, gender, sexuality or place.

Choosing whom to interview, then, involves a targeting of people who are likely to have the desired knowledge, experiences or positionings, and who may be willing to divulge that knowledge to the interviewer. As Susan Smith (1988: 22) has argued: 'any attempt on the part of an analyst [geographer] to enter the life-world of others is above all, strategic … it makes both moral and analytical sense to expose the power relations inherent … at an early stage of the research.' Sometimes researchers will have difficulty in gaining access to an identity group or organization. Here we often encounter 'gatekeepers' who either are in an official position to handle inquiries about an organization or who are powerful figures within a particular group (Hughes and Cormode, 1998). Gatekeepers will often attempt to structure the access of the researcher to others, pointing us towards 'helpful' or 'safe' interviewees who in the judgement of the gatekeeper are appropriate onward contacts. Naturally, there are pitfalls inherent in these decisions such that the researcher

> must retain the leeway to choose candidates for interview. Otherwise there is a grave danger that the data collected will be misleading in important respects, and the researcher will be unable to engage in the strategic search for data that is essential to a reflexive approach. However, gaining access to informants can be quite complex, sometimes as difficult as negotiating access to a setting. Indeed, it may even be necessary to negotiate with gatekeepers before one can contact the people one wants to interview. (Hammersley and Atkinson, 1995: 134)

Sometimes, however, researchers have to work with the interviews they can get, and it is crucial that the context in which knowledge is being produced is fully acknowledged in these circumstances.

'Working with what you can get' also applies to groups of subjects who will not be amenable to direct approaches or who are difficult to get hold of. Here 'snowballing' can be used whereby a small initial group of informants (this might be one person) can be asked to provide interview contacts with friends and acquaintances within a particular social or identity group, thus permitting a chain of interviewees to emerge. Similar issues to those relating to the role of gatekeepers arise here, although they are often less obvious in snowball samples. Where the emphasis is to use interviews to hear different stories, the propensity of people to pass the researcher on to people whom they consider to be like-minded needs to be acknowledged in the intertextualities which underpin interviews.

Finally, in terms of how to prepare for interviews, it is worth noting that we should be prepared for different levels and types of response from research subjects. This may be illustrated in terms of Erving Goffman's ideas about the 'two-sidedness' of people performing occupational roles: 'One side is linked to internal matters held dearly and closely, such as image of self and felt identity; the other side concerns official position, jural relations, and style of life, and is part of a publicly accessible institutional complex' (1968: 119).

Goffman's work suggests that there will be a 'front' and a 'back' to the responses given by interviewees. In the case of interviews with representatives of agencies, the front responses may be the readily available public relations narratives designed for public consumption. The 'back' responses will be more hidden and yet may be the areas of most interest to the researcher in that they may reflect opposition to, or negotiation with, the public face of the front responses. It can also be argued, however, that a much wider range of interviews will involve these public/private distinctions, which can be anticipated in interview preparation. In the interpersonal drama of interviewing, however, such distinctions emerge in a much more messy and complex form and may well defy the notion of clearly defined back or front regions of knowledge and response. Indeed, the active subject will be creatively making meanings during the interview itself, and so the idea of fixed reference points of back and front knowledge should represent instead rather less fixed markers in the critical tension in interviews between rigorous preparation and dramaturgical spontaneity.

Conducting an interview

By using the metaphor of drama to emphasize both interpersonal performance and the development of an plot within interviews, we lay ourselves open to an account of conducting an interview in which the script is tightly adhered to, and the interviewer plays all the powerful roles. For example, Berg (1989: 35) suggests that interviewers must play the role of actor, director and choreographer (see Box 5.6) during the 'performance' of an interview:

> interviewers must be conscious and reflective. They must carefully watch and interpret the performance of the subject. Their interpretations must be based on the various cues, clues, and encoded messages offered by the interviewee. Included in the information these interactions supply may be the communication of a variety of moods, sentiments, role portrayals, and stylised routines, which represent the interviewee's script, line cues, blocking, and stage directions. From this information, interviewers must take their cues, adjust their own blocking, and effect new or responsive stage directions.

This extended use of dramaturgical metaphor perhaps overemphasizes the interview process in which improvisation is at least as important as delivering 'classic' lines. However, it does point us towards a discussion of the degree to which an interviewer can influence the intersubjectivity of the interview by providing a situated context, by employing interviewing skills and by asking the 'right' kinds of questions.

Sometimes the researcher will not be in control of where an interview takes place. The subject may choose where to be interviewed – in a place of paid employment, in the home or perhaps in some neutral location – and gatekeeping individuals or agencies may well dictate the interview location. For example, many interviews in schools take place in broom cupboards marked 'office'. Equally, the researcher will not always be able to choose who else is present during the interview, and inevitably the presence of parent, child or co-worker will alter the interpersonal dynamics of the interview. As an extreme example, it is not unknown for the friendly attention of the pet dog or cat to induce such a strong allergic reaction that an interviewer cannot complete the interview concerned (see Popkin, 1998)!

The 'textbook' interview calls for the interviewer to establish a 'rapport' with the subject. The achievement or otherwise of rapport is

Box 5.6 Roles played during interviews

Interviewers as Actors. Interviewers as actors must perform their lines, routines, and movements appropriately. This means that in addition to reciting scripted lines (the interview questions), they must be aware of what the other actor (the interviewee) is doing throughout the interview. Interviewers must listen carefully to line cues in order to avoid 'stepping on the lines' of the interviewee (interrupting before the subject has completely answered a question).

Interviewers as Directors. Simultaneous to their performance as actors, interviewers must also serve as directors. In this capacity, interviewers must be conscious of how they perform lines and move as well as of the interviewee's performance. Interviewers must reflect on each segment of the interview as if they were outside the performance as observers. From this vantage point, interviewers must assess the adequacy of their performance (for example, whether line cues from the interviewee are responded to correctly and whether avoidance messages are being appropriately handled).

Interviewers as Choreographers. The various assessments made in the role of director involve a process similar to what Reik ... described as 'listening with the third ear'. By using what they have *heard* (in the broadest sense of this term) in a self-aware and reflective manner, interviewers manage to control the interview process. As a result, interviewers, as choreographers, can effectively block their movements and gestures and script their response lines.

Source: Berg (1989: 35)

certainly influenced by whether the location of the interview appears to reinforce the power either of the interviewer or the subject. Rapport is also thought to be influenced by the appearance and demeanour of the interviewer (see, for example, Gorden, 1987). Characteristics of age, gender, race and the like cannot be 'fine tuned' unless a team of different interviewers is involved. It may sometimes be crucial, however, to undertake interviews in a native language in bilingual or multilingual areas. Other characteristics such as style of dress, manner of speech and more general demeanour *can* be adjusted to suit a particular occasion. For example, casual clothes may be essential when interviewing young people on sensitive issues so that the interviewer does not come across as an authority figure. Conversely, the interviewer will perhaps wish to come across as an authority figure when talking to

subjects who have that kind of expectation about what their interviewer should look like (Walford, 1994). Thus a business suit is often worn when interviewing high-ranking managers in industry or commerce.

These attempts to 'fit in' are valid reflections of the wish to establish rapport with the research subject, but they by no means guarantee that barriers to in-depth intersubjectivity will be removed. Many active subjects will 'see through' obvious attempts to establish rapport and will rely much more on the sensitivity displayed by the interviewer during the interview. This in turn begs the question of the degree to which the acquisition and practice of interviewing skills will enable the interviewer to achieve the desired depth and scope of conversation. On the surface, such skills are both demanding and formidable. As Mason (1996: 45–6) points out:

At any one time you may be: listening to what the interviewee(s) is or are currently saying and trying to interpret what they mean; trying to work out whether what they are saying has any bearing on 'what you really want to know'; trying to think in new and creative ways about 'what you really want to know'; trying to pick up on any changes in your interviewees' demeanour and interpret these ...; formulating the next question which might involve shifting the interview onto new terrain; keeping an eye on your watch and making decisions about depth and breadth given your time limits. At the same time you will be observing what is going on around the interview; you may be making notes or, if you are audio or video tape recording the interview, keeping half an eye on your equipment to ensure that it is working; and you may be dealing with 'distractions' like a wasp which you think is about to sting you, a pet dog which is scratching itself loudly directly in front of your tape recorder microphone, a telephone which keeps ringing; a child crying, and so on.

Key skills for interviewing will therefore include: listening sensitively; remembering what has already been said; achieving an effective balance between listening and speaking out (with questions, prompts and responses to questions from the interview subject); sensitivity to unspoken signals, particularly in the area of body language and demeanour; and technical response in tape-recording or note-taking. For some, such skills are artistic and intrinsic; for others they can be taught and learnt mechanistically (Roth, 1966). The presence or absence of these skills will alter the intertextual construction of narrative in an interview.

Another facet of the 'textbook' interview is to ask the right kinds of questions. For example, questions using affective words (that is, words which carry some kind of emotional charge or which are likely to elicit an emotional response) are supposedly to be avoided. For example, according to Berg (1989: 24), 'the word *why* in American culture tends to produce in most people a negative response' and should therefore be avoided. Equally, overly complex or misleading questions are also to be shunned. While all this in theory sounds like good advice, there is a danger here of interviewers being so trained up in demeanour, skills and language that their interpersonal contribution to the interview is as a walking textbook rather than as a sensitive and honest human being. In our experience, interviewers who present themselves as something other than themselves will be relatively transparent to many research subjects. Indeed, it could be argued that the various attempts to characterize the 'textbook' interview represent a conceptual clinging to the notion of interviewing as extracting grains of truth from the vessel-of-answers that is the interviewee. Gaining access to the meanings which subjects attribute to their experiences of their social and geographical worlds may not be a technical issue at all. Critical awareness in interviews will come from acknowledgement of intersubjectivity, which in turn depends on the interviewer giving something of her or his self in the interview. The creative tension between methodological rigour and dramaturgical spontaneity will, therefore, be as much to do with intellectual and personal honesty as to the demands of textbook interviews.

Discussion groups

Introduction

One form of questioning and interviewing that deserves special mention is the group discussion, which has become increasingly popular as a tool of qualitative inquiry in human geography over recent years. Although the term 'focus group' tends to be used generically, Burgess (1996) makes the very important distinction between *focus groups*, which refer to single-group interviews in the tradition and culture of market research, and *in-depth discussion groups*, referring to interactive group interviews in the tradition and culture of psychotherapy. Focus groups have been widely used in contemporary society, particularly in areas of market research, ranging from the testing of particular food products to the preliminary evaluation of policy initiatives from the New Labour government in Britain. An excellent example of the use of focus groups in human geography is the research by Peter Jackson and colleagues designed to investigate the cultural significance of the new generation of men's lifestyle magazines (such as *Arena, GC, Loaded, FHM and Maxim*). Some 20 one-off focus groups were used to prompt interactive discussion in this research (Jackson et al., 1999). For an example of best practice in the use of in-depth discussion groups in human geography we would point to the work of Jacqueline Burgess and colleagues who have made excellent use of in-depth discussion groups to explore the contested meanings associated with nature conservation and environmental values (see Burgess et al., 1988; 1988b; 1991).

To avoid confusion over terminology we prefer to use the generic term 'discussion groups' and then to differentiate between focus groups and in-depth discussion groups as appropriate. Although there is a range of specialist literature which underpins the methodologies deployed in discussion groups,[4] we suspect that for some human geographers the apparent ease of 'doing a discussion group' leads to instances of dangerous misuse. This risk should be reviewed alongside the very positive potential advantages from this technique when deciding whether discussion groups are an appropriate method of inquiry in a specific research context.

In simple terms, the generic idea of discussion groups may be defined as 'a qualitative method involving a group discussion, usually with six to twelve participants, focused around questions raised by a moderator' (Pratt, 2000a: 272). However, as Goss (1996: 113) points out, such groups are more specifically

'a confusion of the focused interview, in which an interviewer keeps a respondent on a topic without the use of a structured questionnaire, and a group discussion, in which a relatively homogeneous, but carefully selected group of people discuss a series of particular questions raised by a moderator'.

It is important therefore to position discussion groups critically among the array of other methods involved. As an *interview* technique, group discussions allow researchers to draw out interaction between participants and make direct comparisons between the experiences and opinions narrated in the group. However, the ability to go into depth with particular individuals is commensurately less than in individual interviews. As a *group observation* technique, discussion groups present the opportunity for researchers to observe very significant interactions between individuals in confined space and time. Indeed, the prompting of interaction by the moderator feeds back into the interviewing process as participants begin to narrate material as a response or addition to what they have heard from other members of the group. However, for some, the discussion group may represent an unnatural setting in which they are less likely to enter into sensitive or difficult discussion.

Just as qualitative methods literatures have tended to specify a 'model interview', so too have 'the common sense and the preferred practices of a few researchers ... been reified into rules of thumb, or myths, ... that specify the ideal form of the group discussion' (Goss, 1996: 113). In an informative account of the techniques of discussion group work, Morgan (1997) identifies particular phases of research and poses key questions about each so as to provide appropriate guidelines for undertaking discussion groups. Two phases are important here, as follows.

PLANNING GROUPS

This involves consideration of how big the groups should be, how many are required, who will participate in the groups and how structured the discussions will be and what level of involvement is envisaged from the moderator. Morgan's rule-of-thumb answers to these questions suggest that groups projects 'most often (a) use homogeneous strangers as participants, (b) rely on a relatively structured interview with high moderator involvement, (c) have 6 to 10 participants per group, and (d) have a total of three to five groups per project' (1997: 34).

He insists, however, that such norms are a point of departure rather than some kind of industry standard. Interestingly, geographers have used a wide variety of methods to recruit and conduct discussion groups (see Holbrook and Jackson, 1996). The Jackson et al. (1999) study on reading men's magazines, for example, deliberately chose to use pre-established groups rather than groups of strangers and completed 20 such groups, each lasting for around an hour.

CONDUCTING GROUPS

This involves questions about the site of the interview, the content of the interview and the role of the group moderator. The site can be in the home or in a 'neutral' venue such as a room in a public building. It should permit the participants to be comfortable but also allow the interviewer to record and moderate proceedings. In terms of content, Merton et al. (1990) have suggested four basic criteria for an effective discussion group: 'It should cover a maximum range of relevant topics, provide data that are as specific as possible, foster interaction that explores the participants' feelings in some depth, and take into account the personal

context that participants use in generating their responses to the topic' (cited in Morgan, 1997: 45).

The role of moderation should include an honest introduction to the topic, a series of ground rules for the discussion, an ice-breaking discussion starter, the use of prompts to continue the discussion and an effective conclusion which debriefs participants and informs them about any further involvement with the transcriptions and resultant narratives that will be produced. In each of these aspects there is considerable room for experimentation and adaptation to particular circumstances which contextualize the research.

Discussion group strategy

It will be clear that in the above discussion we have skated over a number of important conceptual as well as technical issues which need to be resolved critically and carefully in the deployment of discussion groups. However, our main concern here is to suggest, first, that focus group methodologies permit considerable flexibility in decisions about recruitment, group composition and discussion format and, secondly, that many of these detailed decisions tend to fall into place once the *strategy* of the focus group is clear. This is where Burgess's (1996) distinction between focus groups and in-depth discussion groups is of clear significance, as this distinction offers human geographers a strategic choice in the use of discussion group techniques in their research – a choice which may be illustrated from Jacqueline Burgess's own expert studies. She argues that *focus groups* are most advantageous in circumstances where available fieldwork time is short or when research conclusions have to be delivered quickly, such as when urgent policy recommendations are required. For example, she was commissioned in 1993 by the Community Forest Unit of the Countryside Commission to undertake a speedy one-year research study of the social and cultural dimensions of fear in recreational woodlands in the urban fringe (Burgess, 1996). With only five months available for the empirical parts of the research, she combined participant observation of small groups of people who were taken on a guided walk through woodland with a follow-up 90-minute focus group in which their direct experiences and the deeper fears which may have underpinned that experience were discussed. Given her aim to discover how risk and fear associated with woods are socially constructed, she chose single-gender white and black focus groups, recruited though existing contacts and subsequent snowballing. Her experience was that the focus group discussions enabled people to share their views and feelings about the issues, and that the role of moderation was to ensure that the following mental checklist of issues was addressed in the discussion:

1. The pleasures of being in woodlands.
2. Perceptions of risk.
3. Specific features of the wood and how design and management might alleviate fear.
4. Crime and safety issues in woods compared with built environmental settings.
5. Tensions between 'wild' and 'tame' woodlands.

Burgess's report on this methodology stressed that

> These five topics were discussed by all the groups, although the depth with which each was discussed and the sequence in which it was discussed depended on the salience of the topic for the group. Having done the walk together, the groups were much more comfortable with one another than would otherwise have been the case, while the shared experience provided a common focus for discussion in early stages of the group. (1996: 133)

The results of her research illustrated clearly that not only do feelings of fear and anxiety vary between groups of different gender, age and cultural background but also people's fears are socially constructed rather than prompted by the nature of woodlands per se.

In general, the focus group setting is better suited to investigating publicly available discourses than to understanding deep-seated personal and group values. Jackson (2001: 207) emphasizes the need to make sense of the world through 'systems of intelligibility' that are shared by individuals and groups who are part of the same 'interpretative community'. He suggests that focus groups methodology allows researchers to highlight these *discursive repertoires* which characterize the ways in which different groups attempt to make sense of particular texts or issues. Focus groups equally permit analysis of how individuals and groups adopt particular *discursive dispositions* to the identified repertoires.

In-depth discussion groups, on the other hand, are much more suited to research where an intimate and sustained grouping of people allows participants to share their feelings with each other and to create a strong sense of mutual commitment. The group forms its own culture and decides what is to be included or omitted from its discourse. Burgess et al. (1988) argue that this philosophy is particularly appropriate in the study of environmental values and meanings. In a project designated to explore local experiences and local provision of open space in the Greenwich area of London, Burgess and her colleagues recruited mixed-gender groups from three contrasting localities within the area, and an additional group of Asian women with which to explore the needs of minority populations. Groups met in the local community centre for six consecutive weeks, for a 90-minute session. The research team provided both a 'conductor' (moderator) and an observer.

The emphasis in group discussions was to analyse the broad themes which emerged from group analytic processes. Thus Burgess et al. (1988b) identified three important processes:

1. The developmental phases of the group matrix where people first establish and then deepen their relationships by means of increased trust and intimacy.
2. The manifest and latent content of communications, especially where latent meanings reflect individual and collective unconscious feelings.
3. The role of the conductor, with clear boundaries between the work orientation and a therapeutic orientation for the group discussions.

It was from these processes that more systematic themes relating to open space emerged. Three are emphasized in the importance of sensory experience, the rate and scale of change in the urban environment, and differences in values for, and uses of, green spaces in the contrasting urban and rural areas. Thus application of the principles of group analytic psychotherapy allowed researchers to address a human geography problem. A strategy of in-depth discussion, therefore, resulted in very different research findings from those that a focus group would have. As Burgess et al. (1988b: 475) reflect:

> By means of sharing in the history of the group matrix, one witnesses a break down of rational intellectualisation to the discussion of emotions and feelings. Over time, too, as people become closer to one another and feel safe within the group, they are able to talk about their negative feelings for environments and landscapes, as well as their preferences ... In a once-only group, we could have gained an understanding of the consensual values of a sensitive middle-class group which, above all else, valued solitude in open spaces. Only by means of the life of the group were we all able to realise the extent to which 'being alone with nature' is a very small part of highly complex, holistic value systems in which open space is charged with multiple individual and collective meanings.

This strategic distinction between focus groups and in-depth discussion groups will not answer specific questions about what happens if only two people turn up, or how directive we should be when people in the group talk freely about their interests but not about ours. It does, however, provide a very important framework within which to pose key epistemological questions about the nature of investigative research in group settings. Conceptual awareness of what we are looking for – and here the distinction between generally available discursive repertoires and deep-seated personal/group values is crucial – is a necessary first step to decisions about what kinds of group discussion to deploy. Practical methodological queries can also be assessed against this more conceptual framework. A one-off focus group of two people may not be as instructive as a larger group for investigating discursive repertoires and dispositions. However, in-depth discussion with two people could reveal interesting deep-seated values. A focus group that is 'side-tracked' may be problematic whereas an in-depth discussion uses the interests of participants in the development of a group matrix from which a deeper understanding of systematic issues will emerge. Once again we suggest that intellectual honesty and criticality are as important as textbook blueprints in the methodological rigour of this form of work. Technical methodological details matter not so much because they lead to 'successful' research but because they determine the nature of the data being constructed and influence the purposes to which those data can be directed.

Ethics: an important end note

Over recent years it has become increasingly important to emphasize ethical considerations in the conduct of research (see, for example, Penslar, 1995; Smith, D., 1995; Hay, 1998). This trend is to be applauded, not only because of a previous neglect of ethics, particularly when using seemingly objective and remote quantitative methodologies, but also because clear interconnections have needed to be made between ethical issues and the kinds of subjects to which qualitative inquiry has been directed. For example, it has been strongly argued by Mason (1996) that the use of qualitative methods to construct thick descriptions of the lives of individual people inevitably raises more ethical difficulties than the use of other, more quantitative, research strategies in maintaining the confidentiality and privacy of those with a personal involvement in the research.

Janet Finch (1993) also reminds us that these qualitative methods are being applied to very different kinds of people. Accordingly, she reminds us that it is now commonplace in research to distinguish between powerless social groups – where the need to protect the rights to privacy are extremely important – and powerful social groups who are well able to provide their own protection. In this way, an ethical approach to research will need to embrace notions of power and powerlessness, as Finch herself maintains in the context of research into gender: 'because issues of power are central to gender relations, one cannot treat moral questions about research on women as if they were sanitised "ethical issues" divorced from the context which makes them essentially political questions' (1993: 178).

We want to stress, therefore, that the use of interviews and focus groups as discussed in this chapter cannot be assumed to represent a collaboration of equals. While it is possible to imagine that interviewers can in some ways be overpowered by their subjects, it is absolutely essential that human geographers are aware of the power which they potentially hold over their subjects. Thus, some very important questions need to be raised about the ethical conduct of researchers when interviewing and conducting focus groups. For example, the questions asked will sometimes delve into personal issues which could be painful, traumatic or confidential for the subject. An appropriate response may be to avoid such questions or to be free to switch off the tape-recorder and terminate the interview if the subject is upset by the issues raised (see England, 1994). The way in which questions are asked may also be unethical, particularly if the questioning is aggressive, underhand or discomforting. Here there can be an effective imposition of power relations over the subject, not only through the nature of the questioning but also in a broader sense by the way in which a rigid agenda is set for the interview and strong control is established over the resulting 'data'. For example, there is strong evidence to suggest that everyday rules and habits of conversation between men and women operate such that the effects of social power will produce bias in favour of the views of men over those of women (see Fishman, 1990). Moreover, Ann Oakley's (1979) work with women about their experiences of transition into motherhood rejected 'textbook' interviewing, which pointed to a one-way process of gaining information from subjects but not answering their queries. She argued that *engagement* should become a key principle of feminist research (see also Oakley, 1984; 1990).

These social theories of power raise obvious moral and political questions about practising human geography. In terms of gendered power relations, a praxis of 'doing feminist research' (Roberts, 1990) has emerged. There are, however, other axes of social and cultural power which present equivalent challenges for researchers working with issues of, say, race, sexuality, children, disability, poverty and homelessness. Acknowledgement of intersubjectivity in interviews must include reference to the uneven distribution of social power in society as well as to the particular relations of dramatic collusion which occur in the interview itself.

Human geographers have sought to answer these questions about ethics in two significant ways. First, there have been attempts to generate genuine 'research alliances' between researcher and researched, and in particular between interviewer and interviewee. Pile (1991: 467), for example, has suggested that in the past 'geographers have acted as if they stand outside the specific historicity and geography of their subjects; this has enabled them to comment on the reality of the subject's view of their own situation, while not allowing the subject's valid versions of reality'. He draws on psychoanalytic notions of the 'therapeutic alliance' to suggest the possibility of fostering 'research alliances' between the researcher-interviewer and the researched-interviewee. Such alliances entail a process by which the balance of power of the emotional relationship in an interview is transferred from the interviewer to the interviewee, thus creating the interview as a safe space in which deep, mutual and trustworthy interactions and understandings emerge. This outcome permits the subject's version of reality to be aired and valued while the interviewer's project benefits from a more sensitive and critical grasp on the intersubjectivities involved.

It is not necessary to buy into psychoanalytical ideas to applaud the basic sentiments of the 'research alliance'. Indeed, Hester Parr (1998b) has warned against the implications of figuring the research interview through the psychoanalytic lens of doctor (therapist, counsellor)/patient (psychologically damaged

individual) relations. In part she does this through reporting on her experience of interviewing people with mental health problems, many of whom have been through supposedly therapeutic interviews. From this context she argues for a more flexible approach to interviewing which draws on the spirit but not necessarily the letter of good psychoanalytic practice.

The second response to ethical questions has been a discussion about how to construct a widely agreed set of ethical guidelines which can be used to establish appropriate 'research alliance' relations between researchers and the people they work with. For example, four such guidelines are recognized by Hammersley and Atkinson (1995):

1. *Informed consent*: The researcher should inform the researched about the research in a comprehensive and accurate way, and the researched should give their unconstrained consent.
2. *Privacy*: The researcher should resist making things public which are said and done for private consumption.
3. *Harm*: The researcher should avoid negative consequences both for the people studied and for others.
4. *Exploitation*: The researcher should avoid 'using' respondents to gain information while giving little or nothing in return.

We would add a fifth concern to this list:

5. *Sensitivity to cultural difference and gender*: The researcher should be sensitive to the rights, beliefs and cultural context of the researched, as well as to their position within patriarchal or colonial power relations.

The ethnographic turn in human geography has brought with it two particular strands of research which make particular demands on this kind of ethnographic strategy. The first is an acknowledgement of the need for *reflexivity*

Box 5.7 Challenges to 'textbook' ethical guidelines

Informed consent?

Gaining permission for an interview is not a cut and dried event at the start. Even with agreement to conduct an interview, it seems as though the interviewee doesn't really release their permission until they begin to trust you. Equally, their responses are generally more interesting when permissive trust has been established. I certainly encountered instances of Punch's *mutual deceit*, where I was on my best ethical behaviour, often asking questions which were peripheral to my main interest, *until* I was being trusted, and effectively given permission to be more personal and directive. There is no doubt that I was more prone to exaggeration, concealment and other trust-seeking devices during the early part of interviews. This is a really distressing, though not perhaps uncommon, conclusion to reach (see Punch, 1986).

Privacy?

I had many opportunities to take photographs of the homeless people I met, and I sensed that many would have given me permission so to do. Nevertheless, there was something in my moral/ethical map which directed me away from this practice. Was it about the potential transgression of privacy? Was it a judgement about exploiting the subjects of the photographs? Maybe. However, all of these factors rolled into an imagination of how in the future I might give talks on the research. Somehow I knew that I would not want to flash up images of these people in order to impress or stimulate an audience. It was this sure knowledge that worked back into the research practice of not using photography.

Harm?

I had been quite keen to talk to James, and, in the course of a number of fairly superficial conversations, he did express a willingness (even desire) to tell me his story and to put it to tape. In the end, it took three or four different 'attempts' to actually sit down and talk with James, because each time things had worked out such that this would have been rather inappropriate. On one occasion, anxiety over an impending court appearance had just got too much; on another James was struggling over the anniversary of a friend's untimely death; and on two others, I came to the conclusion that had I gone ahead it could have jeopardised James's efforts at the time to quit alcohol. I would not want to pretend that these were necessarily easy decisions to take, nor that the whole process was not extremely frustrating; for in the end it could have meant 'losing' James's story (and there were others that I did lose in this way).

Exploitation?

I had written recently about the dangers of exploitative research tourism, and the unethical nature of 'flip' ethnographies in which academics flip in and flip out of the lives of other people staying just long enough to collect juicy stories. Yet I found myself doing just that in a particular village. Staying for just a weekend I had a chance encounter with two homeless men, and their stories have become integral to our findings. My commitment to them was so small, and the fact that I have a more regular commitment to homeless people in Bristol seemed at the time to be a disconnected excuse for flip ethnography (see Cloke, 1997).

As I entered this area I become aware of 'Tony', who was begging under the main archway. Here, then, was the opportunity to produce an image relevant to my forthcoming presentation. However, I was also aware that the local press had been somewhat hostile to people begging on the streets of this town over recent months. As such,

Box 5.7   (Continued)

I needed to give some thought to how I would approach Tony, introduce myself and the project, and ask permission to take his photograph … I first placed some money in Tony's polystyrene cup… and crouched down next to him. I began making conversation … I asked Tony whether he had heard of our homelessness project … I provided some brief details … how we had been trying to talk to homeless people in the local area and record their stories. He didn't seem that interested … I explained to Tony that I was also keen to collect some visual images to place alongside the stories, and as such wished to take a photograph of him. 'You're not working for the papers are you?', he asked. I stressed that I was conducting research. 'OK, but I'm not giving you my name'. 'That's OK, I don't want your name, just your picture!' We reached an agreement and I moved across the cobbled street to compose my image. It was a hurried shot … I returned across the street, thanked Tony for his help and wished him well.

Sensitivity to cultural difference?

The stronger part of my self-identity is that of a single parent, and my own struggle with the negative representations of my 'group' in the media and wider society. While I have never been homeless, I have, on occasion, had difficulty (as many people do) with trying to maintain my rent payments and provide for my family. These experiences allowed for a certain degree of identification with my interviewees but, in certain situations, this proved problematic – at times it felt as if I was at once the researcher and the researched, perhaps too much of a 'insider' in some ways.

Source: Cloke et al. (2000a)

which has resulted in a conscious analytical scrutiny of the researcher's self. The second is a growing acceptance of Bakhtin's (1986) *dialogics* by which researchers seek to encounter otherness through the potential of dialogue. Together, reflexivity and dialogics suggest research to be a process rather than a product. As England (1994) has argued, research is thereby transformed by the inputs of those who previously have been regarded as the 'subjects' of research. Furthermore, the researcher is now recognized to be an integral part of the research process. Hence the interview and the focus group become unfolding performances, permitting the telling of particular narratives rather than quarries for the truth.

These changes have direct ethical consequences and often serve to problematize any straightforward application of the ethical guidelines listed above. To illustrate these difficulties we turn to an account by Cloke et al. (2000a) of the ethical issues in research with homeless people in rural areas. Extracts from the research diaries of individuals in the research team (Box 5.7) clearly indicate that in each case the five 'textbook' principles for ethical interviewing were challenged by the practicalities of interviewing homeless people. It is clear that standard ethical approaches to the relationship between the 'researcher' and the 'researched' are both turbulent and problematic. The practice of human geography can never be a neutral exercise. As Cloke et al. (2000a: 151) suggest:

> For good or ill, the very act of entering the worlds of other people means that the research and the researcher become part

co-constituents of those worlds. Therefore, we cannot *but* have impact on those with whom we come into contact, and indeed on those with whom we have not had direct contact, but who belong in the social worlds of those we have talked to ... Ultimately such matters are entwined with the need to avoid exploitation of research subjects, and to give something back to them through the research process. These are matters of complex negotiation ... to suggest that a degree of negotiation does not regularly take place over differential ethical risk, in order to garner material with which to achieve certain ends, is to hide behind ethical standards so as to obscure the real-time dilemmas of research.

These issues of research ethics are crucial to the undertaking of the kinds of methodologies discussed in this chapter. In order to deal appropriately with the responsibilities we have in the worlds that we so readily gatecrash

in the name of research, we require more detailed and critical ethical maps by which we can traverse this difficult terrain.

Notes

1 See, for example, Moser and Kalton (1971), Bynner and Stribley (1978), Dillman (1978), Dixon and Leach (1978), Sudman and Bradburn (1982), Rossi et al. (1983), Fink and Kosecoff (1985), Bell (1987), Fowler (1988), De Vaus (1991), Oppenheim (1992), Foddy (1993), Czaja and Blair (1996), Lindsay (1997) and Parfitt (1997).
2 See Downs (1970) as an example of the use of semantic differential scales in retailing geography.
3 See Oppenheim (1992: chs 10 and 11) for a full account.
4 See, for example, Goldman and McDonald (1987), Greenbaum (1988), Kreuger (1988), Stewart and Shamdasani (1990) and Morgan (1993; 1997).

6

Doing Ethnographies

Introduction: what is ethnography and how can it be geographical?

Elsewhere in this book we have begun our answers to such questions by going to Greek roots. And this is quite useful. While geography literally means 'earth-writing', ethnography means 'people-writing' (Hoggart et al., 2002), the kind of 'writing' most often associated with early and/or traditional anthropological fieldwork. Here, Western researchers typically spent a year or more living in far-off, non-Western, small-scale, isolated, rural communities (Jackson, 1983). They learnt local languages, watched and participated in day-to-day activities and (sometimes with research assistants and translators) entered into conversations about these activities with local people. They asked what was going on and why, took notes on what they saw and heard, sketched or took photographs of particular people, places or events, kept tallies, collected objects and anything else that could help to record, make better sense of and/or better represent to others what they were doing and learning. Ideally, such extended, intensive and detailed work would enable the researcher to understand that 'people's' 'worldview', 'way of life' or 'culture', like an inquisitive 'insider' (Herbert, 2000). Ethnography has its roots in European imperialism and in the roles that academic disciplines such as anthropology and geography played in

this (see Asad, 1973; Livingstone, 1992; Gregory, 1996). Subsequent adoptions, adaptations and transformations, however, mean that, today, its practitioners, politics, fieldsites, theoretical content and purposes complicate this stereotype considerably.

This doesn't mean that our opening questions can't be answered. Wherever it is practised, 'ethnography' has common characteristics. First, it treats people as knowledgeable, situated agents from whom researchers can learn a great deal about how the world is seen, lived and works in and through 'real' places, communities and people. Secondly, it is an extended, detailed, 'immersive', inductive methodology intended to allow grounded social orders, worldviews and ways of life gradually to become apparent. Thirdly, it can involve a 'shamelessly eclectic' and 'methodologically opportunist' combination of research methods but, at its core, there *must* be an extended period of 'participant observation' research (Jackson, 1985: 169). Fourthly, participant observation uniquely involves studying both what people *say* they do and why, and what they are *seen to* do and say to others about this. Fifthly, ethnography inevitably involves tricky negotiations between researchers' words and deeds, both within standard field settings and between these and other settings. Finally, therefore, ethnography involves a recognition that its main research tool is the researcher and the ways in which he or she is used to acting in more familiar circumstances *and* learns to act in the often strange and strained circumstances of his or her research settings. Here differently 'theorized' (academically and otherwise) and/or taken-for-granted worldviews, ways of life, self-understandings, relationships, knowledge, politics, ethics, skills, etc., are accidentally and deliberately rubbed up against one another.

Ethnographic findings are not therefore 'realities extracted from the field' but are 'intersubjective truths' negotiated out of the warmth and friction of an unfolding, iterative process (Parr, 2001; Hoggart et al., 2002). In many ways, Steve Herbert argues 'all humans are ethnographers whenever they enter a new social scene; one moves from outsider to insider as one comprehends the world from the insider's point of view' (2000: 556). This may be true, but novice ethnographers must recognize, develop, complement and sometimes unlearn existing attitudes, habits, sentiments, emotions, senses, skills and preferences. A good ethnographer is someone willing and able to become a more reflexive and sociable version of him or herself in order to learn something meaningful about other people's lives, and to communicate his or her specific findings, including their wider relevance, to academic and other audiences.

Ethnography's methods other than participant observation (and what is involved in understanding the data generated through them and in representing that understanding) are dealt with elsewhere in this book. Here, therefore, we focus on issues that geographers have brought into, and got out of, ethnographic research and the participant observation at its heart. Recent geographical reviews have pointed out that, although participant observation is now a standard approach in the discipline's methodological literature,[1] only a tiny proportion of mainstream publications appear to be based on its use. For instance, of the human geography articles published in the *Annals of the Association of American Geographers* and *Society and Space* between 1994 and 1998, only 3.8% and 5%, respectively involved participant observation research (Herbert, 2000). Moreover, only a handful of 'proper', book–length ethnographies by geographers have been published in the past 30 years, and these have usually been revised versions of PhD theses.

This poor showing, according to Herbert (2000), results from a general geographical scepticism about ethnography's merits. Crang (2002) has added more practical explanations including problems of arranging and conducting extended periods of fieldwork around family, work and other commitments, and of funding bodies' requirements for research proposals specifying 'outcomes' which are difficult to predict when using 'a method in which the units and boundaries of inquiry are as little predetermined as possible' (Jackson, 1983: 45). Under these circumstances, it is not surprising that qualitative research in human geography is primarily based on interviewing or, at best, 'short-term ethnographies' (Thrift and Dewsbury, 2000: 424; Crang, 2002).

Ethnography is often cast as entering geography as a response to the dehumanizing effects of the quantitative revolution in the 1960s (see Billinge et al., 1984). So, the story goes: in the 1970s it was used by humanistic geographers keen to put 'real people' back into the discipline and their work subsequently inspired the widespread adoption of qualitative methods during the cultural turn the early 1990s (see Philo, 1991; Cook et al., 2000b). Historians, however, of geography have argued that the sponsorship and publishing of ethnographic studies – i.e. 'first hand accounts of travels and explorations' (Livingstone, 1992: 161) – go much further back. In a 1899 edition of the *Journal of the American Geographical Society of New York*, for example, Titus Munson Coan set out in his 'Hawaiian ethnography' the 'physique', 'mental and moral traits', 'language', 'religion', 'social usages' and 'arts' of the 'brown Polynesian' who was 'everywhere substantially the same – or was before the obliterating hand of European civilisation swept over the finely graven plate of Nature's etching' (1899: 24). Between 1906 and 1910, 0%, 3% and 14% of the articles published in *Geographische Zeitschrift,* The *Geographical Journal* and the *Bulletin of the American Geographical Society* were classified as 'anthropology and Ethography' and 60% of articles in *The Geographical Journal* were categorized as 'exploration and mapping' (Close, 1911: 743; Genthe, 1912). This therefore means that participation, observation and 'ethnography' have, one way or another, comprised a core methodological tradition in geography since the beginning of the nineteenth century (Livingstone, 1992; Anderson, 1999). This tradition also includes numerous women's accounts of 'travel and exploration' which, historically, were not deemed 'proper geography' or 'proper anthropology' by these disciplines' male-dominated professional bodies.[2] The work of 'geographers' as diverse as Mary Kingsley, Titus Munson Coan, Carl Sauer and Bill Bunge could therefore be seen as part of geography's long-running ethnographic tradition.[3] The quantitative revolution 'may, therefore, be seen as an aberration rather than a revolutionary paradigm … that it was claimed to be at the time' (Winchester, 2000: 14).

In this book we want to make a contribution to the kinds of ethnographic research that geographers want to do now. And we want to offer more than a potted method, which just says what ethnography is, what it involves and how to do it well. Instead, we want to outline ethnography as a long-running and distinctive *geographical practice* that has learnt much from, and can contribute much to, wider interdisciplinary debates.

For the past 30 years, we argue below, the exploration of relationships between people, place and space has been the hallmark of geographical ethnography. However, the ways in which these relationships have been theorized, designed into research practice, 'discovered' (or not) in the 'field' and written up (or not) as 'fieldnotes' have changed as the discipline has changed. All sorts of fascinating theoretical ideas, impressive ethnographic work, missed opportunities and thoughts for the future have come and gone.

Currently, participant observation is entering a new phase of popularity, with agenda-setting geographers making fresh calls for its use (e.g., Thrift, 2000b; Smith, S.J., 2001; Hoggart et al., 2002). But these latest 'cutting edge' ethnographies could learn a great deal from older, pioneering studies. There are important connections to be made between humanistic ethnographies from the 1970s and 1980s and the 'new' ethnographies subsequently advocated as geography took its cultural turn. Thus, below, we outline and exemplify key debates about theory and practice within and between these ethnographies, use these to provide some top tips for prospective ethnographers and then finish with a discussion of 'field-noting' – the production of data from participant observation research.

Geography's humanistic ethnographies

Humanism advocates a people-centred understanding of the world and its workings, and argues that humanity must be central to intellectual endeavour (Cloke et al., 1991). Geographers' commitment to these ideals dates from the late nineteenth century and their influence connects the educational humanism of Peter Kropotkin and H.J. Fleure, Paul Vidal de la Blache's school of *la géographie humaine*, Carl Sauer's Berkeley School of cultural geography, J.K. Wright's (1947) 'geosophy' and David Lowenthal's (1961) writings on 'geographical epistemology'.[4] All were concerned with understanding people's minds, hearts, imaginations and their combined, 'properly human', geographies. They questioned relationships between 'the world outside and the pictures in our heads' (Wright, 1947, cited in Jackson and Smith, 1984: 21), and attempted to understand human agency in relation to experiential, societal and physical 'structures'. Some of these ideas were absorbed into the quantitative revolution's behavioural geography, but a self-consciously humanistic geography emerged in the early 1970s as an ethical and philosophical critique of the dominant spatial science tradition in human geography which was 'insensitive to the humanity of both geographers and people under study' (Cloke et al., 1991: 67).

Here, in response to spatial scientists' claimed 'distance' from what they studied, humanistic geographers emphasized the humanity of all researchers and how this unavoidably entangled them with their objects of study. In response to spatial scientists' claimed neutrality, objectivity and ability to work without values, they emphasized the centrality of human values in everyday life and argued that claims to rise above them were not only untenable but also unethical, irresponsible and 'cloaked their complicity in global inequality and exploitation' (Rose, 1993b: 42). Humanistic geographers argued that 'Quantifiers and spatial analysts have failed to make a single important discovery either about mankind or about the human environment' (Prince, 1980: 294). 'Human' geography therefore needed a 'radical reorientation of thought and vision' (Buttimer, 1976, cited in Rose, 1993b: 43). This reorientation involved heated, often acrimonious, debates, with accusations of 'irrelevance' and

'hypocrisy' frequently made on the pages of learned academic journals and books. However, what these debates contributed to geography and to the foci of its ethnographic research was a distinctive and important set of arguments about the inescapable entanglement of people and place.

To those who know their history of Anglo-American human geographic thought, much of this is a familiar story. Readers might therefore expect us to start outlining humanistic geography's main philosophical critiques of spatial science and its attempts to imagine a new, fully 'human' geography. They might expect us to discuss the importance of phenomenology and existentialism, the work of Husserl, Heidigger, Merleau Ponty and Sartre, among others and some key principles picked up from this work. These might include the blurring of positivism's object/subject and fact/value distinctions to take a position where 'one's goals, intentions, and purposes can never be totally isolated from one's experience and knowledge of the world' (Entrikin, 1976: 626) and the belief that people's goals, intentions, purposes, experiences, knowledge and bodily movements were always 'incarcerated' in places as diverse as 'movements, art-works, buildings ... cities' and other entities that 'organise space as centres of meaning' (Tuan, ested in Entrikin, 1976: 626). We might then say something about humanistic geographers' concern to understand the 'human experience of space and place' and the, often taken for granted, 'geographies of the lifeworld'. Readers might then expect us to conclude by discussing a slew of ethnographic studies that grounded and extended these ideas in concrete places and everyday lives. This empirical connection was very rarely made because humanistic geography was, however, a largely reflexive philosophical exercise.

Humanistic geography's primary concern was to understand *the* human experience of space and place, *the* 'geographies of the lifeworld and so on – in other words, *the* essential qualities of the humanity which spatial science had brushed aside. For example, its phenomenological method involved, in principle, researchers being able to 'bracket out' all worldly presuppositions in order that the raw, pure, direct 'essences of man, space, or experience' could be revealed (Tuan, cited in Entrikin, 1976: 628). This was an intuitive, introspective, idealistic, conceptual process which did 'not depend upon empirical verification' (Entrikin, 1976: 629).[5] So while being philosophically rich, humanistic geography's intuitive, introspective, idealistic and essentialist approach – ironically – kept it distant from the people whose humanity it cared so much about (Eyles, 1986).

It seems, therefore, that ethnographic research should, in principle, have been part of this new, humanistic geography but, in practice, was not (Jackson and Smith, 1984). Yet things are not so disappointing as they seem because humanistic geography

made its most significant contribution to human geography not in directing the attention of a few researchers to the deepest phenomenological and existential connections people have with their places, but in sensitising numerous researchers ... to the everyday and yet often quite intimate attachments all sorts of people (and not just philosophically inclined scholars) have to the places that encircle them. (Cloke et al., 1991: 81).

Humanistic geography literature produced only a handful of ethnographies detailing people's attachments to place, but none drew heavily on the philosophies and phenomenological methods outlined above. Only three were discussed in any significant detail: David Ley's (1974) *The Black Inner City as Frontier Outpost*, Graham Rowles' (1978a) *Prisoners of space?* and John Western's (1981a) *Outcast Cape Town*. All three were revised American PhD theses in urban social geography and, as well as being published in their own rights,

were also showcased in edited collections of humanistic geography at the time. These ethnographic chapters often stood out from the philosophical crowd, and gave one reviewer 'greater confidence in the ability of these humanists to open new realms of geographical discovery' (Prince, 1980: 295).

These three ethnographies are now regarded as milestones in geographical inquiry. Two have been 'revisited' as 'classics in human geography' in *Progress in Human Geography* (Ley's in 1998 and Western's in 1999), and all three can be seen as important precursors of geographers' current interests, first, in representation–reality, structure–agency and space–body relations and, secondly, in the *practice* of ethnographic research. They are discussed below both to demonstrate their impacts and influences on the discipline and to illustrate how ethnographic research proceeds and what the finished products can look like.

The Black Inner City as Frontier Outpost

David Ley's (1974) The *Black Inner City as Frontier Outpost: Images and Behaviour of a Philadelphia Neighbourhood* was submitted as a PhD thesis at Penn State University in 1972 and revised and published two years later as an Association of American Geographers' monograph. At the time of its publication it was seen as 'a solid research achievement and a model for further studies' (Palm, 1975: 573). Addressing 'radical concerns about racialised injustice' at the time, Ley's book has been recently described as geography's 'first and in many ways most lasting example' of ethnographic social research and as 'signalling a turning point in the development of [urban] social geography' (Jackson, 1998: 76, 75).

Ley lived in Philadelphia's black inner-city community of Monroe for a six-month period in 1971, and then spent another two and a half years writing proposals for its community association. While there, he participated in community activities, became acquainted with residents, undertook a questionnaire survey of 116 residents and examined numerous newspaper articles and 'city files' (Darden, 1976; Palm, 1998). Behind the scenes, he drew on an 'eclectic range of intellectual sources including a humanistic reading of the French tradition of *la géographie humaine*, an ethnographic interpretation of the Chicago School of urban ecology, combined with an interest in various strands of phenomenology together with a strong influence from behavioural science and environmental psychology' (Jackson, 1998: 75; see also Jackson and Smith, 1984; Cloke et al., 1991). He also drew inspiration from more 'off the wall' sources, such as the travels of explorer James Cook, cybernetics and accounts of Caesar's *Gallic Wars* – the source of his 'frontier outpost' model (Ley, 1998; Palm, 1998).

What this meant, he has written recently, was that the book was 'too conceptually busy' (1998: 79). It was also busy in a methodological sense. Along with the tapes and notes from his participant observation research, he analysed the results of his questionnaire 'using semantic differential tests to determine stress profiles, and factor analysis to derive components of a "stress index"' (Palm, 1998: 77). In the mid-1970s, this *range* of theoretical and methodological approaches was seen as an exemplary combination of 'the rigour of positivist science with the insights of experiential exposition' (Palm, 1975: 572). Twenty-three years later, Palm (1998: 77) described this as 'the highest possible compliment I could [pay] at the time'.

Ley's findings were presented in two parts. In the first – 'The shadows' – he examined dominant 'outsider' representations of black inner-city neighbourhoods in the USA as united and oppositional to mainstream white society. Here, he examined both mass media and academic representations, most often put together 'by white investigators who have never lived in a black community but who nevertheless emerge as so-called authorities on the life and behaviour of blacks' (Darden, 1976: 117), but also by 'members of the city's Black Muslim population' (Jackson, 1998: 76). In this section, Darden comments:

> He concludes that the popular white image of black America is false, for it is grounded in history and American institutions, nurtured by social and spatial segregation, and traditionally reinforced by the white media from which most Americans derive their most important and in many cases only insight into black America. Ignorance and lack of firsthand information have helped white Americans believe in the existence of a monolithic, hostile, organised black community that is preparing for revolution in collaboration with outside agitators … Occasional visits to the inner city will not change these false images. (1976: 116)

The sustained and systematic explorations of an urban ethnographer could, however, change those images. Thus, in the second section of Ley's book – 'The real' – he outlined the findings of his research *within* the community (i.e. his 'insider' perspective). Here, in what remains the most 'compelling' part of the book (Jackson, 1998), he described in rich ethnographic detail a community divided by 'gang territorialization', and how the everyday lives of its people were full of complexity, suspicions, mistrust, disunity, 'unpredictable alliances', 'uncertainty and even chaos' (Palm, 1975: 572).

According to Ley, what characterized this community was not a black oppositional unity but 'individuation' or the 'retreat into self and family as a means of coping with the environment' (Palm, 1975: 573). And this was where the frontier outpost model came in. As Darden (1976: 117) explained, the

> social group, the community association, the street gang – all may be frontier outposts. Each builds up an internal defensive

repertoire, each stockades itself and pursues a policy of individuation, each seeks to aggrandize itself, sometimes at the expense of members of its own kind, each develops a group image, its own symbols and stereotypes.

Representations and realities of the black inner city were therefore poles apart, and this was something which a white outsider such as Ley could only have discovered through developing an insider's perspective. However, as he noted in the book's Afterword, while he prised 'representation' and 'reality' apart in its two main sections, he could not be sure that he had done so in practice.

In 1998 he recounted the 'gropings' of his early weeks of participant observation when he found that his 'socially constituted stock of knowledge … was challenged at every turn by everyday life in North Philadelphia' (p. 79). The role of social scientists, he had written in the book, was to 'search for order in [such] chaos' (cited in Palm, 1998: 78). But this meant that, while his conclusions clearly derived 'from the reality of the area studied', they also derived from:

- 'the postulates of the scholar's conceptual framework, postulates which are probably ethnocentric, or, at best, limiting' (Palm, 1975: 572);
- his presence in the field as 'the neighbourhood exotic' whose British accent 'local adolescents sought to mimic … when we passed on the street'; and, in particular,
- his 'two key informants, a middle-aged woman and an adolescent male, both active in the neighbourhood' (Ley and Mountz, 2001: 239).

Despite these criticisms, however, when Peter Jackson revisited Ley's book in 1998, he argued that it was

justifiably regarded as a classic because of the way it embodied and advanced … trends

[in ethnography and radical anti-racist politics], prefiguring many subsequent developments in the discipline: the concern for social theory, for qualitative research methods, for the politics of representation and the rise of cultural and media studies. (p. 75)

Outcast cape town

While Ley's ethnography was based in a black inner-city neighbourhood in the USA which seemed to be fragmenting itself, Western's (1981a) *Outcast Cape Town* was based in a 'coloured' inner-city neighbourhood in South Africa which was being forcibly fragmented as a result of apartheid's 1950 Group Areas Act. South Africa's two and a quarter million 'coloured' people were

the creation of miscegenation among Whites, who established themselves in South Africa after 1652, moving towards the interior from Cape Town; their slaves, who were imported mainly from Madagascar and the East Indies; and the autochthonous Khoisan peoples, otherwise known as Hottentots and Bushmen, now almost completely wiped out. Today the majority of Capetonians are Coloureds, neither Whites nor Black Africans … Thirty percent of all the coloureds in the republic live in Greater Cape Town, forming about sixty percent of its total population of over one million. (Western, 1978: 298)

Despite the fact that the people of Cape Town 'range[d] along a phenotypic continuum from those who appear to be Nordic Whites to those who appear to be Black Africans' (Western, 1978: 298), the government imposed 'a discrete hierarchy of White-Brown (Coloured)-Black (Bantu) categories through apartheid legislation' (1978: 298). Residential segregation of black South Africans had been in place for decades. The Group Areas Act therefore was aimed at those 'in the middle' – the 'coloureds' whose genetic make-up made most nonsense of apartheid's discrete racial categorizations. Couched often as the clearance of slums adjacent to white districts, neighbourhoods were systematically razed and their coloured inhabitants were relocated to 'new ghettoes ... typically situated several miles from the white towns, with a buffer zone in between' (van den Berghe, cited in Western, 1978: 300).

As reviewers of Western's work noted this was a 'chilling' topic (Pirie, 1984: 259), and his accounts of its history, justifications and effects on the people of Mowbray (his case-study Cape Town neighbourhood) carefully and vividly captured this. Some saw him as arguing 'from the position of an objective observer basing his conclusions upon historical evidence, statistics, and personal interviews' (Entrikin, 1979: 256), while others praised his work as 'no clinically-detached survey' (Pirie, 1984: 259), as 'original, frank ... modulated rather than angry ... earnest ... sincere' (Pirie, 1984: 260) but 'hard hitting' (Jones, 1980: 115). For one, it combined 'serious scholarship with passion in a way that I have never seen them intertwined' and, despite some unusual source materials, should therefore 'be read by every geographer, by every graduate student in geography, and by many nongeographers. The book is a superb example of human and humanistic geography, of painstaking, exhaustive, and explorative research, of penetrating analysis, and of civilised, literate reporting. In short, the book is geography at its relevant best' (de Blij, 1982: 493).

Western had not set out to work in South Africa, let alone in Cape Town. He initially wanted to work in Burundi but was unable to because of the 'African holocaust' of the early 1970s in which 300,000 people were killed (de Blij, 1982). He first saw South Africa in July 1974, 'a time that in retrospect seems the high water-mark of apartheid' and through his experiences there he became 'really impassioned about "politics"' (Western, 1999: 425). He explained, it 'was only when rummaging around apartheid's legacy of empty and gentrifying formerly "coloured" areas of Mowbray that I sensed something wicked had this way come' (Western, 1999: 425).

In 1999 Western mentioned two people who had made a big difference in his work. The first was Robert Coles, whose passionate talk at UCLA in 1973 had been inspiration for Western in the audience. Coles' tales of ethnographic research had fired Western up and made him ask 'Couldn't I, as an inquisitive urban geographer, do such hands-on, please-tell-me-in-your-own-words research?' (Western, 1999: 426). The second was Victor Wessels, a South African 'Unity Movement leader then banned under the Suppression of Communism Act', and one of the first people he met in Cape Town (Western, 1999: 426). As Western recounted, 'He befriended me and challenged me to, damn it, *do* something'. 'Are you indeed a so-called scholar?' Wessels asked him. 'Then write your best contribution to our struggle' (cited in Western, 1999: 426). Through his extended ethnographic research that struggle became 'real' for Western, as what he saw, heard and read profoundly distressed him.

When *Outcast Cape Town* was revisited as a 'classic in human geography' in 1999, David Simon (a geographer who had grown up in Cape Town and been radicalized as a child by the processes described in Western's book) recalled his 'admiration', as a postgraduate, 'at the quality of writing and vividness of the accounts of gross inhumanity and suffering conveyed' (p. 423). For many South African social scientists, this was their 'first exposure to humanistic geography' (Dewar, 1999: 421), and it created 'a sense of frustration, even indignation, on the part of some ... that a USA-based British geographer had come to Cape Town and produced an acclaimed masterpiece in their own back yard' (Simon, 1999: 424).

Like Ley's account of Monroe, Western's account of Mowbray drew on, and 'showcased', the sociologically oriented phenomenology of Alfred Schutz (Entrikin, 1979). Like Ley, he juxtaposed outside representations of Cape Town with the real social geographies on the ground. And, like Ley, his ethnography was methodologically eclectic and busy. *Outcast Cape Town* combined 'every skill at the geographer's disposal – cartographic, quantitative, literary, and visual' and a wide range of sources, 'from Gamtaal poetry to governmental statistics, from questionnaire-generated data to real-estate advertisements in newspapers' (de Blij, 1982: 493). Western even 'indulge[d] in some informal modelling, reconstructing the ideal apartheid city according to the canons of enabling legislation' and recorded 'an 82 per cent congruence between his deductive model for Cape Town and the actual case' (Pirie, 1984: 258).

This multiple methodology allowed him to chronicle 'the effects of separate development philosophies and practices in a city that before the mid-twentieth century was "by far the least racially segregated in Southern Africa, and perhaps in all of sub-Saharan Africa"' (de Blij, 1982: 493). He cast a critical eye over accounts of the pre-apartheid city, the designs that apartheid planners produced for it, the ways in which these plans were put into practice and the effects that this had on Mowbray's removed and resettled coloured population (Jones, 1980; de Blij, 1982; Dewar, 1999). Western's ethnographic research 'meticulously document[ed] their histories, experiences and perceptions' (Pirie, 1984: 258) during and after 'removal'. It revealed how the:

> racially motivated eviction of coloureds [had] induced collapse of community, as well as apprehension, social and private distress, material disadvantage ... fear ... civic unrest ... [and the] humiliating damage which removals and destruction of homes had on self-esteem and security, and the way in which people had to learn again their place on the edges of a callous urban society. (Pirie, 1984: 258)

Western's ethnography therefore had important implications. For South African students, it contained an important political lesson that '"people are living there"; that there are human consequences to the use (and abuse) of power and that impersonal, unseen structural forces are not entirely deterministic. There is a primacy in human agency' (Dewar, 1999: 422). And for geographers more widely, it showed how space could be 'manipulated to preserve and promote a divided society in which whites will be master. A distinctive territoriality cements a racial lifestyle and outlook, entrenching black powerlessness and servile deference' (Pirie, 1984: 258).

Fifteen years later, reviewers could see how it had shown how 'social relations are ... both space-forming and space-contingent' and how 'in remaking the city, apartheid also remade Cape Town's citizens' (Dewar, 1999: 421). Thus, it seemed, Western's ethnography had 'foreshadow[ed] the insights of structuration theory and postmodernism' (Dewar, 1999: 423). Moreover, as Western argued in 1999 he had been an 'accidental humanist'. When he began his research in 1974, the notion of humanistic geography hadn't been properly crystallized. By the time he had finished it, this had fortunately emerged as 'an accepting subdisciplinary niche into which to deliver my now-completed study' (Western, 1999: 426). Not surprisingly, then, he hadn't drawn heavily on its literature. However, precisely because the Group Areas Act uprooted people from places where they had lived for generations and clearly caused so much suffering, the situation (and Western's book about it) could not help but illustrate a core theme within it – i.e. 'the complex and inextricable link between people's identities and places in their community and society' (de Blij, 1982: 493).

Prisoners of space?

Both Ley's and Western's ethnographies are, without doubt, 'classics in human geography'. Both were hugely influential in the subsequent development of urban social geography, with its combinations of political, theoretical and methodological concerns (Jackson and Smith, 1984). However, we want to argue that Graham Rowles' (1978a) *Prisoners of Space? Exploring the Geographic Experience of Older People* is an equally classic humanistic ethnography. Rowles' reviewers were as raving as Ley's and Western's. They praised his painstaking, inductive, 'experiential' approach, his 'fluid', 'intelligent' and 'exciting' accounts and the way he gave 'a sense of holism to individuals within their physical environment' (Stutz, 1979: 473). They praised his engaging, first person, 'novelistic' writing style which took his arguments 'beyond explanation and generalisation' and 'stop[ped] one short, confronting one's perceptions' (Western, 1981: 276, 275). They praised the social relevance of his detailed ethnographic work and believed that it would inspire and allow 'the eventual emergence of more realistic theory and more appropriate planning' that was sensitive to people's lived experiences (Ley, 1978: 356). And one reviewer, singled out Rowles' chapter in Ley and Samuels' (1978) collection as its 'most humanistic and engaging' contribution (Prince, 1978: 295).

Despite these plaudits, Rowles' work does not appear to have had the same disciplinary impact as Ley's or Western's. This is perhaps because he has worked on the substance of his PhD – i.e. the human aging process – in the interdisciplinary field of 'gerontology'.[6] However, the recent resurgence of interest in humanistic geography's core concerns – e.g. geographical knowledges, taken-for-granted, 'performative' everyday lifeworlds and body-space-identity relationships[7] means that it might now be time to revisit what is geography's most direct and detailed humanistic ethnography.

Rowles is interesting because he undertook his PhD in a hot-bed of humanistic geography – Clark University – where he was supervised by Anne Buttimer and studied alongside David Seamon and others (see Buttimer and Seamon, 1980; and Buttimer, 2001). His work grew out of and challenged the principles, abstraction and introspection of this new geography and he believed that its 'introspection and anecdotal or literary evidence in part reflects a reluctance to engage in [what he called] experiential field work' (Rowles, 1978a: 174). Like Ley and Western, he used ethnographic research methods to gain an insider perspective of a group too often represented in simple, abstract, stereotyped terms. However, while they did ethnographies on a conventional neighbourhood or community scale, he worked at a much smaller scale, developing an experiential approach to 'interpersonal knowing' intended 'to eliminate the intellectual separation of observer and observed and thus mediate the distinction between objective and subjective knowledge' (Entrikin, 1979: 256; see also Jackson, 1983). In practice this meant that:

> Unlike conventional participant observation, where the researcher aims to vary his [*sic*] social contacts and maintain a certain detachment from them, interpersonal knowing is predicated upon a deepening friendship between participant and researcher, a relationship which takes time to establish and in which otherwise hidden or suppressed aspects of personal understanding become knowable. Ethically, the objective is one of mutual discovery, as researcher and researched are both engaged in a mutually rewarding project. (Ley, 1978: 355)

Prisoners of Space? was, in a general sense, an exploration of the differences between 'Cartesian and lifeworld space' (Western, 1981b: 276). However, more specifically, it was an ethnographic critique of behavioural suppositions about the constriction of elderly people's geographical lives as a result of the ageing process. Rowles undertook his research in Winchester Street, a working-class neighbourhood in an unnamed East Coast

American city. This was the context for what the book was *really* about: the biographies, geographical experiences and everyday life-worlds of elderly people living there. He initially intended to recruit large numbers of elderly research participants but only five stuck with him – Stan (aged 68), Marie (83), Raymond (69), Evelyn (76) and Edward (80). However, as he later wrote, this 'anxiety provoking [recruitment] 'failure' turned out to be a blessing in disguise' (Rowles, 1978b; 177).

Through developing close, long-term relationships over a three-year period, he was able to appreciate in vivid detail what it was like to be, live and see the world as an elderly person. In his book, one chapter was devoted to his 'interpersonal knowledge' of each person. Two passed away before he finished writing but he discussed draft chapters with the other two before settling on his final versions. Some geographers found this interpersonal, person-centred approach intriguing (Entrikin, 1979), but, for others, it was much more problematic than 'more orthodox participant observation' because, first, the intensity of its commitment meant its participants would be largely self-selecting; secondly, its ethical goals could only be ideals, given the inevitable imbalances between the "co-explorers" concerns, actions and 'public disclosure'; and, finally, some of the details being sought were so intimate that 'the outsider should not, perhaps, have been given such privileged access' (Ley, 1978: 356).

Like Western, Rowles examined people's attachments to places. While these came to the surface in Mowbray through forced detachments, in Winchester Street they emerged from the relationships Rowles developed with these five people. In the early stages of his work this didn't work at all to plan. Marie, for example, wasn't interested in talking to him about the kinds of issues raised in the academic literature on ageing, however carefully he worded his questions. So, he recounted,

Instead, as we sat in her parlor poring over treasured scrapbooks in which she kept a record of her life, she would animatedly describe trips she had taken to Florida many years previously. She would muse on the current activities of her granddaughter in Detroit, a thousand miles distant. She would describe incidents in the neighbourhood during the early years of her residence. Blinded by preconceptions, I could not comprehend at first the richness of the taken-for-granted lifeworld she was unveiling. (1980: 55)

As a result of such 'frustrating' initial experiences, his research became less focused and more time-consuming as he went with the flow of their lives and their ways of thinking about the experiences in which he was interested. He 'hung out' with his elderly participants, one by one, chatting in their homes, going for walks, meeting friends, going for a beer, attending community meetings, whatever they liked to do. This allowed relationships to develop at their own pace. And here he found that hours and hours of interaction, lots of casual chatting and long periods of silence occasionally unearthed striking flashes of insight. Marie's account of her local church's Christmas dinner party, for example, shocked him:

'We had a big banquet. There was over two hundred people, two hundred and forty people. We all bought our tickets. We had a caterer.' 'What about the poorer older people?' I interjected. 'The poor ones?' she replied incredulously. 'There's no more poor. Don't talk about poor people. The old people? There's no more of that. The old people now live better than the young ones because they have pensions. They have good pensions, too. So they all live like millionaires now,' she concluded emphatically. The more I pondered this unsettling exchange the more apparent it became that Marie's image of poverty, framed by conditions during the depression, was completely different from my own. Gradually I came to internalise rather than merely to acknowledge that the world she inhabited was a different 'Winchester Street', not the run-down environment I could view. (1978b: 178–9)

As these relationships developed, Rowles produced reams of notes, photographs, sketch maps and taped conversations with or about each participant. He slowly learnt and internalized the richness, complexity and intimate dimensions of the worlds they inhabited. His conclusion was that these elderly people were attached to Winchester Street via their:

- intimate bodily memories of the dimensions, routes through and physical features of its many interconnected spaces and the almost effortless routine movements this allowed;
- autobiographical memories which connected who they were to where they were through the continual sparking of memories by the 'incident places', mementoes and other bits and bobs of their lives; and
- social networks, near and far, everyday and occasional, experienced and imagined, through which their lives were intertwined with others, alive and dead.

These attachments showed that the stereotype of elderly people's deteriorating physical and perceptual capabilities leading the world to close in on them was a 'demeaning oversimplification' (Rowles, 1978b: 183). In its place, Rowles' work 'emphasise[d] the vibrancy and creativity of the human mind as well as the tyranny of biological, economic, and social constraints' (Ley, 1978: 356). And these agency/structure relations had important policy implications. As Rowles later argued, family, medical and social service professionals should listen to elderly people's requests to 'age in place' because this could prolong their lives. Removing them to nursing homes, for example, severed intimate physical, psychological and social attachments to particular places that could by no means be easily reestablished in a new setting. And, for those who had to relocate, the design of their new homes could be made more sensitive to residents' multidimensional geographical knowledges, experiences and bodily capabilities.

Geography's 'new' ethnographies

The above discussion has no conclusion because these ethnographies foreshadowed much of what was to come as Anglo-American geography took its cultural turn in the early 1990s. 'Cultural turn' is a 'shorthand description for a vast number of different and sometimes incommensurable trends within human geography, united only by their diverse appeals to concepts of culture and to the intellectual field of cultural studies' (P. Crang, 2000: 142). These included the diverse insights of feminism, postmodernism, poststructuralism, 'queer theory', anti-racism, postcolonial theory, psychoanalysis, social psychology, cultural studies, science studies and social anthropology which, together, brought into geography concepts and concerns about difference, identity, embodiment, language, knowledge, texts, discourses, images, communication, value, ways of seeing, agency and structure, power relations and resistance, to name a few (Philo, 2000; Shurmer-Smith, 2002a). Many were not, however, entirely new to the discipline. There were for example, some striking similarities between old humanistic and new feminist geographies. Both attempted to study and understand better:

- how different people understood the world and their place(s) within it;
- the routinization of everyday lives in time and space (including the 'home'); and
- the ways in which embodiments, memories, emotions and feelings tied together places and social/personal identities.

Moreover, in studying these relationships, both refused scientific rationality, demanded researchers' reflexivity and dispensed with dualisms – e.g. 'analysis and empathy, insider and outsider, thought and pleasure, body and mind, individual and context' – which 'are always bound into the power relations of a society' (Rose, 1993b; 48). While humanistic geography left a lot to be desired (e.g. because of its inadequate theorization of power relations and its neglect of gender relations; Cloke et al., 1991; McDowell, 1992a; Rose, 1993b), its biggest influence on a cultural turn which 'sent shockwaves throughout the length and breadth of human geography' (Philo, 2000: 28) was arguably via the widespread adoption of its qualitative research methods (textual, visual and intersubjective; Dwyer and Limb, 2001; Ley and Mountz, 2001). These, enabled

economic geographers [to] 'discover...' [the] embeddedness of local economies in local social practices; political geographers [to become] aware of new nationalisms and notions of identity in boundary formation and exclusion; urban geographers [to] turn ... their attention to lifestyle and ... [to become] enthusiastic about cultural regeneration of cities; ... [and] retail geographers [to become] enthusiastic about sites of consumption, as opposed to patterns of distribution'. (Shurmer-Smith, 2002a: 2)

The now substantial body of work outlining these methods and how to use them well is, therefore, an overdue but vital element of this cultural turn literature.

Geography's cultural turn was also its ethnographic turn because social and cultural geography were brought closer together, in part, under the influence of the new and feminist ethnography emerging from anthropology and sociology.[8] Here, theory, politics, poetics and their relationships with research design and practice were discussed in substantial critical detail. Below we focus on three ways in which these became central to geography's new ethnographies: in debates about, first, categorizing subject communities; secondly, relating abstract theory to everyday life; and, thirdly, being reflexive.

Categorizing subject communities

Literal definitions of ethnography as 'people writing' can be misleading because, originally, these people weren't the kind you might meet in the library, on holiday or in your research. Early ethnographers set out to describe people in the sense of *a* people, *a* culture, *a* nation, *a* community – a *group* of people. Thus 'the assumption of a sociocultural unit, spatially and temporally isolated, is deeply embedded in the conventional framing of subjects for ethnographic analysis' (Marcus and Fischer, 1986: 86). Geographers taking the cultural turn were therefore critical of the way in which ethnography had been used to bolster an understanding of the world as divided into discrete peoples or cultures. But they were also critical of humanistic geography's belief that, beneath the surface in the realm of human essences, all people were the same (Ekinsmyth and Shurmer-Smith, 2002). So how could relationships between group and individual identities and the places and spaces of everyday life be understood? Clearly, ethnographic research methods are well suited to this kind of study. But what do researchers mean when they say they're using them to understand the 'inner workings' of a people, a culture or a community? How are their boundaries defined? By whom? At what stage(s) in the research process? What difference can this make to research findings? And what can answers to these questions mean both for the design of, and data construction within, geography's new ethnographies?

As we have already noted, the ethnographic traditions in geography and anthropology are historically entangled. However, as geography began to take its cultural turn this entanglement took on a new significance. No account of this turn is complete without discussing James Duncan's (1980) critique of the 'superorganic in American cultural geography'.[9]

Duncan's paper was a critique of the theory of culture that, through the work of Carl Sauer and his Berkeley School students, dominated twentieth-century American cultural geography. Sauer, as we said earlier, is an important figure in the history of geography's ethnographic research. He was an advocate of fieldwork in 'primitive rural areas' (Duncan, 1980: 194); he 'supervised some 40 PhD theses, the majority on Latin American and Caribbean topics'; and he conveyed 'to all his students his firm belief in the need for first hand field experience and for learning the language of the people being studied' (Jackson, 1989: 10). So why did Duncan have such a problem with his 'cultural' geography? Why, for some, did Duncan's 1980 paper have such an impact on the discipline, helping to 'open … the way for a "new" cultural geography' (Shurmer-Smith, 1998a: 567)? Why, for others, had his paper made so little difference to this new cultural geography (Mitchell, 1995)? What do we mean by culture? And what difference does this make to the way we, first, understand other related words such as 'people', 'community', 'neighbourhood' or 'place' and, secondly, undertake ethnographic research with or within them?

Duncan's (1980) main critique of superorganicism was that it reified cultures as *things*. He argued that Sauer and many of his students saw culture as existing above, and being independent of, the everyday actions of human beings. *It* acted on *them*. *It* had the power to make things happen. *It* evolved 'according to its own internal logic and presumed set of laws' (Zelinsky cited in Duncan, 1980: 187). Individuals were merely *its* 'agents', 'bearers' or 'messenger[s]' carrying information across the generations and from place to place' (Duncan, 1980: 184). *Its* 'values grip men's [*sic*] minds and force them to conform to its will' (1980: 184). *It* was a 'powerful nearly sovereign primal force [which]

should share star billing in our research and pedagogy, along with geomorphological agents, climatic process, biological process, and the operation of economic laws' (Zelinsky cited in Duncan, 1980: 188). Sauer's was another 'human geography' which seemingly 'had nothing to do with individuals' (Jackson, 1989: 18). At best it implied that people were 'passive and impotent' and the ties that bound them were 'external to them' (Duncan, 1980: 190). Like Coan's (1899: 24) descriptions of 'the brown Polynesian' who was 'everywhere substantially the same', this geography involved the analysis of distinct, bounded, essential, internally created, self-perpetuating and homogeneous cultures. Individuals, at best, were represented as 'modal personality types' (Matthewson, 1998: 570).

The superorganic theory of culture had been brought into geography in the 1920s and 1930s largely through Sauer's association with the Berkeley anthropologists Alfred Kroeber and Robert Lowie. As early as the 1940s, though, it had been more or less dismissed in an anthropology which had changed its focus to address 'questions of how individuals, interacting with other individuals through institutions, create, maintain, and are in turn modified by their environment … [and questions of] how individuals exercise choice, how they are strategists who manipulate the contexts in which they find themselves' (Duncan, 1980: 196). While this 'superorganic' theory came 'under devastating attack and [had] long since been rejected by the vast majority of anthropologists' (1980: 182), generations of American cultural geographers 'ignored the variety of alternative definitions' (p. 182). Instead, they 'continued to adhere uncritically' to this theory in their research (Jackson, 1989: 18–19) and 'promoted it to generations of students through textbooks and undergraduate courses outlining the kind of 'world regional geography … [which] classified the globe's population into culture "worlds", as if cultures and regions were homogeneous and easily bounded entities' (Anderson, 1999: 4). Only in the mid-1970s did this understanding of culture begin to wane as 'a number of geographers … distanced themselves from [this] tradition' (Anderson, 1999: 4).

More recent critics of superorganicism point out that, despite being academically defunct, it continues to maintain its power in popular understandings of culture, and can often sneak back into academic research in a casual, taken-for-granted way (Mitchell, 1995). We are used to hearing talk of British culture, Islamic culture, gay culture', youth culture, black culture, or urban culture. Many people are fascinated by their 'own' and/or 'other cultures', see themselves as moving 'between cultures' and/or as having a 'mixed culture'. Culture is often used to explain conditions and events. All this arguably reflects a tendency towards culturalism: the 'assumption that culture "independently" exists, that cultural distinctions are necessarily real and rooted in the peoples being analysed, and that culture can be used as explanation' (Mitchell, 1995: 108). However, while definitions of culture may be taken for granted, they are not innocent. Take Peter Jackson's (1989: 27–8) critique of its use in a newspaper report:

On 12 May 1984, the *Los Angeles Times* reported that a high ranking official of the federal Department of Housing and Urban Development had sought to explain the poor living conditions of the Hispanic population in terms of their cultural 'preference' for overcrowded housing. Rather than citing poverty as the explanation for their overcrowded condition, or their uncertain political status as 'undocumented workers', the official chose to emphasise their large family size and 'preference' for living with their extended families. He failed to recognise that 'preferences' are subject to material constraint and that it is unreasonable to infer 'preferences' directly from behaviour. Neither can the statement be rejected simply

as a matter of individual ignorance. It is a reflection of institutionalised attitudes and structured inequality. The implication that Hispanic 'culture' explains their poor living conditions, regardless of their material circumstances, is … a culturalist explanation.

The main critique of 'superorganism' and related popular culturalist explanations is that they are 'divisive and dangerous as [they] underpin … unnecessary oppositions and enmities' (Shurmer-Smith, 2002a: 3; see also Massey, 1991b), and 'can represent and reify difference by [both] obfuscating connectedness' and 'over-valorising localism' (Mitchell, 1995: 111). These obfuscated connections are not only those between different, spuriously separate cultures (peoples, communities, nations, etc.) but also between culture, economics, politics and other, spuriously separate realms of existence (Latour, 1993: Thrift, 1995). Categorizations are central to implicit and explicit theories of how the world is organized, how it works and how it can or cannot be changed. But as the 'cultural turn' convinced many geographers, all categories and theories are social constructions which help to establish, maintain, justify, and transform unequal power relations. Western's (1981a) study of the categorization and treatment of South Africa's Cape coloureds illustrated this better than anything.

So how can geography's new ethnographers study a people, a culture, a community or any other social group when the politics of categorization are so dangerous? Let's go back to Jackson's (1989) newspaper article. Imagine yourself as an undergraduate in Los Angeles in 1984. You're fascinated by urban social geography and are planning a dissertation on this. You come across that article in the *Los Angeles Times*. It's topical. It's on your doorstep. It seems to reflect a number of issues you've been reading about in the literature. You're probably not part of this Hispanic culture but you think you could study it using ethnographic methods. But your literature review also involved reading Duncan's (1980) paper. So you're suspicious of the newspaper's description of this discrete, homogeneous, Hispanic culture and what that official said *it* was responsible for. This is a superorganic story so you can't uncritically accept what it says. But you still think it's worthwhile doing research on housing problems in this part of LA and decide to base your study in one, predominantly Hispanic, neighbourhood. To get this going, you know you need to find a way into that neighbourhood, that community, that culture via one of its gatekeepers, ideally a community leader. Once he or she accepts you, you're in. You will be able to move from being an insider in your community or culture to an outsider in theirs.

But doesn't this thinking also reflect that superorganicism you're so worried about? Isn't it reproducing, again, that divisive and dangerous, separate and homogeneous, 'them and us' mentality? The boundary between one culture and another is policed by that gatekeeper, isn't it? But, hold on, if the gatekeeper is a community leader, hasn't everyone gone superorganic? This is confusing, especially after you have spent some ethnographic time in that neighbourhood, being introduced to more people, talking to them and to others, hanging out, getting involved in things, like David Ley did in Philadelphia. Here it becomes very clear that the people most keen to view this community (or culture) as a separate, bounded, homogeneous entity are those who wish to govern, police, represent, create, maintain and/or transform *it*. However, you remember Pamela Shurmer-Smith's question that, in anyone's day-to-day life, 'Which of us shares *every* aspect of our life with even one other person, let alone a group' (2002a: 3)? She's right. The people you speak to seem to have all kinds of affiliations to do with their age, their gender, their financial circumstances, their jobs, their religious beliefs and practices, their politics, the languages they speak, their hobbies, their tastes, their diverse life experiences, whom they choose to (and can't help but) spend their time with, and more besides.

Moreover, these affiliations aren't only with local people or only in the present day. What about those first, second, third-generation Mexican, Guatemalan and Nicaraguan migrants you talked to? It's like that area of London where Doreen Massey lives because:

> while ['it'] may have a character of its own, it is absolutely not a seamless, coherent identity, a single sense of place which everyone shares. It could hardly be less so. People's routes through the place, their favourite haunts within it, the connections they make (physically, or by phone or post, or in memory and imagination) between here and the rest of the world vary enormously. (1991b: 28)

So how can you set out to study a subject community whose existence as a neatly bound entity is, at best, highly suspect? How can you identify a people who may have few concrete boundaries, and whose precise membership may be difficult to gauge? This makes it difficult to research or talk about 'them'.

The answer is that, in an ontological sense, these people-grouping terms (like many others, most notably 'race') are meaningless, but the ways in which they are defined, deployed and contested in real-world situations have profound effects (Mitchell, 1995; Shurmer-Smith, 2002a). So, however suspect, they can't simply be ignored. They must be tackled, head on. Take culture, for instance. According to Don Mitchell (1995: 108), this cannot and should not be studied as a thing which people possess but as a 'powerfully determined idea' bound into specific, grounded circumstances, and specific, contested power relations. Like Duncan, Mitchell draws upon the work of Clifford Geertz. His 'interpretive anthropology' has been held up as an 'attractive alternative to "superorganic" definitions' (Jackson and Smith, 1984: 39) because its:

> approach is essentially semiotic: regarding culture as a series of signs and symbols which convey meaning. Alluding to Weber's idea that man [sic] is an animal suspended in webs of significance which he has himself spun, Geertz argues that culture comprises these webs and that their analysis is 'not an experimental science in search of law but an interpretive one in search of meaning' (Geertz 1973, 5) ... The diverse and contradictory evidence of human behaviour observed in the field is thus recognised for what it is, rather than as aspects of some postulated cultural entity: 'what we call our data are really our own constructions of other people's constructions of what they and their compatriots are up to' (Geertz 1973, 9). (Jackson and Smith, 1984: 38)

In terms of research design and practice this means a number of things. First, researchers could choose to undertake research to challenge stereotypical, popular, culturalist explanations of the living conditions of one categorized and located group (like Rowles' elderly people). Here ethnographic research could be undertaken to understand better the power of categorization in specific circumstances, how this may also be part of specific people's self-categorization and the additional and alternative categorizations in use 'on the ground'.[10] Here, researchers may *start* with culturalist explanations but not necessarily end up with them if the partiality, complexity, contradiction and complications of belonging to any group are both expected and followed through as part of the research design.

Secondly, instead of framing ethnographic research as a process of moving between cultures it could better be framed as a process of moving through, and perhaps extending, existing social networks which, themselves, cross boundaries (sometimes with difficulty) between discursively and physically separated cultures, communities and/or places. Thirdly, it might not only be worth detailing in fieldnotes the networking that is the research process but also to design research which allows the tracing of network relations that help to create the conditions experienced by the people under study. Here researchers could plan to undertake ethnographic studies in more than one place, with people who may think they lead separate lives until, perhaps, that research is done and those boundary-crossing connections can be more clearly seen (Marcus and Fischer, 1986). In addition (and/or contrast), they could study the lives of people who routinely make these connections themselves: on foot, in cars, on planes, on the Internet.[11] More about this later.

From beginning to end ethnographic research is all about networking so this should ideally be done *from the beginning* of a project. Ethnographers, have to learn how to work through networks, to make appropriate connections and to 'go with the flow' when preconceptions come to light and alternative interpretations begin to make more sense. So it is not a good idea to follow the 'standard three-stage ... read-then-do-then-write' model of research design (Cook and Crang, 1995: 19) to prepare for such an unpredictable process. Nipping culturalist thoughts in the bud by networking as early as possible is, therefore, a wise damage limitation exercise.

Relating abstract theory and everyday life

During the 1980s it was not only the retheorization of culture that made a difference to geographical inquiry. Theory, in general, became much more important. Central to this were attempts by geographers like Nigel Thrift (1983b) to bring together two bodies of work that developed largely in parallel during the 1970s. On the one hand, there was the humanistic geography 'which [hoped] to capture general features of place through the specifics of human interaction' and, on the other, was the Marxist geography 'which [hoped] to read off the specifics of places through the general laws of tendencies of capitalism' (Thrift, 1983b: 23). Somehow, they argued, the former's 'voluntarist understanding of agency had to be related to the latter's 'deterministic' understandings of structure (Agnew, 1995). But a fine line had to to trodden because:

> the 'nitty gritty' of everyday life [could not] be presented as raw, unmediated data – the empiricist fallacy, data speaking for itself – nor [could] it be presented through abstract theoretical categories – the theoreticist and idealist trap, the lack of interest in empirical findings ... [What's] best for the relation – data/theory – is the 'surprise' ' ... that each can bring to the other ... [through a] continuous process of shifting back and forth ... between 'induction' and 'deduction'. (Willis and Trondman, 2000: 12; see aslo Ley and Mountz, 2001)

Ethnographic research into everyday life designed to test or illustrate abstract theory or to generate grounded theory by 'bracketing out' that abstract theory therefore keeps theory and everyday life separate (Dwyer and Limb, 2001). Structuration theory was developed in the 1980s better to understand their entanglements, and ethnographic research played an important part in this.

For structuration theory's main advocate, sociologist Anthony Giddens (1984: 16), structure was 'often naively conceived of in terms of visual imagery, akin to the skeleton or morphology of an organism or the girder of a building' and was therefore treated as '"external" to human action, as a source of constraint to the free initiative of the independently constituted subject'. He argued, instead, that there was a duality of structure which placed human agency in a web of cause and effect which was both buried in the routinized, often taken for granted, ' "how to do" and therefore "how to be" messages' in/of everyday life (Thrift, 1983b: 43) *and* stretched out of awareness and control through the 'systemness' of society. Here, structure and agency were effectively the same thing; the structuration process was unfolding and unfinished; people's everyday activities – consciously, intentionally or otherwise – were central to this process; and, therefore, their consequences were countless, often unintended and inevitably folded back into the contexts for future activities.

For ethnographers, structuration theory was interesting because it emphasized:

- how, where, when and with whom people learnt those 'how to do' and 'how to be' messages;
- how these messages were continually reinforced components *and* results of routinized, compartmentalized and channelled 'activity experiences';
- how much of the knowledge in these messages comprised a taken-for-granted 'practical consciousness', which was only rationalized into 'discursive consciousness' when routines were disrupted and/or actiity experiences jarred when moving between locales (Giddens, 1984; Thrift, 1985);
- how much of this knowledge was geographical and bodily, as Graham Rowles had discovered; and
- how participant observation's uniquely long-term, situated and embodied approach (watching, talking *and* doing) could therefore play a vital role in translating practical consciousness into discursive consciousness.

Ethnographic studies reconnecting structure and agency could therefore address important questions about the 'potency of human agency in an unfinished world' and draw attention 'to the prospects for effective action rather than determinacy in social life' (Smith, S.J., 1984: 353). And, through unearthing, putting into words and questioning the taken for granted, they could potentially 'restore human beings to their worlds in such a way that they can take part in the collective transformation of their own human geographies' (Gregory, 1981b: 5).

Despite this potential structuration theory has more or less disappeared from geographical view. One explanation is that it couldn't turn its theoretical principles into new forms of research practice because 'Giddens ... consistently advocated a methodological bracketing that allows for either the analysis of strategic conduct or the analysis of institutions ... [thereby] transpos[ing] the dualism between '"agency" and "structure" from a theoretical to a methodological level' (Gregory, 2000a: 800). Nowhere was this better illustrated than in Paul Willis' (1977) *Learning to Labour: How Working Class Kids Get Working Class Jobs*. This was a landmark ethnography due to its detailed combination of political economic theory and ethnographic research. Nigel Thrift (1983b) used it to illustrate structuration's arguments about the intended and unintended consequences of social action. And, subsequently, it was a springboard for George Marcus' sustained methodological rethinking of structure/agency relations (1986; 1998; Marcus and Fischer, 1986; 1999). For Thrift (1983b: 43), *Learning to labour*

> shows how some working-class schoolchildren build up a resistance to mental work out of their resistance to the authority of the school. Manual work is invested with seemingly opposite qualities to mental work – aggressiveness, solidarity, sharpness of wit, masculinity – and becomes a positive affirmation of freedom. Here resistance has the effect that working-class schoolchildren

> accept working-class jobs through 'free' choice. They willingly embrace their own repression.

Thrift did not criticize Willis' book, despite its resemblance to the kind of functionalist work he was arguing against. *Learning to labour* was an attempt to trace 'the passage of a group of young men from school to work ... [to] show... how their unequivocal rejection of middle-class values represents a symbolic opposition to their *structural* subordination' (Jackson, 1989: 6, emphasis added). Willis undertook ethnographic research with a group of 'lads' and contextualised their lives and choices within abstract arguments about 'class structured society built into the tradition of Marxist theory' (Marcus, 1986: 186). Half the book was detailed ethnographic description and half was dense theoretical argumentation. Agency was studied empirically and structure was theorized. That was the problem.

Marcus' reading of Willis' approach was more critical than Thrift's. For him it simply added a 'recognition of the messiness of real life ... to conventional notions of functionalist order' (1986: 183). It supported that 'unintended consequences' argument only in the sense that 'agents do what they do in their various contexts, and hey presto!, there is neat order in the system' (1986: 183). Willis made 'the lads real, but ... [reified] the larger social system in which they [lived] ... [by ignoring the fact that] what is "the system" for the lads is the other's (the middle class) cultural form' (p. 186). Thus, Marcus concluded:

> [If] the system thoroughly operates by unwitting interdependencies and unintended, but coordinated consequences, ... it would be difficult to sustain the conventional way of representing the dominant class or classes in Marxist theory ... If [Willis] had instead constructed his text on cross-class ethnographic juxtapositions, he would have had to face the full implications of his

unintended consequences more squarely. (pp. 186–7)

This is a powerful and important structure/ agency, theory/data argument. But what would an ethnography that took it into account actually look like? Marcus and Fisher (1986: 91) answered:

> What we have in mind [is the kind of work] that takes as its subject not a concentrated group of people in a community, affected in one way or another by political-economic forces, but 'the system' itself – the political and economic processes spanning different locales, or even different continents. Ethnographically, these processes are registered in the activities of dispersed groups or individuals whose actions have mutual, often unintended, consequences for each other, as they are connected by markets and other major institutions that make the world a system.

Marcus (1986: 171) argued that 'the most obvious views of systems as objects for experimentation with multi-locale ethnographies' were 'markets (Adam Smith's invisible hand) and capitalist modes of production, distribution, and consumption (Marx's version of the invisible hand – commodity fetishism)'. However, after undertaking a review of multi-locale ethnographies some years later, he (1995; 1998) found that Marxism was not the only body of theory that, in practice, could enjoy such a following. Poststructural theory worked just as well.

One of poststructuralism's core beliefs is that dominant categorizations and binary oppositions must be unsettled and resisted in order to create a 'non-oppressive politics' (Pratt, 2002: 626). As Donna Haraway (1991: 177) has argued:

> self/other, mind/body, culture/nature [male/female,] civilised/primitive, reality/ appearance, whole/part, agent/resource, maker/made, active/passive, right/wrong, truth/illusion, total/partial, God/man [and, we would add, global/local] … have all been systematic to the logics of domination of women, people of colour, nature, workers, animals – in short domination of all constituted as others, whose task is to mirror the self.

Working against this, therefore, requires 'new connective modes of theorising' (Pratt, 2000b: 626) and this is where geographers' contemporary fascination with non-representational theory comes in (Thrift 1995, 2000b). This is where:

- the 'good bits' of structuration theory have arguably resurfaced in new, improved forms;
- calls have been made for new types of research 'where "theory" is not "applied" in a particular "context" but where the theoretical and empirical fold into one another in a glutinous mix of text and context, and facticity and fictionality' (Thrift, 1995: 530); and
- 'observant participation' is now being touted as a key means to accomplish this (Thrift, 2000b; Smith, S.J., 2001: 35–6; Hoggart et al., 2002: 258).

We can only give a flavour of the arguments here, in part because they aren't yet adequately developed in the literature. None of human geography's qualitative research texts deals with this in sufficient, substantial detail, and its finished products have tended to be 'unspecific or unclear' about research practices (Davies, 1998: 76).

Like much of the literature we have discussed so far, non-representational theory is primarily concerned with the 'mundane everyday practices, that shape the conduct of human beings towards others and themselves at particular sites' (Thrift, 1997, cited in Nash, 2000: 65). But its advocates have criticized qualitative research in human geography, first, for registering only 'narrow realms of sensate life' (Thrift, 2000a: 3) and, secondly, for relying on a model of consciousness where 'all acts are secondary to the processing of (or even deliberation over) knowledge as representation' (Hinchliffe, 2000: 575). New research practices are therefore necessary to reach beyond what *can* be rationalized into words (i.e. from, or translated into, 'discursive consciousness') into 'realms of sensate life' that, for most people, *cannot*: for example, the workings of the 'unconscious', the body's precognitive practices (e.g. kinaesthesia)[12] and the 'rounded sensory experience' of everyday life (Smith, S.J., 2001: 31; see also Gregory, 2000a; 2000b; Nash, 2000).

However, as shown in geographical studies of embodiment, experience and performance (e.g. Rowles, 1978b; Crang, 1994; Malbon, 1999), it is surprising what non-specialists *can* and *do* turn into words when ethnographies are conducted appropriately, in sufficient detail, with members of carefully chosen 'subject communities', and in the right times and places. Thus non-representational theory could usefully sensitize prospective ethnographers to push the boundaries of research design (by employing, for example, new media technologies which can help to shift attention beyond words) and to consider the experiential side of its methods in a more holistic manner. Going beyond this could, however, be problematic because:

How can ... precognitive body practices be known, or is the effort to understand and communicate abandoned in favour of abstract theorising of the non-representable? Are ethnographic research methods as redundant as textual or visual sources, since they invite people to speak and therefore cannot access the preverbal? What happens to the project of 'giving a voice' to the marginalised, if the concern is with what cannot be expressed rather than what can? ... How does the focus on noncognitive everyday practices position the academic? The energy spent in finding ways to express the inexpressible, ... seems to imply a new (or maybe old) division of labour separating academics who think (especially about not thinking or the noncognitive) and those 'ordinary people' out there who just act. (Nash, 2000: 662)

To avoid turning into another ethnographic dead-end – calling for much but delivering little – geographers need to undertake and publish substantial studies which show how non-representational theory and its observant participation can surprise one another.

To an extent, this has already begun in one area of non-representational theory – actor network theory or ANT – whose most notable full-blown empirical study – Bruno Latour's (1996) *Aramis, or the Love of Technology* – is packed with 'surprises' (see Laurier and Philo, 1999). ANT is, arguably, what structuration theory could have become if Giddens had more than a 'limited encounter with post-structuralism', if he'd been able to think of 'action as [more than] individual and [had more] fully considered the ghost of networked others that continually informs that action' (Thrift cited in Gregory, 2000a: 800). In ANT, agency is far more than the activity of individual human beings. Rather, it is an attribution of 'the ability to act' to 'functional collectives' which can 'include non-human actors and new couplings of human beings and machines' (Gregory, 2000b: 350). It is 'a relational effect generated by a network of heterogeneous, interacting components whose activity is constituted in the networks of which they form a part' (Whatmore cited in Gregory, 2000b: 350). And its structures cannot, therefore, be seen as 'lying somewhere *outside* or *above* the fray' but are, like everything else, 'intricate weavings of *situated* people, artefacts, codes, and living things and the maintenance of particular tapestries of connection across the world' (Whatmore and Thorne, 1997: 288, emphasis in original).

These are perhaps the barest of ANT's bare bones. Fleshed out, it can be a mind-bogglingly abstract labyrinth of ideas. However, in *Aramis* and in other ANT-inspired empirical studies, it is often a more obvious and thought-provoking 'way of see-ing'. Here the dividing lines between agency/structure, theory/data, global/local and plenty of other binary oppositions disappear from view in the mundane, situated complexities of everyday life.[13]

In this kind of study, the research process takes shape after researchers (1) latch on to specific people, things, metaphors, plots/stories/allegories, lives/biographies and/or conflicts; and (2) follow them, see what 'sticks' to them, gets wrapped up in them, unravels them and where that takes a researcher who is (un)able to follow myriad possible leads and to make myriad possible connections (Marcus, 1995; 1998). Here, questions are asked of 'emergent object[s] of study whose contours, sites, and relationships are not known before-hand' but come into being and take 'unex-pected trajectories' through following up the answers to those questions, and the questions that they then raise, and so on, *where possible* (Marcus, 1998: 86, 80). As Jonathan Murdoch has put it, 'We let them [the people, the things, etc.] show us where to look, what material they use in the course of network construction and how they come to be related to others' (cited in Davies, 1998: 77). Theory is therefore '"held lightly" and is made accountable to fieldwork' (Dwyer and Limb, 2001: 11) rather than 'imprison[ing] field data within preconceived positions' (Ley and Mountz, 2001: 236). The big explanations in terms of politics, economics, organisation, and technology always turn up, without fail: "It's politically unacceptable". "It isn't prof-itable". "Society isn't ready for it". "It's ineffi-cient"' (Latour, 1996, cited in Laurier and Philo, 1999: 1054).

What is interesting in these studies is how the connections being made or followed are 'ordered' by their human and non-human participants (including researchers), and how the research process is thereby at once com-plexly empirical and 'theoretical' (in both aca-demic and everyday senses of 'sense-making'). Ian Cook et al.'s (2002) papaya paper, for example, brings together theoretical debates about:

> culture and economy, consumption and production, postcolonialism and globalisa-tion, commodity chains, political economy, actor network theory, multi-local ethnogra-phy, plant physiology, retailing, gender, 'race' and class, labour relations. More besides. [These are] Academic debates which could never be synthesised into a big, complex theoretical position. By me at least. But they do work together in an empirical account tracing the biographies and geographies of what might initially appear to be a discrete thing. In all of its complexities and contradictions.

Similarly, Mike Michael's (2000: 13) study of dog leads 'simultaneously explor[es] the inter-relations between dogs and humans, cultural representations of animals, community, local and central government, the environment and the body'. What this approach does is 'remove [the idea of context] as a subtle safety net for many theorists who will often invoke it as a term akin to social order or social structure' (Laurier and Philo, 2003: 1055). It can, there-fore, be unpopular because it decentres estab-lished radical standpoints like the 'resistance and accommodation framework that has organised a considerable body of valuable research' (Marcus, 1998: 85). What it offers instead, however, is an approach which, first, is 'modest', 'pragmatic', 'recursive' and 'commit-ted to an order*ing* inquiry into order*ing*, rather than to an ordered inquiry which uncovers other root orders' (Law, 1994: 97), and, secondly, enables the 'mega concepts with which con-temporary social science is afflicted ... [to] be given the sort of actuality that makes it possible to think not only realistically and concretely *about* them, but, ... creatively and imaginatively with them' (Geertz, 1973: 23, emphasis in original).

Few of these studies exist and few of them – including *Aramis* – are, strictly speaking, ethnographic in the sense that they have been based on a substantial period of participant observation research. *Aramis*, for example, is participant observation in style but is, in fact, 'a "novelistic" report' based on 'a rich archive of documentary sources (planning and policy papers, publicity materials, consultancy reports, memos, scribbled notes), iconographic sources (photographs, maps, diagrams) and also in-depth interviews with many of the relevant human actors' (Laurier and Philo, 1999: 1050). Such studies are therefore of little use if we want answers to basic methodological questions 'such as how to construct the multi-sited space through which the ethnographer traverses' (Marcus, 1998: 89). Saying that 'they' will 'show us where to look' doesn't help much because:

> How do we know where they are telling us to look without knowing their vocabulary? How do we understand the significance of their materials without using existing categories? How do we understand their relationships without drawing on prior experiences? And, perhaps most reasonably, how do you get an immensely busy community of people to do your work for you? (Davies, 1998: 77)

These questions are rarely addressed in geography's ANT (and after) literature (although, see Davies, forthcoming). But, fortunately, they have been addressed in geography's growing, process-oriented qualitative methods literature. Here, numerous 'warts and all' stories show that research practice always involves working through, extending and then (often) hiding network relations (Marcus, 1998; see also Keith, 1992; Cook and

Crang, 1995; Davies, forthcoming). A key feature of anthropology's cultural turn was how it took behind the scenes, 'corridor talk' about what shaped fieldwork experiences and objects of study, put them on the pages of academic books and journals and turned them into 'up front' debates about the theory and practice of ethnographic inquiry. This was called 'being reflexive', and geographers joined in.

Being reflexive

Geography's cultural turn led many to 'question the constitution of the discipline – what we know, how we know it and what difference this makes both to the type of research we do and who participates in it with us, as either colleagues or research subjects' (McDowell, 1992a: 399–400). The 'social construction of knowledges and discourses' within the discipline 'and the relations of power embedded within them' were hotly debated (McDowell, 1992a: 399–400). In geography's 'house journals', awkward and profound questions were asked:

> What is the relationship between our own background, current position and values, and our own research agenda? How do we know what we know? Through what sort of lenses is our knowledge filtered? Who is included and who is excluded by the social practices and academic subject matter of academic geography? For whom are we writing? And what is, what should be, the relationship between our theories and our politics, between our thinking and action? (McDowell, 1992b: 56)

And this has continued. Why are most researchers 'commonly white, male and middle class, with disproportionately low ratios between researcher and researched for non-whites, the poor, women and so on' (Hoggart et al., 2002: 261)? Why does research so 'often involve those with more economic and cultural capital studying those with less' (Hoggart et al., 2002: 261)? Why do researchers mainly report their findings to people like themselves (Ley and Mountz, 2001)? How can you justify using methods designed to develop sufficient trust for participants to yield sensitive information and then risk betraying this by publicly writing what may 'upset (or perhaps even disadvantage)' them (Ekinsmyth, 2002: 183; see also Ley and Mountz, 2001)? Given its imperial roots, how can ethnographic research (in the 'Third World' or elsewhere) avoid being the 'handmaiden to broader colonialist projects that invent oppressed groups as a means of controlling them' (Herbert, 2000: 562)? So, 'who has the right to write about whom' (Hoggart et al., 2002: 263)? Are there circumstances where researching 'the Other … [is] ethically defensible' (Hoggart et al., 2002: 263)? And, how can 'the Other' be represented anyway, 'when researchers are so thoroughly saturated with the ideological baggage of their own culture' (Ley and Mountz, 2001: 235)?

These questions are awkward and bewildering, and geographical research could easily have ground to a halt under their weight. However, if anything, the opposite has happened. Qualitative studies of 'non-dominant, … socially excluded or marginalised groups' (Smith, S.J., 2001: 25–6) have flourished (Dwyer and Limb, 2001). This may reflect, first, the increasing involvement and/or visibility of members of non-dominant groups in academic human geography and, secondly, critiques of oversimplified, essentialist distinctions between *'the* Self' and *'the* Other' in debates about culture and the politics of difference (Ekinsmyth, 2002). Here, identity has not been seen in such stark either/or categories but as 'partial in all its guises, never finished, whole, simple there and original; it is always constructed and stitched together imperfectly, and *therefore* able to join

with another, to see together without claiming to be another' (Haraway, 1996: 119, emphasis in original).

These new studies seem to have resulted from the political choices by a, perhaps, more diverse, body of geographers committed to feminist, anti-racist, postcolonial and like-minded projects. They have seen their rights complicated by their responsibilities to do certain kinds of research. Many geographers have wanted to recover and centralize … '"marginalised" voices and stories' (Parr, 2001: 181) with the aim of 'unsettling the status quo, or redefining what is relevant, useful and legitimate [geographical] knowledge' (Smith, S.J., 2001: 25–6). Chris Philo's (2000: 29) verdict, at least, is that this has had the desired effect, creating 'a new ambience for human geography that has blown away many cobwebs of convention, conservatism and even downright prejudice'. And qualitative, 'intersubjective' methods have played their part in this because they make it 'easier to equalise power relations' by 'allow[ing] the researched more opportunity to "frame" (i.e. identify salient issues etc.) the research' (Ekinsmyth, 2002: 181) 'without being invasive, colonising and violent' (Johnston, 1997b: 291).

These principles have been a core contribution of feminist ethnography (Madge et al., 1997; Nagar, 1997a; Ekinsmyth, 2002) and were, arguably, central to our three humanistic ethnographies. Rowles' work, for instance, not only provided vivid insights into the geographical lives and imaginations of elderly people but also challenged stereotypes about the ageing process both in American society in general and in the academic disciplines that claimed expertise in this field. This happened because he organized his research around the belief that elderly people knew an awful lot more about the ageing process than he did, and that he therefore had much to learn from them. He used ethnographic methods to do research *with* elderly people, and his data detailed not only what he learnt but also *how and where* it was (re)framed: in conversation, *in situ*, over time and with difficulty.

This ethnographic tradition has continued in geography. However, it has been supplemented by new work that pushes its politics to their logical extreme and/or studies people who wouldn't normally be considered to be marginal. First, more explicitly interventionist projects have drawn upon participatory action, activist and/or emancipatory research methods. Here, researchers do 'not simply [aim] to empathise with research subjects' but, through contributing professional skills and contacts to the struggles of marginalized peoples, aim 'to help to produce changes in their lives' (Dwyer and Limb, 2001: 7).[14] Secondly, it is important to recognize that shining so much light on the relatively poor and powerless leaves in the shade huge parts of the processes that human geographers aim to understand. Thus there is a growing body of ethnographic research with more wealthy and powerful groups.[15] While such ethnographies are often framed as geographical inflections of particular ethnographic traditions, they are also the result of much more practical choices about when, where and with whom research can be done.

To help think through the problems and practicalities brought up in these debates, many geographers have followed other social scientists by becoming much more reflexive. Reflexivity can be seen to take three forms which attempt to 'link the micro-level activities of the lone ethnographer with macro-level processes and imperatives' (Herbert, 2000: 562). First, the power relations structuring the interpersonal relationships that ethnographies are based upon have been scrutinized. Questions have been asked about whom and what researchers have power over, whom and

what has power over them, how their power and status can change as they move through their 'expanded field', and what relative effects research has on the lives of researchers and researched (See Robson, 1994; Cook, 2001).[16] Secondly, geographical institutions, traditions, ways of seeing and the ways they can affect the 'conditions and modes of producing knowledge about other cultures' (Callaway, 1992: 32) have also been put under the microscope. It has been noted, for instance, that geographical traditions and institutional arrangements attract only a narrow range of students, who work through established social and/or disciplinary networks which guide their research in certain directions (see McDowell, 1992b; Sidaway, 1997; Twyman et al., 1999). Finally, wider questions have been asked about how these issues are related to 'a world historical situation in which, as one face of imperial domination, differentially permeable barriers to the movement of people, and to the flow of information, have been erected and have asymmetrically organized contact between the world's people' (Dwyer, 1977: 148).

Calls to be more reflexive in these ways might, again, seem quite daunting. Some argue that too much reflexivity might produce research that is more self- than other-fascinated and thereby discourages the study of issues that really matter, 'out there' in the wider world (Pile and Thrift, 1995). That said, this self-reflection has, however, resulted in new ways of thinking about the planning and conduct of ethnographic research. First, writing about the prehistories of research projects has shown how the personal is the political, that concerns, motivations and passions rooted in people's life experiences play an important role in energizing and directing their research (see Bristow, 1991; Cloke, 1995; Saltmarsh, 2001). Secondly, reflexive writing has shown that, however 'out there' and 'other oriented' a project appears to be, it will typically have taken shape through exploiting and extending networks of which researchers and/or their supervisors are already, or could easily become, part. Research is often done in places where:

- trusted contacts have already been established in local universities, NGOs, and so on (see Robson and Wills, 1994a);
- spaces and/or social groups can be entered through paying a fee (see Cook, 1997; Desforges, 2001);
- a job, work placement, or voluntary position is only an application away (see Crang, 1994)
- no-fee events, campaigns or other activities, provide relatively free access (see Jackson, 1988; Parr, 2000; Anderson, 2002);
- researchers work out that they know people who know people who can help them to develop contacts in seemingly impregnable or distant organisations and places (see Keith, 1992; Davies, 2003; Cook and Crang, 1995);
- researchers choose to work within and through their own family, friendship and leisure networks (see Lakhani in Hoggart et al., 2002; Saltmarsh, 2001); or.

All this reflection has led to a rethinking of what counts as 'the field': where and when 'it' begins and ends, and where and when researchers should consider themselves to be 'on duty' (or not). First encounters in traditionally ascribed fields – neatly bounded and out there – are by no means the only ones that make a difference to research findings. Ethnographic and other research always takes place in an expanded field, where researchers inhabit and move between a number of different locales, people and frames of meaning, and their work inevitably involves complex translations of meaning between *all* these settings.[17] Here, the variety of research participants is much wider than is usually acknowledged. They include not only those people in a researcher's proper subject community but also those who inspire, fund, direct and act as the official audience for his or her work, *and* the friends and family members whose support and/or understanding (or lack of it) can play an important role in getting things done (see Cook, 2001). In addition to the residents of Mowbray, for example, Robert Coles and Victor Wessels were key participants in Western's (1981) expanded fieldwork.

Secondly, moving between locales and relations where norms of speech, appearance, status, ethics, manners and so on may be very different often involves a complex, confusing, anxiety-provoking, and sometimes bizarre process of 'identity management.'[18] Some of this is specific to participant observation as decisions have to be made about how covert or overt this should be (see Hoggart et al., 2002: 256–7). But, research more generally, 'can be a tricky, fascinating, awkward, tedious, annoying, hilarious, confusing, disturbing, mechanical, sociable, isolating, surprising, sweaty, messy, systematic, costly, draining, iterative, contradictory, open-ended process' (Cook et al., in press). It is invariably an intellectually and emotionally challenging experience, and prospective researchers need to be prepared for this. Ideally such research reaches a point where the researcher has the confidence to 'follow impulses, change directions, and co-ordinate with other people … [because he or she understands that local] unpredictability has its distinctive tempo, and it permits people to develop timing, coordination, and a knack for responding to contingencies' (Rosaldo, 1993: 112). But this may not always happen, especially if the research isn't given sufficient time.

So, to conclude, reflexive writing about the process of doing ethnographies can help researchers to explain – to themselves and to others – what they did and why *under the circumstances. This* is where questions of positionality, politics and, indeed, ethics get worked out, often in quite messy ways (McDowell, 1992a). Expanded field *work,* therefore, requires expanded field *notes,* and prospective ethnographers can learn much from the growing body of geographers' reflexive, 'warts and all' accounts of their experiences.

Top tips for prospective researchers

Before we conclude this chapter by discussing the construction of ethnography's prime 'data' – its fieldnotes – we want to take stock of the arguments presented so far and to condense them into some top tips for anyone thinking of doing participant observation research. If you have got this far and still want to do this, we recommend that you do the following.

Read some 'warts and all' accounts

Read such accounts no matter how irrelevant they may seem to what you plan to do and even if you don't plan to use participant observation as a formal research method.[19] Try to read them *before* undertaking your research. They provide tales from the (expanded) field that can be full of ideas, inspiration, confidence, mistakes, warnings, realizations, etc., from which you could learn in planning and conducting your research. If or when *your* research doesn't quite go to plan, it isn't necessarily because it's going badly but because that's what happens in participant observation research. Like those authors, you would have to cope under those circumstances, think on your feet, change the way you do things, try to maintain some focus and live with the fact that you can't be sure exactly how things will turn out.

Read them alongside some 'methodological cookbooks'

Read those texts that don't provide such rich tales from the field but outline the practicalities involved in using a variety of research methods, including participant observation.[20] Despite the inevitable twists and turns of research practice, it's important to set out to do research properly, systematically, rigorously, to have a plan and to know how to do different types of research well. But, if things don't go to plan and/or if opportunities unexpectedly arise in the field to undertake, for instance, participant observation or archive work, you need to know how to do this as well. Take one or more of these books (or their equivalent) into the field, just in case.

Familiarize yourself with the history of ethnographic research in geography

Ethnographic research has been undertaken across a number of disciplines, all over the world and for over a century. So it is reasonable to assume that someone, somewhere, has done research which somehow relates to what you're thinking of doing. As we have shown, calls for geography's latest, cutting-edge, 'poststructural' ethnographies could usefully draw upon some neglected 'humanistic ethnographies' undertaken in the 1970s. Even if their 'conceptual bookends' may seem inappropriate or out of date, the 'thick description[s] of everyday life' that they contain may still be informative and thought provoking (Ley, 1998: 79). It is important to recognize that there is a long-running and distinctive ethnographic tradition within the discipline, so geographers don't *only* have to look elsewhere for inspiration.

Read some results of participant observation research

Have in mind the kind of final product you may want, or be expected, to produce from your research. 'Warts and all' and 'cookbook' reading can help you to picture what your methodological writing might look like. But what about the findings of your research? How could they be presented? How have others presented their findings? What theoretical ideas, methodologies, field settings and findings have they put together? Have they made them appear to be somewhat discrete? Or have they presented them as jumbled up, more or less as they fitted together in the expanded field? Have they limited themselves to the usual combinations of words, pictures, maps, diagrams, etc., on paper? Or have they expressed things differently on and/or off the page? What academic precedence is there for doing this? And, having read this work and decided what works best for you, what *kinds* of data do you therefore want to produce, in what sorts of detail and in what state of separation for your write-up?[21]

Mix up your reading, doing and writing

Open a research diary as soon as you start thinking about your dissertation or your next research project *especially*, but not only, if you're considering using participant observation. It's essential to undertake research that is interesting (to you and to others), relevant (in an academic, popular, policy, etc., sense) *and* 'doable' (in a practical sense). Interesting, relevant research that can't be done, or doable research that is irrelevant and/or uninteresting should be avoided if possible! Interesting, relevant, doable research has to be pieced together over time – through trial and error, false starts, dead-ends and breakthroughs, flights of imagination

and reality bites and, ideally, tentative forays into the field to make contacts and to try things out. Good combinations of theory, methodology and practice should grow and work together.

So why not record in a research diary any interesting and relevant ideas from academic reading, newspapers and other sources and how these have gone down and made sense as you've discussed them with your peers, supervisors or other people? A diary can also be somewhere to have a go at writing fieldnotes before your really need to. These notes could concern the expanding field of your project. They could outline your early contact-making and suggest how your ideas could develop. They could also be experiments in field-noting where you go somewhere and/or do something new and try to finds ways of vividly conveying that experience to an academic audience (fellow students, qualitative methods lecturers or your supervisor; see Warren et al., 2000). So what guidance can we give on this?

Conclusion: field-noting

Until recently, geography's prospective ethnographers had only a limited methodology literature to draw upon when planning their research. In development geography, for example, students were often expected to prepare for fieldwork through 'a good grasp of theory and a well written literature review' but, rarely, through a good grasp of the actual 'rigours of fieldwork' (Robson and Willis, 1994b: 1). This is what they were expected to learn through a '"baptism of fire" in the field – probably mirroring their supervisors' own experiences' (Robson and Willis, 1994b: 1). In disciplines where ethnography is a core methodology – most notably in anthropology – the ways in which 'fieldwork dilemmas arise and are resolved' have been an important part of research training (McDowell, 1992a: 408). In geography, it has not. Here, far more attention has been paid to the products than to 'the *processes* of actually undertaking the research' (Dwyer and Limb, 2001: 2, emphasis in original). Yet field noting is necessary to record and to make sense of key events as they happen, to think about how the research is taking shape, to try to maintain and/or regain its focus and to plan its next stages. These notes may be systematically written in a proper research diary but they may also take other forms: as researchers scribble notes on stray pieces of paper or around the edges of questionnaires; as they write emails and letters home to friends, family and supervisors; or as they continue to fill the pages of a personal diary or journal. Field-noting is not the sole preserve of ethnographers, then, because *all* research involves a combination of participation and observation in projects that change as they proceed, and such changes are invariably noted in some way or other. Proper field-notes *are,* however, the prime data of participant observation research. These are what should enable researchers to give those who read their work that unique sense of 'being there'.

Katy Bennett has recently recalled the problems she had doing participant observation in a British farming household. She wrote about 'the angst of not knowing what to write down, what ought to be written, what's the most important part of a conversation, an observation, an event …[and] the feelings of drowning in voices and "data" and not knowing where to start.' (Bennett and Shurmer-Smith, 2001: 255). Her supervisor simply said: 'Just keep writing' (2001: 255). What sort of writing are we talking about, though? How spontaneous and/or systematic should it be? It's not just a simple matter of collecting or recording experiential data from

the field. Fieldnotes can never be anything other than 'fictions in the sense that they are "something made", "something fashioned"' (Geertz, 1973: 15 see also Willis and Trondman, 2000; Bennett and Shurmer-Smith, 2001; Hoggart et al., 2002). And these are 'fictions' which shape as much as they reflect researchers' experiences of the research process (Wolfinger, 2002). As soon pen is put to paper, or fingers to keys, the recording and interpretation of data blur into one another (Burgess, 1986).

Field-noting is an ongoing sense-making process. It is a process of creative writing based on first-hand experience. It involves attempts to tie together minutiae of theoretical and empirical detail gleaned in and between the different locales of a project's expanded field (Batterbury, 1994). It usually involves more than one kind of writing for more than one audience. Some of what gets written should be suitable for publication in raw or refined forms. Yet significant parts, however 'juicily' relevant to the research, may have to remain private to protect the researched.[22] Moreover, for a researcher often isolated in the field, a notebook may be a place to write things purely for him or herself, to write out the frustrations of being there, struggling to do that research, to let off steam, let rip, write all kinds of intemperate, intolerant, lusty, clumsy, emotional things (Parr, 2001) that could never, ever, be included in a final product but might be easier to cope with once they have been put into words. For many researchers, this self-reflexive, soul-searching, worry-writing may comprise the majority of their fieldnotes.

Whatever the contents, field-noting is often done in stages, starting with hastily written scratch notes jotted down then and there ('inscriptions'), which are then elaborated upon at regular intervals (ideally each day) in more flowing narratives ('descriptions'), parts of which may later be processed into the more polished tales from the field that can appear in finished products.[23] After the initial, highly detailed scenes, characters and roles have been written, field-noting may narrow its foci as researchers mention only changes to these established scenes, etc., and focus in on, latch on to and follow up parts of these (Wolfinger, 2002). Here it is common for researchers to begin to identify patterns and regularities, rhythms and routines, dominant discourses and ways of seeing/doing, to check facts and verify claims made in one time/place/setting in others, and so on (Jackson and Smith, 1984). Identifiable individuals, places and so on often have to be anonymized, usually in the last stages of writing a finished product, but sometimes as the notes are being written if there is a danger they could fall into the wrong hands (see Warren et al., 2000; Dowler, 2001). The sense-making in these notes may be the work not only of the researcher but also of co-workers such as research assistants and translators, and the ways these perspectives come together may also need to be described (Twyman et al., 1999). Particularly where researchers work in field areas radically different from their home contexts, field-noting may involve writing about taken-for-granted aspects of identities and experiences that may become more noticeable but may also be difficult to describe (Punch, 2001).

Field-noting takes up a considerable amount of time, with every hour of participant observation requiring, according to one estimate, two hours of note-taking (Junker cited in Burgess, 1986). This means that a careful balance has to be struck, *in the field*, between the what, when and where of doing and writing. Field-noting (along with other administrative elements of data construction such as filling and sending mail) is not something that should simply be done in the

researcher's spare time (Crang, 1994). Moreover, wherever and whenever it is done, it provokes emotions (especially feelings of guilt that this is a betrayal of participants' trust) as much as it attempts to describe their role in and between research settings throughout the research process (Stacey, 1988; Cook, 2001; Bennett, 2002). Great care has to be taken to ensure that these notes and other bits of data don't get lost or confiscated. Access to a photocopier and/or a laptop computer is therefore essential to make no-extra-time duplicates which can be kept in or sent to a number of safe places (Dowler, 2001). Finally, only the tiniest proportion of these notes end up on the pages of completed works. So, how can good notes be written and what do they look like?

Participant observation research is supposed to provide its readers with a vivid impression of 'being there'. So how might we conjure up a sense of place of the locale(s) in which the research is based and how this might change over the minutes, hours, days, weeks and months of the research? How might we describe the rhythms and routines of those actors who inhabit and/or comprise that setting? How might we move from these observations to describe our interactions in that setting with those actors, initially and as they develop over time? How might we describe not only what we are learning but also the often strange and strained circumstances under which that learning is taking place? How might we describe the difference this makes to the way the research seems to shaping up, where it can and should go, and what power we have over this? And how would we describe how this develops in and between the locales of our expanding fields?

To answer these questions, we echo the guidance given to researchers planning semi-structured interviews: draw up a checklist which reminds you of the issues you should probably cover, in whatever order they come up, but be prepared to 'go with the flow' when the unexpected happens (Evans, 1988). For more observational or substantive field notes, attention could be paid to the following list of prompts:

1. *Space:* the physical place or places; 2. *Actor:* the people involved; 3. *Activity:* a set of related acts people do; 4. *Object:* the physical things that are present; 5. *Act:* single actions that people do; 6. *Event:* a set of related activities that people carry out; 7. *Time:* the sequencing that takes place over time; 8. *Goal:* the things that people are trying to accomplish; [and] 9. *Feeling:* the emotions felt and expressed. (Wolfinger, 2002: 91)

These could be supplemented by more speculative prompts which can help the researcher to think through what might have been going on there, why that might have been the case and how this might be further investigated. For example:

Who is he? What does he do? What do you think she meant by that? What are they supposed to do? Why did she do that? Why is that done? What happens after _____? What would happen if _____? What do you think about _____? [and] Who is responsible if _____? (Lofland and Lofland cited in Wolfinger, 2002: 90)

This is where observational writing begins to blur with participatory writing. So in addition to and/or, wrapped up in, these substantive notes, it is essential to discuss the more reflexive, participatory aspects of the research process. These are the 'methodological fieldnotes': the 'detailed records of the research process, research relations, and research roles', which provide a means both to 'allow the researcher to talk things through; a dialogue with oneself' and to provide the raw materials to 'construct an autobiographical account of the major phases in the research process' in the final report (Burgess, 1986: 58–9).

Methodological footnotes often comprise descriptions of the ways in which research unfolds or takes shape in or between the settings and interactions described in the substantive fieldnotes. And their prompts should relate more to the research process and the expanded field as they take shape in practice. These could include descriptions of:

- The selection of research topics.
- The networking that helps them to take place and shape.
- The ways in which the researcher manages first and subsequent impressions in and between the various social settings and interactions which comprise the field.
- How people seem to be reacting to him or her as a result of his or her age, gender, skin colour, language skills, nationality, politics, dress, etc., and how he or she attempts to work with and/or against this to get the research done.
- The ways in which he or she identifies and (un)successfully tries to build relationships of trust with key informants, what he or she learns from shadowing and talking to them and how others appear to see these relationships.
- How circumstances change and breakthroughs are made.
- How research methods and theoretical standpoints are tailored to match what's possible, what's happening, and what has unexpectedly turned up through the research process.
- The way that field-noting, photography and other forms of data construction have been able to take place and what effects their inscription and description have on the ongoing research process.
- How the researcher becomes part of the scenes he or she is describing, the effects this has on the research process, and on his or her sense of self and his or her findings.
- How all these issues are brought back to, and used in, the institutional context where the research is formally based.

To conclude, in Boxes 6.1–6 we provide a range of raw and polished examples of these notes from published and unpublished geographical ethnographies. These boxes move from the most observational, substantive and public to the most participatory, methodological and private field-noting. They suggest six layers of description, moving progressively from outside to inside perspectives: from locating an ethnographic setting, to describing the physical space of that setting, to describing others' interactions within that setting, to describing researchers' participation in those interactions, to their reflections on the research process, and ending with self-reflections provoked by the research. We ask to what extent they begin to 'take you there', to that drop-in centre, courtroom, street, restaurant, farm, village or bus journey, and to what extent they can place you in the shoes of the researchers who produced them? What level of detail seems to work best as you read these notes? Given the kind of final product you are working towards, do you think it would be better to disentangle your substantive and methodological notes when you produce that final product or when you produce its fieldnotes (see Hoggart et al., 2002)? What could be added to these depictions by using alternative media? All we can suggest is that you have a go at this yourself, *before* embarking on your proper research. Go to an unfamiliar place and think about how you would try to share that experience with someone who had never been there. Go back a few times with a camera, sketch pad, whatever. Get something down, in detail. Hand it over to your likely audience. See what works for you *and* for them. Go back if necessary to elaborate on the thinner parts of your description, etc., until it seems to click.

Box 6.1: The First Layer: Locating an Ethnographic Setting

Parr's 'St Peter's' drop-in centre.

'The drop-in (which I shall call 'St Peter's') is located in Nottingham's inner city, in a district well known for residential hostels, care homes, sheltered housing and day schemes for people with mental health problems. … St Peter's (see Figure 1 [above]) is a drop in located within a church hall and had been established for about 10 years when the research took place'.

Source: (Parr, 2000: 226)

Box 6.2: Layer 2: Describing the Physical Space of that Setting

a) Sketching a setting: Parr's drop-in centre (the re-drawn version).

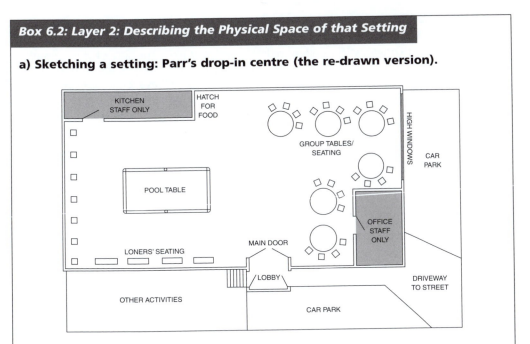

Source: Parr (2000: 230).

b) Describing a setting: Spradley's courtroom:

'There were rows of spectator benches, all made of heavy dark wood, oak or walnut, to match the panelled walls. The rows of benches went for more than twenty-five feet until they met a railing that seemed to neatly mark off a large area for 'official business' … At the right of the area behind the railing were twelve high-back leather chairs behind another railing. A large oak table with massive chairs all faced toward a high lectern *which I took to* be the judge's bench.

Source: Spradley (1980 cited in Wolfinger, 2002: 88).

Box 6.3: Layer 3: Describing Others' Interactions within that Setting

Smith's observations of street policing:

'… one quiet afternoon along the Lozells Road, the orientation towards 'fun' of an ebullient group of black youths eating the fares of 'Dolly's Takeaway' encountered the 'leisured' frame of two police officers approaching from the opposite direction. A chain of subtle gestures was observed as the two parties became aware of each others' approach. An almost imperceptible sequence of exchanges provided enough impetus to transform the episode from one frame (fun and leisure) to another (danger and suspicion). The sequence was as follows. The police officers seemed to stiffen their walk and move closer together. One twitched his hand towards what might have been a radio, and both fixed their eyes on the oncoming group with the consequence that

---

Box 6.3 (Continued)

their subsequent conversation spilled melodramatically from the side of their mouths. The effect was to reframe a leisurely stroll into a display of professionalism. The youths, in turn, exaggerated their usual bounce into what could have been construed as a swagger. They talked more loudly and gesticulated more often. A frame oriented towards fun slipped through into one characterised by ritual demonstration. At the same time, the reserved window–shopping of two elderly white pensioners became an expression of passive solidarity as they linked arms and drew their bags to their bodies. This became, too, an expression of deference, as they crossed the road to remove themselves from the path of the encounter. However, as the groups moved closer and one officer's request for a chip was laughably answered with the flash of an empty packet, a friendly shrug of the shoulders was sufficient to dispel the atmosphere of confrontation and replace it with one of amiable sparring'.

Source: Smith, S.J. (1988: 32–3).

Box 6.4: Layer 4: Your Participation in Interactions in that Setting

Crang's experience of waiting in a restaurant

'An initial panic and despair at not being able to cope with the pace of work, coupled with a demonstrative rushing around and impatience with any delay is eventually replaced by a fragile order, the reduction of customers to ticks on a list. All movement now has a direction, there is none of the earlier aimless congregating; I map out my priorities, I am heading somewhere all the time, and have the next four or five destinations planned out too. Yet this order is vulnerable. Firstly, any interruptions by customers are likely to throw it all into disarray. For that reason I try to control all contacts with tables, for example moving around all my tables to see if there are any queries when I have a spare minute rather than waiting to be called. Secondly, for me that sense of direction and the sheer amount of energy I'm expending mean I need a constant boost; and that is given by tips. If they don't come my motivation collapses. Luckily last night they were pretty good so that feeling of occasional despair as it all gets too much seems worth it'.

Source: Crang, (1994: 683).

Box 6.5: Layer 5: Reflections on the Research Process

Putting plans into action? Cook's experience on a Jamaican papaya farm

'I tell Jim [the farm manager] what I would like to do … which has now become simplified to the impact on the farm and surrounding villages of changes in consumer tastes in the UK. I tell Jim as we leave to mount our bikes again, that I have decided … to make all the contributors to my work anonymous because I don't want to provoke a consumer boycott based on a highly selective reading of the negative aspects of whatever I might find. I emphasised that in my work, I would try to be even handed. I told

Box 6.5 (Continued)

him I'd change his name and a few key identifying features. He said that this had been one thing he had been kind of worried about, and seemed reasonably happy with the arrangement. I think he trusts me, now ...? I said to him at the gate that I thought that my nosing into everyone's business was what each wanted to know and that, hopefully what each gave would be at least equally compensated by what they learned. Of course, there's no way I can guarantee this and it might be dangerous, but we can work with it ... ? We cycled back to the banana field ... We met Philipps, Jim's right hand man ... in his white pickup. Jim told him what I was doing as I had told him and asked whether he would introduce me to the packing house people and tell them what I was doing. I put my bike in the back and we drove there. Walked in (the house had been rearranged for reconcreting the floor), and Philipps called the women one by one, then all at once, and introduced me as being from the Univ of Bristol, interested in the relationship between this and that and would be asking questions about things like their wages and changes in the area. They looked. I thanked Philipps and thought that was that, I'd come back tomorrow and start. He then took me over to the wrapping bench and showed me how they were wrapped. I thought, OK, that's what I'll do tomorrow. But we stayed there for a few minutes like wet lemons, so I started doing it. I was in for the day'.

Source: (Unpublished notes, Wednesday April 1 1992)

Data reflecting/affecting the scene? Robson taking photos in a Nigerian village

'... taking photographs caused a lot of problems and was a very uncomfortable part of the fieldwork. People became demanding that I should come and take their picture and if I did then every time they saw me after that would insistently want to know where their picture was. This led me to only taking my camera out on special occasions and carrying it concealed in a bag so people would not see it and demand to have their picture taken. It was also frustrating because I wanted to be able to take natural pictures. However, the villagers had very set ideas about photography and unless they were unaware that they were being photographed, they would insist on putting on their best clothes and pose rigidly for the camera with very serious expressions, women always with downcast eyes'.

Source: Robson, (1994: 45)

Box 6.6: Layer 6: Self-reflections

Encountering Tanzanian racism on the bus: Nagar's letter 'home'

'At times it pains me so much to have brown skin here [in Dar es Salaam]. I take the bus every day to town and most of the times I am the only Asian riding on the bus. Sometimes I am greeted sarcastically, 'Kem Chho?' (Gujurati for 'how are you?'). I feel angry ... because I can hear and feel the resentment against the brown skin in the voice that is greeting me. And I feel like screaming: 'I am not an exploitative, racist ... *Muhindi* (Asian) from here. Don't look at me like that. I have nothing in common with

Box 6.6 (Continued)

them!' Being resented like that gives me some sense of what an antiracist white person must feel in a black Chicago neighbourhood. But then my position here as an Indian student from India is not quite the same as a radical *Muhindi* from Tanzania either. My problem is that I am a foreigner here, and at times I want to be recognised as such. But anyone who does not know me thinks that I am just an arrogant and aloof *Muhindi* who does not know enough Kiswahili in spite of having lived in Tanzania all her life. It hurts me so much to be resented here … And what really frustrates me is that the nature of my research keeps me away from the *Waswahili* and I just don't have enough time to get absorbed in their culture because I am too busy trying to get absorbed into the Asian communities'.

Source: Nagar, (1997: 215)

Notes

1 See, for example, Cook and Crang (1995), Flowerdrew and Martin (1997), Hay (2001), Kitchin and Tate (2000), Limb and Dwyer (2001) and Hoggart et al. (2002).
2 See Domosh (1991a), Blunt (1994b), Behar (1995) and McEwan (2000).
3 See Merrifield (1995), Gregory (1996), Fuller (1999) and Hoggart et al. (2002).
4 See Buttimer (1978), Ley and Samuels (1978), Jackson and Smith (1984), Livingstone (1992), Johnston (1997b) and Philo (2000).
5 See also Rowles (1978a), Jackson and Smith (1984) and Eyles, (1986).
6 Since 1985 Rowles has held a joint position at the University of Kentucky as a professor in the Department of Geography and in the Sanders–Brown Center on Aging. Findings from his Winchester Street study, and the follow-up study in a rural Appalachian town called Colton, were largely published outside geography in, for instance, the *Journal of Human Development* (1981a), *The Gerontologist* (1981b), the *Journal of Environmental Psychology* (1983) and *The Handbook of Clinical Gerontology* (1987).
7 For example, Nash (2000), Parr (2000), Thrift and Dewsbury (2000), and Smith, S.J. (2001).
8 For example, Clifford and Marcus (1986), Marcus and Fischer (1986), Stacey (1988) and Behar and Gordon (1995); see Cosgrove and Jackson (1987), Gregory and Ley (1988), McDowell (1992a), Gregory (1996) and Shurmer-Smith (1998a).
9 Duncan's (1980) paper was 'revisited' as a 'classic in human geography' in 1998 (see Duncan, 1998; Mathewson, 1998; Shurmer-Smith, 1998a).
10 As well as the humanistic ethnographies discussed above, the culture industry ethnographies of Dydia DeLyser (1999) and Mike Crang (2000), and the small-scale culture war ethnographies of Garth Myers (1996) and Richa Nagar (1997b), are worth reading in this respect (see Herbert, 2000; Crang, 2002).
11 See Clifford (1992), Marcus (1998), Hoggart et al. (2002) and Laurier and Philo (2003).
12 A standard dictionary definition of kinaesthesia is 'the sensation by which bodily position, weight, muscle tension, and movement are perceived' (Anon, 2001: 809).
13 For example, in making a cup of coffee (Angus et al., 2001), using a remote control or dog lead (Michael, 2000), windsurfing (Dant, 1999), working in an office (Hetherington and Law, 2000) or car (Laurier and Philo, 2003), or making natural history films (Davies, 1999; 2000; forthcoming).
14 Sara Kindon's (2003) participatory video work with members of a Maori tribe in Aotearoa, New Zealand, and Duncan Fuller's (1999) work with/for a credit union in Hull in the north of England are notable 'full blown' examples of such work. However, Batterbury's (1994: 81) admission that the NGO lobbying that was not part of his formal research strategy gave him the 'most personal satisfaction' shows how this action research can work on a much smaller, informal scale.
15 This includes work by Michael Keith (1991; 1992) and Steve Herbert (2000) with police officers, by Steve Hinchliffe (2000) with

managers on training courses, by Pam
Shurmer-Smith (1998b) with Indian administrative service workers, by Eric Laurier and
Chris Philo (2003) with mobile sales teams
and by Gail Davies (1999; 2000; forthcoming) with natural history film-makers.

16 Elspeth Robson (1994: 57–8), for example,
wrote how 'in my "normal" life as a PhD
student [in the UK] I play the role of a student, a learner and "subordinate", but in the
field [in the Nigerian village where I worked]
roles were reversed and I found it difficult to
be the teacher, the one calling the tune,
deciding what we should do each day, directing my assistants to do work for me'.

17 See Jackson and Smith (1984), Keith (1992),
Katz (1992; 1994), Madge (1994), Clifford
(1997) and Cook et al. (in press).

18 See Crapanzano (1977), Madge (1994),
Robson (1994) and Nagar (1997a).

19 We recommend starting with Robson and
Willis (1994a), Hughes et al. (2000), Limb and
Dwyer (2001) or Ogborn et al. (forthcoming).

20 Here we recommend starting with Cook
and Crang (1995), Flowerdew and Martin
(1997), Hay (2000), Kitchin and Tate (2000)
or Hoggart et al. (2002).

21 We are referring here to the possibilities
offered by CD or DVD technologies or by
online publishing such as that in the radical
geography journal *ACME* (www.acme-
journal.org).

22 See Burgess (1986), Stacey (1988), Bennett and
Shurmer-Smith (2001) and Dowler (2001).

23 See Clifford (1990), Crang (1994), Cook and
Crang (1995) and Capps (2001).

Geographical interpretations

If human geographers do use data in their research (see the introduction to Part I), it is obvious that we have do something with these data in the course of bending them into intelligible findings amenable to the drawing of conclusions, the offering of speculations and even, on occasion, the tendering of predictions. A few human geographers might suppose that 'the data speak for themselves', implying that once you have gathered or generated the relevant data it is possible simply to 'write up' or 'write out' these data as findings which unproblematically mirror what is happening in the world beyond the pages of the academic text. The majority of human geographers would dispute such a supposition, though, not least because any sensitivity to the complexities of how data are constructed (whether pre- or self-constructed) demonstrates that data can never straightforwardly mirror the world. All the countless complications integral to what data sources entail, the myriad factors integral to how they are first 'made' and then brought together, whether by the researcher or by other people, give the lie to any belief in the innocence, naturalness, transparency and so on of data.

The discussions throughout Part I of our book serve to challenge any simplistic assumptions about both what data entail and what they can reveal. This is not the end of the matter, however, since there is more that also needs to be said about the very definite 'labour' which the human geographical researcher cannot avoid expending in the process of turning data into plausible findings, conclusions, speculations or predictions. In short, attention needs to be drawn to the very substantial amount of work that has to be done on data before they can, as it were, be 'made to speak'. Part of this work does involve being fully cognizant of how data sources have been constructed, as we have emphasized, but there is a sizeable further dimension to this work that comprises a diversity of 'strategies' through which these data sources are *interpreted* by the researcher. The interpretations that thereby result must themselves be seen as accomplishments, as 'made' by researchers through their labours on their data, and this is why we use the slightly odd-sounding phrase 'constructing geographical interpretations' to capture what we see occurring here.

A few existing textbooks on matters of methodology touch upon the work of interpretation, but none of them lays out, explores and critiques as we do below the overall portfolio of different interpretational strategies available to human geographers. We might immediately add that such texts tend to take 'interpretation' as referring to a specific strategy akin to what we term 'understanding' in Chapter 10, whereas we are expanding what we mean by 'interpretation' to the status of an omnibus term encompassing all of the different ways of working on data covered in Part II.

Interpretational strategies

Such interpretational strategies have not always been acknowledged by human geographers, some of whom have been interpreting data without realizing they have been doing so, and this is lamentably true of those researchers who conceive of what they are doing as just 'common sense' or as merely following conventional scientific procedures. In our view, such human geographers *are* most definitely adopting interpretational strategies, ones whose inner logics can actually be specified quite closely and ones carrying with them implications about what sorts of data count as credible and what kinds of findings, conclusions and so on can be arrived at. These geographers *are* interpreting their data, they are working on their data according to frameworks of interpretation which we can readily identify, even if they fail or refuse to recognize that this is the case. Hence in the chapters that follow we will, to some extent, be addressing a range of unacknowledged interpretational strategies, showing how they still perform the task of allowing researchers to 'make sense' of their data.

For many human geographers today, there clearly *is* an awareness of bringing interpretational strategies to bear on their data – even if they might not use this terminology – and such geographers are more or less self-reflexive (see Chapter 1) about the work they are doing in endeavouring to move from data into findings, conclusions and beyond. This being said, the common format in which such issues are debated revolves around the different philosophies and social theories being mobilized to inform research projects: the philosophies which fashion a researcher's prior assumptions about the nature of the knowledge being produced (assumptions about ontology and epistemology), and the social theories which frame a researcher's root vision of how human society is constituted (the inter-relationships between the everyday existence of individuals or groups and the more enduring properties of institutions, systems and structures). This is the terrain of conceptual orientations in human geography, as dealt with in a text such as Paul Cloke et al. (1991) or that by Richard Peet (1998), and there is extensive and often heated debate about the philosophies (e.g. from positivism to humanism to poststructuralism) and the social theories (e.g. from social Darwinism to Marxism to feminism) that human geographers could or should entertain. And yet related to these conceptual orientations – cross-cutting them in ways which are not always as obvious as might be supposed – are numerous different interpretational strategies standing as bridges between abstract conceptual concerns and the data on which researchers work when striving for findings, conclusions and the like. It is upon these interpretational strategies that we now focus, teasing out identifiable strategies for working on data that can be adopted, and urging a greater self-reflection on such strategies that is independent of – or at least not determined by – more overt, we might say more abstract or high-level, philosophizing and social theorizing. In the chapters that follow reference will be made to philosophies and social theories so that concepts of positivism, Marxism, psychoanalysis and such will be mentioned, but the interpretational strategies under the microscope here cannot be 'reduced' to any one of these a priori conceptual orientations. To our thinking, these interpretational strategies do entail something different again: certain stances before data – modes of probing into and manipulating data – that urgently need to be unpacked by human geographical researchers. In this connection, then, we are unquestionably operating on a methodological plane as this is described in Box 1.4.

In the chapters that follow we identify and discuss five interpretational strategies, or perhaps it would be better to speak of three such strategies cross-cut by two interpretational processes or moments (those of both 'sifting and sorting' and 'representing') that cannot but be bound up in the other three. For simplicity, though, we will refer to all five as strategies while admitting that it is only our middle three ('enumerating', 'explaining' and 'understanding') that offer researchers the most obviously coherent – and, as such, often differentiated and perhaps counterposed – modes of strategic intervention in the tackling of data. We have debated long and hard how to characterize these interpretational strategies, what names to give them and what considerations are germane to each, and we realize that the organization and the contents of these five chapters will seem unfamiliar and even strange to many readers. There are also doubtless things that are missing here, stances and modes of working on data about which we do not say enough. Even so, we remain convinced that the strategies identified here do embrace the principal tasks – some being essential and some being contrasting possibilities – that human geographers can and do bring to bear in turning data into findings, conclusions, speculations and predictions. Some of these strategies or at least aspects of them are complementary, open to being combined, while others are likely to be pursued sequentially in a given research project. As already stated, no project can avoid having a 'sifting and sorting' moment, for instance, and no project that is written up in some form can avoid having a 'representing' moment (when the findings, conclusions and the like are written through for an audience of readers or listeners). Similarly, aspects of 'enumerating' and aspects of 'explaining' clearly link into one another on occasion, while the gulf between 'explaining' and 'interpreting' is rarely as wide as certain conventional accounts of methodology imply. This being said, some elements of these different interpretational strategies may be quite hostile to each another, and a simple point is that different, often very different, human interests – ones often wrapped up in very different visions of what human geographical research is *for* in the wider world – fuel the advancing of claims about the vastly superior merits of differing interpretational strategies over others. Nowhere is this more apparent than when those who purport to explain the world (especially if crunching through quantitative data) come into conflict with those who purport to understand the world (especially if immersing themselves in qualitative data). A modest ambition of what follows is thus to question the oppositions and hostilities that sometimes arise in this respect.

From initial siftings to final writings

Once researchers start to encounter data, choosing to examine certain data sets rather than others – as soon as they are selecting certain columns of numbers from a table for attention, choosing to concentrate on certain passages, minutes or entries from a written document, being captivated by certain responses in a questionnaire return or remarks in a taped interview, electing to record certain observations and impressions in a field diary and so on – they are taking the first steps in the interpretational strategy we term 'sifting and sorting'. This strategy is, therefore, intimately bound up in the most basic of dealings with data, and when data are being self-constructed it is probably true to say that the interpretational work of sifting and sorting has already begun. It is a primitive of all human geographical research, then, but in its universality and ever-presence it is almost never subjected to self-reflexive scrutiny: it is almost never brought into the light for special examination and critique. This being said, while the effort to impose order on data cannot but be there from the very outset of a project, we suppose there to be dimensions of the sifting and sorting strategy that become more prominent once the researcher directly confronts the masses of data obtained from the field. It is what happens when the researcher begins to order the data by identifying 'entities' within them that are reckoned to be obvious, significant or meaningful; it is what happens when he or she begins to categorize things, to detect binaries in the data and to draw up hierarchies of more or less important items in the data. No project can proceed without doing work along these lines, even if it is rarely acknowledged that this work is taking place and making a difference to what the data are finally allowed to tell us. Chapter 7 duly tackles sifting and sorting, reflecting quite abstractly, even philosophically, on the issues involved.

The next three chapters cover ground which will probably be more familiar to readers, albeit doing so in a fashion that might seem unfamiliar, disconcerting even. Chapter 8 tackles 'enumerating', arguing that a prime interpretational strategy for many human geographers is one that converts data into numerical forms – one that demands quantification of the world by counting or measuring its contents – and then proceeds to all manner of numerical analyses of these quantitative data ranging from the simplest of descriptive statistics (e.g. calculating a mean) to the most complicated of spatial modelling and geocomputational visualization. Some human geographers might suppose that enumeration is simply working with the facts of the world, believing that the numbers are somehow 'out there' in the world, but most of us would now acknowledge that to pin a number on to something is a decisively human activity which translates that something into a human 'language' wherein it can be technically manipulated statistically, mathematically and computationally. However simple or complex the manipulations, they cannot but insert a distance – we would call this an interpretational distance – between the world (and data extracted from it) and the researcher's findings, conclusions and the like (when efforts are made to translate back from numerical languages into words more obviously referring to tangible phenomena, events and processes around us).

Chapter 9 tackles 'explaining', arguing that many human geographers operate within explanatory frameworks which, albeit in a host of different ways, purport to provide the tools allowing us to answer 'why' questions: Why did that settlement get located where it did? Why did that part of the city end up derelict and riddled with poverty? Why did people in that region develop a strong sense of their own ethno-regional identity and hence wish to secede from their host nation-state? Explanatory frameworks are approaches to knowledge production which think in terms of cause-and-effect relations or, more subtly, of underlying causal mechanisms which exert an influence through differing media – people's psychologies, cultural expectations, economic tendencies – to effect outcomes which might include the location of a settlement, the impoverishing of a city district or the upsurge of dissent in a given ethno-region. The job of 'science' has usually been regarded as that of explanation, particularly if explanation can feed into prediction, and this is as true of conventional science (the hypothesis-testing science that most of us know from the school science room) as it is of more recent post-Newtonian science (which now accepts 'uncertainty' as a feature of explanation) and any number of other intellectual positions which have claimed the mantle of science (from Freudian psychoanalysis to Marxist historical materialism to Husserlian phenomenology). There are, therefore, numerous different versions of what explanation entails and of how it can be achieved and, to a limited extent, these can be contrasted as 'surface' or 'depth' models of explanation, but for us they are all interpretational strategies which researchers operationalize in their efforts to make data yield up definitive answers to the sorts of 'why' questions listed above.

Chapter 10 tackles 'understanding', arguing that many human geographers – notably ones who reject the prevailing models of explanation – prefer approaches where the chief objective is to ascertain the meanings which are held in the hearts and minds of people central to the human situations under study. In contrast to Dilthey and Weber's construct of *Eklären* (i.e. 'explanation'), many human geographers favour their construct of *Verstehen* (i.e. 'understanding'), which is all about the excavation of meanings, values, beliefs, hopes and fears which course through what are occasionally called the 'thought-worlds' of people who are the researched (whether they be princes or paupers, company excutives or the socially excluded) (e.g. Gregory, D., 1978a: 133). For these geographers, the research task is basically complete once the researcher has obtained some understanding of the meanings, values and the like figuring in the thought-worlds of the researched. Drawing inspiration chiefly from the humanities and cultural studies, the tendency is to depend upon highly qualitative methods, like in-depth interviewing or participant observation. And so the anticipation is that, through varying forms of encounter, empathy and dialogue, the researcher can come to understand key dimensions of how the researched think (and feel) about their lives and how they then act – what they do, whether routinely or as one-off occurrences – as decision-makers in their everyday lives at home, in the street, at work or whatever. This being said, in more sophisticated versions of understanding, consideration is given to the broader discourses (ideas, beliefs, opinions) that frame what research subjects think (and feel) about their worlds, in which case the researcher's understanding does not lodge solely in the idiosyncrasies of individual people encountered in the research process. Just as there are different models of explanation, so there are different models of understanding, and there are also different levels of work involved that include addressing both highly conceptual problems (as in rarefied debates about hermeneutics) and mundanely technical ones (to do with how to 'code up' qualitative data sources in the hunt for clear patterns in the realm of meanings, values and beliefs). Once again, though, to our minds these are all facets of an overall interpretative strategy which is understanding: they are all directed to how the data can be shaken to reveal the fundamentally meaningful qualities of those portions of humanity that we purposefully meet or accidentally stumble across in our inquiries.

Integral to some of our endeavours when we are gathering or generating data is a process which can rather grandly be called 'representing'.

Whenever we write notes on a document found in an archive, scribble lines on the context and proceedings of an interview or more formally write up our observations and impressions in a field diary, we are starting to represent our data in textual form. Indeed, in many cases the data *are* this textual form because we have nothing else to carry back with us to our offices and computer suites. What we write and how we write it at this stage is, therefore, already part of the interpretative process that we call 'representing' geographical texts, and it should be evident that what we are doing here is also very much embedded within the sifting and sorting covered in Chapter 7 (at this very preliminary stage these two interpretative processes go hand in glove). Yet what principally concerns us here is less these initial acts of writing through data, important though they are, and more what happens when human geographers – in tandem with subjecting their data to some combination of sifting, sorting, enumerating and understanding – seek to produce in their written output findings, conclusions, speculations and perhaps predictions to be presented to an audience. This audience could be other academics reading a journal paper; it could be students reading a textbook; it could be policy-makers reading a report; it could be people who have been researched reading a report back; or it could even be an audience hearing about the research at one-step remove, perhaps the general public reading a summary prepared by a journalist for a newspaper article. A host of considerations weave through these acts of representation, these re-presentations of peoples, lives, space and places which have been the focus of study, and one way of quickly grasping what is involved is to think about how the people who have been researched might react to representations of them. But more than this, because we are in no doubt that acts of representation (i.e. constructing geographical texts) comprise a maze of different possible interpretative possibilities bound up in the diverse conventions, styles and rhetorics of writing available to the researcher. Chapter 11 duly tackles this amorphous task of representing, reflecting quite widely, often critically, on the issues involved.

7 Sifting and Sorting

Sitting down with your data

So far in this book we have dealt at length with the many ways in which geographical data are constructed, whether by the researcher or by others whose labours he or she then draws upon. Whatever the route by which the data have been constructed, a point arises – although it may well not preclude further generating or collecting of data – when the researcher 'sits down' with all or some of his or her data and starts to puzzle out exactly what to do with them. This can mean quite literally 'sitting there', perhaps in something of a daze, surrounded by piles upon piles of materials: by box-files of photocopied official sources (minutes, policy statements, structure plans); by results from questionnaire surveys; by photographs, mental maps and snippets of poems; by tapes and notebooks carrying thousands of interview words; by field diaries brimful of participant observations; and the like. (And it might help at this point if you were to look at the discussion of the concepts of 'induction' and 'deduction' as outlined in Box 7.1.) While some attempt might have been made to impose a measure of order on this chaos of materials as it was constructed, perhaps by keeping together seemingly similar sorts of data (say, all the official sources) or perhaps according to where the data were gathered and from whom (say, from the planners in a particular local authority planning department), to all intents and purposes this *is* still a chaos which needs to be brought into some kind of order before it can be tackled further.

There are all manner of decisions to be made in this respect: deciding that several field observations should be categorized together as exemplifying (say) a particular type of 'settlement layout' or 'settlement social life'; deciding that several passages from several official planning documents should be categorized together as exemplifying (say) a particular variety of interventionist policy or ideological commitment; or deciding that certain statements from a transcribed interview should be categorized together as exemplifying (say) certain themes such as 'belonging', 'nostalgia' and 'alienation'. In each case decisions have to be taken about what comprises a clearly delimited field observation, documentary passage or interview statement, and then additional decisions have to be taken about what are basically 'like' and 'unlike' observations, passages and statements worthy of being grouped together into categories. It is usually these categories that the researcher will utilize in subsequent interpretations, but he or she will often wish to refer back to the original observations, passages and statements (and the sensitive researcher will try hard to retain some alertness to the nuances within such individual pieces of data).

Box 7.1: Induction and deduction

While the philosophical ground here is contested, a standard distinction is drawn between these two basic approaches to the conduct of academic research, each of which implies a fundamentally different stance before the data constructed (whether collected or generated) in a project. One geographer (Lindsay, 1997: 7; see also the concise statements in Walford, 1995: 4) suggests that *inductive reasoning* 'works from the particular to the general', while *deductive logic* 'works from the general to the particular'. Induction depends upon careful examination of a substantial amount of 'raw data', seeking to discern in them patterns and regularities which can be interpreted as being of some generality, importance or meaning. It is commonly supposed that induction depends on statistical procedures to establish the presence of such patterns and regularities, although our argument throughout Part II of this book is that there are various strategies (quantitative and qualitative) which can be deployed to do this. Deduction entails prior specification of theories, models and laws which can be used to account for the details found in 'raw data' as derived from studies of particular phenomena. Again, it is commonly supposed that deduction depends upon statistical procedures to compare expected outcomes (from the theories) with observed eventualities (what actually takes place), but our argument here is that there are various strategies (quantitative and qualitative) which can be deployed to do this. The usual thinking, moreover, is that only induction requires close attention to data – to 'letting the data speak for themselves' (Gould, 1981) – whereas deduction does not need to bother overly with the niceties of data. In geography, regional geographers have been regarded as inductivists obsessed by data, while spatial scientists have been regarded as deductivists little troubled by them (a claim with a measure of justification; see Chapter 1). A similar distinction might also be detected between humanistic geographers with an inductive feel for data and Marxist geographers with a deductive line on theory, but there is hardly a clear-cut divide here, and the picture becomes even more muddled in relation to feminist, postcolonial, poststructuralist and postmodernist geographies (see Chapter 1 and later in this chapter). Our own view is that such distinctions are invariably overdrawn and that most research tends to contain both inductive and deductive moments in a constant, if uneven and sometimes unappreciated, dialogue. Moreover, we believe that virtually all human geographical research projects, wherever we might wish to position them on the inductive–deductive continuum, display a sustained engagement with data. And this means that they then have no choice but to 'sit down' with their data and to begin sifting and sorting them in the manner which is the concern of this chapter.

It is worth adding that perhaps the most clear-cut example of this kind of activity in human geography today arrives when researchers are undertaking what is called the 'coding up' of textual evidence (notably transcripts of tape-recorded interviews), since it becomes so obvious here that the researcher is seeking to discern an order in – and in effect to impose an order upon – the original chaos of data. We return to the activity of coding in Chapter 10 when discussing what we describe as the 'artisanal' approach to understanding qualitative data (see also Crang, 2001; Jackson, 2001). The basic point, however, is that the researcher simply cannot avoid having to do a lot of preliminary work on the chaos of his or her data as the first stage of moving into the interpretative part of his or her project. It is this work that we refer to here as 'sifting and sorting', and which Mike Crang (2001) imaginatively refers to as 'filed work' (as opposed to 'field work'), because it cannot but involve a process analogous to that of sorting out paperwork ready to be filed in the appropriate drawers and dropfiles of a filing cabinet. If a researcher's attempt at 'constructing geographical interpretations' is going to be at all successful, there needs to be a lot of this careful sifting and sorting. It simply has to happen in a research project, it always has to happen and it should therefore be reflected upon quite consciously. The fact is that little attention is usually given to this activity, partly because it does seem like – and probably often is – such a thoroughly dull aspect of the research process. As Crang (2001: 215) puts it, the 'relative silence' about what is entailed in sifting and sorting is unsurprising given that it 'feels like an unglamorous almost clerical process' (which returns us to the hardly very exciting image of office filing). The purpose of this chapter is, none the less, to examine this absolutely crucial yet rarely acknowledged or discussed part of the research process. In so doing, we aim to show that it is precisely *not* a simple aspect of the process, even if some of the practical actions involved appear straightforward and not warranting much comment. Indeed, we acknowledge that it is actually riddled with taxing conceptual issues which cannot be assumed away by a reflexive human geography.

Entitation

This sifting and sorting of data – combing through them to ascertain exactly what is there; laying them out, rearranging them, classifying them, putting individual items of data into boxes and lists; perhaps going on to detect patterns in the lists and boxes – is not a part of the research process which receives the attention it should. It would be wrong to imply that this deficiency has never been noted, though, and here we reproduce a useful observation taken from a book on 'systems analysis' in geography. Given that we are wary

of systems analysis as an approach to human geography, for reasons relevant to this chapter (see Box 7.2), we have altered this quotation to remove references to 'systems'. Even so, the sense remains apt for our purposes:

> The [initial] phase of [data interpretation] involves two steps: the identification of the basic components or entities of which a [situation or problem under study] is composed; and the definition or description of these components in terms of measurable properties or state variables. This phase is of enormous importance but it has not always been given the attention that it would warrant. The process of identifying entities has been termed entitation; the process of measuring entities has been styled quantification. ... Entitation is vastly more important a stage in [data interpretation] than quantification: Gerard (1969) demonstrated this point with the example of primitive astronomy where stars (basic entities) were grouped into constellations which we now know were, in reality, meaningless. (Huggett, 1980: 29)

The term *entitation* (see also Langton, 1972; Chapman, 1977; Gregory, 1986) usefully captures something of what we think is entailed in the preliminary ordering of data, a task that results in the basic entities relevant to a research project being detected, delimited and described, and perhaps also seen in basic patterns relative to one another. Another term for this chapter could hence have been 'geographical entitation'. In the quotation it is suggested that entitation precedes quantification, the numerical measurement of the entities so revealed, and in Chapter 8 (on 'enumerating') we pursue something of what this arguably more familiar part of a research project (namely, quantification) can involve. But in this chapter we agree with the quotation that many human geographers have conflated entitation and quantification (or enumeration), and have thereby regarded the attribution of numerical values to items of data as *the* primary stage in the interpretation – or, as they would be more likely to say, the

Box 7.2: Systems analysis

Considerable confusion surrounds the introduction of thinking about 'systems' in the geographical literature. One branch of this thinking, derived from the 'general systems theory' of Ludwig von Bertalanffy (1968), has seen geographers seeking for fundamental properties (usually spatial ones) which feature in many different but seemingly analogous worldly systems such as the hydrological system, the ecosystem, the regional system and the settlement system (e.g. Chorley, 1964; Coffey, 1981). Less grandly, a variety of 'systems theory' has sprung up in the discipline which conceptualizes the world in terms of interlocking 'systems', wherein a system is taken simply as an identifiable portion of reality which 'consists of a set of entities with specifications of the relationships between them and between them and their environment' (Wilson, 1986: 476). Put in this manner, a great many geographers could be cast as systems thinkers (and one argument is that we are all in effect systems thinkers), but few explicitly utilize the specialized notions (to do with 'inputs', 'outputs', 'feedbacks' and the like) which have been developed to describe the supposed workings of systems by the likes of Chapman (1977), Bennett and Chorley (1978) and Wilson (1981b). These notions are sometimes referred to as the tools of 'systems analysis' (Huggett, 1980), although in recent years the latter term has been increasingly been restricted to a 'mathematical approach to the modelling of systems' (Hay, 1994: 615). Numerous objections might be raised to systems thinking in its many guises (e.g. Kennedy, 1979; Gregory, 1980), not least because there is a tendency here to conflate the 'real' and the 'system': to suppose that the world really is a bundle of interacting systems, thereby forgetting that this is only a thoroughly *human* picturing of this world, which can lead to simplistic attempts at trying to 'control' these imagined real systems. Moreover, by the lights of the current chapter, there are grave dangers associated with the basic impetus of systems thinking to 'put things into boxes': to suppose that it is a quite straightforward exercise to identify distinct entities which can be named, specified, measured and inserted into the 'boxes and arrows' depiction of a typical systems diagram (see the example illustrated in this box).

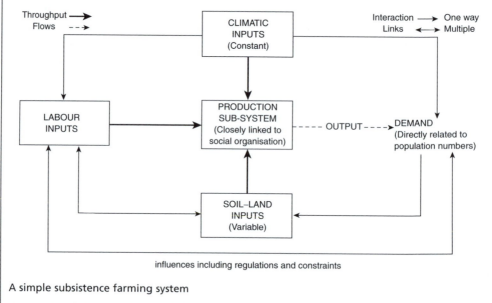

influences including regulations and constraints

A simple subsistence farming system

Source: Carr (1987: 121, Figure 13.1)

analysis – of geographical data. This is obvious from what are arguably the standard remarks made by geographers about the sifting and sorting of data, as we briefly review in a moment. The problem here is that attributing numerical values *presupposes* that the researcher has decided in advance exactly what *are* the basic entities or perhaps patterns between entities warranting enumeration. It is a view which overlooks the very real conceptual labour that goes into entitation, effectively regarding it as automatic or unimportant, or as somehow folded into the more dramatic moment of fixing the numbers to the bits of 'stuff' under scrutiny.

These points about entitation do not only hold in relation to enumeration. Indeed, we would insist that in any human geographical research project, however quantitative or qualitative in its practical execution (see Box 1.7), there must be a phase of entitation which provides a preliminary ordering of the data upon which subsequent interpretative strategies operate. By the latter we could be talking about the utilization of a conventional statistical test or a fancy GIS procedure on numerical entities, but we could also be talking about textual or iconographic interrogations of non-numerical entities entailing words and images (a policy objective in a planning document, a statement in an interview transcript or a photograph in an archival collection). The principle is the same: the entities must have been extracted from the chaos, the mix, of materials comprising what is sometimes called the 'raw data' (see Box 1.3). What we therefore do find strange is the absence of general claims about entitation, and we suspect that this is because many researchers take it as too self-evident a stage in what is being attempted to warrant comment. Such a belief emerges from conventional assumptions about the 'nature' of the data, of the real-world situations behind the data, which are being subjected to a preliminary ordering. The data, and hence the realities,

are not seen as problematic; or, rather, the assumptions underlying how these 'data of the world' are to be sifted and sorted – badgered into boxes and lists – are rarely examined at all critically because they lie so fundamentally at the heart of much taken-for-granted knowledge production and consumption in Western academia.

If this sounds like a big claim, it is meant to be, for it genuinely strikes us that what is involved in entitation runs up against extremely thorny philosophical dilemmas about the positing of identities, binaries and hierarchies. We are not proposing that human geographers can somehow resolve these dilemmas – far from it – but we are recommending that we should become aware of them so as to be more circumspect and even humble about the grounds on which entitation occurs in any (and every) research project. In the course of developing this line of argument, we will take a look at various matters which are usually viewed in a rather different way in the more philosophical literatures of human geography.

Standard remarks

In so far as geographers have thought much about the preliminary ordering of data, rather than taking them for granted, they have tended to do so in terms of what James Lindsay (1997: 22–9) calls both the *provenance* and the *types* of data available to them. In terms of provenance, he refers to the differing sources from which data have been derived, offering a distinction between 'primary' data (generated by the researcher), 'secondary' (generated but not interpreted by somebody else) and 'tertiary' data (already interpreted and available for consultation in existing texts and journals). We have some problems with the starkness of these distinctions, as should be apparent from our detailed examination of precisely these matters throughout this part of the book (see also Box 1.3), but we agree with Lindsay that the preliminary ordering of data must always be highly sensitive to the provenance of these data: to the when, the where and the why of their construction. In a simple sense, this means that the researcher must never lose sight of which items of data have been gleaned from the field through his or her own labours, which have been trawled from the surveys and reports of others, and which have already come to him or her preinterpreted from the library. What operations can legitimately be performed on these data, what kinds of inferences and conclusions can plausibly be drawn from them, depend intimately on such matters of provenance and construction. But these matters, having already been addressed previously in the book, are *not* our main focus here.

When turning to types of data, Lindsay begins by considering the oft-cited distinction between qualitative and quantitative data, a distinction that we are seeking to blur in various ways even if it continues to have meaning for us (see Box 1.7 and Chapter 8). But Lindsay introduces this distinction in the course of turning to a more sustained examination of data in terms of their potential to be quantified (or enumerated), and he thereby echoes what have perhaps been the most typical observations ever made by geographers about the preliminary ordering of data (e.g. Silk, 1979: 6–7; Gregory, S. 1992: 136). This exercise merely divides up data under the four standard headings of nominal, ordinal, interval and ratio (see Box 7.3), and explains the extent to which it is possible to apply numerical values to, and then to use statistical testing procedures on, the different types of data identified. While there are varieties of statistics which have been developed for use on data which are merely nominal or ordinal (and geographers have contributed usefully to the development of 'categorical data analysis' appropriate for such data; see Wrigley, 1985;

Robinson, 1998: ch. 6), the common ambition is still to find ways of converting such data into forms where they can be treated numerically and thereby subjected to more familiar forms of measurement and manipulation. Unsurprisingly, most credence has been given to data capable of measurement on interval and ratio scales, which is thus data which can be most easily made suitable for the numerical work favoured by many geographers (everything from the most basic of descriptive statistical testing, much of which depends on being able to calculate arithmetic means, through to the complexities of multivariate statistical analyses and beyond). It was not so long ago that a methods course in geography, even one in human geography, would quickly dispense with any data that could not be subjected to this toolbag of statistical-mathematical techniques. As indicated, in Chapter 8 we consider at greater length what is involved in the various faces of 'enumerating'.

At this point our concerns are rather different, however, in that we wish to dig beneath what is really being presupposed in such treatments of the sifting and sorting of geographical data. Returning to Lindsay for a moment longer, it is significant to note that he couches his discussion of data types very much in terms of 'measurement'. He claims that 'all data are obtained by some sort of measurement, although at its simplest the measurement might amount to no more than allocation to a class' (1997: 25–6). By his latter comment he means something like choosing to allocate a data item (e.g. a listing of someone's occupation in a historical document as 'carpenter') to a broader class or category ready for further interpretation or analysis (e.g. 'woodworkers'). None the less, it is revealing that he talks of measurement in this regard, given Richard Huggett's argument above that measurement equates with quantification (an impulse to enumerate phenomena) and that quantification is itself an act

---

Box 7.3: The standard view of the basic data types

1 *Nominal* This implies data that fall into categories or groups, such as landforms, land-use types, occupational categories, diseases or nationalities. These cannot be measured, but their frequencies in given areas can be counted.
2 *Ordinal* In these cases the relative magnitude of the phenomena can be known, so that they can be ordered from largest to smallest, even though actual values are not available. This is often because of difficulty in obtaining exact measurements, either due to instrumentation problems or because of limited access to more detailed data.
3 *Interval* With interval data the actual magnitudes of the phenomena are available, such as slope angle, rainfall, production figures or population.
4 *Ratio* These include proportion or percentage measurements, such as voting percentages, relative humidities or chemical concentrations, as well as a variety of indices, such pH values. These ratio data may have finite upper or lower limits, which obviously affect the way they can be analysed statistically.

Source: Gregory, S. (1992: 136)

---

which necessarily *comes after* the prior act of entitation. In other words, we would argue that Lindsay's examination of types of data (the main part of his chapter on 'data handling' in human geography), basically misses out the stage of entitation which cannot but occur – even if taken for granted, even if done commonsensically – *before* arriving at the stage where the researcher can sort out his or her data into piles distinguished as nominal, ordinal, interval and ratio. And this is notwithstanding Lindsay's own admission that 'if data are not naturally given objects, then we have to recognise that they are chosen and shaped in a way which might have unconscious as well as conscious elements' (1997: 22).

Our contention is that, even for a piece of data about some phenomenon, event or aspect of the world to be designated as one of the seemingly simplest (nominal) type, a decision must already have been taken to *name* that phenomenon, event or aspect as one thing rather than another (as a 'carpenter' who can also be classed as a 'woodworker' but not as a 'textile worker'). In Huggett's example, for instance, a decision has been taken to identify a small pinprick of light in the night sky as a star which might be given a name

(e.g. the North Star) and also classed as part of one given constellation rather than another (e.g. the Big Dipper). All this has occurred prior to any assessment of what type of data are involved, let alone of how they might be altered to render them amenable to numerical measurement and manipulation. A further implication of Huggett's example is that medieval astronomers viewed their classification of a star to a constellation as a meaningful scientific act, whereas modern astronomers would suggest that this is less meaningful than identifying a given star in terms of the precise portion of deep space where it is located (which may be nowhere adjacent to any of the other stars in its constellation as it appears to us on earth). The basic lesson is that different decisions may be taken in the naming, classifying and, in effect, entitation of a given star, and in the process all manner of differing assumptions may be made about both its 'nature' and, interestingly, its time-space relations. All this occurs in the preliminary ordering of data pertaining to that star, before questions kick in about the type of data involved, and yet we probably take too little notice of what is happening in the process: we fail to register a whole series of moments in the sifting, sorting and entitation which then *does* have such an influence on any subsequent interpretations of the star as provided by academics.

Spatial referencing

The astronomy example also hints at another issue which we need to note before exploring the deeper problems bound up in sifting, sorting and entitation – namely, that of the spatial referencing of data. The medieval astronomers identified stars according to their position in the geography of the night sky and hence gave priority to constellations as ordering devices, while modern astronomers think in terms of the precise locations of stars relative to those nearby in the particular sectors of deep space which they occupy. For geographers the space in question tends to be earthly rather than celestial, but the issue of spatial referencing is central to the concerns of the discipline, both in its historical development and still today (see Chapter 1). So much of the geographer's data directly describes spatial units of the earth's surface (e.g. specifying the dimensions of length, breadth and area possessed by given units) or more indirectly describes aspects of particular spatial units (e.g. prevailing religious beliefs or stated voting intentions of human populations living in that unit). Much work by regional geographers was precisely concerned with the task of identifying meaningful areal units and of deciding which data sets most appropriately serve to represent the character of the regional entities so delimited (e.g. Hartshorne, 1939), while spatial scientists expressly equated the wider scientific task of 'classification' with the narrower geographical task of 'regionalization' (of allocating individual pieces of data to given areal units while also deciding upon realistic typologies of regions).[1] Even in the more unusual of more recent projects in human geography, virtually all the

data consulted retain some element of spatial referencing. Thus a humanistic geographer might be talking about what creates feelings of security, well-being and belonging for people living in a given place (e.g. Ley, 1974; Relph, 1976), or a feminist geographer might be talking about differences in the 'gender division of labour' between one locality and the next (e.g. McDowell and Massey, 1984). The latter geographers may be less concerned than either regional geographers or spatial scientists about the exact configuration of the regions and spatial patterns involved, being more concerned with the human meanings and political-economic dynamics of the situations under study, but the spatial referencing (the embeddedness in place or the variability between localities) remains absolutely crucial to the inquiry.

For virtually all human geographers, therefore, sifting and sorting their data continues to include some reference to space, place, location and environment. We realize that by raising these issues we are entering into a minefield of debate about what exactly geographers understand, and should understand, by such terms and notions. Right at the heart of any geographical study assumptions *will* be made which, even if not reflected upon at all explicitly, cannot avoid carrying with them an enormous freight of philosophical reasoning about such terms and notions. Even the seemingly simplest of sifting and sorting operations – choosing to sort out data on wage rates by named standard planning regions or choosing to sift out data on people's feelings by types of environment (e.g. run-down inner city or idyllic lowland village) – presupposes a whole array of understandings about what are relevant spatial references, about the likely intertwining of spatial and social processes across geographical scales, about the 'absolute' or the 'relative' properties of space itself and so on. This is not the occasion to begin debating such issues[2], but we do wish to flag them as ones which press upon the practising of human geography.

What we should also underline is the extent to which there is a long tradition within the discipline of thinking quite technically about what one writer terms 'geocoding', which 'can be defined as the process by which an entity on the earth's surface, a household, for example, is given a label identifying its location with respect to some common point or frame of reference' (Goodchild, 1984: 33). At one level geocoding widens into the broader issue of spatial co-ordinate systems, the specification of (x, y) co-ordinates for a given object (the attribution of standard grid references or latitude and longitude, for example), which means that the discussion of geocoding potentially covers the efforts of cartographers, surveyors, remote sensors and various other 'topographic scientists'. Indeed, it could be argued that cartography is a primary ordering device for geographers which should be discussed further here (for an outline, see Blakemore, 1994; Box 7.4), particularly since there was a time when many geographers declared that 'if it cannot be mapped, then it is not geography'. We would not echo such a declaration, although we do support those who are seeking for new, different and more experimental means of mapping phenomena (e.g. emotions and power relations) which have hitherto been regarded as too 'ephemeral' (see Dorling, 1998). We would also note criticisms of how a cartographic obsession dovetails with a prevalent 'occular-centrism' in the discipline's history (see Chapter 1), and at the same time suggest that most procedures integral to standard map-making exercises inevitably run into the tricky problems of fixing (id)entities which we will examine shortly.

Much the same is true of the most recent inheritor of this technical tradition, geographical information systems (GIS), wherein highly sophisticated computer-assisted techniques are deployed to take 'geographic information, [which] in its simplest form, is information which relates to specific locations' (Martin, 1996: 1), and then to store this information in a database which both preserves the spatial referencing and renders it amenable to further manipulation (in seeking to detect inter-relationships within the data). Many see GIS as the future of the discipline (e.g. Thrall, 1985), and we would agree that GIS offers an important new tool for human geographers to utilize in the stages of collecting, storing and exploring large spatially referenced data sets. There are thus compelling reasons for why we should treat GIS further (for an outline, see Goodchild, 1994; Box 7.5), particularly given the key role of GIS in sifting and sorting data, but we are reluctant to accord it greater prominence both here and elsewhere in the book because serious criticisms can be voiced about its conceptual limitations and practical applications.[3] And, once again, GIS arguably intensifies the tricky problems integral to the fixing of (id)entities.

What happens when we put things into boxes and make lists

In effect, we have already been discussing the implications of listing and boxing things during the previous chapters. Indeed, we have already been taking seriously what we are calling entitation, as when noting the differing ways in which the raw data of UK unemployment can be configured and reconfigured to tell different 'stories' about the UK economy (see Chapter 2). Decisions taken in the entitation of unemployment, in giving it shape and possible meaning, have duly been unpacked, and attention has been paid to the 'particular understandings of reality' possessed by different agents and bodies involved in the processes of economic data collection, manipulation

---

Box 7.4: Cartography

Cartography can be defined as a 'body of theoretical and practical knowledge that mapmakers employ to construct maps as a distinct mode of visual representation' (Blakemore, 1994: 44). It is commonly assumed that one of the prime tasks of the geographer is map-making, and at least one well-known history of the discipline (Livingstone, 1992) has a book cover depicting 'the geographer as map-maker'. Many textbook introductions to the discipline concentrate heavily on map-making and thus on the diverse and often quite complex knowledges and skills involved in what is defined above as cartography. In an old textbook such as Bygott (1934), 'mapwork' and 'practical geography' are taken as virtually synonymous, and the focus is basically on how to translate 'real earth' (terrains, settlements, etc.) into 'paper landscapes' (representations of that earth, together with often less visible human constructions like political and administrative boundaries). Attention is thereby given to everything from surveying to projections to map-reading to understanding different kinds of maps (at different scales, with different projections, made by different agencies). A more recent textbook such as Lindsay (1997) still includes a whole chapter on 'maps and mapping techniques' although here it is acknowledged that much map-making – particularly of high-quality maps – is done by specialists (by cartographers who would not necessarily call themselves geographers). It is also recognized that the maps which geographers themselves produce today tend to be representations of particular bodies of spatially variable data – e.g. population profiles, industrial location, income level, quality of life, etc. (Dorling and Fairbairn, 1997; Dorling, 1995b; 1998) – rather than 'base maps' attempting to convey the 'sense' or 'ground truth' of overall physical and human environments. These geographers' maps of today are produced in the context of particular research projects and they become tools in helping to answer given research questions and perhaps (when used more inductively) to frame new ones. In this regard, the technical quality of the cartography becomes of less importance than what the maps can demonstrate, how they can be interpreted and what new possible avenues for inquiry they may open up. Numerous critiques are now directed at standard assumptions about the value neutrality of cartography, the root argument being that all maps are made for given purposes which determine exactly what is shown and how it is shown (Harley, 1988; 1989) and, in this connection, maps must be seen as official (or occasionally non-official) sources *to be examined* as part of given research projects. Thus, how maps are constructed in given situations, for instance by a Nazi *geopolitician* or by a local planning department, becomes something that must itself be subjected to research (see Chapters 2 and 3). What is perhaps less reflected upon is how constructing maps, choosing what is or is not an object (entity) fit for (some kind of) spatial representation, is central to the sifting and sorting of 'raw data'; and much cartography is thereby paradigmatic of the thorny issues to do with the attribution and fixing of identities that are central to this chapter (see also Kitchin and Tate, 2000: 156–64).

---

and (re)presentation. These are the sorts of understandings which we are urging human geography researchers to critique, although it is important to distinguish between those assumptions which are, as it were, taken for granted (because they are reckoned to be so commonsensical) and those which emerge more wittingly from the professional–political interests of the agents and bodies involved in producing, formatting and making available such data. And such a distinction applies as much to researchers themselves when constructing their *own* data as it does to researchers interrogating the data constructed by others, since from the outset of a research project researchers will be making myriad

---

Box 7.5: GIS

Geographical information systems (GIS) can be defined in various ways, some accenting more the technological elements, some the data transformations and some the applications (Kitchin and Tate, 2000). At bottom, though, GIS entails a combination of computer hardware, software and 'liveware' (the researchers) which creates a 'system for capturing, storing, checking and displaying data which is spatially referenced to the earth' (Chorley Committee, 1987, cited in Kitchin and Tate, 2000: 164). For many, the use of GIS – sometimes talked about in terms of a broader sweep of geographical information technologies (GIT) – offers the way forward for the discipline as a whole, providing an arsenal of technologies and data-handling skills, ones specifically applicable to spatially referenced data, which can then be the discipline's distinctive contribution to a broad constituency of social and environmental disciplines. For others, as indicated in the main text, there are many problems inherent in the kind of 'geography' that GIS entails and in the uses to which GIS tends to be put (cf. Clark, 1998; Flowerdew, 1998). This is not the place to discuss the workings of GIS nor of specific GISes (different systems) since such specialist texts are already numerous (e.g. Martin, 1996; Longley et al., 1998a; Kitchin and Tate, 2000: 164–87), but it is appropriate in the context of this chapter to note the following:

> Two major traditions have developed in GIS for representing geographical distributions. The *raster* approach divides the study area into an array of rectangular cells, and describes the content of each cell; while the *vector* approach describes a collection of discrete objects (points, lines or areas), and describes the location of each. In essence, the raster approach 'tells what is at every place' and the vector approach 'tells where everything is'. (Goodchild, 1994: 219, emphasis in original)

As the latter comments make clear, GIS is indeed fundamentally about scrutinizing spatial data sets in such a manner as to ascertain exactly what 'things' are present within the graticules of a preset spatial grid or exactly where are specific 'things' (identified and identifiable objects or, perhaps better, shapes) relative to one another. Such an effort of establishing identities and fixing them spatially, in many ways merely an elaboration of the time-honoured map-maker's craft, comprises the foundation of all GIS work. Again, then, and like cartography, the fashion in which GIS operates on raw data is indeed highly germane to the deeper, even philosophical, concerns of this chapter.

---

decisions about data which, more or less self-consciously, cannot but feed into the entitation of the phenomena, events and aspects under study (a key message, if not quite couched in these terms, throughout Chapters 5 and 6).

It is absolutely vital to explore the more conscious decisions involved here, whether taking into account the political orientations of governments, the aesthetic preferences of photographers, the personal motivations of researchers themselves or whatever: all these dimensions *will* play upon the sifting, sorting and entitation of data. But the chief concern for the rest of this chapter is with the commonly much more hidden, indeed taken-for-granted, dimensions of these acts. As stated above, therefore, this does lead us to tackle some quite deep-seated, ultimately philosophical questions to do with identities, binaries and hierarchies. Before tackling such questions directly, though, let us begin with two brief cameos from the geographical literature – one discussing an example of entitation in action and the other entailing a particular line of thought about such acts of entitation – from which we will draw out points to inform the more abstract reflections that follow.

Occupations in historical-geographical perspective

An important topic for research concerns the character of occupations (the jobs of work people do), and human geographers have often asked about the differing occupational profiles associated with different places and regions in given parts of the world. The historically minded of such geographers have also asked about how these local and regional compositions have changed through time and, in particular, interest has centred on the emergence in eighteenth- and nineteenth-century Europe of specialist 'industrial' districts wherein large numbers of people shifted into occupations ranging from basically craft-based pursuits (e.g. furniture-makers) to being factory hands on early production lines (e.g. textile workers in the 'dark satanic mills'). The data available today on occupations from something like the British Census are certainly not as straightforward as many might suppose, and there are difficulties with deploying the 1990 Standard Occupational Classification (Dorling, 1995b), but the data from periods prior to the 1841 census – when detailed occupational recording began – are more inconsistent again, patchy both sectorally and spatially, and hence resistant to easy interpretation. In a monograph explicitly addressing this matter, Paul Glennie (1990) discusses at length the problems involved with such data and, in so doing, explores the sorts of qualifications already voiced in Chapter 2 about using official sources (in his case pre-census English population listings, tax returns, parish registers, probate inventories and the like).

More specifically, Glennie stresses the problem that the entities which we might wish to extract from the data (the occupations

themselves) do not have the simple, coherent and stable qualities that the researcher would ideally require. While it may be true that there 'are few occupational terms whose general meaning remains [entirely] elusive' (Glennie, 1990: 14), such that the basic designation of 'bricklayer' can probably be understood by us today in a manner akin to that of an eighteenth-century citizen, it is also the case that 'even quite familiar designations encompassed a range of labour relations and forms of work' (1990: 14). In other words, a bricklayer in the eighteenth century might go about his (or her?) business rather differently from a bricklayer today, but so too might an eighteenth-century bricklayer in Lancashire be doing somewhat different things set within a somewhat different set of contractual relations from a fellow bricklayer in Kent. Furthermore, significant differences may exist between how individuals themselves might classify their *own* occupation and how this might be defined by 'social superiors' such as the gentry, the clergy, session clerks or overseers of the poor; and another consideration is that the occupations of women, even formal paid occupations, remained largely invisible to such authorities. The occupations that 'materialize' out of the historical data, therefore, reflect a host of assumptions, decisions and exclusions, ones made by both those in the past who recorded occupations and researchers in the present who adopt or adapt these occupational classifications.

There is nothing overly surprising about these claims, but what we wish to emphasize is Glennie's further line of argument about what is involved in talking about occupations, and in being prepared to identify particular named occupations with both clarity and certainty:

[To] have an 'occupation' *in its modern sense* is a comparatively recent idea. 'John Smith is a tailor' in the twentieth century carries a rather narrower meaning than in, say, 1600 or 1800. It is not so much that the term covers several possible relations of production ... , rather what differs is the exclusivity of activity embodied in modern definitions ... The counting of occupations is important, but mere tabulation of occupations amounts to tacit acceptance that 'an occupation is an occupation is an occupation', regardless of context. (Glennie, 1990: 11, emphasis in original)

What is evidently in Glennie's mind is the fact that few people in premodern England possessed anything like one single occupation and, instead, tended to be involved in a range of different occupations, such that for some of the time they might be farming (either on their own small-holding or as a waged agricultural labourer) whereas at other times they might be cotton-spinning in their attics or furniture-making in their yards. These would be the intersecting activities of the 'proto-industrial household' (Medick, 1976; Gregory, 1982), ones involving all household members, including women and children whose work in this respect rarely figures in the archives, and ones which often slipped into the realms of an informal economy which was again unlikely to be fully recorded. The key issue is that the determination to force the data into a mould which speaks of definite occupations, conveying the impression of clearly delimited and exclusive entities akin to how we tend to think of such things today, is itself something that must always be acknowledged. This is not to say that the researcher should not speak of such occupations – indeed, Glennie is prepared to do so in constructing a general typology of pre-census occupational classifications – but it is to insist that we can never completely forget about how we are carving up the data, and in a sense the reality behind the data, prior to any effort of naming, enumeration and interpretation. And his recognition of how we in effect envisage singular occupational identities (person A is in occupation X; person B is in occupation Y), identities posited as

devoid of overlap or ambiguity, allows us to foreshadow the more philosophical questions at issue here.

A geographical detour into set theory

In a little-known book on 'the mathematics of geography', Keith Selkirk (1982: 5) considers the principles underlying much geographical thinking and begins by suggesting that we start our studies – or should start our studies – through a careful sorting out of what might be termed raw data into *sets*:

A set is a collection of distinguishable objects, for example the set of active volcanoes in South America or the set of inhabited islands in Indonesia. It is necessary to have a criterion by which we know whether any object is included in such a set or not. In the above examples we need a clear definition of the word 'active', and to distinguish whether 'inhabited' includes seasonal habitation or not.

There is a formal notation which attaches to 'set theory', talking about 'elements' or 'members', 'universal sets', 'subsets' and so on (and also developing a related mathematical notation; see Selkirk, 1982: 5), but the key point to underline is the assumption that items of data can be relatively straightforwardly allocated to sets in a process which effectively entitizes them as certain varieties of phenomena rather than others. In terms of occupations, for instance, the aim is to allocate particular individuals to particular occupations in as clear a fashion as possible or to allocate particular specialist occupational classifications (e.g. bell-making) to more overarching occupational sets (e.g. engineering) in a similarly clear fashion. While we may not conceive of what we are doing in this vocabulary, there can be no doubt that we are in effect striving to specify readily identifiable sets (particular sets such as bell-makers or more general sets such as engineers), and then to allocate individual pieces of data to the appropriate sets. The properties of such sets are often visualized in terms of Venn diagrams, themselves an intriguing translation of complex conceptual processes into the abstract spaces of diagrams made up of lines and shapes comprising borders and boundaries. Two examples of Venn diagrams conveying simple geographical information are provided in Figure 7.1, and it might be argued that such diagrams offer a useful way of achieving a preliminary ordering of geographical data prior to any further work being conducted on these data (see also Abler et al., 1971: 152–5).

It should be obvious, as can be seen from the Venn diagrams, that thinking in terms of sets does not necessarily commit human geographers to supposing that data can be allocated to only one set or another set and never simultaneously to two or more sets. In fact, it is quite possible to conceive of the *intersection* of two or more sets, with the intersection containing those elements reckoned to be properly members of more than one set (as in Figure 7.1). This means that in set terms we can identify an individual person who is simultaneously a farmer and a spinner or to identify an occupation (e.g. a railway buffet attendant) that could be allocated simultaneously to different occupational classifications (e.g. a worker in both services and transport). None the less, at root the logic of set theory does conceptualize the situation very much in terms of phenomena either being included in a set or excluded from it, and the further suggestion is that the principles of set theory become rather meaningless if the researcher starts wanting to allocate certain phenomena to four, five or more sets (the implication being that the analytical simplicity inherent in thinking in set terms then disappears, leaving us back with the chaos of reality). Selkirk also argues that conventional set theory does not

handle well the situation where all the elements of a set can be assigned to one or other of two disjoint subsets, the example which he cites (and which is perhaps *not* as obvious as he suggests; see below) being that of the set of all human beings and its two disjoint subsets of females and males. Selkirk refers to such a situation as *dichotomous* and notes that dichotomous divisions are the basis of many classification systems (a consideration to which we will return).

Interestingly, he then makes the following observations with respect to Venn diagrams, as well as introducing another possibility for how we might be more flexible in our envisaging of sets:

> The Venn diagram has a number of disadvantages ... Firstly, Venn diagrams are not easily drawn for more than three sets ..., and secondly they emphasise the inclusion–exclusion aspect at the expense of the idea of dichotomising a set into two parts. An alternative which seems little known to geographers is the Karnaugh map. (1982: 7)

The latter alternative, the Karnaugh map, confronts the problem that many phenomena can (and should) be allocated simultaneously to more than one set, deploying a system which positions individual items (pieces of data) in different sets relating to segmentations of both the horizontal and vertical axes of a grid (for an example, see Figure 7.2). What we have here, then, is another way of translating conceptualizations into the abstract spaces of a diagram – one that, its rather ungainly rectangular quality notwithstanding, is actually more flexible than is the conventional Venn diagram. The possibility of alternative visualizations – of 'other' ways to arrive at conceptual maps – is something we will revisit shortly, albeit in a somewhat different fashion. At this point let us merely underline the extent to which set theory does tend to proceed in terms of an 'inclusion–exclusion aspect', referred to in other literatures as an 'either/or logic', and also the

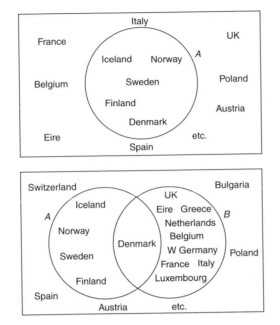

Figure 7.1 Two Venn diagrams used by Selkirk: one showing the set of Scandinavian nations as a subset of the set of European nations, and the other showing the overlapping sets of Scandinavian and (the then) EEC nations

Source: Selkirk (1982: 6, Figures 1.1 and 1.2)

extent to which the conventional diagrammatic representations seemingly reinforce this approach to entitation. This is ultimately a philosophical issue, and it is now to such philosophical reflections that we (have no choice but to) turn.

Identities

The basis of what we are calling entitation (in any project) lies in the attributing of *identity*. It entails being open to the raw data of the world as they come to the researcher, whether through his or her own efforts or in a form mediated by others, and it highlights the process whereby the researcher decides more or less consciously to *identify* the component phenomena, events or aspects seemingly existing in (and constituting) that portion of reality relevant to the study. It involves the crucial link between the verb 'to identify' and

the noun 'identity', two similar words possessing similar etymological roots, with the act of identification establishing the identities of the things reckoned to be present and worth discussing. The usual process of identification, of establishing the identities of what are to be discussed, is one which proceeds by delimiting a phenomenon as a singular, coherent and stable entity (and thus we also speak of 'entitation'), perhaps giving it a proper name (e.g. Merrybank Farm) or depicting it as a particular type of phenomenon (e.g. a sheep farm). Even when we are working with less tangible phenomena such as statements in an interview transcript, our ambitions are still conventionally to delimit them in such a singular, coherent and stable manner. Thus we would want to identify a given statement as a clear instance of, for example,, somebody's expression of 'topophilia' (love of place), 'topophobia' (hatred of place), 'neighbourliness', 'alienation'

229

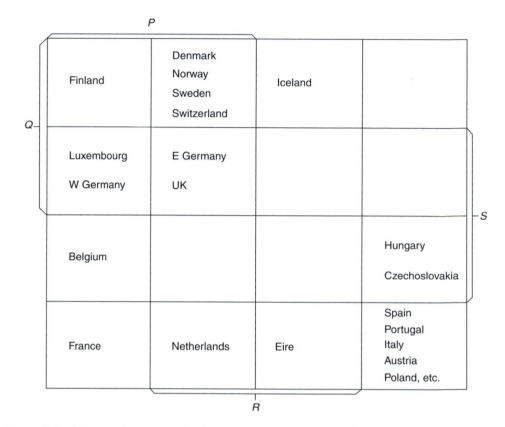

Figure 7.2 A Karnaugh map showing European nations divided on four categories: *P* = per capita income exceeds $1000 per annum; *Q* = over 300 daily newspapers per 1000 population; *R* = over 68 years life expectancy at birth for males; *S* = over 15 miles of railway per 100 square miles

Source: Selkirk (1982: 8, Figure 1.8)

or whatever.[4] When working with more tangible phenomena, the goal is normally to attribute not just a name but also a number, and thereby to give numerical values to a named phenomenon (e.g. Merrybank Farm has a land-holding of 75 acres) or to a depicted type of phenomenon (e.g. sheep farms are ones with over 70% of their income derived from sheep; sheep farms in the district comprise 60% of all farms; or sheep farms in the district employ 50% of local male employees). In order for enumeration (or quantification) to occur, it is normally essential for singular, coherent and stable identities to have been etablished, since numbers can

only really be fixed at all meaningfully to identities boasting these properties.

Thinking generally, though, we would suggest that in any given research project many entities are being ascribed with such well-defined identities to facilitate further interpretation of them and of how they apparently relate to one another. The upshot is a conceptual 'mapping' of the phenomena concerned, creating what might be termed an 'entity map' which effectively positions these phenomena in a stabilized 'space of identities', and – while this may sound like an odd way of putting things – such a map *is* being generated (it cannot help but be generated) in every conceivable research project. It is the outcome of the sifting and sorting which we are talking about here and its production is the unspoken objective of this crucial but unacknowledged activity.

To begin thinking about the notion of identity, and to begin criticizing how it is normally treated by geographers and other academics, we propose to take as our guide the human geographer Gunnar Olsson. This is someone who is often regarded as a rather maverick figure within the discipline, someone who has forsaken the conventional expectations of how to proceed (indeed, of how to practise) as a human geographer, but we would counter that in text such as *Birds in Egg/Eggs in Bird* (1980; see also Philo, 1984) he quite brilliantly confronts the issue of identity and its problems. Here is one set of claims which he makes about identity:

> The problem of *identity* is to recognise something as the same again. In conventional reasoning the equality sign is in the definition that *a* and *b* are identical if *a* and *b* have exactly the same properties. Identical words refer to identical things. The same labelling word must always refer to exactly the same phenomenon. What is called one thing is never called another. The advantage is that others will know what I hold as identical because they can see what I point to. The disadvantage is that, while categories are frozen, the world is not. (1980: 65b, emphasis in original)

This passage lays out the common supposition that the identity of a phenomenon resides in, or at least depends upon, its singularity (each thing being clearly delimited as 'one', and then as different from any other thing), its coherence (each thing having given properties) and its stability (each thing remaining the same from one moment to the next). Another coupling of key words therefore surfaces: that of 'identical' and 'identity'. The form of 'entity map' implied is one possessing entities which are clearly identified in their own pristine individuality, perhaps

named as A or B, and they are pictured as wholly separate with stark boundaries running between them which fix certain 'stuff' in one part of the map (e.g. in A) and other 'stuff' in another part (B). The connection back to the Venn diagrams mentioned above should be obvious, and we can reference literatures which carry a wider examination of such boundary-based 'topologics' and their implications (e.g. Reichert, 1992). Mention might also be made here of what Marcus Doel (1999) critiques as the 'pointillism' of conventional geography, particularly quantitative geography, residing in the discipline's obsession with 'One': with the wish always to be able to specify, delimit and count on the simplicity of the one identifiable, set-off and enduring entity, event, process or whatever as clearly distinct from any and every other one in the world (Doel, 2001).

Despite this critical tone, it must be acknowledged that Olsson appreciates the benefits of such a straightforward approach to identity in that it *is* the basis for people to communicate about phenomena, to share knowledge about what they agree – if tacitly – to regard as the *same* phenomenon with a definite identity. He also suggests that the ultimate grounding for agreeing on identity consists of people physically pointing at the same entity, in which case he hints at a grounding which is given less by pure logic and more by the contingent, embodied act of pointing (a suggestion with resonances is explored further in Olsson, 1988; 1991a). 'Trust' between people regarding the attribution of identity is thereby taken to be more about pointing, showing and demonstrating (see also Olsson, 1995) than about the ethereal logical deliberations of philosophy, an observation that hints at the variability of the 'real world' which requires more the gestures of everyday life than the abstractions of philosophy to create the preconditions for successful human communication.

This is probably to run into deeper waters than we need for our book, but it does flow from the critical claim that Olsson makes about the problem with the conventional approach to identity: that arising because, 'while categories are frozen, the world is not'. Thus, to return to the earlier example, any attempt at imposing a single occupational classification on a citizen of eighteenth-century Britain, whether by an authority at the time or by the researcher in the present, is to risk steamrollering over the fact that he or she may really have done a number of different jobs. Moreover, it risks obscuring the possibility that what the citizen did as (say) an 'inn-keeper' could vary considerably from what someone else given the same title might do in another part of the country. But even something as simple as identifying Merrybank as a 'sheep farm' might not be so straightforward: in an earlier decade the farm might have concentrated more on arable farming; in the present it might devote half its attention to sheep and half to a diversified activity such as running a farm zoo; over half its employees might work full time in the farm zoo but over half its income might still derive from the sale of sheep and wool; and many other complications can easily be envisaged. The identities postulated by the categorizations of 'inn-keeper' and 'sheep farm' are indeed frozen, implying a simple, obvious and unvarying character (in both time and space) which the things of the world almost never possess. In other words, while the world beyond academia is a thoroughly messy place, one which most of us would acknowledge from everyday experience as far from being marked by any singularity, coherence and stability, academics spend much of their time striving to impose just such a definitive quality on it with their neat namings, depictions and 'mappings' of the entities which they purport to find there. The sifting and sorting thus end up generating a picture of the world which is too tidy,

an artificially ordered approximation which cannot fully 'mirror' or 'correspond to' the phenomena, events and aspects which are there, present and happening, in everyday reality. The complaint that our models and theories are too neat and tidy, and hence too simplified, compared to the messiness of the world is actually quite commonplace (they are the life-blood of 'postmodernism'; see Box 7.6; see also Cloke et al., 1991: ch. 6), but Olsson's philosophical point is that such a neatening and tidying-up is integral to our routine assumptions about the identities of phenomena: it is indeed endemic to the very foundations of Western thought, for '[b]anned as outcasts are … fuzziness and indeterminacy' (Olsson, 1980: 62b).

Another 'outsider' geographer, David Sibley (1981b; 1998), has long complained about the obsession with order, or at least with a certain orthodoxly geometric conception of order, which has marked the efforts of many human geographers over the years (most notably spatial scientists). Drawing on an 'anarchist' critique of conventional scientific reasoning, he objects to approaches which insist on trying to fit awkwardly shaped chunks of reality into neat boxes, a process requiring the edges of the real 'stuff' to be planed off and the remains shoehorned into the simple (often symmetrical, sometimes elegant) 'grids and networks' of standard intellectual imaginings. Arguing along very similar tracks, a number of human geographers inspired by 'poststructuralism' (see Box 7.7) have also begun to question the usual procedures and outcomes of how the discipline has gone about its sifting and sorting of data. In the course of an innovative paper recording a staged conversation between a spatial scientist (here called a spatial analyst) and a poststructuralist, Deborah Dixon and John Paul Jones (1998a; see also Natter and Jones, 1993; 1997) outline what they refer to as the 'epistemology of the grid' which underpins

> ### Box 7.6: Postmodernism
>
> 'Postmodernism' has quickly become a much maligned and misunderstood umbrella term for what should probably be best regarded as a *stance* taken by many academics, geographers included, to fundamental questions about what the human world *is* (ontological questions) and how 'we' can acquire knowledge *about* it (espistemological questions). It cannot be regarded as a coherent theory, model or approach, despite occasionally being presented as such. While many writers explore the character of what they describe as the 'postmodern' epoch in which we presently live, one marked by the complexity, hyperactivity, hybridity and mutability of economies, polities, societies and cultures (e.g. Harvey, 1989a), the stance of 'postmodernist' thought is one that effectively mirrors – and argues that our thought should mirror – this *real* 'chaosmos' by refusing *conceptual* certainty: by refusing any simple 'grand theories' or 'meta-narratives' (big stories) about the human world and its workings, whether these be supposedly scientific laws, humanistic emphases on human agency and spirit, Marxist accounts of capitalist determinations and socioeconomic structure, psychoanalytic excavations of repressed desires, or whatever.[5] The watchword of postmodernism as a way of understanding is hence *difference*: an alertness to every line or scrap of difference which prevents things, peoples, situations, etc., ever being truly the *same* as one another, and which thereby insists on taking seriously the 'jagged edges' of particularity that most other intellectual stances before the human world wish to smooth away. This is only to give the briefest sketch of postmodernist thought and, in practice, there are now countless debates about its genealogies, claims and (de)merits (e.g. Benko and Strohmayer, 1997), but it is worth providing such as sketch here because in many respects the very basis for this chapter – with its querying of conventional approaches to the problem of identity obsessed with finding sameness (between things) rather than celebrating differences – lies squarely in the postmodernist critique.

the research of the spatial analyst (and which they suggest is most formally propounded in a paper such as Berry, 1964; see Figure 7.3):

> By our estimation, your [approach] is grounded in a particular way of knowing, one that depends upon, and propounds, the 'epistemology of the grid'. The grid is at once a procedure for locating and segmenting a complex, relational and dynamic social reality. It works to create a systematical horizontality that stabilises both objects and the concepts associated with them – *i, j* – such that both can be rigorously investigated using linear cause-and-effect systems of logic [see also Chapter 9]. The grid segments social life so that it may be captured, measured and interrogated. (Dixon and Jones, 1998a: 251)

Moreover, Dixon and Jones stress that 'the grid['s] … powers of segmentation fashion borders and supervise inter–relations among objects and events in space and time' (1998a: 251).

Writing chiefly against the assumptions built into spatial science but also against the stabilities of identity postulated by 'structuralism' (see Box 7.7), Dixon and Jones go on to highlight the importance of paying sustained attention to *context*: to the specific circumstances framing a particular phenomenon, event or aspect, as anchored in the peculiarities of a given time and space. Because the 'stuff' of the world varies so much from one context to the next, from one portion of time–space to the next, researchers need constantly to be revising the names, categories and concepts which they deploy to isolate, identify and then interpret this 'stuff': their orderings need to shift accordingly; their sifting, sorting and entitation cannot remain frozen. They must be ever vigilant to the possibility that what was meant by 'costermonger' in eighteenth-century London was not identical

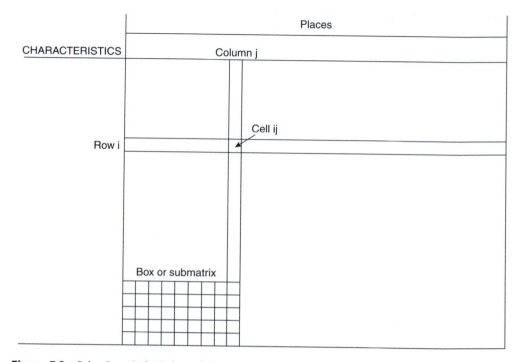

Figure 7.3 Brian Berry's depiction of 'the geographical matrix'. 'A row of this matrix presents the place-to-place variation of some characteristic or a spatial pattern of some variable which can be mapped. Each column contains a "geographic fact": the value assumed by some characteristic at some place' (1964). Berry also introduced time into the picture by imagining an array of geographical matrices in temporal sequence.

Source: Berry (1964: 6, Figure 1)

to what it meant in eighteenth-century Liverpool, for instance, or that what was meant in the eighteenth century by the very notion of 'occupation' is not the same as what we tend to mean by it today.

Dixon and Jones thus conclude that 'context stabilises meaning, but never permanently' (1998a: 256), and they advocate a 'deconstructive' stance (see Box 7.8) alert to how the identities of phenomena (along with associated terms, concepts and disciplines) which have been established with reference to one particular context constantly need to be reassessed with reference to other contexts. It is a stance whose imperatives are at once 'to lineate and de-lineate, to fix and unfix' (Dixon and Jones, 1998a: 255); or, in a vocabulary more closely bound up with feminist

geography, the aim is never to let anything 'settle' (Rose, 1991; 1993b).[8] Allowing such unfixity into research, permitting identities to be unsettled, introduces an openness, a fluidity, into our thinking which is very much hostile to the 'epistemology of the grid'. It runs against the normal assumptions and procedures of practising human geography, suggesting that the entitation part of a research project must always look to the context for guidance, to the temporal and spatial specificity of the subject-matter under study, rather than to any preset grids of expectation.

Binaries and hierarchies

In the course of his reflections on identity, Olsson makes a number of further points

Box 7.7: Structuralism and poststructuralism

'Structuralism' is an intellectual movement emerging from de Saussure's linguistics, Piaget's psychology and Lévi-Strauss's anthropology, and at root its insistence is that everything occurring in the human world – including the seemingly most 'human' of phenomena such as 'our' language and communication, cognition and comprehension, myth and taboo – is ultimately determined in its form and function not by individual human beings making conscious decisions which reflect their own (or any special features of) humanity but by anonymous 'structures' (of language, cognition, myth, etc.) which effectively lie beyond our individual knowing and control. These structures, conceived as relations between different entities identifiable within language, cognition, myth, etc., are then said to 'structure', to impose limits and possibilities, on what is actually said, understood, done, etc., by particular people in particular instances. Such actual happenings are hence regarded as *surface* manifestations of *deeper* structures, and the researcher's task as a structuralist is therefore to ascertain the contours of these underlying and not immediately visible structures. While the linguistic, psychological and anthropological versions of structuralism have occasionally influenced geographers (e.g. Tuan, 1972), the main route whereby structuralism entered the discipline was through an engagement with Marxism during the 1970s. At this time various notions to do with the determining influence of capitalism, a 'structural realm' comprised by the dynamics of capitalist accumulation, crisis and contradiction, were imported in a manner which many duly termed 'structuralist'.[6]

'Poststructuralism', closely related to postmodernism (see Box 7.6), continues the structuralist insight that the human world is 'made' by structures, where these are conceptualized as relations of all sorts in which human beings are caught up (and through which 'we' are all constituted as particular sorts of human beings with particular attributes, aptitudes, attitudes and so on). Yet whereas structuralism tended to suppose that there are ultimately definite entities which can be identified and between which possible relations can be established, poststructuralists are much more sceptical about there being any such 'pre-existing' entities which are not in turn themselves merely configurations of different relations (economic, political, social, cultural, discursive, psychical, spiritual, etc., all as cross-cut by axes of human difference such as class, gender, race, ethnicity, sexuality, (dis)ability, etc.). Any stabilities with which academics have traditionally dealt, even those named in the rigorous structuralist texts of de Saussure, Piaget, Lévi-Strauss and others, are here thrown into question. This is not necessarily to reach the point where there is said to be nothing but chaos in the human world (as in extreme postmodernist positions), but it is to accept that everything under study has to be subjected to the most uncompromising of *relational* analysis (see also Massey, 1999b; Thrift, 1999), in the course of which attention must always be directed at the specific time–space contexts in which the relations of relevance – whatever precisely these may entail – become 'grounded', 'earthed', 'real' in the whirls and twists of everyday human geographies. This, then, is the tricky conceptual terrain from which Dixon and Jones (1998a) develop their powerful attack on the 'epistemology of the grid' which is a central to the arguments of this chapter about identities and their (un)fixing (as integral to the sifting and sorting of raw data).[7]

which are germane to our discussion of sifting and sorting. In the course of talking about 'laws of identity' as linked to categorization, he writes as follows:

These are themselves embedded in the law of the excluded middle. Everything is identical to itself and nothing is identical to anything else. Nothing is itself and not itself at the same time. And yet inherent

---

Box 7.8: Deconstruction

Sometimes regarded as the methodology of postmodernism (e.g. Dear, 1986; 1988), 'deconstruction' is first and foremost a 'technique' for analysing texts – primarily philosophical and literary works – and as such it was introduced by Derrida in his 1967 text *Of Grammatology*. As Poole (1988: 206) explains:

> Derrida showed that, by taking the unspoken or unformulated propositions of a text literally, by showing the gaps and the supplements, the subtle internal self-contradictions, the text can be shown to be saying something quite other than what it appears to be saying. In fact, in a certain sense, the text can be shown not to be 'saying anything' at all, but many different things, some of which subtly subvert the conscious intentions of the writer.

This technique for teasing out the incoherencies, limits and unintentioned effects of a text – an exercise that scuppers any simple notion of being able to discern the 'true' meaning of a text (a notion that privileges the author over the reader) – has had a great influence, notably upon Anglo-American literary theorists. In this guise deconstruction commonly works by 'tracking the traces of oppositional elements' within a text, and there is a 'two-step process' through which 'one first reverses and then displaces the oppositions (eg. male/female, culture/nature or subject/object)' (Pratt, 1994: 468). This reference to isolating and then disrupting binaries (oppositions) is very pertinent to claims further on in this chapter. What has also taken place is that the principle of deconstruction has been extended to the analysis of the contradictions lying in the human world beyond written texts: in all manner of cultural products, social settings and power struggles. As a result, many researchers, human geographers included, now talk about 'deconstructing this, that or the other' when they are not so much pursuing a line of questioning directly informed by Derrida as simply highlighting aspects of a given theory or real-world situation that do not appear to 'add up' or 'fit together' all that well. See also notes on 'the deconstructionist' in Chapter 10.

---

in this stringent definition is nevertheless an element of dialectics [see below], for it is only by learning what something is not that we know what it is. Built into the positive is its own negation. (1980: 62b)

In a related statement he declares that in 'conventional thought, all categories stem from the principle of the excluded middle, which says that nothing can be one thing and its opposite at the same time' (1980: 36b). Here he is underlining the assumed singularity of identity, the straightforward oneness and non-contradictoriness of the phenomena being identified, but in so doing he raises a second

feature of conventional Western reasoning. This is one which tends to see things in terms of binary oppositions between supposedly dichotomous or mutually exclusive 'pure essences', not as condensations of many different phenomena, events and aspects fusing together in an overlapping fashion. This is not just to reiterate the finding that conventional reasoning prefers to hold things apart, assigning them to different and separate positions in the 'entity map' (see above); it is also to indicate its tendency to fix identities by setting up opposites which in a sense define one another. Olsson is perfectly aware of this tendency, elsewhere termed the role of the 'constitutive outside' (Rose, 1993b; 1995), which arises when what are identified as the characteristics 'inside' a given phenomenon (e.g. health, sanity, normal behaviour) are defined by those deemed to be 'outside' it (e.g. sickness, insanity, deviant behaviour). Although it is not necessarily the case that these binary oppositions will prioritize one of the two terms or identities over the other, it is often the case in practice that they do so. '[B]uilt into the positive [of a term or identity] is its own negation', states Olsson (1980: 62b), which means that written into the heart of what is taken as good about a phenomenon, what is valued about it, are suppositions about the bad, devalued, unwanted and even despised elements reckoned to be typical of its opposite.

Arriving by a different route from Olsson, binaries have become a central focus of recent debates in feminist geography, which echo Olsson's earlier work while extending it in significant new directions. Indeed, a rich vein of substantive research by feminist geographers from Linda McDowell (1983) onwards has examined a variety of pervasive worldly practices, notably spatial practices, which reflect a simple binary logic separating off public and formal 'realms' from private and informal 'realms'. As the WGSG authors state near the start of a chapter addressed to this issue:

> Binary categories often suggest the existence of discrete spaces, that is spaces associated with different types of activities ... [They] often depend on the drawing of sharp lines between the two halves within the binary category. These lines are like boundaries, or fences; they are put up between the two sides which comprise the binary category. So, for example, the public/private distinction works by establishing a boundary between the public and private sphere, and by drawing a line between public and private space. (1997: 112)

In effect echoing Olsson's claims, or those of Sibley or Dixon and Jones, about how the conceptual fixing of identity goes hand in glove with 'conservative' practices defensive of the existing social order, these feminist geographers show how a deeply worrying translation occurs from the conceptual 'binarizing' of women and men into the creation of everyday gendered geographies. As is well known, women tend to be portrayed as 'naturally' one kind of entity (caring, emotional, intuitive) and men as another (calculating, rational, cerebral), which relates to women being expected to perform certain kinds of role (nurturing mother, carer, counsellor) and men others (providing father, wage-earner, disciplinarian). Notwithstanding the countless scramblings of these gendered distinctions in ordinary life, there is still a clear link into prevailing assumptions about it being men who populate the public and formal spaces of society (conducting the important affairs of economy and state) while women remain in the private and informal spaces (managing the routine tasks of child-minding, home-making and house-keeping). And such assumptions still have an influence on what actually happens in the grounded geographies of who (out of women and men) is found where and doing what. As a result, '[f]eminist geographers continue to critique the gendering of dichotomous categories used to create space and places, the boundaries which these set up, and the ways in which boundaries work to define some people (and practices) as being "in bounds" while others are located "out of bounds"' (WGSG, 1997: 114; see also Cresswell, 1996). In so doing, these authors insist that researchers should endeavour to ensure that the basic ingredients stirred in at the outset of a project, those through which sense begins to be made of the raw data, are not ones which unwittingly reinforce stereotypical gendered binaries. More positively, they propose that researchers should be purposefully 'illustrating that the boundaries drawn between binary categories are frequently more *blurred* than the dichotomy constructs them as, very often because women and men struggle to resist them' (1997: 115, emphasis in original).

Yet there is more to say here, since there is a line of feminist critique which emphasizes the centrality of binary opposites to conventional Western thought and insists that a great many such opposites, even if it is not immediately obvious, are heavily freighted with gendered meanings and implications. The WGSG authors provide a clear summary of what they refer to as the problem of 'dualisms':

> A dualism is a particular structure of meaning in which one element is defined only in relation to another or others. Dualisms thus usually involve pairs, binaries and dichotomies, but not all pairs, binaries and dichotomies are dualisms. What makes dualisms distinctive is that one of the terms provides a 'core', and it is in contrast to the core that the other term or terms are defined. Thus dualisms structure meaning as a relation between a core term A and subordinate term(s) not-A ... The reason some feminist geographers have paid attention to dualistic ways of constructing knowledge is that dualisms are very often gendered and hierarchised, so that the core term A is masculinised and prioritised, and the subordinate term(s) not-A are feminised. (1997: 84–5)

Intriguingly, in his account of set theory in geography, Selkirk commented on the prevalence and importance of dichotomous thinking and proposed that the most obvious example of such a dichotomy is that between women and men (see above). Feminists agree that this particular dichotomy has been prominent within Western thought, and they argue that it has functioned as something of a 'master' dualism whose effects have rippled out to provide an underlay of expectations and even

prejudices present in the dualistic foundations of both academic and common-sense knowledges. In the process they have exposed the manner in which these knowledges become structured by oppositions between attributes coded as masculine (and in the process deemed as important, good and valued) and attributes coded as feminine (and in the process downgraded as trivial, relatively bad and not worthy of being valued). Gillian Rose (1993b: esp. 66–77) traces such oppositions in some detail, noting their installation within the engine-room of Western thought by male scholars effectively casting intellectual activity in their own self-satisifed and self-congratulatory *self*-image, and she offers a vehement critique which aims to dismantle these oppositions in the course of laying out new possibilities for a non-dualistic human geography whose tactic 'is suggested by the notion of other fields of knowledge beyond A/not-A' (Rose, 1993b: 85; see Box 7.9 for materials covered by Rose in her exploration of gendered dualisms).

Probably the most controversial proposal to emerge in this respect would involve questioning what Selkirk (and indeed virtually all the rest of us) do take as so self-evident: namely, the dichotomy between women and men. Instead, the WGSG authors speculate that

> The notion that we all fit into either a male or female body is just that, notional. It cannot be sustained, either over time or space. Simply put, there is nothing 'natural' about 'male' and 'female' bodies. There is nothing natural about everyone being forced into one sex or the other. (1997: 195)

However disturbing this claim may be to many readers, there is mounting evidence to suggest that the obviousness of biological sex can no longer be taken for granted, and that both sex and gender are much more entangled, much less dichotomous, than is usually assumed. There are many versions of woman,

Box 7.9: Gendered dualisms

The following are Gillian Rose's specification of gendered dualisms, including part of a poem which she borrows from Helene Cixous (where the first term in each line is coded as 'male' and the second term as 'female') and four other listings (with the same codings of first and second terms). Rose's argument is that in each coupling the 'masculine' term is conventionally regarded as a 'superior' personal attribute, academic quality, social condition, cultural space or whatever:

Activity/passivity,		modern	– postmodern
Sun/Moon,		deep	– shallow
Culture/Nature,		seminal	– playful
Day/Night,		great	– fecund
		thrusting	– titillating
Father/Mother,		penetrating	– veiled
Head/heart,			
Intelligible/sensitive,		time-	– humanistic
Logos/Pathos		geography	geography
		public	– private
culture	– nature	transparent	– opaque
city	– countryside	social	– body
space	– nature	knowledge	– maternity
culture	– primitive	rational	– emotional
		space	– place
theory	– empirics		
general	– specific		
abstract	– concrete		
nomological	– contextualizing		

Source: Rose (1993b: 67, 74, 75)

female and feminine, and there are many versions of man, male and masculine: all of us are almost certainly complex hybrids of bodily features, social behaviours and cultural attributes which are conventionally allotted to the non-overlapping boxes marked 'woman' and 'man', 'male' and 'female' or 'masculine' and 'feminine', but which in truth are always mixing together in unpredictable ways within individual bodies, minds and personalities. For feminists and feminist geographers, destabilizing these binary opposites – or even the thought experiment which contemplates destabilizing them – is a vital manoeuvre in challenging the tyranny of dualistic thought and in throwing up new possibilities for thinking about the sifting and sorting of identity. The implications for the more mundane practices which are ostensibly the focus of this chapter may not be all that apparent, but if such a challenge *were* to strike a decisive blow against the standard binary 'inclusion–exclusion' or 'either–or' logics of how to impose order on raw data, the impacts for all sifters and sorters would be immense. And there are one or two immediate implications as well, which will be covered shortly when we draw this chapter to a close.

Before doing this, though, let us conclude this section by briefly mentioning how the specifying of binary opposites in which certain things are valued over others often becomes elaborated in the further ordering device of the hierarchy. The WGSG authors discuss 'the implicit hierarchies' (1997: 114) which are set up in the process of valuing some phenomena while devaluing others, and in so doing they hint at the underlying principles of hierarchies wherein entities are more or less explicitly listed or ranked from those which are reckoned to be most important or fundamental through to those which are reckoned to be unimportant and trivial. (And there is a clear connection here to the so-called 'depth models' of explanation which we review in Chapter 9.) The assumptions which feed into such ranked listings should be reflected upon at some length since what might seem like an innocent act of getting raw data into a neat order prior to the real work of interpretation will actually have a decisive impact upon what interpretations are finally provided (constructed). To give a stark example, a Marxist geographer coding up an interview transcript may look for statements which can be named as to do with 'class', and will more or less deliberately put all these together under this heading at the top of his or her list of things to take into account. Other statements which might be coded under other headings may also be noted, for instance 'gender', but will then be placed rather lower down the list. A feminist geographer looking at the same transcript might arrive at similar codings but will almost certainly reverse the order so that statements under the heading of 'gender' are listed higher in his or her interpretative hierarchy than are those under the heading of 'class'.[9]

Geographical scale is often used as a basis for hierarchical order in the sifting and sorting stage of a project. Factors reckoned to be of relevance from an inspection of the raw data are allotted to levels identified as 'global', 'national', 'regional' and 'local'. There is then often some prior supposition on the part of the researcher that one of these scales is more significant in explaining a given situation (e.g. an international war, a separatist movement, a fire in a factory) than are any of the others (Smith, N., 1993). This supposition appears quite logical, although it must be acknowledged that human geographers of different persuasions do tend to give priority to one scale over another whatever the subject-matter under study (e.g. some Marxist geographers always prioritize the workings of global capitalism; some humanistic geographers always prioritize the hopes and fears of people living in their local neighbourhoods). In line with the broader trajectory of our reasoning, though, it should be added that many human geographers now urge us to avoid prioritizing one geographical scale over another and propose instead that we should always strive to see the intersections of influences emanating from different scales (particularly the articulations of the global and the local, giving what is sometimes referred to as the 'glocal'; see esp. Massey, 1993; 1994: chs 6 and 7). More generally, there is a powerful argument which insists that human geographers, and indeed other academics, should endeavour to avoid the deceptive seductions of hierarchical thinking, or at least to become much more self-aware than hitherto about the urge to list and to rank things according to their supposed worldly importance and hence explanatory relevance (again, see Chapter 9). And once again there are some immediate implications which will be mentioned presently.

Alternatives and recommendations

Much more could be said in this chapter about what happens once researchers start to sift and to sort their data, and about how what initially may seem quite self-evident tasks, such as the naming of entities and the allocating of them to boxes and lists, are really quite problematic and rooted in a host of questionable assumptions. To be sure, these are activities and assumptions deeply entrenched in the conventions of Western thought and action, and as such they are mostly taken for granted and rarely exposed for critical attention. Yet with the rise of new approaches both within and beyond human geography, notably feminism (see esp. McDowell and Sharp, 1997), poststructuralism (see esp. Doel, 1999) and postmodernism (see esp. Strohmayer and Hannah, 1992; Benko and Strohmayer, 1997), such taken-for-grantedness *is* now being extracted and subjected to intense questioning. The foundations of knowledge, both academic and common-sense, are thereby being subjected to a searching critique which to many may seem esoteric and irrelevant but which, none the less, has far-reaching implications for some of the apparently most mundane of our practices as human geographers. Indeed, as should be clear from the discussion throughout this chapter, what might usually be regarded as the simplest, most uncontroversial part of research − the basic sifting and sorting of data, the preliminary ordering in which entities are identified, named, boxed and listed − is really one of the most problematic. What occurs here in terms of the attributing of identities, the specifying of binaries and the positing of hierarchies can no longer be treated as self-evident but must become a further element of the research endeavour about which we are self-critically aware. But what at bottom can this mean?

For many of the human geographers mentioned in this chapter, there is a need to keep on worrying away at these foundations in the hope of being able to formulate alternative ways of thinking about how we order knowledge (about what we take knowledge to 'look like' in the first place). This has certainly been Olsson's ambition for two decades, commencing with his earliest attempts to work with alternative systems of logic (termed 'many-valued' and 'fuzzy logics'), which transcend simple 'either–or' assumptions in preference for an alternative 'grammar of change and action' able to grasp certainty and ambiguity simultaneously (Olsson, 1980: Portfolio 2). Few geographers have followed this lead although those closest to having done so are the likes of Stephen Gale with his 'fuzzy set' theory (1972; and note that Olsson describes Gale as someone 'who was certain about ambiguity' − 1980: n.p.) and also Stan Openshaw with his concern to deploy artifcial intelligence (AI) in tandem with 'fuzzy logic, fuzzy systems and soft computing' (Openshaw and Openshaw, 1997: esp. ch. 10; Openshaw, 1996; 1998). Olsson himself quickly moved on from these technical solutions, however, and even in his 1980 text he framed them with a more 'dialectical' vision derived from Marx (see Box 7.10) wherein 'every category includes both itself and its opposite' (Olsson, 1980: 36b). Matters of identity and binaries were therefore to be recast in terms of 'dialectical change', and it is worth quoting Olsson further in this connection:

> It was exactly these forces of dialectical change that Marx tried to capture by using batlike words in which he could see both birds and mice at the same time. Thus, he allowed his categorial frameworks to be in constant flux because he realised that the things he was talking about were in constant flux. Put differently, he conceived of his seemingly inconsistent words as really being consistent, because the world he saw was consistently inconsistent. (1980: 12b)

With dialectics he still sought to find some way of disrupting the usual obsession with fixing things in sealed boxes, and thereby to

---

Box 7.10: Dialectics

Boasting a heritage stretching back into Antiquity, 'dialectics' is a 'logic of reasoning' which operates differently from standard logical reasoning that is rooted in data derived from sensations, and which is thereby positioned as subservient to the supposedly identifiable, stable and coherent objects in the world. Whereas '[s]tandard or analytic logic … is rigid and abstract, a matter of fixed connections and exclusive oppositions', it is suggested that '[d]ialectical logic sees contradictions as fruitful collisions of ideas from which a higher truth may be reached by way of synthesis' (Quinton, 1988: 225). Hence, rather than being anchored in an apparently constant world where fixed ideas attach to clearly delimited phenomena, dialectics embraces the possibility of a more dynamic, changing and fluid world at the same time as 'freeing' the ideas inspired by this world to enter into higher-level alliances, affiliations and syntheses productive of wholly new insights about human life on earth.

Marx took on board such dialectical logic, arriving at a deep perception of how phenomena are constituted through their relations (and see various remarks throughout this chapter about relational thinking), and developed a broader understanding of human history – his 'dialectical materialism' – in which every society contains within itself the 'seeds' of another society that it might ultimately become (or, more precisely, it encompasses within itself contradictions whose intensification and eventual resolution will tip that society over into being something quite different and new). Nobody has done more than Harvey (esp. 1995) to bring Marxist dialectics into geographical theory, emphasizing that '[d]ialectical thinking prioritises the understanding of processes, flows, fluxes and relations over the analysis of elements, things, structures and organised systems' (1995: 4). Moreover, he states that '[e]lements or "things" … are constituted out of flows, processes and relations' (1995: 5), and that it is vital to appreciate the sheer 'heterogeneity' of things in that they are always themselves constituted through 'internal relations' holding together contradictory forces or tendencies within 'smaller' component things. The implications for thinking about issues of identity, binaries and hierarchies are legion.

---

see A and not-A as at once inter-related and mutually constitutive. This manoeuvre connected him to Marxist geographers such as David Harvey, for whom Marx's dialectics was the preferred starting-point in sifting and sorting the data of the world,[10] but he still found himself feeling that such an orientation was too prone to collapsing the dialectical moment into rigid binaries. The upshot is that he began to turn increasingly to the exploded logics and free-flowing creativities of surrealism – 'every time I returned, I saw more surrealism' (Olsson, 1980: n.p.) – and this has meant that, while striving to produce a new

'invisible cartography' freed from the iron geometries of conventional thought,[11] he has increasingly turned for inspiration to the poetic and 'dreamlike' writing of surrealists like Joyce wherein 'truth emerges when identities are violated and opposites unified' (1980: 47e).

We realize that it may seem odd to encounter such notions in a book about practising human geography, and we are fully aware of the criticisms which may be levelled at Olsson (e.g. Billinge, 1983; Sparke, 1994), but we are still prepared to suggest that researchers could do a lot worse than allowing such 'crazy' notions to swirl around their heads at the stage when they begin seeking to impose order on their materials. Just sitting there with your data, knowing that a respected human geographer such as Olsson has contemplated the possibilities of very 'other' ways of ordering knowledge, even ways which in a conventional sense seem entirely *dis*ordering and remote from the normal expectations of 'science', will potentially be a very liberating (if unnerving) experience. But it is not just Olsson's example which might be pondered, for many other respected human geographers have been equally challenging of standard orderings, none more so perhaps than the feminist geographers who have queried the common dependence on dualistic constructions of knowledge anchored in an equation of what is deemed 'superior' with what is deemed 'masculine'. Even so there is a difficulty about leaving readers with the impression that in their own individual research projects they should emulate the examples of Olsson and certain feminist geographers, and hence devote their energies to challenging the very pillars of Western thought. While we do wish to encourage a critical edge, one that does recognize these pillars to be somewhat more unsteady than is commonly supposed, we appreciate that for much of the time researchers will have little choice but to accept the normal procedures of sifting, sorting, boxing and listing, as predicated on standard assumptions about identity, binaries and hierarchies. But does this therefore imply that most of our discussion in this chapter is irrelevant to the everyday practising human geographer? To some extent we have to be honest and say 'yes', but this is not entirely the case because we can offer some concrete proposals about a critically self-reflexive sifting and sorting which *can* be of immediate relevance.

It is with a listing of such proposals that we shall now conclude the chapter. First of all, it is crucial to *be self-critical about the identities that you identify*. In most pieces of research, we *will* want to sort out our raw data in such a manner that we can identify definite things within them, whether as nameable entities (e.g. Merrybank Farm) or specifiable categories (e.g. sheep farms). More specifically we *will* want to take particular statements (whether from documentary sources, interview transcripts, field diaries or whatever) and code them up under clearly distinguishable headings (e.g. 'class', 'gender', 'community', 'disability'). We must be aware of what we are doing in this respect, though, and give considerable thought to the act of entitation involved. We must remember that we could always be identifying entities and categories differently from how we are doing, and that it is not somehow 'natural' or automatically correct to be identifying certain ones rather than others. The secret is to be self-critical about our identified entities and categories and always regard them as 'open questions', as things that might need to be revisited, renamed and reclassified, even abandoned, as the project progresses. We should endeavour never to let them achieve the solidity, the frozen quality, which is all too typical of much social-scientific research.

Secondly, it is crucial to *be alert to the messiness of identity*. Following from the above, we must acknowledge that even seemingly self-evident entities and categories, such as a named place, an occupational grouping or supposed fundamentals like 'woman' and 'man', may not be quite as straightforward as might be supposed. There may be tricky theoretical debates which suggest this to us, as indicated earlier, but it may be that within the details of our raw data this also becomes apparent. A simple spatial designation like

'Glasgow' can turn out to be quite complex, for instance, in that a researcher may find that what is being called 'Glasgow' in different sources and by different people refers to territories of differing sizes with differing central points and differently drawn boundaries. The potential variability of an occupational grouping like 'carpenter' (or even of the very notion of 'occupation') has already been discussed and could easily be detected from empirical data, as Glennie shows us. The potential variability of the entities 'woman' and 'man' has also been discussed and, while it is unlikely that much of our data will call for us to question the continued use of these categories, it is not impossible that it might (as Lewis and Pile, 1996, found when studying gender performances at the Rio Carnival, Brazil). The implication is that we should be prepared to accept that identified entities and categories are almost always going to be less singular (more overlapping), less coherent (more fragmented) and less stable (more changeable) than is commonly expected of the identities upon which research is based. This does not mean that we should stop working with them as the basis for further interpretations, but it does mean that we must consistently add qualifications about their messiness (if not sometimes foregrounding the fact so as to question the orthodoxies of others).

Thirdly, it is crucial to *be alert to the contexts of identity*. Following from the last two points, we must ensure that anything which we do name or classify is meaningful to the particular context of our research: in other words, that it is appropriate and relevant to the people, situation, time and space under study. Ideally this means that the entities and categories at which we arrive should be ones recognizable and understandable by the people in the given situation, time and space. This cannot and need not always be the case since there will be occasions when the researcher has to operate with names and classes imposed very much from outside the situation (perhaps when studying prehistoric human geography for which virtually no 'written' records remain, or perhaps when purposefully attempting a highly conceptual account dependent on terms derived from prior theories and models). None the less, our usual preference would be, as far as possible, for striving to use the words, phrases and discourses of situated people themselves – whether derived from a planning document, an interview, a casual conversation, even an overheard remark – in order to guide us in identifying entities and categories pertinent to the subject-matter of the research. As an aside, this is a strong claim now coming into geography from scholars influenced by ethnomethodology (e.g. Laurier, 2001; Laurier and Philo, 2003), who give priority to closing the gap between the terminologies of academics and the 'lay' vocabularies through which situated peoples cannot but think and, in effect, perform their own worlds. Moreover, from the previous chapters of the book, we can note that there are many ways of constructing data which preserve the subtle texture of what is meaningful to people in the historically and geographically specific contexts of their everyday lives, work, play, loves and hates. It should be added, of course, that this stance should not preclude us from being critical of the entities and categories favoured by the people under study: they are certainly not beyond criticism, even if this must be done with a contextual sensitivity, and their views can be commented upon judgementally from the standpoint of the researcher (see Chapters 1 and 12).

Fourthly, it is crucial on occasion to consider *scrambling binaries and hierarchies*. In following the previous recommendations, we should also aim to avoid falling unthinkingly into specifying dualisms or adopting hierarchies. In the spirit of what we have just said about identities, this is not to suggest that we need to purge all traces of binaries and hierarchies from our research since this is probably impossible anyway (given just how pivotal are binaries and hierarchies to Western thought). But it is to suggest that we are self-critical in this respect as well, and that we do our best to spot binaries and hierarchies emerging from the preliminary ordering of our data. If we begin to figure that contrasts between the city and the country might be germane to the research, or those between highlands and lowlands, *Gemeinschaft* and *Gesellschaft*, Fordism and post-Fordism and so on, then we are definitely bringing binaries into the heart of our research. It is therefore incumbent upon us to ascertain whether or not such contrasts are meaningful to the particular study being conducted, perhaps asking if they are ones registered as significant by the people involved; and we must beware of unwittingly prioritizing one element of a binary over the other in a fashion according it with value while dismissing its opposite as valueless (which would be to fall into the dualistic trap with its likely gendered underpinnings critiqued by feminist geographers). This being said, it may be that the people under study think in terms of binaries, and quite possibly in terms of dualistic and gendered binaries, in which case our task is to report on this finding while retaining a critical distance alert to the sociospatial inequalities perhaps resulting from grounded actions informed by such binaries. Finally, the same principles apply to hierarchies. From the moment we start to list things which we reckon to be relevant to our research, the very ordering of the list has the potential to become an unexamined ranking of what is considered more or less important, determining or of explanatory value. We must therefore explore the relevance of our listings to the people under study, noting their sense of what is or is not important, and keeping their rankings (as far as these can be discerned) as a check upon our own. That we should do this self-critically, but also with a critical eye on the largely taken-for-granted hierarchies harboured by our research subjects, can presumably now go pretty much without saying.

Notes

1 See, for example, Philbrick (1957), Grigg (1965; 1967), Abler et al. (1971: ch. 6) and Semple and Green (1984).

2 See Blaut (1961), Harvey (1969), Sack (1980), Gregory and Urry (1985), Gregory (1985; 1994), WGSG (1997: esp. ch. 1), Massey (1992; 1994; 1999a; 1999b) and Thrift (1996; 1999).

3 See Goss (1995b), Pickles (1995), Curry (1998); but see Clark (1998) and Flowerdew (1998).

4 This is precisely what we are attempting to do in most versions of 'coding up' transcripts and other textual sources, as explained further in Chapter 10; see also Crang (1997) who uses the phrase 'sifting and sorting' precisely in this connection.

5 See Gregory (1989b), Cloke et al. (1991: ch. 6), Strohmayer and Hannah (1992); see also Chapters 9 and 10.

6 For good surveys of structuralism in geography, noting its links to Marxism, see Gregory (1978a: ch. 3; 1994) and Peet (1998: ch. 4).

7 For further guidance on poststructuralism in geography, see Dixon and Jones (1996a), Natter and Jones (1993; 1997), Peet (1998: ch. 6) and Doel (1999).

8 Interestingly, elsewhere Jones utilizes a numerical modelling technique called the 'expansion method' as a formal spatial-analytic vehicle for coping with the variability of context; see Jones and Casetti (1992).

9 Of course, attention *should* be given to the total numbers of statements under each heading, in which case a simple numerical measure can be deployed to ensure that

the priorities in the understandings of the interviewee are not completely over-ridden; see what we say about numerical 'contents analysis' in Chapter 8.

10 See Harvey (1973: esp. ch. 7; 1995; 1996); see also Castree (1996).

11 See Olsson (1991a; 1991b; 1994); see also Philo (1994).

8

Enumerating

Enumeration and human geography

In this chapter we turn to the use of numbers in the practising of human geography. Numbers are a fundamental facet of our everyday lives. They convey information to us; they underwrite key categorizations of size, standard, price, importance, progress and so on; they are used to present significant evidence to us in favour of, or in opposition to, some opinion, policy, product, way of life or place; and sometimes they are offered as proof of some underlying generalization. Thus we buy clothes of a stipulated size and we buy albums from numbered charts. Our schools, universities and hospitals are measured in terms of their performance and our housing values and our incomes are taken as a reflection of the standard of the area and the taxability of the people who live there. It is inevitable, therefore, that human geographers will be informed by, and in turn will seek to inform, this numerical world using processes and practices of enumeration.

For some human geographers, quantitative analysis can be represented as a modus operandi for addressing the nature of major social and political problems in their own terms. A recent critical review by Johnston et al. (2003) charts national-scale problems such as failing schools, unemployment, ill-health and hospital waiting-lists alongside international-scale problems associated with living standards, famine, AIDS and environmental degradation, and contends that such problems are difficult to assess, let alone respond to, without qualitative analysis. They argue that

> The identification of these and many other problems – their what, where and intensity, all of which have to be addressed before one can turn to why and then how to remove them – involves measurement. Although many of the problems involve alleviating (at least) the situations of individuals, nevertheless they can only be attacked by focussing on population aggregates, such as famine sufferers, or the unemployed, or the long-term sick in particular areas – without committing the ecological fallacy. Knowing where such people are involves first measuring the symptom … and then identifying other characteristics of either the people themselves or the places they live in, so as to isolate potential causative factors for their conditions and then analyse the impact of attempted cures (aggregate social research involving the ability to handle large data sets). (2003: 158)

The use of enumeration to interpret human geography data reflects in some ways an uncomplicated story by which ever more sophisticated quantitative techniques have been developed to analyse numerical spatial data and to construct and test mathematical models of spatial processes. There is another sense, however, in which the development of quantitative geographies has been implicated in a rather tense and often dichotomous relationship between quantitative and qualitative geographers undertaking their quantitative and qualitative analyses, sometimes in opposition to each other. Thus alongside the evolving sophistication of quantitative analysis, there has come a proselytizing zeal to present enumeration as *the* way of doing things, matched only by the 'reverse swing' of those who would dismissively regard qualitative analyses as *the* way. It would be very easy to characterize, or even satirize, the relationship between quantitative and qualitative geographies as a debate between two entrenched, mutually antagonistic camps. Such a debate will, however, be variously constructed and perceived. Our team of authors includes individuals who have encountered trenchant opposition to the introduction of qualitative methods courses into the curriculum of particular universities. On the other hand, when reflecting on the 'two entrenched, mutually antagonistic camps', Ron Johnston (pers. comm.), a quantitative researcher of high repute, reflects:

> the more I think about it, the more I feel that this is not the case, and that there is an asymmetry: few 'quantifiers' dispute the validity of the general questions being posed by the 'qualifiers' … on the other hand, many 'qualifiers' deny the validity of the questions the 'quantifiers' ask, and brand all of the latter as logical positivists.

In a recent theme issue of *Environment and Planning A* on 'reconsidering quantitative geography', Philo et al. (1998) recall how the emergence of the quantitative 'revolution' in the early 1960s was itself flavoured by an antagonistic divide between the older qualitative and new quantitative approaches. They quote Floyd, writing in 1963, as an illustration of how this new geography was received at the time:

> Faced with the intellectual immensity of the geographer's goal, a number of young workers … would have us eschew the integrative, global philosophy of the subject, time-honoured as it is, and turn our attention to sterner, more 'masculine' things: a rigorous wrestling with restricted data and areas utilising the 'new' scientific methods, quantitative ways of analyses, statistical and mathematical tools and models, and so forth. The old ways often lead to a never-ending trail of 'trivia, boredom and ennui', the new presents challenging, brain-cracking problems in combinational topology, scaling, theories of central place location, nearest neighbours, the journey to work and linear (or at least quasi-linear) programming. (p. 15)

As Philo et al. point out, Floyd's recognition of the excitement and potential associated with emergent quantitative geographies was tempered by a strong concern that these innovations would signal the death knell of the existing mainstream position of qualitative methodologies, to the extent that 'the non-quantitative geographers are politely if firmly relegated to the lower order of workers within the geographical hierarchy' (Floyd, 1963: 15).

The speedy dominance achieved by quantitative approaches in geography has often been accompanied by self-representations of considerable machismo:

> Quantitative geography is here to stay. The revolution is over. It is not a 'new' geography in the sense that it alters the nature of the discipline itself; but only in that it offers new techniques – for the solution of old problems as well as new ones – techniques which have brought a new rigour into geographical thinking, and have made possible the exploration of new fields. (Hammond and McCullagh, 1974: xi)

On the other hand, this 'new rigour' has been held by others to be at the expense of understanding *meanings* and explaining *phenomena* due to the uncritical conceptualization of geographical objects and subjects. As Dey (1993: 3) suggests: 'The growing sophistication of social science in terms of statistical and mathematical manipulation has not been matched by comparable growth in the clarity and consistency of its conceptualisations.'

Gould's (1979) reflections of geography in the 1960s and 1970s suggest that many early borrowings of statistics by geographers were inappropriate and that the quantitative 'revolution' which at first seemed to liberate human thought was in danger of being later transformed into dogma and myth. Subsequently, the quantitative project in geography has become subject to increasing criticism. In an influential paper, Barnes (1994) points to two strands of criticism. First, and perhaps most obvious, are criticisms from Marxist and humanistic geographers about the applicability of mathematical knowledge to the representation both of social and political conflicts, and of the significant characteristics of human action. Secondly, Barnes focuses on the *internal* blind spots and dead-ends of mathematics as a structured system of knowledge. He sees mathematics as a prime example of *logocentrism* – a term from Derrida (see, for example, 1991) which describes a belief in a fundamentally ordered world that can lead to a further belief that there is a fundamental level of knowledge beyond which we need not go. What is being suggested here is that quantification in human geography tends to present the world in an ordered way which can be accessed using numeric data and understood by using these data both to describe and infer characteristics of the human world, and to provide evidence for or against hypotheses and laws by which that world can be understood. It is important here to discriminate between different underlying approaches to quantification. For some, the privileging of observations represents an empiricist approach while others (whose testing of repeatable observations leads to theory construction) may take a more positivist approach (see Box 8.1). Yet others use quantitative techniques to test non-positivist theories

(see, for example, Plummer and Sheppard, 2001). Albeit underpinned philosophically in these different ways, the quantitative mind-set, once adopted (and sometimes it becomes deeply ingrained in the intellectual psyche), makes it difficult to see beyond the numeric and problematic to accept that there are other (non-numeric) bases of knowledge which can be equally relevant to the broad project of understanding the human world. In these terms, quantification can be viewed as a (rather partial) attempt to impose a rather particular form of order to the world.

Qualitative movements have emerged in challenge to these quantitative orthodoxies, suggesting that some matters of concern to human geographers cannot simply be counted or measured, and therefore deploying a range of methodologies which focus on intensive research of a non-quantitative kind. Perhaps because of the strength of numbers of quantitative practitioners in human geography, and the consequent difficulties that others have sometimes encountered in gaining legitimacy (in postgraduate theses, teaching, research funding, etc.) for their qualitative methodologies, there has been rather a defensive posturing from qualitative geographers who have sometimes seemed unwilling to sanction any numbers-based approaches to their subject. Hence quantifiers have been often dismissive of qualifiers, and vice versa, and, in this context, enumeration is therefore positioned as a game played by quantitative geographers.

Some commentators now argue that the quantitative take-over from the qualitative, as commented on by Floyd, has been reversed. By 1973, Harvey had pronounced that 'the quantitative revolution has run its course, and diminishing marginal returns are apparently setting in' (p. 28) and, as the millennium turned, Johnston (2000a: 131) was moved to declare that 'Spatial analysis isn't dead (or even dying), therefore, so why is it being committed

---

Box 8.1: Empiricism and positivism

Empiricism

Gregory (2000e) describes empiricism as a philosophy of science which privileges empirical observations over theoretical statements. Empiricism assumes that observed statements are the only way of making direct reference to the real world and that such statements can be judged as true or false without reference to theory.

Positivism

Positivism builds on these empiricist foundations but, additionally, insists that scientific observations have to be repeatable, that the generality of observations leads to the formal construction of theories which when verified empirically can be regarded as scientific laws. In turn, these scientific laws can be unified progressively and integrated into a single system of knowledge and truth (see Gregory, 2000f).

---

to the discipline's past by many of those concerned with human geography's future?'

The worry now in some quarters is that the increasing dominance of qualitative methodologies may signal the death knell of the previously mainstream position of quantification, representing an increasingly difficult 'contest for resources, power and influence within the discipline' (Johnston, 2000a: 131) and certainly marking a 'downturn' in the fortunes of quantitative geography during the 1980s and 1990s (Graham, 1997; Johnston, 1997a). Fotheringham et al. (2000) suggest five issues which set such a downturn in context (Box 8.2). They argue that many of the criticisms traditionally levelled at quantitative geography no longer apply. For example, not all quantifiers can be assumed to be positivists:

> just as some quantitative geographers believe in a 'geography is physics' approach (naturalism) which involves a search for global 'laws' and global relationships, others recognise that there are possibly no such entities ... the emphasis of quantitative analysis in human geography is to accrue sufficient evidence which makes the adoption of a particular line of thought compelling. (2000: 5)

Neither, it is argued, should they be presented as being driven by any deep-rooted political or philosophical agenda: 'For most of its

practitioners, the use of quantitative techniques stems from a simple belief that in many situations, numerical data analysis or quantitative theoretical reasoning provides an efficient and generally reliable means of obtaining knowledge about spatial processes' (2000: 4). Such a recognition of diversity among quantitative practitioners is both significant and, in our view, very helpful. We want to recognize a myriad of entry points into human geography – those places, events, readings or interests which spark an individual's imagination to go deeper into the subject. It is clear to us that quantitative methods can be, for some people, just such an entry point. For some researchers and students it is the compelling association with mathematical modelling, statistical analysis and computing which is the unambiguous point of departure from which the journey through the terrain of human geography begins. Indeed, the selective nature of that terrain will often be mapped out by decisions about what kinds of subject-matter provide 'good fit' with the desired methods. It seems likely in these circumstances that quantitative geographies, or perhaps a search for order which underpins some of them, will be the dominant factor informing judgements about what geography is and what it should be.

Box 8.2: Reasons for the 'downturn' in quantitative geography

1. A disillusionment with the positivist philosophical underpinnings of much of the original research in quantitative geography and the concomitant growth of many new paradigms in human geography, such as Marxism, postmodernism, structuralism and humanism, which have attracted adherents united often in their anti-quantitative sentiments.

2. The seemingly never-ending desire for some new paradigm or, in less polite terms, 'bandwagon' to act as a cornerstone of geographical research. The methodology of quantitative geography had, for some, run its course by 1980 and it was time to try something new.

3. A line of research that appears to be better accepted in human geography than in some related disciplines is one that is critical of existing paradigms. As quantitative geography was a well-established paradigm it became, inevitably, a focal point for criticism.

4. As part of the broader 'information revolution' which has taken place in society, the growth of geographical information systems (GIS), or what is becoming known as geographical information science (GISc), from the mid-1980s onwards, has had some negative impacts on quantitative studies within geography. Interestingly, these negative impacts appear to have resulted from two quite different perceptions of GISc. To some, GISc is seen either as the equivalent of quantitative geography, which it most certainly is not, or as the academic equivalent of a Trojan horse with which quantitative geographers are attempting to reimpose their ideas into the geography curriculum (Taylor and Johnston, 1995; Johnston, 1997a). To others, particularly in the USA where geography has long been under threat as an academic discipline, GISc has tended to displace quantitative geography as the paramount area in which students are provided with all-important job-related skills (Miyares and McGlade, 1994; Gober et al., 1995).

5. Quantitative geography is relatively 'difficult' or, perhaps more importantly, is perceived to be relatively difficult both by many academic geographers, who typically have limited quantitative and scientific backgrounds, and by many students. This affects the popularity of quantitative geography in several ways. It is perceived by many students to be easier to study other types of geography, and their exposure to quantitative methodology often extends little beyond a mandatory introductory course. It deters established non-quantitative researchers from understanding the nature of the debates that have emerged and which will continue to emerge within quantitative geography. It also makes it tempting to dismiss the whole field of quantitative geography summarily through criticisms that have limited validity rather than trying to understand it.

Source: Fotheringham et al. (2000: 1–3)

However, it is also important to recognize that a single methodological impulse – whether quantitative or qualitative – need not occupy an all-embracing position in the personality of human geography. Practices of enumeration, statistical analysis and spatial modelling can also form part of a wider portfolio of techniques open to human geographers. In this case, a particular topic, issue, subject or question may be the entry point which sparks the geographical imagination and the selection of suitable technique(s) with which to practise human geography in this area can be dictated by the nature of that subject–matter.

Indeed, we would strongly suggest that quantitative and qualitative analytical techniques need not be viewed as discrete choices, and there are now signs of a growing interest in analyses which span the quantitative–qualitative divide. Although there are geographers who continue to use mathematics in a fundamentally logocentric way, there are many others who have sought in different ways to explore the potential interconnections between enumeration and other interpretative methodologies (see, for example, Fielding and Fielding, 1986; Fielding and Lee, 1991). For example, essentially quantitative geographers (see Wilson, 1974) have suggested that multilevel modelling (see below) supposedly represents a way of using quantitative approaches to address some of the contextual and conceptual issues being raised in highly politicized and 'subjective' arenas. A recent move in this direction has come from Openshaw (1996; Openshaw and Openshaw, 1997). Starting from an unpromising re-articulation of the quantitative ('hard')–qualitative ('soft') divide in which soft methodologies are recognized as having 'good' and 'bad' traits (Box 8.3), Openshaw opens out some interesting possibilities for translating 'soft' material into a form which is compatible with computational methods derived from artificial intelligence. His advocacy for using 'fuzzy logic' not necessarily in a numeric way represents a small but potentially significant step towards interconnecting the quantitative and the qualitative.

From a very different starting point, some users of qualitative methodologies (see Dey, 1993) have suggested that the growing importance of *computer-based* qualitative analysis, dealing for instance with enumerating the uses of different metaphors and the links in discourse, will bring qualitative and quantitative analysts closer together. Yet others, often using a realist philosophy of science, foresee the linking of causal mechanisms and the variety of concrete events determined by those mechanisms in terms of quantitative enumeration (see Foot et al., 1989). One

further area of interconnection which has considerable potential is where mathematics is used *critically*, recognizing that it is a language just like any other. Researchers such as Sibley (n.d.), Pratt (1989) and Sheppard and Barnes (1990) have rejected the idea of mathematics as a logocentric project yet have used numeric data in an exploratory way as 'a technique for organising one type of information, information that needs necessarily to be complemented by a more relational, contextual understanding, as well as more abstract theoretical development' (Pratt, 1989: 114).

Many of the arguments rehearsed so far in this chapter illustrate the potential for defensive divisiveness in human geography's methods. If you are quantitative then you are presumed to be anti-qualitative, and vice versa. While such stereotypes are no doubt given expression in lecture theatres, tutorial rooms and coffee houses throughout the international community of human geography (and certainly the wider community of geography, in which physical geographers often have a strong contribution to make on these issues), we want to argue that this need not be so. We view enumeration as one of a range of available interpretative strategies for human geographers. Moreover, we agree with Philip (1998) that there is considerable value to be gained from scrambling the supposed quantitative–qualitative divide and focusing more often on multiple methodologies which deploy different techniques and different stages of a research process, and which revisit the supposed binary relationship between objectivity and subjectivity.

In the remainder of this chapter we present a brief account of some of the methods of analytical enumeration available to human geographers. We do not seek here to present a detailed manual of individual techniques. There are many other texts which perform this task. Neither do we shrink from introducing critical debates around the intellectual context of particular forms of enumeration

Box 8.3: The goods and bads of 'soft' methodologies in human geography

Possibly good aspects

- They are a means of coping with the task of understanding complex human systems.
- They are concept rich and theory rich.
- They strengthen links with other social sciences.
- They focus on developing a social and political dimension.
- They are extremely flexible and non-constraining on the topic of study.
- They emphasize the importance of social context.

Possibly bad aspects

- They are ignorant of the vast amounts of computer information that exists.
- They neglect information technology.
- They engender severe computer skill deficiencies.
- They are intolerant of non-conformist research.
- They exaggerate the problems of geographical information systems and quantitative approaches.
- They involve no understanding of science.
- They involve a denial of science.
- They focus both on the uniqueness of the world and on interpretations of the world.
- No consensus-based research agenda is possible.
- They are more strongly linked to fringes of other social sciences than to geography.

Source: Openshaw (1998: 325)

since the deployment of this 'critical edge' is also part of the portfolio of skills for practising human geography. We do, however, regard these forms of analysis as potentially powerful and useful tools with which to make sense of many kinds of human geography data.

Describing, exploring, inferring

Following the quantitative revolution of the 1960s, the analysis of numeric data using statistical techniques became *the* orthodox means of practising human geography and, in the USA, for example, this orthodoxy has continued to be dominant in many educational curricula and research programmes. In the UK, as in parts of the US system, the growth of

qualitative geographies has presented a challenge to any notion that quantification is *the* orthodox methodology, but numeric analysis remains a significant part of the human geographer's toolkit. In the remainder of this chapter we explore some of the manoeuvres available to researchers wishing to analyse and interpret numeric data. We begin with perhaps the most straightforward practices: summarizing data through the descriptive use of statistics; exploring data through visualization; and inferring aspects of the population from samples – and it is worth noting at this point that this statistical pathway should not be equated merely with solely quantitative research strategies. Although by definition the interpretative techniques offered by enumeration require the data to be in the form of numbers, it is certainly the case that such interpretation can be implemented on data (for example, words, visual images, sounds) which have been categorized or otherwise transformed into numbers, as well as data constructed in numerical form. As such, enumeration is a valid option for interpreting many of the different types of data construction discussed earlier in this book, and not just for the more obvious constructions of official data, social research data and questionnaire interview data. Indeed, several aspects of more qualitative research strategies are open to interpretation by enumeration. For example, Krippendorff's (1980: 7) account of content analysis suggests that its history has been firmly rooted in a 'journalistic fascination with numbers', and basic techniques such as word counts and textual content analyses will certainly transform data constructed in non-numeric form into numbers which can then be interpreted by enumerative methods. Moreover, although it is often the case that human geographers will be focusing on the singularities of particular texts rather than looking for patterns within them, such singularities are likely to be embedded in images or language which will probably contain a number of implicit classifications and comparisons. As Dey (1993) argues, classification should not be viewed merely as the stuff of structured questionnaire interviews; rather it may be the aim of various strategies of qualitative research to recognize and specify the comparisons and classifications used by the subject of that research. Thus enumeration is often bound up with classification (see Chapter 7):

> Enumeration is implicit in the idea of measurement as recognition of a limit or boundary. Once we recognise the boundaries to some phenomenon, we can recognise and therefore enumerate examples of that phenomenon. Once we know what a 'school' is, we can count schools. Indeed it may be hard to describe and compare qualities entirely without enumerating them. (Dey, 1993: 27)

Having established that statistical description, exploration and inference can be applied to many different types of data, we now broach one of the most thorny issues in enumerative human geographies – that of how different types of statistical manoeuvres should themselves be categorized. The approach of many textbooks is to divide groups of statistics into their usual analytical function. In this, way, we are presented with 'descriptive statistics', 'inferential statistics' and so on. This approach can lead to considerable confusion over whether there are separate descriptive and inferential statistics. In our view there are not. All statistics can be used to describe, and all can be used to make inferences. Thus a simple measured mean can be used to *describe* the central tendency of the data set but can also be used to *infer* the characteristics of the mean of the wider population from which the data set has been drawn. Similarly, a regression equation can be used to describe the relationships in a set of values, as well as to infer relationships more widely if the data are from a sample. Bearing these potential confusions in mind, it is important to reiterate the importance of appropriate sampling strategies (see Chapter 5). Some of the most important statistical devices in human geography are those which focus on sampling error and yet, at the undergraduate level, it is often easy to under-emphasize their role (for example, in dissertation research).

In our account of enumerative techniques we have chosen to discuss processes of using statistics to 'describe' and to 'infer' rather than seeking to identify 'descriptive' or 'inferential' statistics per se.

Using statistics to 'describe'

There has been a regular flow of textbooks over the years which have sought to introduce and expand the subject of statistical methods in geography.[1] Although there have been some interesting innovations (see Wrigley and

Bennett, 1981; Bailey and Gatrell, 1995; Rose and Sullivan, 1996, for example), this body of textual material suggests a well-trodden core pathway of statistical interpretation. It is usual to point out first that in most instances geographers do not have access to all the cases for which data are, in theory, available. Although most official data such as population censuses and employment statistics *are* supposed to be comprehensive, in practice it is often necessary to study only part of the data set, selecting samples of data with which to work (see Chapter 5). Having acknowledged the partial nature of the data being used, the next stage in statistical methodologies is to describe and analyse the characteristics of these samples of data. Here we are introduced to statistics which provide measures of central tendency, the variability to data and the nature of the overall data set. These statistics can be presented numerically or graphically and they perform two main tasks.

SUMMARIZING DATA

The statistical toolkit provides measures of central tendency (mean, median, mode) and measures of dispersion (standard deviation, coefficient of variation) which help to summarize large data sets. Establishing central tendency is a useful but potentially problematic process. As an illustration, Table 8.1. draws on a major survey of 'Lifestyles in rural England' (Cloke et al., 1994), part of which attempted to measure poverty in the rural areas of England. Surveys of 250 households were carried out in each of 12 study areas, and data on household incomes were analysed to produce mean and median values for each area. These different 'average' values were then used to suggest levels of affluence and poverty in the rural areas concerned. The table highlights the results of three attempts to provide an indicator of those households whose income levels might be described as being in or on the margins of poverty. Such indicators are by nature

fraught with potential definitional problems (see Cloke et al., 1995a; 1995b; Woodward, 1996) but, leaving these aside in this present context, the first two indicators make use of averages to represent households in this way.

The first identifies those households whose income falls below 80% of the mean household income in all 3000 households in each study area surveyed, and the second substitutes the median value for the mean. As can be seen from the table, not only is the range of percentages of households with incomes below 80% of average incomes markedly different according to whether the mean (higher values) of the median (low values) is used but the rank ordering of different case studies also varies according to whether the mean or median is used. For example, Shropshire has the fifth highest number of households in or on the margins of poverty according to the mean indicator but the eleventh highest according to the median indicator, whereas West Sussex is tenth using the mean indictor but second using the median indicator. Such variations in the use of 'average' statistics render the simple interpretation of such data extremely problematic. The distribution of incomes in particular study areas, and especially the ways in which the distributions deviate around the mean, would require close attention before any real sense could be made of what the statistics are describing.

EXPLORING DATA

Statistical procedures can also be used as part of a more inductive framework of analysis by presenting data in an appropriate form for suggesting hypotheses about underlying processes, and thereby suggesting fuller modelling practices which could be employed to investigate these processes further. Such exploration often uses visual devices – scatterplots, boxplots, histograms, pie charts and so on – to represent frequencies and to suggest patterns, trends and processes which can be followed up using other techniques.

Table 8.1   Indicators of households in or on the margins of poverty in 12 rural study areas

Mean*		Median**		Income support***	
North Yorkshire	61.9	Devon	47.1	Nottinghamshire	39.2
Northumberland	61.1	West Sussex	45.8	Devon	34.4
Notttinghamshire	58.8	Essex	44.4	Essex	29.5
Essex	53.3	Cheshire	43.6	Northumberland	36.4
Shropshire	51.4	Suffolk	42.6	Suffolk	25.5
Devon	50.0	Nottinghamshire	41.2	Wiltshire	25.4
Wiltshire	49.3	North Yorkshire	40.5	Warwickshire	22.6
Warwickshire	48.4	Warwickshire	38.7	North Yorkshire	22.0
Cheshire	46.2	Northumberland	37.5	Shropshire	21.6
West Sussex	45.8	Wiltshire	36.6	Northamptonshire	14.8
Suffolk	44.7	Shropshire	35.1	Cheshire	12.8
Northamptonshire	43.6	Northamptonshire	34.5	West Sussex	6.4

Notes:

*in or around margins of poverty: < 80% mean
**in or on the margins of poverty: < 80% median
***in or on the margins of poverty: < 140% income supplement entitlement

Source: Cloke *et al.* (1994: 94)

As with the task of summarizing data, such representation and exploration are both fundamental to some research tasks and yet open to variable interpretation, either due to lack of knowledge or by design. To illustrate, we present in Box 8.4, a dated but still interesting use of the exploration of frequency statistics, drawn from Brown et al. (1992). Taking data at a world scale on the electrical generating capacity of nuclear power plants, the analysis performs several simple interpretative devices in order to offer different facets of description. A figure of 326 000 megawatts capacity for the year 1991 has some interest in its own right but it is enhanced by being placed in the context of the capacities recorded in previous years. Not only does this allow a comparison with the early years of nuclear power generation to show the dramatic increase that has occurred over the 40 years concerned but it also affords a year-on-year comparison which informs us that, between 1990 and 1991, nuclear generating capacity fell for the first time since commercial nuclear power was introduced.

Graphic display of these data (often by bar charts as well as lines) helps to interpret the pattern concerned, although in this case the recent downward turn is hardly discernible in their Figure 1. In their Figure 2 we are invited to interpret the trend in nuclear generating capacity in terms of the concomitant trend in the number of construction starts for nuclear reactors. It may have been assumed that the seemingly inexorable rise in generating capacity was associated with a continuing flow of new reactors. However, this frequency distribution shows a context of peak construction in the 1970s since when the number of starts has been subject to a jagged downward trend. Such processes of providing numerical context for the original frequency distribution could have gone further, of course, perhaps illustrating the number of reactors shut down over the period, the relative performances of coal-fired or gas-fired electricity generation, the geographical distribution of generators and so on. The authors themselves introduce these and other factors in their summary interpretation which provides an excellent example of well-crafted description. Here we are presented with an informative and convincing account based on enumeration of 'facts'. It remains description, however, because there remain so

Box 8.4: Describing frequencies: the example of the electricity-generating capacity of nuclear power plants, 1950–91

Between 1990 and 1991, total installed nuclear generating capacity declined for the first time since commercial nuclear power began in the fifties. There were 421 nuclear plants in commercial operation in January 1992, fewer than at the peak in 1989. These plants supplied 17 percent of the world's electricity and had a total generating capacity of 326,000 megawatts – only 5 percent above the figure reported 3 years earlier. (See Figure 1.)

World net installed electrical generating capacity of nuclear power plants, 1950–91

Year	Capacity [megawatts]
1950	0
1951	0
1952	0
1953	0
1954	5
1955	5
1956	50
1957	100
1958	190
1959	380
1960	830
1961	850
1962	1,800
1963	2,100
1964	3,100
1965	4,800
1966	6,200
1967	8,300
1968	9,200
1969	13,000
1970	16,000
1971	24,000
1972	32,000
1973	45,000
1974	61,000
1975	71,000
1976	85,000
1977	99,000
1978	114,000
1979	121,000
1980	135,000
1981	155,000
1982	170,000
1983	189,000
1984	219,000
1985	250,000
1986	275,000
1987	298,000
1988	311,000
1989	321,000
1990	329,000
1991	326,000

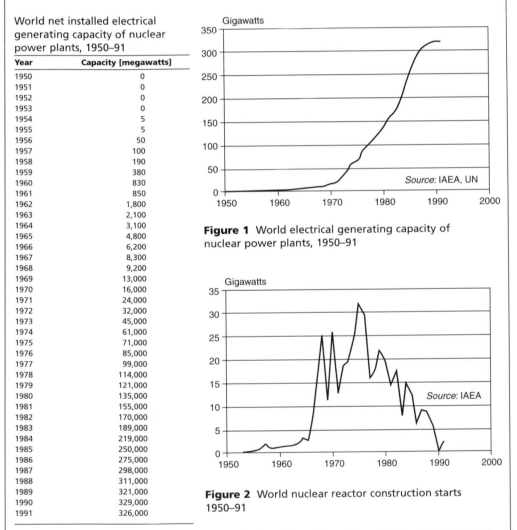

Figure 1 World electrical generating capacity of nuclear power plants, 1950–91

Figure 2 World nuclear reactor construction starts 1950–91

Sources: R. Spiegelberg, Division of Nuclear Power, International Atomic Energy Agency, Vienna, Austria, unpublished printout, March 18, 1992; United Nations, Department of International Economic and Social Affairs, *Energy Statistics Yearbook, 1950–1974* (New York: 1976); Greenpeace International, WISE-PARIS, and Worldwatch Institute, *The World Nuclear Industry Status Report: 1992* (London: 1992).

Source: Brown et al. (1992: 48–9)

Box 8.4 (Continued)

Forty-nine nuclear plants are under active construction worldwide, a quarter as many as a decade ago (see Figure 2), with a combined capacity of 39,000 megawatts. Many of these are nearing completion, so in the next few years worldwide nuclear expansion will nearly stop. It now appears that by decade's end the world will have at most 360,000 megawatts of nuclear capacity – only 10 percent above the current figure. This is less than one tenth the figure predicted for the year 2000 by the International Atomic Energy Agency in 1974.

Some 76 reactors, with a total generating capacity of 17,000 megawatts, have already been retired, after an average service life of less than 17 years. As technical problems continue to crop up at many older plants, retirements seem likely to accelerate. Dozens more could be closed in the next few years, nearly cancelling out reactors coming on-line.

In Western Europe, nuclear expansion plans have been stoppped everywhere but in France. And even the French program is now in jeopardy due to rising public opposition and the $38-billion debt of the state utility. Only six plants are under construction, and just two have been started since 1987. The United Kingdom has one final plant being completed. Canada and the United States together have just three under construction. No new plants have been ordered in the United States in 14 years, and it has been nearly two decades since a plant was ordered that was not later cancelled.

Japan still has an active nuclear construction program, but there too, public opposition is rising and utility executives are considering alternatives. In the Third World, operating nuclear plants generate 18,000 megawatts – 6 percent of the world total. Many are well over budget, behind schedule, and plagued by technical problems. As a consequence, only a handful of orders have been placed in the Third World in the past decade.

Recently, nuclear programs in Eastern Europe and the former Soviet Union have also begun to come unglued. With the arrival of democracy and with some 300,000 people now being treated for radiation sickness, a torrent of public criticism has been unleashed, focusing on the failure of nuclear plants to meet western safety standards. Scores of nuclear plants have been cancelled in Bulgaria, Czechoslovakia, Hungary, Poland, Russia, and the Ukraine. As concern about deteriorating equipment and discipline grows, pressure mounts to close those that remain.

This international trend away from nuclear power is propelled by the two serious accidents at Three Mile Island and Chernobyl, rapid cost escalations, and rising concern about a healthy environment. Many people are worried about the danger of accidents and a continuing failure to develop safe means of disposing of nuclear wastes. Opinion polls in most countries indicate lopsided majorities against the construction of more reactors.

In addition, nuclear costs have risen to the point where nuclear power is no longer competitive with many other energy sources. Not only coal plants, but also new, highly efficient natural gas plants and new technologies such as wind turbines and geothermal energy are less expensive than new nuclear plants.

The old market niche that nuclear power once held is nearly gone. Since 1988, nuclear advocates have tried to use concern about global warming as a reason for reviving the industry, but they have had no impact so far, as reactor orders have continued to dwindle. Nuclear power is an expensive way to offset fossil-fuel-fired power, and several hundred plants would have to be built in order to reduce carbon emissions significantly. Given the current economic and political state of the industry, a major revival seems unlikely.

many other questions about context, intentions, meanings and processes in which these facts are embedded. To use one brief example of this, some understanding of the *authorship* of this piece on nuclear energy provides a context for the facts presented. Brown et al. (1992) are writing on behalf of the Worldwatch Institute, an environmental pressure group. In the Foreword to their book they express a clear desire for their descriptive analysis to be interpreted in particular ways:

> We hope that this volume will meet many needs: that national governments will use it to compare their efforts to reduce carbon emissions or soil erosion with those for the world as a whole; the industrial firms will use it to compare the share of recycled materials in their output with that of the rest of the world; and that individuals who are helping to establish bicycle-friendly transportation systems will be interested in the global trends in the use of bicycles and automobiles. More generally, we hope that not only environmental activists but concerned citizens everywhere will want to find out whether population growth is slowing, how rapidly CFC production is being phased out, or whether the world's nuclear arsenal is shrinking, as disarmament agreements promised. (1992: 12)

Thus the presentation of descriptive enumeration of nuclear power plants may be interpreted as being part of this environmentalist framework, and the statement in Box 8.4 that 'several hundred plants would have to be built in order to reduce carbon emissions significantly' suggests that the commonly used link between nuclear power and environmental policy is being rejected, and therefore that the authors have an underlying opposition to nuclear power. In fact the display of these data begs many other hypotheses (for example, relating to the costs of nuclear energy and the reduced strategic need for military nuclear materials) which would require further investigation and analysis.

Using statistics to 'infer'

Using statistics, with their associated exploratory manoeuvres, to perform descriptive functions is often regarded as the foundation level of analysis, to be built on by strategies of hypothesis–testing so as to infer either difference or association. Statistics are used for inferential purposes in two main

ways. First, statements can be made about the population from which a sample has been drawn and analysed. Secondly, statements can be made regarding the likelihood of something being observed in the data set being analysed with no reference to a wider population. Most usage adopts the former approach, which has led to the innovative idea of inference as an *exploratory* data analysis as well as a confirmatory analysis.

Inference is commonly viewed as a valuable process of enumerative analysis in its own right and as a necessary progression from descriptive parameters. As Gregory (1992: 137) argues:

> Useful as this [descriptive parameters] may be in a limited sense, it often fails to ensure maximum scientific benefit from the data collection that has been carried out. Ideally, the items forming the sample, and the sample itself, should be as representative as possible of the total data set from which the sample has been selected. If this representativeness can be ensured, then – within certain limits – the characteristics of the sample can be used to infer the characteristics of the total data set, i.e. the population from which the sample has been drawn.

It is worth noting here that there is a clear discursive bias in favour of going further in enumerative interpretation than simple description. Description is 'limited' and 'fails'. Inference gives more 'scientific benefit' and it makes the most of the process of data collection. Gregory is certainly not alone in using such a discourse. The logical step in almost all accounts of statistical methods in geography is to progress smoothly from descriptive parameters to inferred estimations and then to inferential tests.

Use of statistics for inference varies according to differing types of data (nominal, ordinal, interval, ratio) and in addition may be 'parametric' – that is, where the sample data are assumed to belong to an overall population of data which accord to a normal frequency distribution – or 'non-parametric' where no such assumption is necessary, although the results of all inferential tests (including non-parametric ones) are held to be 'reliable' only if they are applied to statistically 'unbiased' samples (see Ebdon, 1985: 17). This does seem to be rather a big 'if' in the case of much human geography data. Indeed, spatially distributed data will often violate many of the assumptions which need to be met if parametric statistics are to be used for inferential purposes. This is fully realized by the leading proponents of quantification and has often inspired a quest for thoughtfulness and sensitivity in the whole trajectory of spatial statistics (see Haggett et al., 1977).

The nub of inferential tests is the establishment of a null hypothesis which suggests that the difference occurring in the sample data is there by chance and not because of any 'real' difference or relationship existing in the overall population of data. Interest particularly arises when the null hypothesis can be rejected with a degree of statistical confidence (using significance levels such as 95%, 99%, 99.9%, etc.) in favour of the alternative hypothesis. The chi-squared, Kolmogorov–Smirnov and Mann–Whitney U-tests are common non-parametric tests used for inference, while the F-test or Student's t-test are equivalent parametric tests. There are a series of manoeuvres which need to be understood in this process. Clearly, the notion of hypothesis implies assumed reasoning: x is different from y; x is the same as y; x is related to y; x is explained by y. In this way, inferred relationships or differences can easily become *explanatory* in the minds of researchers. Huck and Sandler (1979), who themselves favour the use of statistics for inference, comparison and relationship, have produced a book full of cases where explanations for events based on hypotheses can be challenged because of the existence of 'rival hypotheses' which belie the original explanations given (an illustration is presented in Box 8.5), and Huff (1973) provides a series of examples where statistically 'proven' hypotheses hide spurious associations between different data.

One further issue which is crucial to the use of statistics for inference is that of the *spatial* characteristics of data and analysis. As human geographers will usually be interested in understanding spatial processes, the spatial location of the data being used becomes an important component in the analysis. Some statistical procedures effectively discard the spatiality of data by regarding input information as strings of numbers which can be reshuffled regardless of where they are located spatially. Other procedures clearly identify the spatiality of the data. For example, the presence of spatial autocorrelation – 'the presence of spatial pattern in a mapped variable due to geographical proximity' (Hepple, 2000: 775) – is widespread in data which are mapped out geographically. Spatial autocorrelation can be analysed to suggest similarities and differences between adjoining areas, although care needs to be exercised in the definition of an appropriate scale of neighbourhood unit in such analysis.

Considering the spatial characteristics of data, then, adds both to the potential utility of statistical inference and to the potential problems encountered in their use (see Box 8.6). As with more qualitative strategies of analysis, rigorous technical expertise is the key to useful output. This is particularly so when extending inferred differences to situations where statistics are used to describe the relationship between two or more variables measured from a particular sample of data. The range of bivariate or multivariate techniques is wide but is often held to be *the* key area of interpretation in geography. As Gregory (1992: 138) again puts it:

In most geographical problems, factors and causes do not operate singly, but rather in intricate interaction with one another. It is the unravelling of such interactions and the evaluation of the relative importance of each contributory element, that often provides the focus and the very *raison d'être* of the geographical enquiry.

Box 8.5: Huck and Sandler's illustration of a rival hypothesis: newspaper advertising

Newspaper advertising I: the hypothesis

Each year, millions of dollars are spent on newspaper advertising. Obviously, the people who pay for the ads feel their financial outlay is worthwhile. In other words, there is a presumed cause-and-effect relationship involved here, with the newspaper advertising being the cause while subsequent increased sales constitute the effect. But do the ads really bring about, in a causal sense, more consumer purchases? Recently, two researchers conducted a little study designed to answer this simple yet important question.

The setting for this study was a small town in northern Illinois (population about 3000). The subjects were 142 female customers who regularly purchased items at a grocery store. A list of 28 items appeared in the local newspaper for four consecutive days prior to the day of the study. Each of these 28 products was advertised at a reduced price, and the purpose of the study was to determine whether this advertising made a difference in sales of these items.

Following the four days of advertising, the data of the experiment were collected. The procedural aspects of the study on this fifth day were as follows. As each subject came through the checkout counter, the clerk examined her purchases to see if any of the advertised sale items were included. If one or more of the 28 items were about to be bought, the clerk was instructed to ask whether or not the consumer had read about the sale items in the newspaper ad. Although several of the subjects came back to the store a second (or third) time on the data-collection day of the study, responses from each consumer were recorded only for her first time through the checkout counter.

Results of the study indicated that all 142 subjects purchased one or more of the 28 advertised items. Ninety-nine of the subjects stated that they had read about these items in the newspaper while 43 subjects admitted that they had not. Based on these figures, the researchers concluded that 'reading the newspaper advertising seemed to increase purchase of advertised items more than not reading the paper' (Peretti and Lucas, p. 693). To probe the data further, each subject's socioeconomic status (SES) was assessed by the store clerks (who knew the customers quite well) through an instrument called the Index of Status Characteristics. Using this SES information, a group of lower-class consumers was compared with a group of middle-class consumers. Results indicated that a significant difference existed between the two groups, and in the words of the researchers, 'advertising tended to affect lower-class consumers' buying more' (p. 693).

Based upon the data that were collected in this study, can we conclude that a cause-and-effect relationship has been established? Does newspaper advertising lead to increased purchasing behavior? And does the alleged effect really exist more for lower-class consumers than for those in the middle class?

Newspaper advertising II: rival hypotheses

There are, in our judgment, three problems associated with this study that potentially invalidate the conclusions. The first problem has to do with the question that was asked of subjects as they came through the checkout counter. The second has to do with the personnel used to collect these answers. And the third problem has to do with the honesty of responses.

Box 8.5 (Continued)

The main problem with this investigation relates to our inability to tell how many of the subjects would have bought a sale item even without the newspaper ads. Surely some of the subjects made plans to visit the grocery store on what was actually the fifth day of the study and read the local newspaper during one or more of the four preceding days. Although these subjects responded affirmatively to the question about having seen the newspaper ad, this does not necessarily indicate that the ad caused them to go to the store or to buy any of the 28 sale items. Very possibly, they were planning to visit the store to purchase (among other things) a subset of the 28 'critical' products *before* they saw the advertisement. Did they see the ad before shopping? Yes. Did the ad cause them to buy advertised products? Possibly not.

To get around this problem, a different sort of question could have been asked. Instead of inquiring whether or not the consumer had read about the sale items in the newspaper, the clerks could have asked, 'Would you be buying these sale items if they had not been advertised in the paper?' Or, 'Did our newspaper ads about these sale items cause you to purchase the items you have selected?' However, we feel that an alternative strategy would have produced more valid data and also made the questioning unnecessary. Without too much difficulty, the printed advertisement could have taken the form of an insert and been included in a random half of the home-delivered newspapers. A record could have been kept concerning who got the ads and who did not, and then the data could have been collected regarding the presence or absence of the 28 sale items in the market basket on day five of the study.

The second problem of this study relates to the personnel used to collect the data. These were the store clerks at the checkout counter. They examined each subject's purchases to see if any of the 28 advertised items were present, and if any were they asked their question about having read the newspaper ads. However, since the store clerks were also the ones who assessed each subject's socioeconomic status, we wonder whether the clerks' knowledge of customer SES might have biased the way in which they interacted with the subjects as the subjects came through the checkout counter. Possibly more (or less) time was spent talking with the lower SES subjects *because* the clerks knew they were in this category. We feel that the personnel used to classify subjects into SES categories should not have been the ones to collect the data regarding the effects of the newspaper ads.

Finally, we wonder about the honesty of responses to the question: 'Did you read about these sale items in the newspaper ads?' A significantly greater number of lower-class subjects than middle-class subjects answered affirmatively. This almost gives the impression that lower-class consumers read the paper (or at least the ads about sales) more than middle-class consumers. Maybe this is true. However, a rival hypothesis is that middle-class consumers are embarrassed to admit that they are buying items simply because they are on sale. A partial (but not complete) way to get around this problem would have involved a cut-out coupon in the paper that had to be presented at the checkout counter before the consumer could get the sale price.

[*Note:* Problems of establishing causes and effects are by no means limited to survey-based methodologies or to techniques of enumeration. Honesty is equally an issue for qualitative geographies.]

Source: Huck and Sandler (1979: 66, 184)

> **Box 8.6: Problems encountered in the use of inferential statistical analyses in human geography**
>
> 1. *The modifiable areal unit problem.* In the analysis of aggregate spatial data, the conclusions reached might be dependent upon the definition of the spatial units for which data are reported.
> 2. *Spatial non-stationarity.* Relationships might vary over space making 'global' models less accurate. Global models will also hide interesting geographical variations.
> 3. *Spatial dependence.* To what extent is the value of a variable in one zone a function of the values of the variable in neighbouring zones? Strong spatial dependence can affect statistical inference.
> 4. *Non-standard distributions.* Examine the nature of the data being analysed. Not all data are normally distributed! Consider experimental methods of statistical inference if data have an unusual distribution.
> 5. *Spurious relationships.* Make sure that any relationships identified are meaningful and are not caused by a variable or variables omitted from the analysis.
>
> *Source*: Fotheringham (1997: 167)

Measuring the correlation of two variables can have a descriptive use but, more usually, tests of significance are carried out to see whether variations in one variable 'cause' those in another (but note here that these are tests of *co-variation* and not of *cause* – see also Chapter 9). The product–moment correlation coefficient is the usual parametric test here, and the Spearman rank correlation is the non-parametric equivalent. Correlations stem directly from regression analyses where the dependence of one variable on another is measured. Regression statistics are used to produce trend lines but also to encompass non-linear relationships, and they allow for the prediction of relationship between variables. Other, more complex multivariate techniques such as principal components analyses, factor analysis and cluster analyses use multiple relational statistics to sort out particular components or factors in the data set and indicate how strongly particular variables relate to these components (see Johnston, 1978, and, for an example, Pattie and Johnston, 1993). There is often, however, some ambiguity as to what these components or factors actually represent, particularly in factor analysis where data are rotated in different ways to find the optimal configuration of variables or where different time periods of analysis are being dealt with.

The use of these multivariate techniques offers human geographers explanatory power in their interpretation of data. However, these tests can be highly manipulative and open to significant user error (in relation to whether data are really fit to meet the required assumptions) and to considerable user discretion (for example, in relation to which of the many different ways to construct a regression line or decide which factor-analysing technique is chosen in a particular case). Moreover, there is a strong sense in which the researcher is never quite sure what is causing what he or she is measuring, even though allegedly causal explanations are being offered by the technique concerned. As Marsh (1988: 225), an exponent of correlation and regression analyses, points out:

> It is one thing to declare confidently that causal chains exist in the world out there. It is quite another thing, however, to find out what they are. Causal processes are not obvious. They hide in situations of complexity, in which effects may have been produced by several different causes acting

together. When investigated, they will reluctantly shed one layer of explanation at a time, but only to reveal another deeper level of complexity beneath.

Some human geographers will now reject entirely the notion of cause-and-effect chains, and others will want to stipulate that many causes of events, circumstances and practices in the contemporary world are not easily measurable in numeric form, and can therefore be invisible to these statistical techniques. Others will merely argue for a more careful and rigorous approach in pursuit of statistical explanatory power. These issues of the authority of enumeration will be revisited later in this chapter.

Modelling spatial processes

Much of the pioneering quantitative research in human geography in the 1960s and 1970s stemmed from a desire to test hypotheses about the spatial ordering of points, lines, areas and flows, the patterns for which were suggested by economic themes (for example, relating to central places and industrial location). The seminal work of Peter Haggett (1965) set the tone for much which was to follow and, as such, emphasized points and flows and paid far less attention to areal patterns. In a recent review, Johnston (2000a: 127) has suggested that this rich heritage has led to the energy of the quantitative revolution being used in particular ways, and he identifies other applications which have not been so widely exploited:

> Geographers have made much more use of statistical methods for the spatial analysis of data than of mathematical procedures for modelling spatial processes, though this is less true of physical than human geographers. The books by Bennett and Chorley (1978) and by Wilson and Bennett (1985) introduced a wide range of procedures for the spatial analysis of spatial systems, complex and otherwise, but changes in the philosophical orientation of much human geography have seen them little used.

Johnston's analysis of the relative underdevelopment of spatial modelling in human geography is an interesting one. It might be argued (cf. Fotheringham, 1998) that there are two main constraints on making progress with quantitative research in human geography – the intellectual power which articulates how spatial processes work and how they can be modelled, and the computational power to test, refine and run such models. Initially it was the computational power which lagged behind. It is worth remembering that as late as the early 1980s many geography students were learning about quantitative methods by using hand-held calculators, and that even the subsequent computerization was for a considerable time restricted by the boundaries of classic software packages such as SPSS and Minitab. Yet as computational power has increased, the philosophical swings towards, first, political economic theory and, secondly, the postmodern and poststructural underpinnings of the 'cultural turn' have directed the gaze of increasing numbers of human geographers away from an adherence to techniques of mathematical modelling. Those who have been turned on by the power and thrill of computing have typically forged a geocomputational trail (see below), with the propensity to model being directed by the task of developing existing procedures of spatial modelling. Thus the intellectual power directed at modelling spatial processes has typically been vested in a relatively small group of researchers, often with links outside geography (for example, with mathematics and econometrics) as well as inside (focusing, for example, on 'modellable' topics such as health, voting, urban processes and so on).

In these circumstances it is unsurprising that the cornerstone of progressive attempts to model spatial processes has been the general linear model, which effectively extends the relational use of statistics discussed earlier in this chapter. What we explore here, then, is how the general linear model has been developed and how interesting new directions, especially those relating to modelling the local, the complex and the chaotic, have opened up yet greater opportunities for human geographers to practise using this genre of approach and technique.

Developing from the general linear model

O'Brien (1992: 3) has defined generalized linear models as 'a class of linear statistical models which share some important mathematical characteristics' of which three are particularly significant:

1. They can be reduced to a common mathematical form: where y = a 'response' component and μ = a systematic component to be estimated from the data.
2. They investigate the behaviour of the response by linking it to 'explanatory information' drawn from survey data using a form of linear, or additive, structure.
3. They usually draw the error component from exponential probability distributions.

Techniques contributing to the general linear stable of modelling include multiple correlation and regression, multivariate analysis of variance, principal components analysis, factor analysis and discriminant analysis. These have been used in a range of contexts, including gravity models, optimization models and network models. As Johnston (1978) admits, researchers have increasingly realized that, to a considerable degree, human geography applications tend to violate one of the basic assumptions of the general linear model – that concerning independence among the observations. During and since the 1970s, therefore, there have been considerable efforts to produce more sophisticated versions of the general linear model which address these

violations of key assumptions. For example, Openshaw's (1984) seminal work on the modifiable areal unit problem (MAUP) demonstrated that aggregation and scale effects were important, separate but often interdependent influences on observed relationships. In other words it is possible to get whatever correlation you want when ecological data sets are created from individual data sets or when, as for example in the census which aggregates individuals into areas, the ecological data set has to be taken as given.

An important development was also achieved by Wrigley (1976), whose realization that standard regression procedures should not be used on closed number sets – such as percentages or proportions – (because such usage can produce nonsensical results such as percentages below 0 and above 100!) stimulated more sophisticated analyses using, for example, logistic regression. Other developments focused on the issue of autocorrelation (Cliff and Ord, 1973), much of which is assumed to be spatial in nature, but the fact that adjacent areas have similar values (for example, residual values from a regression) does not automatically mean that a spatial process is causal or even involved. It is clear that some human geographers continue to be concerned with the shortcomings of techniques such as linear and spatial regression models, and therefore with the shortcomings of the general linear model itself (Haining, 1990).

An excellent illustration of these concerns is provided by Fotheringham et al. (2000) (Figure 8.1). They use 1991 census data to map owner-occupation levels of housing in Tyne and Wear (a), noting that the highest rates are at the periphery of the county while the lowest rates are in the central area. Then they map male unemployment rates for the same area (b), noting that the highest rates are in the centre. These problems suggest a linkage between owner-occupation and unemployment and a linear relationship is indicated

(a)

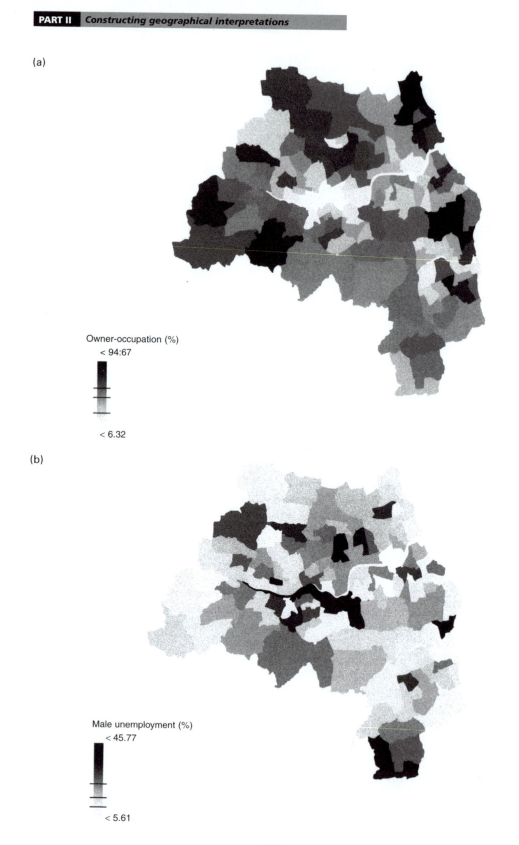

Owner-occupation (%)
< 94:67

< 6.32

(b)

Male unemployment (%)
< 45.77

< 5.61

(c)

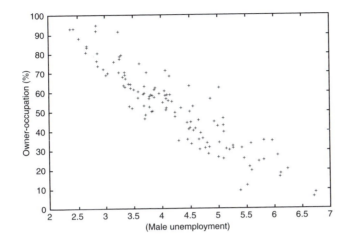

(d)

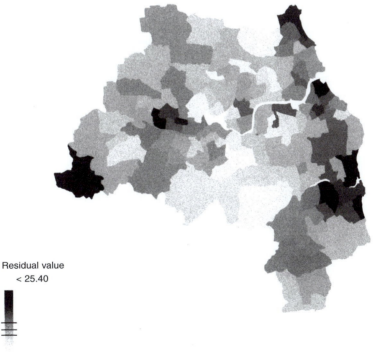

Residual value
< 25.40

< –20.31

Figure 8.1 Problems with linear and spatial regression models

Source: Fotheringham et al. (2000: 163–6)

by a scatterplot of the square root of male unemployment against owner-occupation (c). However, a map of the residuals of a regression model (d) clearly suggests that these residuals do not vary randomly over space but seem to occur in localized clusters. Moreover there are sudden changes between low and high values. The model therefore transgresses the assumption that each observation is independent of the others. Any assumption that male unemployment is a good predictor of owner-occupation is (by this evidence) unsafe, and additional modelling developments are required. While some human geographers have sought these developments within the general linear model, others have preferred to develop alternative models.

One example of this search for alternatives is the development of dynamic linear modelling, as in the field of space–time forecasting models, which Burrough (1998: 166) defines as

> a mathematical representation of a real-world process in which the state of a location on the Earth's surface changes in response to variations in the driving forces. Any system for modelling space-time processes must include procedures for discretising space-time, and for the computation of new attributes for the spatial and temporal units in response to the driving forces.

This time-series, autoregressive moving average has been used extensively in physical geography, addressing issues such as catchment hydrology, slope stability and so on, using PCRaster programming. However, the space–time forecasting model has been less successfully employed in human geography, perhaps because its 'black-box' structure deals less well with the complex spatialities and socio-economic behaviours typically reflected in social science issues.

A second illustration of the search for alternative models is the move into dynamic non-linear modelling, as in the example of spectral analysis (see Bennett, 1979). Spectral analysis is a technique which models the oscillations in a time-series using calculations of the relative importance of different frequency bands. It can be applied to regional economic cycles and to diffusion models (for example, in epidemiology). Here, cross-spectral analyses between different time-series permit a multidimensional appreciation of how, say, measles diffuses during different epidemics in different places, allowing the development of more general predictive models. As with the example of dynamic linear modelling, however, this 'alternative' has not displaced the continuing dominance of previous general linear models. There are, however, two further sets of initiatives in modelling spatial processes which are demanding of very serious attention in terms of the future of quantitative human geographies.

Modelling 'the local'

The first of the new directions in spatial modelling stems from the recent focus in spatial analysis on identifying and accounting for differences rather than samenesses across space. An emphasis on difference has pointed researchers towards statistical techniques which are aimed at the local scale. In a review of these moves, Fotheringham (1997) notes an increasing tendency to question the assumptions inherent in regression analyses about the stationary nature of relationships over space. He discusses four types of research which are making a particular contribution to modelling the local:

1. *Local point pattern analysis* (see, for example, Boots and Getis, 1988), which involves the development of exploratory methods for detecting spatial point clusters.
2. *Local measurement of univariate spatial relationships* (see, for example, Ord and Getis, 1995), where the measurement of spatial association inherent in data permits the modelling of how the values of an attribute are spatially clustered.

3. *Local measurement of multivariate spatial relationships* – the measurement of more complex relationships in their local form, using either 'expansion' (Jones and Casetti, 1992) or 'multilevel' (Jones, 1991) modelling methods.

4. *Mathematical modelling of flows* (see, for example, Fotheringham and O'Kelly, 1989), which involves the building of spatial interaction models which can be applied to migration, residential choice, retail and recreational behaviour and so on.

Some of the implications of these concerns with 'the local' can be seen in the example of multilevel modelling. For instance, Jones and Duncan (1996) compile a list of the necessary requirements for modelling approaches which take the notion of place seriously and which allow for an examination of difference (Box 8.7). They argue that the use of multilevel modelling can detect 'effects' or differences which operate at different scales. For example in a study of immunization uptake among children in Britain, an important policy-related factor is that general practitioners are rewarded differentially by reaching targets based on immunizing certain percentages of children. Jones and Duncan (1996: 94) argue that 'multilevel models, by modelling at the individual and clinic levels simultaneously, are able to get some purchase on which practices are performing well given their client groups, and thereby permit the development of a contextualised measure of performance'. This type of modelling (which has a range of applications in health, education and political geographies) does seem to offer considerable scope for a more spatially sensitive form of spatial analysis in future quantitative geographies.

Modelling 'the chaotic'

The second set of initiatives draws on theories of chaos and complexity. Theoretically, there has been an upsurge of interest from human geographers in ideas of chaos (see Prigogine, 1984). Wilson's (1981a; 1981b) research showed the nature and importance of the *constraints* to solutions of flow patterns at both the origin and the destination of those flows. He argued that an entropy-maximizing approach was a better mechanism for modelling constrained flows than the unconstrained regressions that had been used previously. With the acknowledgement of constraints has come an awareness of the potential for chaos. Mathematically, chaos suggests relationships between variables which demonstrate no sense of order in either magnitude, space or time. Such attributes have a direct application in physical geography (for example, in the modelling of some aspects of climate change) but, as yet, are largely untapped in human geography modelling contexts. Even so, ideas of the chaotic appeal strongly to those human geographers who are interested in concepts of deconstruction and difference, and so there are prospects for quantitative initiative in this area.

One example of the complexity of the sub-chaotic in human geography comes from Batty and Longley's (1994) modelling of fractal cities. Fractal characteristics are reflected in objects with an irregular spatial form, the irregularity of which is repeated geometrically across different scales. As Batty and Longley (1994: 4) suggest:

Cities have quite distinct fractal structures in that their functions are self-similar across many orders or scales. The idea of neighbourhoods, districts and sectors inside cities, the concept of different orders of transport net, and the ordering of cities in the central place hierarchy which mirrors the economic dependence of the local on the global and vice versa, all provide examples of fractal structure which form the cornerstones of urban geography and spatial economics.

This acknowledgement of alternative geometries, or even in the case of chaos theory disordered non-geometries, is opening up interesting new lines of inquiry for spatial

Box 8.7: Requirements for spatial modelling which takes 'place' and 'difference' seriously

Contextual differences
Modelling must recognize that 'people make a difference and places make a difference' so that people are not reduced to statistical aggregates and places to generalizations. In a contextual model of health, the setting in which the behaviour literally takes place is seen as crucial for understanding. For example, in studying tobacco consumption we need to consider simultaneously the micro-scale of individual characteristics and the macro-context of local cultures in which people live.

Place heterogeneity
The effects of context can potentially be very complex, with relationships varying in different ways. For example, there may be places where people of all ages have a high consumption of tobacco, whereas in others it is the young who have a relatively high rate, but in others again it is the elderly. This age differential may also be accompanied by complex differences for other individual characteristics such as gender and class. If this is the case, it is not possible to have a simple 'map' of smoking for consumption will vary according to who you are in relation to where you are.

People/place interactions
Modelling must also take into account the possibility that individual differences may 'interact' with context. For example, a person of low social class may act quite differently according to the social class composition of the area in which he or she lives. To put it in another way, between-place differences need to be examined in relation to the social characteristics of individuals in combination with social characteristics of places.

Individual heterogeneity
It is not only the higher-level contexts that must be allowed to vary in their effect, for individual differences may be too complex to reduce to an overall 'average'. Thus people of low social class may not only smoke more on average but they may also be more (or less) variable in their consumption. Heterogeneity between people as well as between places must be expected and modelled.

Changing people, changing places
Changing contexts and changing behaviour are important too, so that the approach must be able to handle longitudinal data and time (as well as place) as context.

Inter-related behaviours
It is important that differing but not unrelated behaviours are not considered separately but modelled together. Any meaningful model of health-related behaviour would need to consider smoking alongside alcohol consumption, for example.

Quality and quantity
The behaviour and actions of individuals have both a qualitative aspect (occurrence – does it occur?) and a quantitative element (amount – how much, how often, how many?). Both these need to be considered simultaneously. For example, the variables which are closely related to whether a person smokes or not may reveal nothing about the number of cigarettes smoked. There may be places where there are a few people who smoke, but those who do so smoke heavily; an average figure would be very misleading.

Box 8.7 (Continued)

Multiple contexts
It is likely that there will not be a single context but many. For example, smokers may be influenced not only by the milieu in which they live but also by the work environment.

Source: Adapted from Jones and Duncan (1996: 81–2)

modelling. However, as is evident from the reflexive acknowledgement of 'nets', 'ordering', 'hierarchy' and so on above, the driving force for many existing applications remains the desire to model the order of disorder and the capacity of GIS to manipulate vast quantities of data into these models. As is discussed in the next section, the acknowledgement of difference in more recent spatial models tends to be driven by the ability to compute large data sets rather than by any rapprochement between the philosophical underpinnings of spatial analysis and the qualitative cultural turn.

Geocomputation

From the early 1980s onwards, the availability and affordability of increasingly sophisticated and powerful computing technology have created a whole new world for enumerative analysis in human geography. Not only did these technological changes aid the continuing use of spatial modelling but they also prompted the creation of geographic information systems (GIS) which set up integrated computerized processes for handling and analysing geographic data (see Chapter 7). Much has been written about the origins and development of GIS[2] but, in essence, what began as the computerization of mapped information very quickly became a toolbox of procedures for the input, storage, manipulation and output of geographical data.

As Martin (1997: 216) suggests, the early development of GIS was quickly wrapped up in the needs of commercial users:

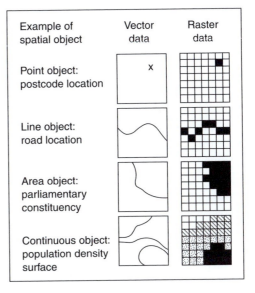

Figure 8.2 Vector and raster data structures

Source: Martin (1997: 217)

In commercial use, one of the most important roles of GIS is as an organising structure for extremely large quantities of operational information. Such inventory applications tend to require compact data storage and rapid retrieval, but do not place particularly heavy analytical or statistical demands on the system. For this reason, much of the development effort in GIS (and the corresponding literature) has been directed towards ways of holding and querying geographical data more efficiently.

Two main spatial organizing systems emerged (Figure 8.2). *Raster* systems work on a framework of rectangular cells and input information on the basis of each cell. This relatively continuous modelling of space has been favoured in physical geography applications. *Vector* systems use points, lines and areas to input the location of discrete objects using map co-ordinates. Social science applications of GIS have generally found vector systems to be more useful to their needs.

From these simple beginnings, the technology of GIS has matured, providing the basis for sophisticated modelling of spatial distributions, flows, networks and hierarchies. However, as Longley (1998: 3) has stressed, GIS provides the environment in which spatial processes can be explored, but not the guiding principles for that exploration: 'In itself, GIS provides little guidance as to how best spatial problems might be solved. It is clear that big technical and epistemological questions remain about the ways in which selected facets of geographical reality are discarded or emphasised in the process of model-building.'

Thus, the computer has been transformed from being an adjunct to the research environment to being *the* research environment (Goodchild and Longley, 1998). It is the *coupling* of the input, throughput and output capacities of GIS with analogous models for analysing, visualizing and modelling data associated with spatial patterns and processes that has brought about the most significant application of GIS in human geography. The term 'geocomputation' has been used to convey this coupling of technology and analytical purpose (see Longley, et al., 1998b). Here the intensive computational environment of GIS is used for research-led problem-solving purposes and thus creates intellectual space for continuity with the longstanding concerns of quantitative geography. So Macmillan (1998), for example, regards geocomputation as being founded on the analytical philosophies and scientific methods which have marked the established traditions of quantitative human geography. This view is not universal, however. For example,

Couclelis (1998) finds geocomputation to be lacking in systematic connection with spatial theory or, indeed, with computational theory – she regrets that the GIS environment has produced a grab-bag of tools rather than a more 'revolutionary' unifying epistemology.

The reasons for these different viewpoints are clear. On the one hand, most users of GIS environments will, implicitly or explicitly, adopt a 'scientific' research design involving hypothesis-testing using data analysis, and employing rational, ordered reasoning. However, the GIS world has become populated by 'an uncoordinated proliferation of digital datasets (many of which are collected by unorthodox or even profoundly unscientific means)' (Longley, 1998: 7). The rich diversity of digital representation in GIS can thus lead to epistemological conflict with the 'answers', 'truths' and 'spatial processes' which they help to generate. Thus the fruits of geocomputation can prompt debate about whether quantitative traditions are being enhanced, or diluted, in GIS environments. We illustrate these claims in three areas of applied geocomputation.

GIS and spatial analysis

In another of his reviews of quantitative human geographies, Fotheringham (1998) sees the role of geocomputation as a crucial guiding force for innovation. In his view, computation is driving the form of analysis undertaken rather than just providing a convenient environment for applying independent techniques. One benefit of geocomputation is therefore an emphasis on variation and differences rather than similarities. The processes and relationships under analysis are not regarded as homogeneous, and the geocomputational ability to analyse small-scale spatial variations has thus enhanced the ability to study differences in broad spatial and social processes. As a result, geocomputation has boosted *exploratory* (rather than confirmatory) techniques and provides a mainly inductive instrument by which the representation of data can lead to new hypotheses about the processes which produce the data.

Fotheringham (1998) outlines a number of important geocomputational applications in human geography, of which two are particularly relevant here.

SPATIAL INTERACTIONS[3]

Geocomputation has permitted unidirectional and bidirectional interactions between the GIS and spatial analytical routines. Thus, for example, data derived from a GIS can be used to undertake a statistical analysis such as regression, and the resulting statistics (for example, residual values) can be returned to the GIS to create fuller maps (see, for example, Figure 8.1). Such integration has become increasingly dynamic, creating a seemingly seamless movement between mapped data and analysis. The use of Windows software, for example, permits the review (or 'brushing') of scatterplotted data in one window and a simultaneous map-referencing of those points in another. These improved exploratory techniques do not, of course, guarantee any greater insights into the spatial processes concerned, but the ability to move between analytical and mapping software in Fotheringham's (1998: 286) view produces 'a reasonably high probability of producing insights that would otherwise be missed if spatial data were not analysed with a GIS'.

Another geocomputational advance in spatial analysis has been in addressing the modifiable areal unit problem. The use of areal data at different scales (for example, census data at enumeration district, ward and district levels) is often problematic because the results of analyses can be sensitive to the particular spatial units in which the data are presented. The flexibility provided by raster- and vector-based systems has presented one route (but not the only route) by which those problems can be addressed and overcome.

ARTIFICIAL INTELLIGENCE[4]

Geocomputation has prompted human geographers to begin to work with artificial intelligence systems, the ultimate aim of which is 'to produce computer programs which can solve problems for which they have not been programmed' (Fotheringham, 1998: 289). These developments have involved applications of *heuristic searches*, using point pattern analysis to identify particular clusters of points in particular locations, and *neurocomputing*, where neural nets are trained to represent a priori connections in simulation processes. There has also been development of evolutionary computing techniques in order to represent the growth of objects (for example, cities) (Batty and Longley, 1994).

GIS and geodemographics

Geocomputation has also been at the core of an important phase of applied modelling of locational analysis in human geography over the last two decades or so. Using longstanding ideas about locational modelling alongside aspects of operational research and computational theory, a group of human geographers has set about using GIS environments to fulfil the commercial needs of a range of business and service providers (Longley and Clarke, 1995). The success of this approach is illustrated by GMAP – a consultancy company established by the University of Leeds in 1987 and subsequently taken over into majority ownership by the American marketing company, Palk, in 1997. Members of GMAP and others re-pioneered geographical expertise in modelling the best locations from which to maximize sales to particular target groups. In so doing they built up a more general expertise and interest in geodemographics.

Box 8.8: ACORN geodemographic classification

Category A: 'thriving'

1 Wealthy achievers, suburban areas
2 Affluent 'greys', rural communities
3 Prosperous pensioners, retirement areas

Category B: 'expanding'

4 Affluent executives, family areas
5 Well-off workers, family areas

Category C: 'rising'

6 Affluent urbanites, town and city areas
7 Prosperous professionals, metropolitan areas
8 Better-off executives, inner-city areas

Category D: 'settling'

9   Comfortable middle-agers, home-owning areas
10 Skilled workers, home-owning areas

Category E: 'aspiring'

11 New home-owners, mature communities
12 White-collar workers, multi-ethnic areas

Category G: 'striving'

13 Older people, less prosperous areas
14 Council house residents, better-off homes
15 Council house residents, high unemployment
16 Council house residents, greatest hardship
17 People in multi-ethnic, low-income areas

Source: Clarke (1999: 580)

Clarke (1999) suggests a number of methodological advances relating to geodemographics. First, the location of customer types has required a profiling of geographical areas in terms of the key customer segment types within them, using classifications such as ACORN (Box 8.8). During the 1990s, increasingly sophisticated systems linked population demographics with commercially obtained lifestyle information. Secondly, GIS environments have also been used to assess retail site location through the geocoding of populations within catchment areas and the calculation of demarcated ('buffered') travel times around the store location. Population catchment estimates can then be translated into sales estimates, taking account of other competing stores in the area. Such calculations are not without problems, however, with both the definition of catchment areas and the modelling of competition being subject to considerable debate which, in turn, has sponsored new developments in interaction and flow modelling.

In some ways geodemographics raises in practice the theoretical tensions (discussed earlier in this chapter) between the scientific

Figure 8.3 The outcome of the 1997 General Election in England

Source: Upton (1999: 401)

philosophy of research and the sometimes uncoordinated proliferation of digital data sets. Many of the commercially derived data imported into geodemographic modelling are patchy, incomplete and therefore 'unscientific' in pure terms. It also rests on the (somewhat arguable) theoretical assumption that 'you are what you consume'. Nevertheless, there is little doubt that geodemographics have assumed an important and applied role in contemporary analysis by enumeration.

Visualization

Geocomputation has also used GIS to explore new ways of displaying, viewing and 'reading' quantitative data that represent more complex spatial processes and systems. To begin with,

human geographers have begun to develop new forms of cartography (Dorling and Fairburn, 1997). For example, customized population cartograms have been used to create more realistic and reliable understandings of demographic distributions (Dorling, 1995a), and new diagrammatic ways of representing the complex relationships between variables have enhanced the graphical display of data. Figure 8.3 illustrates a mapping of the outcome of the 1997 General Election in England. Instead of a conventional spatial patterning by constituency area, constituencies are represented by circles, the size of which is proportional to the area of the constituency. The visualization of electoral data in this case permits a holistic view of the spatial distribution of party control and the types of

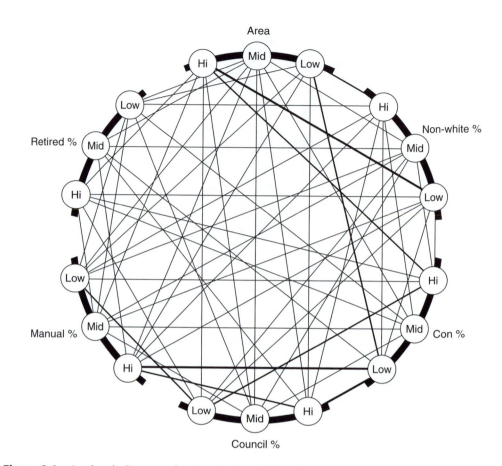

Figure 8.4 A cobweb diagram of sociological variables influencing voting

Source: Upton (1999: 412)

constituencies represented. Figure 8.4 uses the outer circle to display 18 values, three each for six variables relating to voting in that election. These values are then linked using a cobweb diagram, with the thicker lines corresponding to stronger relationships. The room for computer-driven experimentation here is limitless, with multivariate data being represented by anything from triangularization of space to the variable features of a frowning or smiling face situated at each data point.

The cognitive processes inherent in these new cartographies are clearly underpinned by particular statistical modelling of data. Visualization becomes much more difficult when dealing with hypervariate data than

with univariate or bivariate data and, as Fotheringham (1999: 604–5) stresses: 'there would appear to be a trade-off between presenting a clear image...and understanding what that image means in terms of the original data. The problem is not new in multivariate data analysis, and the same issue arises in principal components analysis, factor analysis and discriminant analysis.' Visualization, then, merely reflects the wider epistemological and interpretative issues of the techniques which underlie it.

Two other initiatives in visualization are also significant in this context. First, the principles of the new cartography have focused research attention on the possibilities of developing realistic virtual spaces in which the data-rich GIS can be encountered in different ways. Batty et al. (1998), for example, have shown how immersive and remote environments can be built up using GIS such that single and multiple users can interact in a variety of ways. They illustrate these ideas using four of their different developmental projects:

> a multi-user Internet GIS for London with extensive links 3-D, video, text and related media; an exploration of optimal retail location using a semi-immersive visualisation in which experts can explore such problems; a virtual urban world in which remote users as aviators can manipulate urban designs; and an approach to simulating such virtual worlds through morphological mapping based on the digital record of the entire decision-making process through which such worlds are built. (1998: 139)

There are obvious commercial and education advantages to these kinds of virtual visualizations. The modelling aspect is again an enhancement which is only as effective as the underlying models and techniques permit it to be.

The second initiative which displays the potential of geocomputational visualization is the development of 'wearable' computing. Clarke (1998) has shown that the availability of fully independent mobile computing and developments in human–computer interaction are leading to important new territories of visualization. Hyperinteractive devices will permit the conversion of physical movements, sounds and perhaps even thoughts into digital input streams. Return displays might be envisaged which simulate vision and sound. Once again we see the potential for technology to be the driving force of science. Wearable computing could open up a whole new suite of questions to be answered through geocomputation. Equally, the potential for the loss of privacy

or even, worse, invasions of the human consciousness is also apparent. As Clarke (1998: 136) speculates:

> I dread the day when I am woken from a sound sleep by a noisy, flashing advertisement projected onto my retina urging me to download a new free Web-browser, one that I cannot turn off without focussing on a dark grey 'Decline' button hovering at the far range of my peripheral vision. Nevertheless, as a Professor, I look forward to teaching students directly while walking in the mountains or on the beach, and to sitting again at a desk devoid of a massive, noisily fan-cooled, cathode ray tube.

The authority of numbers?

The questioning of science

Throughout this chapter we have attempted to convey a sense of the growing sophistication of quantitative analytical techniques in human geography. The use of such techniques now ranges from relatively straightforward exploratory analysis using descriptive enumerations and statistics, recently aided by new and interesting forms of visualization in GIS environments, to highly complex modelling, including sustained attempts to discover and exploit the spatiality of data. Earlier in the chapter we emphasized the potential for varying interpretation, and even misrepresentation, when using the simplest of descriptive techniques. Such potential difficulties in no way undermine the legitimacy of descriptive analyses of enumeration, but they do point to the likelihood that the deployment of ever more sophisticated statistical modelling will be accompanied by an increasing risk that the underlying enumeration can be lost sight of during iterative mathematical manipulation.

The principal challenge to any presumed supremacy for quantitative methods in human geography has, however, taken the form of a critique of the positivist philosophical roots of quantification. That is, it is not the use of numbers per se which has been objected to; rather the framework of thinking which dictates the assumptions and interpretative directions which so often characterize numerical analyses. Putting this argument simply, the principal language of science is mathematics; science in human geography has been underpinned by the flawed philosophy of positivism; therefore, mathematics carries with it the flaws of positivism and should be rejected accordingly (see Barnes, 1996). Such a critique is inexorably linked with the adoption of other philosophical frameworks in human geography. Thus Cosgrove (in Gregory, 2000d) regards the modelling of spatial science, with its privileging of abstraction and generalization, as an intrinsic expression of the modernist impulse. It follows that the practice of quantification is seen as incapable of responding to the challenges of 'difference' posed by poststructuralist philosophies. Dixon and Jones (1998a) follow up this ontological challenge with an epistemological one. Poststructuralism, they note, calls into question any assumption that the 'real' can be separated from the 'represented' – an assumption on which the epistemology of model building relies.

For Sibley (1998), the nub of the question is about the search for order. He notes that practitioners of spatial science and GIS exhibit considerable anxiety about the notion of disorder and base their research approach on searching out order. Even the more recent turn towards enumerating difference, as noted earlier in this chapter, has tended to emphasize the order of disorder. Sibley argues that the turn towards poststructuralist philosophies has posed significant questions about this emphasis on order. He especially highlights that practices whereby human geographers claim to adopt value-neutral approaches and 'just look' for

order in data are not innocent practices. Indeed, the radically non-innocent nature of the search for order raises unsettling implications for the process of spatial modelling. The upturn of these critiques has been a significant shift by many human geographers away from the quantification of spatial science, thereby reinforcing a quantitative–qualitative divide.

It is worth noting at this point that much of this critique of the philosophy underlying quantitative practices has failed to establish whether geographers using quantitative methods were ever actually wedded to positivism (with its attendant philosophical and sociocultural baggage) as a framing philosophy, and even if they were whether positivism was the main justification for their use of mathematical analysis. It may well be, for example, that quantitative practices rely more on broad philosophies of empiricism than on the more specifically focused philosophy of positivism. This very significant point opens up a wider debate about the social and cultural contexts in which quantitative techniques are used – a debate which has been particularly drawn out by Trevor Barnes (1998) and Les Hepple (1998). Barnes argues that correlation and regression analyses – kingpins of the general linear model – emerged from very particular social and cultural contexts, and that these contexts were embued in the very nature of the techniques:

> Correlation and regression were the first of the inferential statistics, but they too were designed to advance a particular social and ideological agenda, eugenics. As a result they were not a set of value free, neutral techniques, but carried with them the assumptions, conditions, and wider problematics of the intellectual movement in which they arise. As correlation and regression analyses became black boxes, they moved out of biometrics and into other fields, including eventually human geography in the mid 1950s... Because statistics is a

social and cultural product, its framework cannot necessarily be transported to a different context without inconsistencies occurring. (1998: 220–1)

We should note here that Barnes is not claiming that regression analysis is somehow an inherently racist technique. However, he *is* emphasizing that regression originated in a particular intellectual context (eugenics) and therefore carries with it certain background assumptions (the need for independent sampling, large numbers of samples, etc.) which 'facilitate control and power over certain bits of the world' (1998: 220).

Hepple debates with Barnes the extent to which the emergence of regression and statistical inference has been constrained by these specific sociocultural assumptions and contexts. He argues that the technical form of these analytical procedures was not significantly constrained by specific biometric assumptions and contexts. More widely, he argues that 'quantitative and cultural approaches are both necessary, shedding light in different ways on social life' (1998: 232).

Ground Truth

These questions about the connections between philosophy and practice in quantification, and about the way in which quantitative analytical techniques carry with them important sociocultural baggage, prompted intensified debate in the era when GIS came to the fore. The early exchanges tended to be fundamentalist in tone. For example, Taylor (1990: 211–12) chastens quantifiers in human geography who have ignored and by-passed legitimate critiques of their positivist underpinnings and their empiricist superstructure and sought to capture the rhetorical high ground of the subject by swapping 'knowledge' for 'information':

Knowledge is about ideas, about putting ideas together into integrated systems of

thought we call disciplines. Information is about facts, about separating out a particular feature of a situation and recording it as an autonomous observation ... The positivist's revenge has been to retreat to information, and leave their knowledge problems – and their opponents – stranded on a foreign shore. But the result has been a return of the very worst sort of positivism, a most naïve empiricism.

In reply, Openshaw (1991: 626) offers a somewhat imperialist vision of GIS:

GIS can provide an information system domain within which virtually all of geography can be performed. GIS would emphasise an holistic view of geography that is broad enough to encompass nearly all geographers and all of geography. At the same time it would offer a means of creating a new scientific look at geography, and confer upon the subject a degree of currency and relevancy that has, arguably, been missing.

An important clarification of a more sophisticated critique of GIS came with the publication of *Ground Truth* edited by John Pickles (1995). *Ground Truth* set out to tackle three sets of issues:

1. The impact of geography on the transformation of data-handling and mapping capabilities provided by GIS.
2. The ideas, social practices and ideologies which have emerged with these new forms of data-handling and spatial representation.
3. The situating of GIS as a tool of, and an approach to, geographical information in the context of late twentieth-century transformations of capitalism.

GIS, then, is viewed in *Ground Truth* as

a tool to protect disciplinary power and access to funding; as a way of organising more efficient systems of production; and as a reworking (and rewriting) of cultural codes – the creation of new visual imaginaries, new conceptions of earth, new modalities of commodity and consumer, and new visions of what constitutes market, territory and empire. (1995: viii)

Interestingly, some of those seeking to debate *Ground Truth* immediately complain about the non-comprehensibility of its language. For instance, Flowerdew (1998: 299), with reference to the above extract among others, reflects that 'With considerable effort, I can work out approximately what these sentences mean (I think) but it is hard to imagine GIS boosters, who perhaps need to hear the underlying message, struggling to understand them, let alone be convinced'.

This focus on language can be compared, somewhat ironically, with the list of reasons for the 'downturn' in quantitative geography in Box 8.2, and specifically with the claims therein that the 'difficulty' of quantitative geography 'deters established non-quantitative researchers from understanding the nature of debates'. The coding of quantitative geography as 'difficult' and of *Ground Truth* as 'over-complex' seems here to reflect ideological representations and could easily be reversed. More importantly, the issue of language seems to provide an excuse for some quantitative geographers (Flowerdew clearly excepted) to eschew any meaningful engagement with the *Ground Truth* project.

In *Ground Truth* there is a series of deep concerns over the impacts of unmediated technical practices in GIS. Three examples illustrate these concerns. First, Curry (1995) argues that human geographers need to take serious account of conflicting stances towards *ethical* issues in GIS. There are ethical concerns relating to the expertise involved. GIS users often know how to operate the system but not how it was conceived or how it works in detail. User expertise might therefore be thought of as meeting ethical requirements, despite the fact that the systems in use may well have been developed in the entirely different ethical contexts of military and intelligence

use (see Smith, N., 1992). There are also ethical concerns about how the analysis of large data sets tends to regard people-places as an appropriate category for interpretation, thus contributing to the dissolution of real individuals. Clearly such concerns need also to be linked with an opposite risk in which GIS can be used (explicitly or implicitly) for surveillance purposes, thus perpetrating invasions of privacy.

Secondly, Goss (1995a: 161) reflects on the ability of GIS to manipulate behaviour. As he notes, GIS

reduced the complex multidimensional character of human identity to a set of correlated variables which are then aggregated into abstract constructs or clusters based on place of residence; these clusters are then reified into ideal types of 'consumer neighbourhoods' used to predict consumer behaviour, which is in turn reduced to a simple scale representing the likelihood of purchasing a particular commodity.

As Goss points out, GIS can thus become part of a self-fulfilling prophecy as social identity is defined in terms of particular clusters of consumer goods consistent with idealized lifestyles, and commercial interests will then try to sell these identities back to consumers in the form of complementary consumer goods. Not only do such processes go a long way towards constituting social subjects but they can also direct the gaze towards particular (wealthy, high-status) subjects who are of greater commercial interest, thereby diverting the gaze from other (impoverished, socially excluded) subjects.

Thirdly, Harris et al. (1995) question the relationship between GIS and democracy. They argue that the use of GIS for research, planning and project assessment, particularly in developing nations, has generally turned out to be a 'technicist legitimisation of the historical power relations associated with traditional developmentalism' (1995: 196). A more radically democratic alternative is to establish a participatory application of GIS in which an appreciation of local knowledge, community needs and specific social histories can be incorporated into the developmental process. Harris et al. (1995) report on their experience of attempting to introduce just such a participatory application in the context of post-apartheid rural reconstruction in South Africa. Their conclusion is stark: 'In traditional South Africa, it is likely that GIS will reinforce traditional market-based and technicist approaches to policy formation to the detriment of alternative restructuring strategies: this process is inherently undemocratic' (p. 217).

The control over data collection, analysis and interpretation at the scale now offered by GIS environments remains heavily top-down and elitist. There *are* ways of imagining a more democratic and localized set of inputs but, in the pragmatics of GIS use, in both the public and private sectors, such democratic aims are likely to be superseded by technocratic, commercial and bureaucratic concerns.

Quantitative responses

The responses to issues raised in *Ground Truth* and analogous debates suggest three strategies for taking forward practices of enumeration in human geography. The first – to ignore the emergent critique altogether – is reflected by Clark's (1998: 305) extraordinary use of masculinist war discourses to describe the *Ground Truth* attack on a productive profession of quantifiers:

> The attack on the complacency and flawed thinking of GIS launched by Pickles (1995) in *Ground Truth* is an academic Blitzkrieg of unusual power and ferocity. It strikes at the core of a productive profession by using weapons (arguments) from another world, in much the same way that the Panzer divisions rolled East to impose a new order on Europe's agricultural and industrial heartland. Both attacks were pre-emptive and were fuelled by a mission rather than a measured and reflective desire to contemplate alternative models. However, while Blitzkrieg appeared initially to be immune to resistance, it is amazing that this most radical critique of GIS social implications has failed to raise even a whimper, let alone an enraged outcry, from most of the practitioners of GIS. Indeed, in a very real sense the attack mounted on *Ground Truth* has failed completely to engage the enemy! The tirade has gone unheard not because the fundamental message was irrelevant or wrong but because the messenger spoke a foreign language and spoke in an empty room.

This kind of viewpoint suggests that an adherence to a positivist philosophy of geography will valorize objectivity as a goal and take a benign view of technology as the means of reaching that goal.

A second response might be to adhere to quantitative methodologies on grounds other than a philosophical attachment to positivism. Many quantitative human geographers would now agree with Flowerdew (1998: 292) when he suggests that while 'quantitative geography has lost and largely abandoned the philosophical debate, quantitative work is still important'. Earlier in this chapter we discussed Fotheringham et al.'s (2000) view that there are important differences in philosophy across the protagonists of quantitative geography. For example, the philosophical tenets of pragmatism – what counts as knowledge is determined by its usefulness – were developed by Rorty (1979), who proposed a conversation between the mind and the world with no fixed end-points, strict rules or necessary logic. Pragmatism can and does lead to a use of quantitative methods to provide the answers required by particular end-users. Spatial analysis and modelling therefore provide ways of meeting the demands of the commercial marketplace, the government decision-maker and the research council anxious to be seen to be relevant to the needs of the nation. In these contexts numeric analyses are perceived as useful (and sometimes even vital), so applied human geographers supply their information in this form.

It is difficult to know whether quantitative human geographers have *really* given up on positivistic philosophies, despite claims to have set aside the philosophical debate. For example, Flowerdew himself is reluctant to veer from 'value-free' research (1998: 293) and suspects that most GIS scholars and practitioners are reluctant to embrace alternative social theories which might, of course, point out the links between 'value-free' research and positivist philosophy. The suspicion remains that although some quantifiers recognize the place for other, more qualitative, forms of research (see, for example, Jones and Duncan, 1996), most will not have rejected the principal facets of scientific method in their own work, even if they feel the need to explain these in terms of 'the best way of doing things' rather than of positivist philosophies.

A third response has been to incorporate parts of the critique into new forms of quantitative research. An important illustration of this comes from the Radical Statistics Group which was formed in 1975 as part of the radical science movement associated with the establishment of the British Society for Social Responsibility in Sciences. The group reflects concern about the mystifying use of technical language to disguise social problems as technical ones; the lack of control by the community over the aims of statistical investigations, the way these are conducted and the use of the information produced; the power structures within which statistical and research workers are employed and which control the work and how it is used; and the fragmentation of social problems into specialist fields obscuring connectedness (http://www.radstats.org.uk). Such concerns have prompted campaigns for better 'official' statistics and a better understanding of statistics. The group is particularly interested in demystifying the

selective use of social statistics to support the policies and status of those in power. They also aim to introduce the kinds of statistics needed 'to understand and improve life' (Dorling and Simpson, 1999b; see Box 8.9). The approach reflected by the Radical Statistics Group acknowledges that statistics are a product of something and therefore are never neutral. Such statistics

> cannot be ignored nor can they be substantially changed unless society itself is changed. People who wish to make statistics more open ... more democratically linked to the priorities of local and other communities, less dominated by government and trade, are allies of anyone who wants to change society in these ways. (Simpson and Dorling, 1999: 415)

Here, then, is a very significant group of quantitative social scientists who advocate the explicit introduction of political values into their work. This permits them to address issues relating to ethics and democracy as part of their research design. It *may* also make them more amenable to more qualitative practices which offer 'evidence' of the politically directed 'targets' of their research.

Individual responses

We began this chapter by emphasizing that techniques of analytical enumeration provide potential powerful and useful ways of practising human geography. We also suggested that quantitative and qualitative techniques need not be viewed as discrete choices. It will be evident from the account presented in this chapter that a deep and sometimes uncommunicative divide has often separated practitioners of the quantitative and the qualitative, with each questioning the philosophy, language, assumptions and usefulness of the other. Many will wish to unsettle some of the more dichotomous views of quantification

> ### Box 8.9: Examples of statistical deficiencies in Britain
>
> - There are few statistics on gender, and children are generally treated as properties of their households. This is a relic of an era when women and children were 'goods and chattels'.
> - Most of the work done by people at home, including looking after children, shopping and cooking, is not measured by either economic or social statistics. Only paid work is measured in the System of National Accounts and, until recently, we had no official time budget statistics.
> - Unemployment statistics do not measure the number of people who want jobs.
> - Health statistics do not measure health or care needs.
> - Housing statistics do not measure homelessness or housing need.
> - Poverty statistics do not measure the number of poor people.
> - Economic and transport statistics do not measure the environmental and social costs of new roads or industrial activity.
> - Until the 1990s one of the few sources of statistics on 'ethnic minorities' were the 'mugging' reports of the Metropolitan Police, but there were almost no statistics on racially motivated assault.
>
> *Source*: Dorling and Simpson (1999b: 4)

and qualification as being, at least sometimes, a defensive and occasionally arrogant posturing which unthinkingly rejects the potential for using together both numeric and non-numeric data construction and data interpretation. This position can be founded on concepts developed by Geertz (1973) and Denzin (1978) who differentiate between 'thick description' and 'thin description'. According to Denzin (1978: 33), *thick description* includes 'information about the context of an act, the intentions and meanings that organise action and its subsequent evolution'. Interpretation here, then, focuses on the context, the motivation and meaning, and the process in which action is sedimented. *Thin description* focuses more on facts about, and characteristics of, the research subject. We do not rule out the possibility of using forms of enumeration as thick description – witness Prigogine's (1984) use of quantitative catastrophe theory which has become the foundation of significant strands of modern social theory. However, we do take a collective view on the place of enumeration in thin and thick description – a view which is based on our experience as *users* of enumerative methods in our research rather than exponents claiming to be experts in the field.

Our approach is that enumeration in human geography is best seen as a form of thin description, capable of identifying certain characteristics and patterns of data, but incapable of describing or explicating the meaningful nature of social life (see Chapters 9 and 10). There are two key points to emphasize here. First, 'thin description' should not be viewed as a pejorative or derogatory term. It describes a perfectly legitimate research function, usually involving large numbers of items of information. Indeed, thin description can often achieve a high level of impact when publicized in numeric form. For example, the data in Table 8.1 were used to made very public statements about rural poverty in Britain. When, however, similar attempts were made to use qualitative narratives to influence public agencies about the plight of the rural poor, there was much more resistance to the ideas being advanced (Cloke, 1996). So there is no reason why 'thin' and 'thick' descriptions cannot be used sympathetically together. Indeed, in practice they often are, without any explicit acknowledgement of that fact. Secondly, thin description, although normally applied to numeric data, can in principle equally well be applied to non-numeric data and is therefore available as part of an interpretative strategy for what would generally be described as qualitative research, as well as for the more obvious quantitative work.

Clearly, then, there are individual decisions to be made about whether the role of enumerative analysis can or should go beyond thin description. Tests of inference, comparison and relationship (and the models which are founded on them) are clearly embedded within a particular scientific logic of 'explanation'. Some readers will want to pursue this form of analysis and, in so doing, will need to be aware of the philosophical and technological assumptions which they are endorsing in these practices. Others will view these analytical techniques as attempting the task of thicker description without sufficient recourse to the context, meaning and process which can be delivered using more qualitative methods. At a descriptive and exploratory level, we see no necessary incompatibility between the outcomes of quantitative and qualitative data analysis. Indeed, fruitful collaboration may be envisaged between these outcomes. Collaboration beyond this point, however, will be more restricted, partly because technical choices will often dictate different subjects for research, and partly because collaboration beyond this point will be weighed down by the considerable philosophical and intellectual baggage which currently accompanies quantitative and qualitative approaches.

Notes

1 See, for example, Cole and King (1968), Hammond and McCullagh (1974), Smith (1975), Norcliffe (1977), Gregory (1978), Johnston (1978), Silk (1979), Matthews (1981), Unwin (1981), Clarke and Cooke (1983), Ebdon (1985), Williams (1986), Haining (1990), O'Brien (1992), Walford (1995), Robinson (1998) and Fotheringham et al. (2000).
2 See, for example, Martin (1996), Chrisman (1997), DeMers (1997) and Longley et al. (1998a).
3 See, for example, Bailey and Gatrell (1995).
4 See, for example, Openshaw and Openshaw (1997).

9   *Explaining*

The complexity of explanation

We will begin this chapter by returning to one of the two geographers who featured in Chapter 1. According to Linda McDowell, the *raison d'être* of geography 'is the explanation of difference and diversity' (1995: 280). Not many would wish to argue with either part of this statement. Geography is concerned with difference and diversity, its own distinctiveness as a discipline resting on its enduring interest in the uneven development of both the social and the natural worlds. As an academic pursuit, it also seems indisputable that geography should be concerned with explaining these differences, as well as with charting and describing them. To this end it seems entirely reasonable that human geographers are involved in producing explanations for the geography of social events and processes – accounting, in other words, for where, why and how things happen rather than just pointing out when and where they take place. This is what we would all expect of academic research – it is undertaken in order to explain events and processes. At first sight, then, the statement that the rationale of geography is to explain difference and diversity seems fairly unproblematic.

On closer inspection, however, a number of problems do arise. These are largely due to the extremely complex nature and meaning of the term 'explanation'. A hint of the difficulties can be conveyed by considering the way we use the term in day-to-day life. We 'explain' to passers-by how to reach a particular destination: a largely descriptive form of explanation which simply involves the transfer of relatively accurate information. We 'explain' to friends and colleagues why we acted in a certain manner: a developmental form of explanation in which we attempt to account for the development of particular sets of social actions and activities. We also 'explain' why we didn't take a particular course of action: often invoking a comparative form of explanation by asserting what would have happened if we had undertaken that course of action. We even 'explain' why we are going to undertake a particular course of action in the future: a predictive form of explanation which is tied up with predicting the outcome of a set of social events that have not yet happened. We usually pay no heed to these different types of explanation, and there is usually no need to – social conventions and accepted forms of communication will normally ensure that each different meaning of the term 'explanation' is clearly conveyed and understood.

In human geographical research the meaning of the term is more restricted and is usually confined to a developmental form of explanation – that which attempts to trace and account for the 'development' of processes and events (see Mason, 1996: 127) – although comparative and predictive forms of explanation are also found to a lesser extent. There are also well-established procedures and principles that academics use when they attempt to explain something. Unfortunately this does not lessen the complexity surrounding the term because different researchers will work with different understandings of what constitutes an adequate form of explanation, and they will use different procedures in order to provide one. In his book on social science methods, Sayer (1992: 232) states that the question 'What is a good explanation?' is one of the most common yet most exasperating for a methodologist to be asked. As he points out, the answer first depends on what kind of social activity or process is being investigated and, secondly, on what the researcher wants to explain about it. What the researcher wants to explain about it is in turn linked to how he or she conceives of the relationship between more abstract conceptual and theoretical issues and empirical research. Thus there is no single route to a good explanation, and the path taken will depend on what needs explaining as well as on what the researcher understands to be the components of a good explanation. This chapter will take us through the different forms of explanation commonly used by geographers and will also highlight how the researcher can help to select those which are more suitable for his or her own particular piece of research.

Explanation through laws: geography as spatial science

In his book examining the development of human geography, Chisholm opened the chapter on 'Theories of spatial structure and process' with the claim that 'Explanatory theories may be regarded as the apex of the scientific pyramid, the goal to which we strive' (1975: 122). This statement is instructive

for three reasons. First, it assumes that human geographers will seek to follow a particular version of the 'scientific' form of explanation. Secondly, it states that we should derive explanation from theoretical statements. Thirdly, it implies that we should all follow this one prescribed route. These three claims are, of course, linked, and the 'scientific' form of explanation, drawing on universal theoretical statements or laws, came to have a huge influence on the discipline of human geography from the mid-1960s onwards.

Prior to this, explanation in geography was largely concerned with applying intuitive understanding, based mainly on description, to a large number of individual and seemingly unique cases. The notion held throughout the first half of the twentieth century that geography was an idiographic science, concerned with studying the unique and particular, militated against any search for generalized forms of explanation. It was felt that, because geographers were largely concerned with studying particular places, they could search only for particular understandings. This view was challenged in the 1950s, most forcefully by Schaeffer (1953), who argued that geography should be nomothetic and seek to provide general laws about spatial organization. This call gradually gained ascendancy and, for much of the 1960s, geographers looked for ways of formulating and applying such laws.

They did so by working within a positivist framework of inquiry. For the positivist, predictive and explanatory knowledge comes from the construction and testing of theories, which themselves consist of highly general statements expressing the regular relationships that are found in the world. In this process, scientific theories consist of universal statements whose truth or falsity can be assessed through observation and experiment (Keat and Urry, 1975). Those theories which were empirically verified would then assume the status of scientific laws. The full transfer of scientific method into human geography came in 1969 with the publication of Harvey's book *Explanation in Geography* where he argued for the primacy of deductive theoretical forms of explanation. Under this form of explanation, the event or process to be explained is deduced from a set of initial or determining conditions and a set of general laws – given Law L, if initial conditions C and D are present, then event E will always occur. The event E is 'explained' by reference to the laws and the initial conditions. There is no logical distinction between the explanation and the prediction of any given event. An explanation simply refers to an event which has happened, and prediction to one in the future. In his book Harvey outlined a number of explanatory forms in geography which used this deductive form of reasoning. Which is applied depends on the kinds of questions being asked and the kinds of objects under investigation, but they all share the methods of hypothetico-deductive science whereby explanation comes from the formulation of theories which yield testable hypotheses. By testing these hypotheses we can discover the regularities of the universal laws which govern patterns of events. The events are then 'explained' by reference to the universal laws. As long as a correct set of initial conditions is specified, these laws can be drawn on to provide explanations. This kind of reasoning dominated human geography in the 1960s and most of the 1970s and generated the search for law-like statements of order and regularity which could be applied to spatial patterns and processes. Hence the succession of models which appeared in geography over this period – for instance, Christaller's model of settlement hierarchy, Alonso's land-use model, Zipf's rank-size rule of urban populations and Weber's model of industrial location. All were an attempt to use law-like statements in order to explain and predict spatial outcomes.

However, a fundamental problem with this type of explanation is that what causes something to happen has nothing to do with the number of times or the regularity with which it happens: as Sayer points out, 'unique events are caused no less than repeated ones, irregularities no less than regularities' (1997: 115). And as he goes on to point out, even showing that an individual event is an instance of a universal regularity or law fails to explain what produces it. The inadequacy of positivist argument arises from confusing the provision of grounds for expecting an event to occur with giving a causal explanation of why that event occurred (Keat and Urry, 1975: 27). This problem was compounded in geography by the fact that much of the work carried out within 'spatial science' was actually statistical rather than deductive. And as Guelke (1978: 44) has pointed out:

> The use of the word 'explain' in a statistical sense is most misleading, because in such a sense it carries no connotation of cause. Yet in everyday language it often has such causal implications. In consequence, and in spite of disclaimers, geographers have seen statistical analysis as an end in itself, when such 'analysis' is basically a mathematical description.

Thus, geographers working within a positivist scientific framework tended to confuse description with explanation and looked to explain events via statistical regularity rather than causal process (see also Chapter 8). Despite this, this method of explanation was predominant for most of the 1960s and 1970s and has left a huge legacy within the discipline. A more recent example of its application can be found in an article by Bennett and Graham, published in 1998 and interestingly entitled 'Explaining size differentiation of business service centres'. In this case, the 'explanation' is given in purely statistical terms. As the authors themselves put it, 'our purpose is to demonstrate statistically the general pattern of differentiation of groups of centres of differing sizes' (1998: 1477). But demonstrating a statistical pattern, in this case a hierarchical one of five stages, does not explain that pattern. The factors which produced it are not interrogated at all, and analysis stops once the pattern has been identified. What takes over is intuition; the authors state that 'the five levels can be interpreted in an intuitive way' (1998: 1478), but even this intuition results only in the five levels of business centre being labelled as national (and international) foci, regional centres, secondary regional centres, shadows and local centres. Again, there is no explanation of these differences, merely a descriptive labelling process. What is instructive is that the authors felt that such description amounted to explanation – once their statistical tests had yielded regular results, they felt the term explanation could be applied – when what actually had been provided was a very detailed statistical description.

Explanation as causation

In contrast to this emphasis on mathematical order and statistical description, Sayer (1992; 1997) advocates a causal approach to explanation. In this view, an explanation amounts to statements about what actually causes an event to happen, and an adequate causal explanation requires the discovery of relations between phenomena and of some kind of mechanism which links them. To produce an explanation we therefore need a knowledge of the underlying structures and mechanisms that are present and of the manner in which they generate or produce the phenomena we are trying to explain.

Sayer uses the example of manufacturing shift to explore the difference between the two types of explanation, pointing out the difference between the view that empirical observations themselves can generate explanations and the opinion that they merely act as a guide to identifying some of the underlying processes at work (1997: 117–19). In order to make the distinction clearer, he draws on research which examined the tendency for metropolitan areas to lose manufacturing employment in the 1960s and 1970s, while small towns and rural areas enjoyed gains. One explanation that was sought for this phenomenon explored the influence of government grants on business location by looking at whether the areas of employment decline had assisted or non-assisted area status. The hypothesis that there was a link between business success and grant aid was rejected, for non-assisted areas such as East Anglia experienced employment growth as well as assisted areas such as Northumberland, and the larger urban areas within assisted areas suffered decline in the same way as their non-assisted counterparts. Thus, because the regularity was not found and the universal statement could not be made, the presence of government grants was rejected as a potential source of explanation.

But as Sayer points out, there is another way of positing such a link which does not rely on universality but, instead, looks at mechanisms and processes. This would explore the effect of regional aid on employment growth and decline by thinking through which kinds of firms might make use of such aid and in what circumstances. Thus instead of positing, and looking for, a simple regularity between incentives and employment change, we would now be looking at the different circumstances of individual firms and asking what kinds of firms would find regional grants attractive and why. This recasting of the question requires a different type of research in order to provide a different type of explanation. Research would be less concerned with gathering large-scale sets of data which could be statistically tested and more concerned with investigating the strategies and responses of individual firms through interviews and questionnaires. We would, of course, expect more complex answers to emerge – in some cases we might find the influence of incentives on location and in others we wouldn't – depending on the type and nature of the firm under investigation. Moreover, even if there was a positive influence this need not necessarily translate into employment gain: some firms may well use regional development grants in order to buy new machinery which might make workers redundant, resulting in overall employment loss. This more messy conclusion is a reflection of the complex nature of the problem under investigation. It also emerges from seeking a different type of explanation – one based not on the identification of a regularity of occurrence but on searching for the processes and mechanisms which cause an event to happen.

Intensive and extensive research

This in turn has implications for the type of research methodology adopted and the type of research questions asked. Remaining with the above example (from Sayer, 1997: 118), if we ask what determines the location of manufacturing investment, the answer will almost certainly be that several different processes do, depending on the particular firm in question. In our research we might seek to find which processes are most important in a limited number of cases. This is known as 'intensive' research and it, focuses on a small number of case studies in order to find the causal processes and mechanisms behind a particular event. But as Sayer (1997: 118) notes, 'discovering how a mechanism works does not tell us how widespread that mechanism is: for example in how many instances were regional incentives influential?' In contrast, the type of research which may lead to an answer for this

type of question is termed 'extensive' – instead of seeking causal mechanisms it asks about the 'extent' of certain phenomena. The confusion between the two types of explanation we have considered stems from this difference. The two types of research actually provide answers to two very different kinds of inquiry. The problem is that 'this second kind of question regarding distribution and extent is expected to answer the first regarding causation' (1997: 118). This is not accidental or the result of sloppiness on the part of the researcher but, as we have seen, it is built into the very foundations of scientific method where evidence of regularity (or extent) and statistical description are put forward as 'explanations' of particular events. However, when looked at from a perspective which views explanation in terms of causation, while the search for regularities and generalizations may contribute answers to questions regarding the quantities and distributions of phenomena (extensive questions), it will throw little light on what produces them (intensive questions).

Exploring this distinction between intensive and extensive research reminds us that the type of explanation which can adequately be given is highly dependent on research design and methodology. In other words, it is important to appreciate in advance the type of explanation being sought or the data which are generated from the research will not have been constructed in an appropriate form. As set out in Table 5.1 in Chapter 5, the differences between intensive and extensive research, which at first sight might appear to be rather superficial ones centred around the scale of research being undertaken, are in fact quite fundamental differences of research design, technique, method and explanation.

As we noted in Chapter 5, each is actually addressing a different type of research question, and from this follows a different content and form to the analysis. Because intensive research is concerned to identify causal relations and connections, the groups it studies will be those whose members are thought to relate to each other, structurally or causally. Specific individuals or 'actors' will be studied because of the properties they possess (such as a being an owner-occupier or being homeless, or being unemployed or being a managing director) and the relations they have to others (such as being in a position of authority or being a receiver of a certain service). In contrast, extensive research will be concerned to find patterns of regularity and distribution and, as a result, will focus largely on taxonomic groups whose members share similar formal attributes but who need not necessarily be connected to one another at all beyond this. The research methods will also tend to be different. Extensive research will favour relatively large-scale and standardized methods such as formal questionnaires, while intensive research will focus more around less formal and more interactive methods, such as in-depth interviews or participant observation. However, as we noted in the previous chapter, the roles of the two types of research can be seen as complementary rather than competing, and in assessing which to use we should be mindful that both may well have a place in the same research project. Preliminary work of an extensive nature may well be necesary in order to set a broad context (the level of home ownership in two different areas of a town, for instance) and intensive work can then be used to identify causal processes and relations within this (why particular people do or do not buy houses in these two places). The important point to stress in this chapter is that, while extensive research may be used to identify broad trends and patterns, it will be less helpful in terms of explanation. This is because the relationships it identifies are those based on similarity and formal correlation rather than those based on causal connections.

The search for a revolution in geographical explanation

On one level, the discussions initiated by Sayer around descriptive versus causal explanation were of great assistance in clarifying the position over the form and nature of explanation in geography, and they helped to formalize the growth of causally based explanation during the 1980s. On another level, however, they deepened a complex set of debates over precisely which elements of social activity should be incorporated into causal explanation. These debates had begun in the early 1970s as doubts began to emerge over the adequacy of supposedly 'scientific' explanations rooted in positivism.

These doubts were at once conceptual and moral. They were perhaps expressed most forcefully by David Harvey in his 1973 book *Social Justice and the City*, where he charted his own personal journey away from positivism and set out an alternative frame of reference for understanding social and geographical change. The initial reasons given for distancing himself from the positivism which his own book *Explanation in Geography* (1969) had done so much to establish were methodological and conceptual. Harvey explained that he turned away from positivism and towards a Marxist-inspired analysis because he could 'find no other way of accomplishing what I set out to do, or of understanding what has to be understood' (1973: 17). Among the reasons given for this shift were that the scientific form of explanation used within spatial science has a 'tendency to regard facts as separate from values, objects as independent of subjects, "things" as possessing an identity independent of human perception and action, and the "private" process of discovery as separate from the "public" process of communicating the results' (1973: 11–12). According to Harvey, each of these tendencies carried implications for the adequacy of the explanation that could be given by a positivist human geography.

To these academic concerns underpinning the shift towards a different type of explanation Harvey also added moral and ethical imperatives. As he wrote in *Social Justice and the City*:

> There is an ecological problem, an urban problem, an international trade problem, and yet we seem incapable of saying anything of depth or profundity about any of them. When we do say something it appears trite and rather ludicrous. In short, our paradigm is not coping well. It is ripe for overthrow. The objective social conditions demand that we say something sensible or coherent, or else forever…remain silent. It is the emerging objective social conditions and our patent inability to cope with them which essentially explains the necessity for a revolution in geographic thought. [Our task] does not entail…mapping even more evidence of man's patent inhumanity to man…the immediate task is nothing more nor less than the self-conscious and aware construction of a new paradigm for social geographic thought. (1973: 129, 144–5)

The search, then, was on for the concepts and categories which could help accomplish such a revolution in geographic thought and explanation.

Explanation through abstraction

Having set out the critique, Harvey immediately began to develop his own preferred 'explanatory schema' (1989b: 3). In order to do this he turned to the writings of Karl Marx. As he explained in *Social Justice and the City*, 'the most fruitful strategy at this juncture is to explore that area of understanding in which certain aspects of positivism, materialism and phenomenology overlap to provide adequate interpretations of the social reality in which we find ourselves. This overlap is most clearly explored in Marxist thought' (1973: 129). In a series of extremely influential writings stretching over the next three decades, Harvey continued to explore Marxist thought in an effort to develop this 'area of understanding'. One reason for the influence of Harvey's work, aside from its intellectual richness, is the clarity which he brings to the task of explanation and understanding. In all his work he is at pains to chart a precise route linking theory and concept with empirical findings.

The key to this journey lies in the notion of abstraction. This refers to the practice of isolating in thought particular aspects of the object under study. This is done in order to simplify some of the complexities of the world we are researching, at the same time as identifying key or critical aspects of that world. The Marxist-inspired methodology Harvey draws on utilizes the process of abstraction in distinct stages. In order to show how these stages proceed, Harvey (1989b: 8–10) examines the various elements which make up a typical daily meal. As he points out, we can happily consume these without knowing anything of the conditions which led to their appearance on our table – indeed, this is usually the case. If we think hard we might remember where we purchased them and how much they cost. But this is about as far as we would get in terms of understanding the geographies behind the production of this meal. Yet in terms of understanding, and indeed explaining, the exchange of money for a packet of cornflakes or a loaf of bread is almost the very end of the process rather than the beginning. Hidden behind this simple exchange is a whole host of social relations and technical conditions which have helped to govern how the commodities on the table were produced, traded and consumed.

To expose these relations and conditions we use the process of abstraction: isolating in thought particular aspects of the object under study. In this case we can begin to view the food not simply as part of a meal but as a commodity, bought by us but produced elsewhere by manufacturers using machinery, labour power, raw materials, technical knowledge and money. The concepts used to describe this process – buying, selling, production, commodities – are abstractions in that they are describing partial aspects of the food on the table. Issues to do with taste and smell, and enjoyment, and the gender relations behind who cooked the meal, and who washes up, are being put to one side as we concentrate on isolating those concepts which can help us explain the food's production. (There is, of course, a strong case to be made, even at this point in the analysis, for not setting these things aside at all and for arguing that these issues are actually bound up with the production of the food, just as much as those processes which we have traditionally defined as 'economic'.) In this way Harvey argues that we can reduce the 'complexity of everyday life to a simple set of…representations of the way material life is reproduced' (1989b: 9).

At this stage in the argument the abstractions used are fairly straightforward and refer to everyday processes expressed in everyday language. But following the process of reasoning used by Marx, Harvey now takes things further and introduces a further round of abstraction, this time using non-observable and somewhat specialized concepts to relate the production of our particular meal to the production of food within a capitalist social system as a whole. He uses concepts such as surplus value, class relations and means of production to situate and integrate the individual case within a much larger whole. In a similar manner, the American geographer Michael Watts has recently written of how he sometimes brings an oven-ready chicken into his undergraduate class and asks students to identify 'this cold and clammy creature which I've tossed upon the lectern' (1999a: 305). After

some five minutes of the students crying out 'it's a chicken', 'it's a dead bird', 'it's a fryer', Watts reveals how he then solemnly declares that it is in fact none of these but, instead, is a bundle of social relations encompassing, among others, farmers, immigrant labourers, transnational feed companies, processing plants, fast-food companies and the biotechnology industry (Figure 9.1 depicts the key actors involved in the production and consumption of the chicken). Here we see the role of abstraction in moving beyond the immediate object – the chicken – to isolate and identify particular aspects of its existence. Watts goes on to chart how the chicken industry in the USA moved in the 1940s and 1950s from the mid-Atlantic states to develop large-scale integrated production complexes across the US south as part of a global industry now facing competition from Brazil, China and Thailand. He concludes that 'you start with a trivial thing – the chicken as a commodity for sale – and you end up with a history of post-war American capitalism' (1999a: 308).

Interestingly, when Marx first elaborated this method of reasoning in his book *Grundrisse*, he uses a geographical example concerning the distribution of population. In order to understand how population is distributed between places, and between different sectors of industry, Marx points out that it would seem correct to begin with an analysis of the population itself: with what we might call the surface appearance. However, he goes on to argue that on closer examination this route proves false because our view of the population will be incomplete unless we include the classes of which it is composed. In turn, these classes will be 'an empty phrase' unless we are familiar 'with the elements on which they rest. E.g. wage labour, capital etc.' (1973: 100) These latter in turn presuppose exchange, division of labour, prices, value, money. Marx paints this as a movement towards ever thinner abstractions

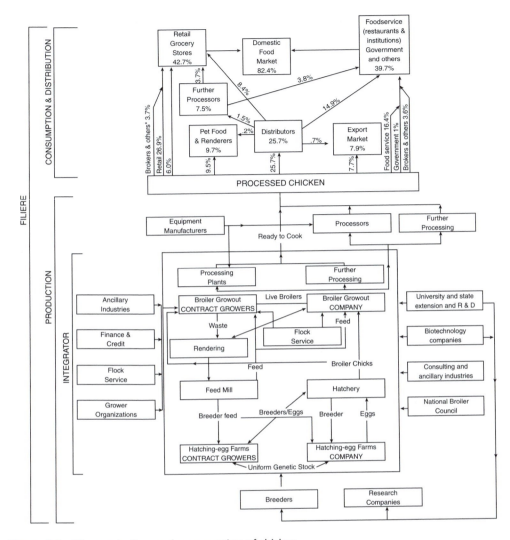

Figure 9.1 The production and consumption of chicken

Source: Watts (1999: 310)

until we arrive at what he terms 'the simplest determinations' (p. 100). From here, however, the journey has to be retraced until we arrive at the population again. In other words, once we have taken our object of study apart and isolated key relationships and components through abstraction, we have to put it back together again. This time, however, it appears not as an incomplete surface picture but 'as a rich totality of many determinations and relations' (p. 100).

To return to the example drawn from Watts, we started with the simple and somewhat trivial oven-ready chicken and, via the process of abstraction, we finished with the rather complex commodity circuit set out in Figure 9.1. The ready-to-cook chicken at its centre now appears as the totality of many determinations and relations. We have thus made a double movement – from the original empirical object of study to the abstract concept, and from the abstract back to the concrete, with concrete here being understood as the unity of many diverse aspects. Harvey argues that we can use such movement to 'help us understand all kinds of surface occurrences that would otherwise remain incomprehensible' (1989b: 10). He goes on to argue that the proof of this line of reasoning is found in its usage – 'explanatory power becomes the central criterion of acceptability' (p. 10).

It should perhaps be stressed at this stage that the procedure of abstraction is by no means limited to Marxist-informed research. Research on gender relations might draw on abstractions developed in feminist research concerning patriarchy. That on the service sector might use abstractions about bureaucracy originally devised by Weber. Work on political geography and the state might draw on abstractions connected to notions of power and governmentality derived from Foucault. Whatever the source of the abstraction, the way in which such explanation proceeds by relating the surface appearance of an empirical event to the actual processes taking place behind it is summarized in Figure 9.2. The figure shows how Sayer envisages the relationship between the abstract and the concrete, and between causal mechanisms and structures and the events they produce. It represents a complex social system in which the activation of particular mechanisms produces effects (or events) which may be unique to a particular time and place. But with different circumstances and conditions the same mechanisms may produce different events and, conversely, the same kind of event may have different causes. The use of abstraction permits the analysis of objects in terms of their 'constitutive structures' as parts of wider structures and in terms of their causal powers. In order to explain any particular event we need to undertake concrete research to examine what happens when these combine.

Figure 9.3 develops the information presented in Figure 9.2 in order to clarify the relationship between different kinds of research and different kinds of explanation. It shows how the researcher can examine mechanisms, structures and events in different ways to produce different types of account. Thus *abstract theoretical research* deals with structures and mechanisms in an abstract sense, and events are only considered as possible outcomes. *Concrete research* of an intensive nature does study actual events, treating them as phenomena that have been caused by specific mechanisms and structures (themselves uncoverable through abstract research). By contrast, the method of *generalization* is undertaken through research of a more extensive nature, where the researcher seeks to establish regularities and patterns at the level of events. Most positivist law-like explanations actually consist of generalizations made at this level of the research process. It is also possible to add a fourth type of *synthesis research*, which attempts to combine abstract and concrete research findings with generalizations covering a wide range of events. Interestingly, Sayer notes that 'research of this kind is especially common in…geography' (1992: 236). This is because as a discipline geography has a tradition of seeking to understand variation across a range of events as well as a history of examining the causes of specific events. It is concerned with both the general and the particular, and, often, with the interconnection between them.

Some of the major problems in explanation arise when the capacity and purpose of these different types of research are misunderstood. In particular, as Sayer points out, 'researchers often *over-extend* them, by expecting one type to do the job of the others' (1992: 238, emphasis in original). This can lead to over-ambitious claims being made on behalf of the explanatory power of specific pieces of research. In particular, abstract research can be over-extended when it is used to explain events directly, without relevant empirical

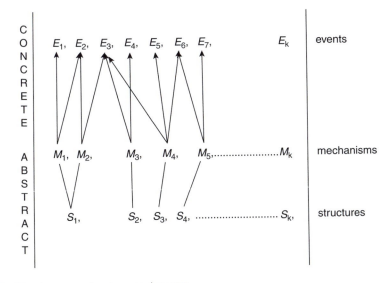

Figure 9.2 Structures, mechanisms and events

Source: Sayer (1992: 117)

research. On the other hand, concrete research can also be over-extended when the findings related to a particular event are the subject of overgeneralization or when the explanation of that event proceeds without reference to the broader sets of structures and processes that have brought it about (Cloke et al., 1991: 154). Sayer concludes that the key requirement with this type of explanation is to be able to conduct theoretically informed concrete research so that events on the ground can be properly linked via abstraction to the mechanisms and processes which produced them. In this section we have set out the central role that abstraction might play in this process.

Explanation and subjectivity

If we return to Figure 9.2 we can see that Sayer essentially stops the research process at the level of events. He does, however, contend that there may be a case for adding a fourth

level to the vertical dimension 'to cover meanings, experiences, beliefs and so forth' (1992: 116). After consideration he concludes that, since these can form structures, function as causes or be considered as events, they have already been taken into account in this explanatory schema. There are others, though, who would dispute this, and the recent history of geography would bear witness to the fact that many believe that events remain unexplained unless we are able to account for what people understand, mean or intend by their actions. In these cases explanation proceeds through an engagement with subjectivity.

Human geography's encounter with the subjective realm of meanings and experiences gathered pace in the 1970s. Like the Marxist-inspired work which underpinned explanation through abstraction, it was also fuelled by a dissatisfaction with positivist forms of explanation. In this case the critique focused on positivism's neglect of human agency and its tendency to reduce the complexity of human emotions and feelings to 'little more than dots on a map, statistics on a graph or numbers in an equation'

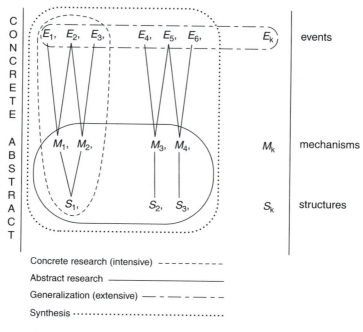

Concrete research (intensive) – – – – – – – –

Abstract research ————————

Generalization (extensive) — - — - — - —

Synthesis ·······································

Figure 9.3 Types of research

Source: Sayer (1992: 237)

(Cloke et al., 1991: 69). According to Entrikin (1976: 616), spatial science presented a view of behaviour which was 'overly objective, narrow, mechanistic and deterministic'. In contrast, and as a reaction, those geographers involved in this particular form of critique studied 'the aspects of [people] which are most distinctively "human": meaning, value, goals and purposes'. This concern with human awareness, consciousness and creativity was soon labelled as 'humanistic geography' (Tuan, 1976a).

In terms of explanation, those geographers working under the humanistic banner subscribe to Schutz's view that 'all scientific explanations of the social world can and, for certain purposes, *must* refer to the subjective meaning of the actions of human beings from which social reality originates' (cited in Gregory, D., 1978a: 125, emphasis in original). Thus the 'scientific' form of explanation derived from the natural sciences and followed by spatial science is seen as inadequate, not

only because human geographers examine a world which is meaningful to its inhabitants but, crucially, also because those meanings enter into the very constitution of that world (Barnes and Gregory, 1997: 356). In other words, the key to our understanding 'is to be found somewhere within ourselves' (Gregory, D., 1978a: 124). The task of humanistic geography is not to impose its own external frame of reference on to the world but to inquire into and understand the frames of reference drawn on by the social actors themselves. In this sense explanation becomes contextual as well as causal through a search for the meanings of particular actions (see Daniels, 1985). As Schutz puts it, the basic question becomes 'what does this social world mean for the observed actor within this world and what did he [*sic*] mean by his acting within it'? (cited in Ley, 1977: 505).

Under humanistic geography, then, the issue of explanation becomes heavily tied to an understanding of the meanings, perceptions and motives of social actors. As Schutz implies, however, there is actually a dual sense to the term explanation as it is used in humanistic geography – 'what does this mean *for* the actor' and 'what did he mean *by* acting' in this manner. In some cases, then, research is concerned with uncovering the antecedents to an action: with uncovering what was meant *by* an action or with ascertaining what meanings and motives lay behind an action. This is somewhat akin to explanation as causation, although the cause is not uncovered through abstraction but through exploring the meaning of the event for those involved. In other cases, however, the humanistic geographer will seek to explain the meaning of something *for* those involved – and in this sense the 'something' is not confined to a social action but could equally refer to a place perhaps, or a landscape, or a cultural event.

We will use some examples to show how particular research within the field of humanistic geography can be placed at different points along this continuum of explanation. At one end of the continuum, closer to the notion of explanation as description, we can situate the work of Edward Relph. In his pioneering book *Place and Placelessness*, Relph explicitly rejects a form of explanation based on models or on abstraction. As he states (1976: 6–7): 'My purpose…is to explore place as a phenomenon of the geography of the lived-world of our everyday experiences…I do not seek to…develop theories or models or abstractions…My concern is with the various ways in which places manifest themselves in our experiences or consciousness.' Hence Relph was concerned to demonstrate and describe a range of different reactions to, and relations with, place. He uses concepts such as 'rootedness', 'rootlessness', 'insideness' and 'outsideness' to chart these reactions, which he views as intensely subjective and personal. This is no easy task and, as Relph

points out, a human geography then based on a positivist form of explanation was ill-equipped to tackle it. In his words, 'any exploration of place as a phenomenon of direct experience cannot be undertaken in terms of formal geography…It must instead be concerned with the entire range of experiences through which we all know and make places' (1976: 6). As well as shifting geography's view of place – from 'a set of externally observable features found at a particular site' (Barnes and Gregory, 1997: 293) to subjective centres of experience and consciousness – Relph was also helping to redefine its attitude to the explanation of place. This shifted from 'scientifically' collecting and analysing a set of 'facts' about a particular place towards charting and describing people's experiences of different places and examining the meaning of those places for those people.

Although Relph's work concentrated on describing the meaning and experience of place, and on recovering the character of everyday geographical encounters, others have used experiential description as the first stage in more complex accounts of social action and behaviour. Such work tends towards a causal view of explanation, and one of the foremost practitioners of this type of humanistic geography is David Ley (see also Chapter 6). In his path-breaking monograph on life in Philadelphia's black inner-city, Ley set out to 'order and understand the world' around him (1974: 4). He, too, was concerned with experiences and meaning but only as a step on the road to understanding social activity. In his words, the researcher

has to discover the salient environment which prompts decision making and behaviour, the environment as perceived and experienced, and its forces which provide rules for human action…Only by coming to grips with the experiential environment,… only by exploration of the topographies of meaning, may we uncover regularities in behaviour. (1974: 4)

Ley explored these meanings, perceptions and experiences through fieldwork in inner-city Philadelphia. From the outset he also argued that such research would involve a different form of explanation from the positivism which was then dominant within spatial science. As he put it: 'To understand a decision-making world, a more holistic paradigm is required, one which…proceeds from the individual to the aggregate, and which recognises inductive understanding as no less appropriate than deductive prediction…To understand the environment as perceived… is an invitation to more subjective methodologies' (1974: 9). In this case, these subjective methodologies were built around a six-month period of participant observation (see Chapter 6) in which Ley 'walked the streets, attended meetings, belonged to a church' and was involved in joint activities with the local community which 'ranged from helping with homework, and a bake sale, to social evenings at the bowling alley, to informal "raps", and two out-of-town summer excursions' (1974: 17–18). Through these methods Ley hoped to build up a picture of the complex system of black inner-city life and of the role of the external environment in that system.

His findings were rich and detailed (see also Ley and Cybriwsky, 1974). They led to the proposition that space and environment both have a range of ascribed meanings and that, crucially, these meanings contain cues which prescribe appropriate forms of behaviour. In order to examine this, Ley looked at the behaviour of inner-city street gangs and at where residents felt safe and unsafe. Figures 9.4 and 9.5 show the gang territories, as defined by the location of graffiti and the pedestrian routes taken by residents. Ley found that residents differentiated their own neighbourhood from others primarily through behavioural rather than physical properties. Descriptive indicators such as housing quality, traffic density and land use were used to differentiate

one area from another less often than behavioural indicators such as gang activity, violence and quietness. He concluded that when people ascribed properties to particular spaces, prior to decision-making or to movement, the perceived meaning of an environment was more important than its purely physical characteristics.

For Ley, then, 'actions are intentional and purposive, they have meaning, but access to this meaning requires knowledge of the motives and perception of the actor, his [*sic*] definition of the situation' (1977: 505). Thus a full explanation of such action hinges around an understanding of the motives and perceptions of those involved – around an appreciation of their 'definition of the situation'. Taylor (1971) has drawn attention to a problem with such a procedure. He agrees that, ultimately, 'a good explanation is one which makes sense of the behaviour'. However, he points out that 'to appreciate a good explanation one has to agree on what makes good sense; what makes good sense is a function of one's readings; and these in turn are based on the kind of sense one understands' (cited in Gregory, D., 1978a: 145). This in turn means that it is difficult to draw a clear distinction between subjective understanding and subjective explanation, and that what we judge to be an adequate interpretation of social action will always be dependent on the constructs which are embedded in our own frame of reference: in how we make the translation between the first-order intentions of the actors involved and what we regard as adequate second-order representations of those actions (Gregory, D., 1978a: 145). This concern has always surrounded the more subjective forms of explanation. As Gregory points out, overcoming it would mean that geography would 'have to dismantle the oppositions between subject and object, actor and observer, and emphasise the mediations between different frames of reference' (1978a: 146). We next turn to a form of explanation

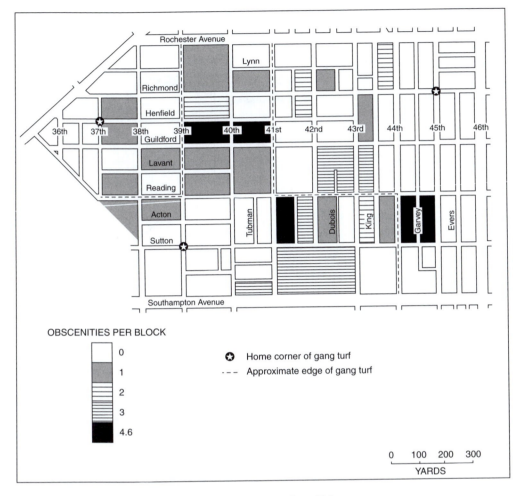

Figure 9.4 Gang territories defined by the location of graffiti

Source: Ley (1974: 218)

which has attempted to do just this, by explicitly acknowledging the enormous difficulties involved in adequately representing, or perhaps more accurately re-presenting, the actions and motives of others.

Explanation and practice

In contrast to those types of explanation which seek to uncover the causes of events, whether through abstraction or through uncovering the meanings and perceptions of those involved,

some geographers have turned to a mode of thought which explicitly rejects any search for an underlying causal process. This is known as 'non-representational' thinking for it is based on the premise that we can never hope 'to re-present some naturally present reality' (Thrift, 1996: 7). This in turn is because it is our practices, or performances, which constitute our sense of the real. Thus, reality is not sitting somewhere as a pre-given, waiting to be uncovered, but is constantly constructed through social practice and social activity. According to such a view events 'have an

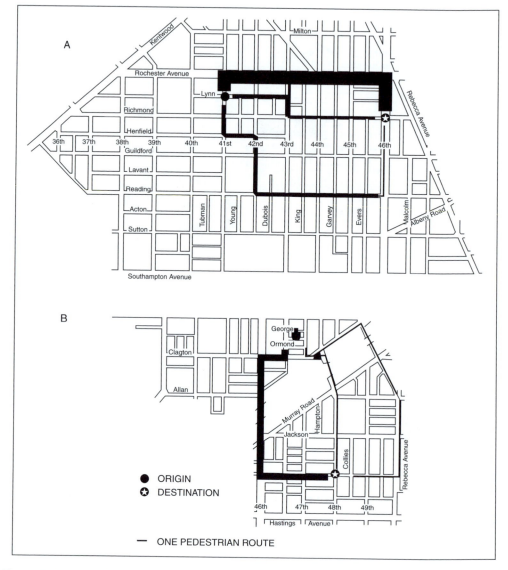

Figure 9.5 Pedestrian routes

Source: Ley (1974: 224)

imminent quality that means they are shaped as they are made' (Hetherington and Law, 2000: 127). This understanding of events inevitably leads to a 'rather different notion of explanation', one which prioritizes social practice and is more 'concerned with thought-in-action, with presentation rather than representation' (Thrift, 1996: 7). As such, explanation and

understanding are not 'about unearthing something of which we might previously have been ignorant, delving for deep principles or digging for rock-bottom, ultimate causes as [they are] about discovering the options people have as to how to live' (Thrift, 1996: 8). In summary, 'non-representational theory is an approach to understanding the world in terms of *effectivity* rather than representation; not the *what* but the *how*' (Thrift, 2000d: 216, emphasis in original).

One strand of non-representational thought which considers *how* social practices are formed, maintained and changed is actor-network theory. Sometimes known as the 'sociology of translation', actor-network theory was initially developed within the sociology of science (see Law, 1986; 1991; 1994) and then taken up within human geography during the 1990s.[1] A closer consideration of the theory, and the way it has been used in geography, will help to clarify the distinctive approach taken by non-representational theories towards the issue of explanation. Actor-network theory gains its name from the central tenet of its approach, which uses the notion of a network to consider how social agency is constructed and maintained. Unlike causal explanation, where entities such as the mode of production or classes are seen as the causes of events, actor-network theory views such structures, and the interests they serve, as a set of effects arising from a complex ordering of network relations. And unlike a subjective form of explanation (which prioritizes the intentional activity of human subjects), actor-network theory views agency as an effect distributed through a heterogeneous arrangement of actors and intermediaries, both human and non-human – such as texts, technical artefacts, money and machines. Thus actor-network theory understands events and effects as processes which are generated by the interaction of human and non-human agents operating through a collective network.

In order for the network to produce, or order, social effects it must be stable. Such stability is in turn achieved through 'translation': defined as 'The methods by which an actor enrols others' (Callon et al., cited in Woods, 1998: 323). According to Woods (1998: 323), these include 'defining and distributing roles, devising a strategy through which actors are rendered indispensable to others, and displacing other entities into a forced itinerary'. Through translation, then, an inter-related set of entities (both human and non-human) is enrolled into the network in a manner which standardizes and channels their behaviour in the directions desired by the enrolling actor (Murdoch, 1995: 747–8). Effectivity, or that which produces effects, springs from the interactions of the various entities involved in the network. Power is therefore associative and invested in relations and connections rather than in the entities themselves. What matters in producing change and promoting activity is the manner in which the entities are brought together. As a result the researcher should turn away from 'laws and regularities' and instead search for 'exchanges and interferences, connections and disconnections…In other words we need to look for different kinds of topologies, based on communication and connectedness' (Thrift, 2000d: 221–2).

Woods (1998) has used such an approach to research rural conflicts operating around land use and the right to hunt deer with hounds. He takes the example of the attempt made by Somerset County Council to ban such hunting, which itself must be viewed 'as part of a broader longer-running political contest in the county: between a conservative, largely agricultural, elite which has traditionally dominated the local power structure and which has close connections with hunting, and an emerging liberal "elite" drawn largely from the professional middle classes' (1998: 324). Woods uses concepts and methods drawn from the actor-network approach to examine the five key sets of entities enrolled by the anti-hunting campaigners in their actor network (see Figure 9.6). The continuation of stag-hunting is positioned as an obstacle to the realization of each of their goals, and thus the banning of hunting is identified as an 'obligatory passage point' which all the entities must support. In opposition, a pro-hunting network developed (see Figure 9.7)

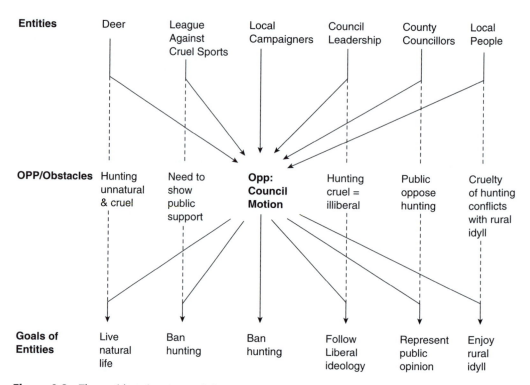

Figure 9.6   The anti-hunting network (OPP = obligatory passage point)
Source: Woods (1998: 329)

which consisted of a further set of human and non-human entities (some of which were redefined through their enrolment into the second network, such as the deer and the county councillors).

Woods goes on to chart the different elements involved in the enrolment and stabilization of each network, through four 'moments':

1 *Problematization*, where the goals and problems of each entity are defined by a potential translator in a manner which presents the translator's objective as the solution to the other entities' problems.
2 *Intressement*, whereby entities are attracted into the network by blocking other possible alignments.
3 *Enrolment*, where the roles of the actors in the network are clarified and distributed.

4 *Mobilization*, when the interests are enacted (1998: 323).

He shows how, despite an initial success in passing an anti-hunting motion, the anti-hunting network was eventually defeated when a second pro-hunting network emerged including two new entities – the judiciary and lawyers. This proved a stronger and more stable network than the first pro-hunting one, and the inclusion of legal entities proved crucial in effectively 'invalidating the definition of the county council as an entity with the power to ban hunting on its land' (1998: 334).

Woods's account is meticulously researched and presented in telling a set of stories which together trace the political and social conflicts operating in Somerset around the issue of hunting and land use between 1993 and

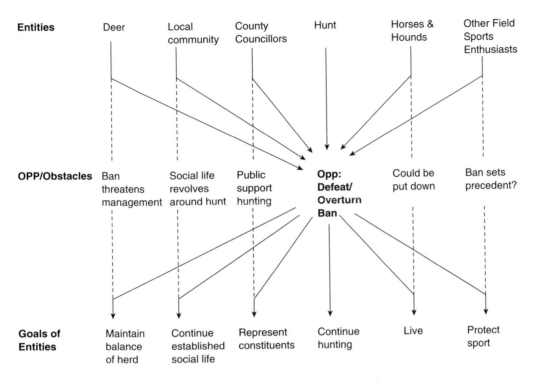

Figure 9.7 The pro-hunting network (OPP = obligatory passage point)

Source: Woods (1998: 332)

1995. That he chooses to set these stories and narratives in an actor–network framework is significant because in such an account the telling of the story and the presentation of the narrative become the explanation. As Murdoch puts it: 'We must explain by using the descriptions of network construction, and not by recourse to some underlying historical logic' (1995: 731). Thus, for the actor–network theorist there is an elision between description and explanation: the one literally becomes the other, and it is only by thoroughly describing the inter-relations operating within the network that the researcher can hope to explain their effects (Murdoch, 1995: 752). As Latour puts it:

> The description of socio-technical networks is often opposed to their explanation, which is supposed to come afterwards...Yet nothing proves that this kind of distinction is

necessary. If we display a socio-technical network we have no need to look for additional causes. The explanation emerges once the description is saturated. (1991: 129)

This image of explanation through saturated description is certainly beguiling. If we follow it, says Latour,

> we will never find ourselves forced to abandon the task of description to take up that of explanation...There is no need to go searching for mysterious...causes outside networks. If something is missing it is because the description is not complete. Period. Conversely, if one is capable of explaining effects of causes, it is because a stabilised network is already in place. (1991: 129–30)

According to Latour, then, when we have fully described our network we have at one and the same time written our explanation. But as Woods points out there are problems with this, and certainly with transferring such an approach from the sociology of science into human geography (1998: 338). This is because a key difficulty remains in knowing which entities should and should not be included as actors in the network – with drawing, in other words, the boundaries of the network to be described. As Woods puts it, 'there are thousands of entities that might be considered to have an influence upon the network. Where would we draw the limits or do we accept the network as infinite?' (1998: 337). To a certain extent, then, we are left with a similar problem faced by those forms of explanation which stress subjectivity: how do we faithfully translate and describe the thoughts, perceptions and meanings of others? This concern grows when we also consider that any description of an actor network, however complete, can tell only the agent's narrative, excluding alternative stories from those who are not enrolled in the network or those of other entities less able to communicate their own narratives.

Thrift (2000d: 215) has written that through the use of non-representational theory we can move away from positions 'which assume that a developmental account tells all, towards a "history of the present" (Foucault, 1986)'. As we have just noted, the problem still remains of which history of the present to write and from which (or whose) perspective? In an effort to overcome these twin concerns, those involved in non-representational work have retreated somewhat from the network metaphor and from the 'panopticism of actor-network theory…where the analyst looks down and counts the poor and huddled masses of scallops and machines' (Callon and Law, cited in Woods, 1998: 337). Instead, non-representational theory has prioritized a focus on 'practice' and on 'performance'. Recognizing that in some senses the notion of an actor network was itself representational, researchers have come to concentrate

on the event itself rather than on the network behind it. As Thrift points out this has significant implications for academic research, for 'what counts as knowledge must take on a radically different sense. It becomes something tentative, something which…is a practice and is a part of practice' (2000d: 222). Drawing on the work of Gibbons et al. (1994), he goes on to note that this implies that we are entering a third type of knowledge production (1997: 143; 2000d: 223). Type 1 systems of knowledge production were disciplinary and homogeneous and were rooted in the 'modern' search for certainty and fixity. Those 'geographical laws' produced by spatial science can be seen as the result of such systems of knowledge. Type 2 systems of knowledge production, by contrast, are trans-disciplinary and heterogeneous. They have been produced by a wider range of work which draws on less certain metaphors to do with 'complexity', 'transience', 'flexibility', 'adaptability' and 'turbulence'. Actor-network theory with its origins in the sociology of science and its stress on relations and connections can be said to represent this type of knowledge.

According to Thrift we are now entering a third type of knowledge production system – one which stresses the ways in which the world is conducted through activity. It is a type of knowledge which privileges 'the embodied non-cognitive activity in which people engage' (Thrift, 1997: 138), and one which shifts the interest towards issues of emotions, desires and imaginations. It is a realm which examines the host of complex 'real-time encounters through which we make the world and are made in turn' (1997: 138). Such a view would see the geographies around us as for ever imminent, as constantly in use and essentially incomplete, and as always in the process of becoming. If investigating this world requires a new type of knowledge, it certainly implies a new form of explanation. Or perhaps it implies a break with the notion of explanation altogether. Its stress on imminence and becoming rather than on certainty and completeness would lead us to question whether we can ever fully 'explain' anything. Perhaps those who would follow such inquiries are better placed to search for an understanding of events rather than for an explanation of them. It is to these considerations of geographical understanding that we turn in the next chapter. Before we do so, however, we want to provide a brief conclusion to our thoughts on explanation.

Concluding comments: from the explanation of geography to the geographies within explanation

We have covered much ground since we began this chapter, initially reviewing human geography's attempts to formulate and use laws as part of positivist forms of 'scientific' explanation, before moving on to examine various types of postpositivist explanation – whether through abstraction, subjectivity or non-representational theory. By way of a conclusion to this chapter on explanation in human geography we want to draw out one of the key implications the discipline itself has had to face as it has made this journey.

As we have noted throughout the chapter, as part of its move beyond spatial science human geography has necessarily engaged with a variety of social and cultural theories in an effort to establish new forms of explanation. This engagement has left its traces on both parties: as Dodgshon points out, 'It is crucial to any understanding of how social theory has colonised human geography to realise that the discovery, like all true discoveries of one culture by another, was mutual' (1998: 5). Gregory characterizes the process of

these mutual discoveries as a double movement in which

> a form of social theory was projected into human geography as a way of meeting a series of supposedly 'internal' concerns, and then, as a direct consequence of this conceptual mapping, the a-spatiality of the original formulation was itself called into question. It is those return movements that have widened the discourse of geography still further. (1994: 111)

Critically for our purposes in this chapter, this broadening of the geographical discourse through mutual discovery has had implications for the content, as well as the nature, of explanation. We have moved from a position where the prime object of human geographical research was to explain the geographies which were uncovered around us, to one where those geographies become part and parcel of the explanation of a range of other processes.

This shift was explored in an influential set of essays published in 1985 entitled *Social Relations and Spatial Structures,* jointly edited by a geographer (Derek Gregory) and a sociologist (John Urry). In their introductory essay, the two editors conclude that 'spatial structure is now seen not merely as an arena in which social life unfolds, but rather as a medium through which social relations are produced and reproduced' (Gregory and Urry, 1985: 3). It is this view of space as constitutive of social relations which has underpinned the dialogue between human geographers and those working elsewhere in the humanities and social sciences. As this dialogue has proceeded it has changed the notion of what geographers should explain, as well as the idea of how they should explain it. At times it is an encounter which has shifted our thoughts away from explanation and more towards understanding. Indeed, as we have seen, for some scholars the division between understanding and explanation is difficult to sustain and the two meld into each other. It is to a continuation of this particular aspect of the dialogue between geography and social theory that we now turn, in inquiring further into the nature of geographical understanding.

Note

1 See, for instance, Murdoch and Marsden (1995), Thrift (1997), Murdoch (1995; 1997; 1998) and Woods (1998).

10 Understanding

Introduction

So far in this part of the book we have traced a path from the initial stages of interpretation (sifting and sorting), through the so-called 'thin description' of enumeration, to the analytical mode of interpretation known as explanation. To oversimplify somewhat, each of these procedures can be seen as steps in a broadly *scientific* enterprise. Although the parallel is not exact, 'sifting and sorting' brings to mind the categorizing activities of the naturalists of the eighteenth and nineteenth centuries – the exercises in classification of a Linnaeus or a Darwin. Enumeration, with its emphasis on quantification, generalizable descriptions and statistical testing, sits squarely in the 'geography as science' corner. Similarly, explanation (whether of a positivist, critical rationalist, structuralist, Marxian or critical realist kind) usually involves a commitment to some idea of science – understood (at minimum) as a process of logical reasoning supported by empirical evidence. That said, throughout our earlier chapters we have tried to question some of the conventional scientific assumptions about what can be achieved when we sift and sort, enumerate and explain.

None the less, the scientific aspirations of much human geography are well known and reflect the two views of the subject that were dominant in the Anglophone literature until at least the mid-1970s. One view held that geography was a unified discipline and that it should be considered one of the natural sciences. As Derek Gregory (1978a: 19) points out, this (at least implicit) belief in the natural scientific status of the subject unites the supposedly 'new' geography of the 1960s with 'the ideals of its Victorian forebears'. Another view is that physical and human geography are, in certain important respects, epistemologically distinct but that human geography can still be seen as a science, albeit this time as a *social* rather than a *natural* science. Without wishing to gloss over the fundamental differences between these two views, it is interesting that they both share a commitment to the practice of science (even if they disagree about what that means).

In this chapter we want to move away from this scientific terrain to focus on some rather different approaches to geographical inquiry that have grown in popularity in Anglophone human geography over the last 30 years to the point where they have become the primary modes of interpretation in many parts of the discipline. These approaches, which we are grouping together under the heading of 'understanding', derive from the arts and humanities as much as from the social sciences. What they share is an interest in and concern for the *meanings* that attach to the human geographies of the world, and a belief that the production and communication of meaning are two of the defining attributes of the human condition and by extension of our *human* geographies. These approaches to interpretation are based in part on the conviction that enumeration and explanation can take us only so far; that there is a central element of what it means to be human that is left untouched and unexplored by scientific methods, whether of a natural-scientific or a social-scientific kind. Arguably, therefore, in our practices as human geographers we reach a moment in our attempts to make sense of the data before us where we have to do something that is not really central to, even if perhaps on occasion still conformable with, the practices of enumeration and explanation.

The point – so obvious in some ways and yet so lost on many researchers – is that, put a touch cryptically, *people are not rocks*. As far as we know, rocks on a slope or any natural objects studied by, say, physical geographers do not possess an 'interior' realm wherein they are aware, conscious of and making judgements about their surroundings. As far as we know, these rocks do not have the capacity to monitor their local environment nor to make decisions about how to act within that environment, choosing to do one thing rather than an other or to do nothing. People clearly do possess such an interior realm, however, and our suggestion is that this realm cannot but be *meaningful*, full of basic meanings charged with emotions regarding the immediate conditions in which people find themselves (ones of comfort, discomfort, love, hate, fear, disgust, anger, passions of all kinds), but probably also containing more sophisticated meanings carried in their more or less developed notions, opinions, ideas, ideologies and worldviews. Some of these meanings may be very much taken for granted, indeed barely even acknowledged, by the people involved; others of these meanings, conversely, may be constantly reflected upon and available to be discussed in detail and even recorded in written,

sung, acted or otherwise represented forms (see Chapters 3 and 4).

We realize, therefore, that we are sandwiching together many different aspects of the mental states possessed by human beings, not all of which many scholars would accept under the blanket term of 'meanings'. None the less, all that we wish to underline is that understanding this 'stuff', these interior realms shot through with meanings of one sort or another, almost unavoidably becomes a concern for human geographers wishing to take people seriously as people complete with these messy interior realms replete with meanings (see Cloke et al., 1991: ch 3). Thus, figuring out what spaces, places, environments and landscapes *mean* to people, or trying to understand the *meanings* that people in given situations acquire, elaborate, share and perhaps contest in relation to activities of geographical consequence (whether locating a factory, opposing the siting of a landfill site, choosing where to go on holiday, whatever): these are fundamentally different research practices of interpretation from those documented so far in our book.

Before turning to a discussion of what is involved in the practices associated with understanding meanings, it may be helpful to contrast the concerns of this chapter with those of the two previous ones in a little more detail. First, part of our focus here is on what the anthropologist Clifford Geertz (following Gilbert Ryle) calls 'thick description' (Geertz, 1973: 3–30). Indeed, in an early plan for our book this chapter was entitled 'Geographical thick description'. We will discuss Geertz's application of this idea in more detail below, but, briefly, the concept of thick description emphasizes the complex layers of meaning that can attach to what are often apparently simple social behaviours. This richness is typically only apparent to 'insiders' in a given social situation who are able to pick up on the variegated subtleties of meaning because of their intimate shared knowledge of the social conventions, language and informal codes being used for communication within a specific social group. In this approach the work of the anthropologist (or by extension the social or cultural geographer) involves becoming an insider (at least on a temporary basis) in order to learn the conventions through which the deeper meanings are conveyed. Understanding in the sense of 'thick description' can thus be contrasted with the 'thin description' that, we suggested in Chapter 8, was the realm of enumeration. This is not to decry enumeration but it is to recognize that the principal strength of enumerative methods is their ability to provide coverage in breadth rather than to develop the richly descriptive depth of the Geertzian approach. We should perhaps note in passing that the idea of social worlds having this meaningful depth is itself not without its critics, a point to which we will return later.

Secondly, we can contrast the idea of *understanding* with that of *explanation* elaborated in Chapter 9. The distinction between *verstehen* (understanding) and *erklären* (explanation) is most closely associated with two German thinkers: philosopher Wilhelm Dilthey (1833–1911) and sociologist Max Weber (1864–1920) (see Weber, 1962). If *erklären* equates with the practices noted in Chapter 9, *verstehen* is used 'to denote *understanding* from within by means of empathy, intuition, or imagination, as opposed to knowledge from without, by means of observation or calculation' (Burke, 1988: 894, emphasis added). The distinction can be loosely mapped on to a number of others, including the distinction between subject(ive) and object(ive), or that between 'lifeworld' and 'system'. Jürgen Habermas (1972) usefully distinguishes between *empirical-analytic* and *historical-hermeneutic* forms of knowledge, and he argues that the production of knowledge is never neutral but rather always linked to particular social interests. In broad terms, the empirical-analytic form of knowledge is concerned with an object world ('system') and is associated with the goal of attaining technical control over that world. By contrast, the historical-hermeneutic form of knowledge is concerned with a subject world ('lifeworld') and is associated with the goal of aiding mutual understanding between human subjects occupying this social world. For the purposes of our discussion of different modes of geographical inquiry, we can also see explanation as concerned with discovering what produces (leads to/causes/generates) a particular phenomenon, whereas understanding treats the phenomenon under investigation as itself a *source* of meanings (ideas/concepts/relationships) that need to be interpreted.

In this chapter we start our discussion of the practices of understanding in the specific sense of the interpretation of meanings, a tradition of inquiry also known in some circles as 'hermeneutics'. The word hermeneutics comes from the Greek work *hermeneus* ('an interpreter'). The term was originally used to refer to the interpretation of biblical texts but was brought into philosophy in the nineteenth century by Dilthey (who was also the first to use the ideas of *verstehen* and *erklären* in the senses discussed above). The original emphasis of hermeneutics as the interpretation of (particular kinds of) written texts is one that has been retained in geography, as in other disciplines which are concerned with written language, and as a result at various points in what follows we will couch our discussion in distinctly textual terms, thinking about 'authors', 'texts' and the 'reading of texts'. By this textual model, authors wish to convey meanings of various kinds through their texts, employing all manner of 'means' to do so spanning from the very words themselves to countless stylistic devices, and the meanings written into their texts are potentially available for readers to discern, more or less faithfully in line what the authors had intended, and perhaps then to inform the lives and actions of these readers. (Many now critique the simplistic assumptions about how meaning is produced and circulated that inheres in such a model, as we tackle below when talking about 'the deconstructionist'.)

Dilthey broadened the task of hermeneutics, however, so as to include the interpretation of the meaningful character of not just texts per se but of all manner of human actions, intentions and institutions. This activity, striving for a *verstehen* or understanding of the overall human world, was held to be the appropriate method of the *Geisteswissenschaften* ('sciences of the [human] spirit'), while the *Naturwissenschaften* ('sciences of nature') were left to tackle the more familiar goal of explanation.[1] More broadly conceived, therefore, hermeneutics has to do with the recovery of the meanings that are presumed to be present in all kinds of human situations, whether expressed in verbal language or not. This partly reflects the widespread (if not always unproblematic) extension of the concept of text to include a range of expressive human products and activities, such that (for example) paintings, films, landscapes and even social life in general come to be defined as texts for methodological purposes. Again, in our discussion below, such problems regarding this extension of the textual metaphor will never be far below the surface, although we fully support the claim that understanding must be about far more than just careful reading of actual texts and must entail an encounter with meanings proliferating anywhere and everywhere in the social world.

Seven modes of understanding

So far, then, we have established that 'understanding' involves the recovery of the meanings present (or presumed to be present) in written texts, human utterances and in other kinds of human artefacts and activities. Human geographers have adopted a range of different kinds of practices that relate to the idea of understanding. In this section of this chapter we duly consider seven different 'modes of understanding' by looking at the practices of seven different kinds of researchers or *figures*: 'the critic', 'the artisan', 'the ethnographer', 'the iconographer', 'the conversationalist', 'the therapist' and 'the deconstructionist'. Each of these figures treats the problem of understanding somewhat differently, adopting rather different techniques and approaches, and giving rise to rather different kinds of human geographical research and writing as a consequence. Taken together, though, they provide a repertoire of practices through which human geographers can engage with the meaningful qualities of human life in the social world. This being said, the modes of understanding indexed by these seven figures do overlap, are not entirely exclusive of one another and can be to some extent complementary.

What we might also say by way of preliminary clarification is that, broadly speaking, the first four figures all adopt a similar definition of meaning as something that inheres in the text/situation under study, and that the meanings involved can be extracted or recovered through the application of appropriate skills or expertise. The next two figures extend this approach to meaning, emphasizing the extent to which the specific meanings upon which the researcher works cannot but be generated in the context of interactions *between* the researcher and his or her research subjects (the people at the heart of the human geographical situation under study). The last figure adopts a more questioning attitude to the very possibility of meaning itself, however, and in so doing reflects some of the most challenging frontiers of contemporary human geography wherein much that has been routinely taken for granted by those attempting to understand the human realm is pulled out for critical reassessment.

The critic

Our first figure is 'the critic' or, as we are also tempted to term it, 'the man of letters'. We use the gendered term deliberately to emphasize the specifically masculine subject position involved. The archetype that we have in mind is that of a particular kind of literary critic, one perhaps personified above all by Sir Arthur Quiller-Couch (1863–1944). Quiller-Couch was a novelist, poet, essayist and literary critic and Professor of English Literature at the University of Cambridge. In his inaugural lecture at the university on 29 January 1913, Quiller-Couch set out the principles that he thought should guide his work as a critic. Foremost among these is the view that the purpose of studying any great work of literature is to discover the meaning *as intended by the author*:

> I put to you that in studying any work of genius we should begin by taking it *absolutely*; that is to say, with minds intent on discovering just what the author's mind intended; this being at once the obvious approach to its meaning ..., and the merest duty of politeness we owe to the great man addressing us. We should lay our minds open to what he wishes to tell, and if what he has to tell be noble and high and beautiful, we should surrender and let soak our minds in it. (Quiller-Couch, 1916, emphasis in original)

This emphasis on 'authorial intention' involves a number of assumptions. First, it assumes that the author has a clear, coherent and conscious intention. Secondly, it assumes that he or she is able to express it in words. Thirdly, it assumes that the meaning of the text is the same as the meaning intended by the author, and that the text is thus a neutral medium for the transmission of the author's message. This in turn takes it for granted that the meaning of the text is not influenced by the social and political context in which it is written, and nor is it influenced by the nature and limits of language itself (except in so far as these lie under the control of the author). As we shall see, all these assumptions have been challenged by more recent literary theorists, especially by those writing since the 1960s.

The other aspects of Quiller-Couch's approach revealed in the quotation above are the emphases on the 'greatness' or 'genius' of the author, an apparently ineffable and inexplicable quality in *his* insight, and on the supposed nobility and beauty of literary art. This faith in the essential goodness of literature is even clearer in a later passage in the same lecture:

> [B]y consent of all, Literature is a nurse of noble natures, and right reading makes a full man in a sense even better than [philosopher Francis] Bacon's; not replete, but complete rather, to the pattern for which Heaven designed him. In this conviction, in this hope, public spirited men endow Chairs in our Universities, sure that Literature is a good thing if only we can bring it to operate on young minds. (Quiller-Couch, 1916)

One of the difficulties with Quiller-Couch's approach is that, while the discovery of the author's intended meaning is seen as the primary goal of reading, it is very unclear how, in practical terms, such discovery is supposed to take place. It took another Cambridge critic, I.A. Richards (1893–1979), to set out an approach to 'practical criticism' that for the first time provided a clear methodology for the recovery of the meanings that were assumed to lie embedded in literary texts. Richards's method is set out in his book *Practical Criticism* which, since its publication in 1929, has been very widely used in English-speaking universities to teach the techniques of literary criticism. Richards advocated interpretation based on 'close reading' of texts. He argued that the reader should develop judgements about the literary merits of a work without being influenced by knowledge of the status (or even the identity) of the author, nor by concerns with the social and political context in which a text has been written.

A further development of close reading came in the work of another Cambridge literary critic, F.R. Leavis (1895–1978). Leavis (and the journal *Scrutiny*, which he founded) took a strongly moral approach to literature, combining Richards' practical criticism with a judgemental style that won him many enemies in the literary and academic worlds. According to Leavis the role of the literary critic was not only to interpret the meaning of a work but also to come to a critical evaluation of its artistic merit and moral worth. He was thus concerned with what he saw as the social and cultural benefits of reading great literature, and he had strong (and unorthodox) views about which literary works really merited the label 'great'. Leavis stressed values such as spontaneity, 'authenticity' in interpersonal relationships and a 'reverent openness before life', and he then judged literary texts according to whether or not they promoted such values. Moreover, Leavis thought that making such judgements in an effective way required the critic him- or herself to have these personal attributes, and hence to have a kind of cultivated sensibility through which great literature could be appreciated. His approach was thus unashamedly *elitist*, a far cry from later development in cultural studies stressing the importance of both popular culture and everyday ways of understanding.[2] It also focused on supposedly universal human values and, in that sense, can be seen as seeking to transcend the individual author and his or her intentions and to recover the more universal meanings revealed in particular ways in different texts. In geography the work of some early humanistic

geographers on geography and literature comes close to the approach of the critic or man of letters (see, for example, the contributions to Pocock, 1981a). Indeed, the interpretation of imaginative literature was central to the development of humanistic geography more generally. For Pocock (1981b: 10–11):

> It is the deliberately cultivated subjectivity of the writer which makes literature literature and not, say, reporting. It is the work of the heart as well as the head; it is emotion, often recollected later, perhaps 'in tranquillity', when an earlier stimulus is reworked and given expression, in the manner perhaps that the painter in his studio may develop his quickly pencilled field sketch … The truth of fiction is truth beyond mere facts. Fictive reality may transcend or contain more truth than the physical or everyday reality.

In Pocock's respect for the 'cultivated subjectivity of the writer' and his claim that fiction reveals a 'truth beyond mere facts' it is not difficult to see an echo of the approach of a Leavis or a Quiller-Couch.

The artisan

Like the man of letters the 'artisan', our second figure, works through a detailed and close reading of texts. In this case, however, the goal is to understand the world as it is revealed in the everyday experiences, encounters and utterances recorded in written texts such as interview transcripts, letters, diaries and notebooks (see Chapters 5 and 6). Whereas for the man of letters the purpose of close reading is to uncover what are presumed to be universal truths about the human condition inhering in the text, for the artisan close reading involves a detailed understanding of the points of view, subjective experiences and 'local knowledges' (see below) of specific individuals and groups located in particular contexts in time and space. For the artisan, therefore, the goal is not so much to transcend the author by recovering universal human meanings but to stand in the place of the author and to see the world through his or her eyes

We have chosen to use the notion of *artisanal* practice to highlight those techniques, skills and procedures that can, in principle, be acquired by any interested and conscientious researcher. In other words, they do not require the researcher to have some rather mysterious and especially refined sensibility as seems to be the case with the man of letters. For the artisan, while understanding undoubtedly requires hard work, it is a craft that can be learnt rather than a talent with which one is born (or not, as the case may be). Of course some practitioners will be better at it than others and, as with any craft, skills improve with practice and experience. That said, the basics are relatively straightforward and it is an approach that can reasonably be adopted in, for example, an undergraduate student's research project. This is not in any sense to disparage the artisan – on the contrary, we absolutely want to stress that this is a highly skilled activity but also that there is no reason why even the most inexperienced researcher should not begin to acquire and practise the skills involved. There are a number of different kinds of techniques that can be employed to gain the kind of empathetic understanding that the artisan seeks. Perhaps the best known and most widely applied (not least in human geography) is the approach to qualitative analysis developed by the American sociologist, Anselm Strauss (1916–96), and his erstwhile collaborator, Barney Glaser. Strauss called his approach 'grounded theory' to emphasize that a social researcher's concepts about the world ('theory') should always develop from and have their roots – and thus be 'grounded' – in the concepts voiced by the people whose lives and activities are being studied. Analysing

qualitative data, such as a set of interview transcripts, can seem like a very daunting task. Grounded theory provides a step-by-step procedure that both makes the task more feasible and aims to ensure that the conclusions drawn by the researcher are validly supported by empirical research evidence.

There are a number of different versions of grounded theory (for example, the approaches of Glaser and Strauss diverged after their early collaboration). However, at the heart of the process lies the activity of 'coding'. Coding begins with an initial careful reading of the text being analysed, which could be a newspaper article, an interview transcript or a focus group transcript. The next step is a version of 'sifting and sorting' the raw material of this text (see also Chapter 7), working through the text quite systematically, identifying sections of it (typically phrases, sentences or paragraphs) that are relevant to the research and tagging these with shorthand labels or 'codes'. For example, in response to a question about attitudes to the environment, an interviewee might say 'environmental conservation is very important to me'. This phrase might be labelled with the code 'conservation'. The same code can then be applied to other text segments that touch on the same issue. Another phrase about the risks associated with climate change might be given the code 'risk' or 'climate change' or both. When the whole text has been coded the researcher will have generated a list of codes and a set of associated text segments or quotations (Box 10.1 gives an example of a coded segment of an interview from the work of the geographer Meghan Cope). At this stage the artisan is already in a position to use the text in new ways. For example, it would now be possible to pull out all the quotations that share a particular code so that the key themes of the text can be identified. Alternatively, one could find all the quotations that have been coded with *both* 'risk' *and* 'climate change' to isolate an interviewee's comments about the risks of

---

Box 10.1: Coding an interview

Text	*In vivo* codes/ description	Analytic codes
Q: What do you expect to see happen as people hit [welfare] time limits or are cut off from programs?		
A: Yeah, the safety net thing. I see them coming into that. I suspect, and the big picture here if you look at the [welfare] population approximately a third of the individuals can probably get work on their own with a little bit of help, another third need some fairly intensive services, basically all the services that the employment and training and other human services community can provide them, and then I believe – and this is my own opinion – that there is probably another third or so that really are going to have a very very hard time of things after the five year limit. They may be in a situation where they're really not going to be able to work. There are some people who have serious problems if you want to call them problems, if you want to label them as such, that are going to prevent them from working.	safety net who can work who needs services who is unable to work	issue of 'creaming' – early success of those ready to work role of job training and support services 'hard to reach' population respondent is sensitive to 'labeling'

Source: Interview conducted in July 1997 by Meghan Cope with a staff member of a social service organization in Buffalo, NY

Source: Cope (2003: 453)

---

climate change from those concerned with other kinds of environmental risk.

Codes can be derived from terms used in the text (in the case of an interview transcript that will mean the actual words spoken by the interviewee). Such codes are called *in vivo* or 'emic' codes. In the example above 'conservation' is an emic code because it is a word

314

taken from the text itself. By contrast, 'etic' codes are words or phrases that do not appear in the text itself but are chosen by the researcher because they refer to his or her research questions or conceptual framework. Thus, if our hypothetical interviewee had spoken of being 'worried about global warming' without using the words 'risk' and 'climate change', 'risk' and 'climate change' would be etic, not emic, codes. It might be argued that the truly *grounded* grounded theorist should use only emic codes as a means of keeping his or her coding (and thus the subsequent development of his or her theories and concepts) as close as possible to the sense and spirit of the interviewees' worldviews. In practice, however, many commentators question whether the distinction between 'emic' and 'etic' codes can really be sustained. Although emic codes may adopt the words of the research subjects, it is still the analyst who selects which specific words are to be used out of the hundreds spoken in the course of a typical interview, and such a selection cannot but be influenced by the worldview, interests and existing knowledge and concepts of the researcher. Thus it is probably best to see coding as a constant movement between the concepts of the researcher and the interviewee, and it is only a short step from here to the 'dialogical' model of understanding embodied in the figure of the 'conversationalist' (see below). Moreover, coding usually entails an iterative approach in which an initial analysis provides the basis for a return to the field (see Chapter 6) for further research. This might involve further interviews with the same person or others, which then provide more material for interpretation.

Having undertaken this initial phase of coding, which is sometimes referred to as 'open coding', the next stage consists of identifying the links between the codes and categories and combining them through so-called 'axial codes' that express different kinds of relationships (such as X 'is an example of' Y, or X 'is caused by' Y, or X 'contradicts' Y). One way to present these relationships is in a diagram or 'code map'. Cook and Crang (1995) provide an example of a code map taken from Mike Crang's research, which is shown in Figure 10.1. The unpolished nature of the map (with hand-written labels and various deletions and additions) reveals its role as a tool of research in progress. The final stage of coding is 'selective coding' in which one or more core codes are identified to form the focus of the researcher's own written study, and it is when all other codes and subcodes are systematically related to these selective codes that one is said to have generated 'grounded theory'.

What grounded theory offers is a detailed and empirically rich interpretation of the social situation under investigation deeply anchored, so it is claimed, in the understandings of the participants as revealed in interviews or other accounts. It is artisanal in the sense that it provides a set of procedures that offer clear guidelines for practice, without being overly rigid or prescriptive. Understanding human geography through the strategies of textual coding is thus not mechanical and it does require the creative input of the researcher, but it is not mysterious. On the other hand, it can be somewhat laborious, which is why for many practitioners an obvious development of these techniques has been the use of computer software to automate some of the more routine aspects of the work (van Hoven, 2003). Software packages typically allow the uploading of texts into what is effectively a specialized database, and then provide a graphical interface to allow on-screen coding. Thus text segments can be selected using a mouse and codes assigned from a menu. One of the most useful features of such programs is their capacity for rapid searching, allowing quick retrieval of phrases or longer quotations based

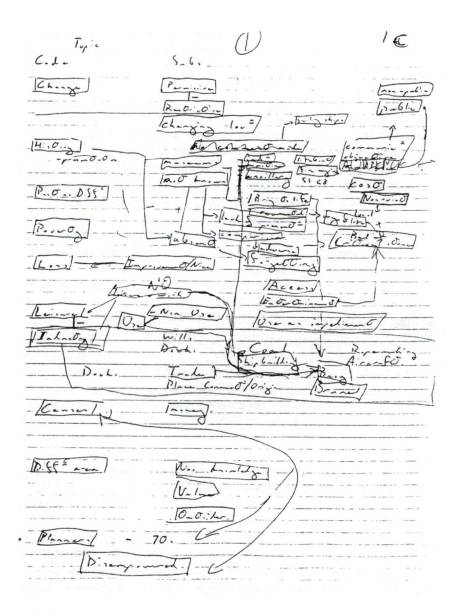

Figure 10.1 A 'code map'

Source: Cook and Crang (1995: 89)

on searches for individual codes or code combinations (such as the 'risk' and 'climate change' example mentioned above). Other functions might allow various forms of content analysis (working out how often certain terms or codes appear, for example), more effective memoing or appending of notes about a specific portion of interview text, and the organization of collaborative research by enabling remote access by several researchers to the same store of textual materials and the sharing of code lists and results.

Although now a little dated, the only comprehensive comparative survey of the various packages available (as well as advice about their use) is that provided by Weitzman and Miles (1995). More recently, Weitzman (2000) has outlined the main advantages and disadvantages of using computers in this way and has summarized the five main types of packages available:

1 *Text retrievers* that identify all the instances of a particular word or phrase in a text or texts.
2 *Textbase managers* that organize and store texts in a database and allow them to be sorted according to various criteria.
3 *Code-and-retrieve programs* that allow the researcher to apply codes to sections of text in a similar way to the manual process described above and then to search the text by code.
4 *Code-based theory builders* that have additional functions to support the development of theoretical propositions and hypotheses based on searching large amounts of coded text.
5 *Conceptual network builders* which facilitate the construction and display of networked visualizations of codes, rather like the code map in Figure 10.1.

While the use of computer software can undoubtedly assist with the frequently labour-intensive work of coding (and some of the additional functions, such as network displays of code relationships, are well worth exploring), the development of computer-aided qualitative analysis has been seen by some researchers as a rather mixed blessing. Crang et al. (1997), for example, while generally supportive of the use of these packages, counsel against fetishizing the technical capabilities of software tools over the substance of the analysis.

Inevitably the primary focus of this kind of analysis has been textualized information of various kinds and it works best in relation to fairly specific research questions in relation to clearly defined social settings. However, not all qualitative research is of this form. Researchers seeking a more holistic understanding of the complex flows of social practices that make up the 'way of life' of a whole community are more likely to adopt the participant observational methods discussed in Chapter 6. The record of such observational research is typically a field diary or field notebook. Of course, in principle a researcher could 'code up' his or her own notes or diary entries following the process outlined above. We suspect that in practice most field notebooks are not analysed in that way, though, but rather are treated as the first drafts of the 'thick descriptions' espoused by the anthropologist Clifford Geertz. It is to Geertz's approach that we now turn through a consideration of our third figure: the 'ethnographer'.

The ethnographer

The third figure we wish to consider is that of 'the ethnographer', although we should immediately acknowledge that the artisan and the ethnographer are very closely allied in terms of their approach to the domain of meaning. In practice, the two figures can end up appearing much the same in their close attention to the meanings circulating in everyday human situations, and a number of human geographers today would probably claim to be moving between the two possibilities (in deploying artisanal tools to achieve ethnographic effects). There are important differences, however, and tracing these helps to clarify different steps that can be taken on the way to the goal of understanding. While the artisan is arguably more inspired by the interpretative garb of modern sociological practice, taking on board the notions of grounded theory and the tools of coding, the ethnographer is more likely to trade on anthropological senses of the researcher's intuitive 'sensitivity' to what it is that events and phenomena end up meaning in the lives of particular people. While the artisan is more concerned with the meanings held by people regarding *specific issues*, such as their feelings about the closure of a school, the persecution of asylum-seekers or the problems of water supply, the ethnographer will usually be more concerned to reconstruct, as far as possible, the *overall worldview* – the broader sets of meanings about love, life, death, history, politics and so on – possessed by an identifiable human grouping occupying a definite place in the world.

Thus, whereas the artisan will examine data from documents, interviews and other sources for clues about a relatively restricted set of meanings, albeit often recognizing how such meanings are framed by more wide-ranging beliefs, the ethnographer will want to pull together as many threads of data as possible – often derived from conversational and practical interactions over long periods with particular people in a particular place – so as to discern something of the entangled patterns of meaning constitutive of shared and relatively all-encompassing worldviews. In the textual lexicon, it might be reiterated that the artisan is aiming to stand in the place of the author, hoping to triangulate the meanings held by a variety of individual authors relative to an event or phenomenon such as the opening of a superstore or the marketing of a city. Particular meanings here attach quite clearly to particular authors, and the approach is quite atomistic, with the artisan peppering his or her own account with the words of different individuals apparently voicing distinct and sometimes contradictory meanings (and the artisan is always on the look out for cleavages within the domain of meaning). Instead, while respecting the individual and sometimes repeating the words of named individuals, the ethnographer is happier with the goal of weaving together the stories told by many different authors – by many different folks in the setting under investigation – with the goal of encompassing these many authors and many stories in one (relatively) seamless narrative account of their shared worldview (their collective framework for making sense of the world around them, human and natural).

In this latter approach it is almost as if the whole span of everyday life within a situated time and place becomes regarded as a 'text', a rich layering of stories pregnant with meanings one on top of the other, and that the task of the ethnographer is to 'read' this text, to decipher its bundle of constituent meanings. This is then done less by identifying tensions between meanings than by registering deeper resolutions whereby they do ultimately cohere (if imperfectly) in some fashion. It is here, therefore, that we encounter one of the most influential routes into thinking about social life as text: one that has clearly shaped a host of developments within contemporary human geography (arguably more so than other routes into the social life as text metaphor).

Such an approach does still throw emphasis on the ethnographer as the privileged person able to perform the skilful act of deciphering the mesh of meanings, and in this regard there are overlaps with the role adopted by the man of letters. This being said, the ethnographer allows him- or herself to be thoroughly immersed in the everyday hurly-burly of peoples and places, something that the man of letters would abhor, and the ethnographer is in no doubt that it is only through being so immersed – by being a character in the text of social life, not divorced from it or transcendent of it – that he or she will be able to appreciate the 'local' constellation of meanings dictating why some events and phenomena are reckoned meaningful and others meaningless. At this point, then, we should recall that in Chapter 6 we have already discussed at length the ethnographer, exploring the techniques of immersion, of participant observation and field note-taking, through which he or she constructs a wealth of data available for subsequent interpretation. Putting aside the not insignificant point that the ethnographer cannot but commence the task of interpretation long before returning from the field – he or she will constantly be trying to make sense of the situation in which he or she is participating simply to get by, not to make mistakes, have one's 'cover blown' and the like – what we need to spell out now is greater detail about what the ethnographer's mode of interpretation entails.

As an example of a prime exponent of the mode of understanding we are tackling here, we will mention the well-known social anthropologist Clifford Geertz. It is actually Geertz who popularized the term 'thick description', borrowed from Ryle, which we have already utilized in this chapter and elsewhere in the book. In many respects this notion – accenting the need to achieve a real 'thickness', a richness or substance, in the 'description' (note, not explanation) provided – goes to the heart of how the ethnographer approaches the meaningfulness of the human realm. In a chapter arguing for 'an interpret[at]ive theory of culture', Geertz writes as follows:

> The concept of culture I espouse ... is essentially a semiotic one. Believing, with Max Weber [see above], that man [sic] is an animal suspended in webs of significance he himself has spun, I take culture to be those webs, and the analysis of it to be therefore not an experimental science in search of law [a standard explanatory approach: see above] but an interpretative one in search of meaning. (1973: 5)

The 'search for meaning' is hence predicated upon semiotic (see Box 10.2) assumptions about culture residing in 'webs of significance'. Such webs are spun from meanings attached to the occupants and goings-on in the surrounding world, the attributions of significance and insignificance, of meaningfulness or meaninglessness: and all these are

the ingredients of what Geertz wishes to unearth in his fieldwork, usually immersed in the everyday worlds – we sometimes say lifeworlds – of particular peoples in particular places (of 'others, guarding other sheep in other valleys' – Geertz, 1973: 30). He presents these said ingredients as the food of the ethnographer.

Geertz stresses that ethnography includes the tools of 'selecting informants, transcribing texts, taking geneaologies, mapping fields, keeping a diary, and so on' (1973: 6), but that it should never be reduced to such tools nor to the techniques of, for instance, coding up interview transcripts. This means that he distinguishes the ethnographer from the artisan in a manner akin to that rehearsed above:

> [Understanding], then, is sorting out the structures of signification – what Ryle calls established codes, a somewhat misleading expression, for it makes the enterprise sound too much like that of the cipher clerk when it is much more like that of the literary critic – and determining their social ground and import. (1973: 9)

The reference to the 'literary critic' repeats that alignment with the man of letters also mentioned above, and in this vein Geertz characterizes the apex of the ethnographer's art as being the accomplishing of 'thick descriptions' which provide a 'reading' of everyday lifeworlds as 'texts' full of meanings. As he expands:

> ethnography is thick description. What the ethnographer is in fact faced with ... is a multiplicity of complex conceptual structures, many of which are at once strange, irregular, and inexplicit, and which he [sic] must contrive somehow first to grasp and then to render ... Doing ethnography is like trying to read (in the full sense of 'construct a reading of') a manuscript – foreign, faded, full of ellipses, incoherencies, suspicious emendations, and tendentious commentaries, but written not in conventionalised graphs of sound but in transient examples of shaped behaviour. (1973: 9–10)

Box 10.2: Semiotics

Semiotics is the study or interpretation of signs and symbols. The founder of semiotics is usually regarded as Ferdinand de Saussure (1857–1913). Saussure argued that a sign consists of two elements: the 'signified' and the 'signifier'. The signified is a concept or an idea and the signifier is the form taken by the sign (for example, the word 'tree' is a signifier and the concept of a tree that the signifier points to is the signified; together they make up the sign for 'tree'). Note that the signified (the concept 'tree') is not the same as the physical specimen of a tree growing in the ground. Semioticians use the term 'referent' to indicate the physical object the sign relates to.

One of Saussure's key insights was that signs are arbitrary. That is, there is no automatic or inevitable connection between a particular signifier and a signified. Thus two people might use different signifiers (e.g. 'tree' and 'bush' to indicate the same idea and to refer to the same physical specimen). For people over the age of 45 the sign 'wicked' usually refers to things that are evil. By contrast when today's teenagers say 'wicked' they usually mean 'excellent'. This 'arbitrariness of the sign' means that the meanings of words and images are not transparent and self-evident: this is one reason why they require interpretation. Signs can convey meanings in two ways, through 'denotation' and 'connotation'. Thus the word 'green' *denotes* the colour that comes between yellow and blue in the rainbow spectrum, but it *connotes* a whole set of other ideas such as 'envy', 'nausea' (as in green face), 'springtime' (green shoots), 'environmentalism' (as in Green Party), 'naivety', 'Ireland' and so on. Here the interpretation of meaning is more like using a thesaurus than a dictionary. Authors have a lot of freedom to choose between many different possible words to describe or explain the same phenomena. The denoted meaning may be the same in each case but the connotations may be quite different and even contradictory. Moreover, no author can think through all the possible connotations of every word and phrase before making a selection, so texts are always open to multiple interpretations – many of which may not have been intended by the author.

Semiotics also stresses the relational character of meaning. The meaning of any sign is constituted through its differences from all the other signs in the system. We learn the meaning of 'green' by coming to understand how it differs from other colours. People with certain kinds of colour blindness cannot understand 'green' in the same way because they cannot see all the differences that give 'green' its meaning.

According to some forms of semiotics, the production of a text involves the 'encoding' of meaning in the form of a pattern of signs, while the interpretation of a text involves the 'decoding' of those signs. However, because of the arbitrariness of the sign and the influence of social context on processes of encoding and decoding, no interpretation corresponds perfectly to the meanings encoded during textual production. Communication thus involves the transformation of meaning rather than its simple transmission. When an interpretation is written down, meaning undergoes a further transformation, or 'recoding'.[3]

As it happens, Geertz clearly does attend to 'graphs of sound', to the words and stories uttered by his field informants and acquaintances. As a result, the metaphor of reading that which is *not* straightforwardly written yet which still clearly means so much – a text of conduct, movement, gestures, struggles, anger, laughter and tears, together with the spoken products of an oral culture – is one that has since come to pervade much of contemporary human geography. We provide some more flavour of Geertz's mode of understanding in Box 10.3, where we follow his encounter with the Balinese cockfight, but let

Box 10.3: Clifford Geertz and the Balinese cockfight

Early in April 1958 Geertz and his wife arrived in a Balinese village and were soon greeted to the unnerving spectacle of a large and illegal cockfight being held in the public square to raise money for a new school:

> Attempting to meld into the background [Geertz and his wife] watched the Balinese villagers watching the cocks fighting in the ring. All of a sudden, the local police arrived, armed with machine guns and looking menacing, and immediately the gathering dispersed: People raced down the road, disappeared head-first over walls, scrambled under platforms, folded themselves behind wicker screens, scuttled up coconut trees … Everything was dust and panic. (1973: 414–15)

The Geertzs ran too, ending up tumbling into the courtyard of another fugitive, whereupon 'his wife, who had apparently been through this sort of thing before, whipped out a table, a tablecloth, three chairs and three cups of tea, and we all, without any explicit communication whatsoever, sat down, commenced to sip tea and sought to compose ourselves' (1973: 415).

One of the policemen marched into the yard, to meet fierce rebuttals from the Geertzs' unexpected host, who declared that all three of them had been there all the time drinking tea and discussing matters of culture. The next day, news of this event had travelled throughout the village, rendering the Geertzs an interesting novelty for teasing – locals mimicked their awkward running style – but also leading them to be accepted and then allowed into the everyday rituals of village life in a manner that might not otherwise have been the case. Moreover, as Geertz elaborates:

> It [this event] gave me the kind of immediate, inside-view grasp of an aspect of 'peasant mentality' that anthropologists not fortunate enough to flee headlong with their subjects from armed authorities normally do not get. And, perhaps most important of all, … it put me very quickly on to a combination of emotional explosion, status war, and philosophical drama of central significance to the society whose inner nature I desired to understand. By the time I left [Bali] I had spent about as much time looking into cockfights as into witchcraft, irrigation, caste or marriage. (1973: 417)

As a consequence, therefore, Geertz devoted much energy to reconstructing the meanings that the Balinese villagers derived from and possessed about the cockfights, partly by hearing their own countless stories about such cockfights in the past, but also simply by attending and carefully observing similar events in the present. As well as noting the double entendre of 'cocks and men' which is obvious in English but also present in the Balinese village worldview, he arrived at the following generalizations about the place of the cockfight in what might be termed 'local knowledge' (Geertz, 1983) or even the local 'economy' of meaning:

> What sets the cockfight apart from the ordinary course of life, lifts it from the realm of everyday practical affairs, and surrounds it with an aura of enlarged importance is not … that it reinforces status discriminations (such reinforcement is hardly necessary in a society where every act proclaims them), but that it provides a metasocial commentary upon the whole matter of assorted human beings into fixed hierarchical ranks and then organising the major part of collective existence around that assortment. Its function, if you want to call it that, is interpret[at]ive: it is a Balinese reading of Balinese experience, a story they tell themselves about themselves. (1973: 448)

Box 10.3 (Continued)

The reference to the cockfight as a story is, of course, telling in that it plays upon the idea of social life as text, although it is interesting that Geertz suggests that it is not just him – the clever anthropologist – reading this story, it is in effect a story that the Balinese villagers are telling themselves to affirm their own identity. We are certainly in a sophisticated hermeneutical realm here, one where meanings are being produced not only by words but also by actions, and one where meanings are circulating in various directions (between the villagers themselves and between them and the visiting professor of anthropology). By examining in microscopic detail the cockfights and by seeking answers to simple questions – who owns a fighting cock? in what order do owners fight their cocks? who is on-looking and who is gambling? – Geertz is indeed able to advance a 'reading' of the cockfight that can stand as a prime exemplar of the ethnographer's mode of understanding. We might also add that attention to the geographies of the cockfight, to its staging in public space and to the bodily spaces of who is positioned where during the event within the square, can further enrich the account that is being given.

us now briefly trace the movement of his ideas into the practising of (certain strains of) human geography.

Following James Duncan's (1980) criticism of an older way of thinking about culture in human geography, one associated with Carl Sauer (see Chapter 1) wherein Culture appears as this undifferentiated 'super-organic' entity somehow existing apart from the transactions of everyday life, the search for alternative conceptualizations of culture almost inevitably took some geographers towards Geertz's take on culture as a web of meanings in constant affirmation and change. As Peter Jackson and Susan Smith (1984: 39) put it, '[f]ollowing Duncan's (1980) critique …, an interesting debate arose in which Geertz's interpretative anthropology was alluded to as an attractive alternative to "superorganic" definitions'. Miles Richardson (1981: 285) duly follows Geertz in supposing that culture resides in how people *talk* to each other, thereby sharing with one another 'in some incomplete way' their 'feelings about the past', their 'thinking about the present' and their 'plans for the future'. Talk is not narrowly conceived by Richardson, however, for he supposes it to include not only spoken words

but also written words, bodily gestures, physical activities, mime, dance, ritual and the manipulation of material objects. Through such expanded talk, people build up shared meanings about the world around them and sometimes beyond, and Richardson underlines that talk is indeed intersubjective – shared between separate human subjects – and thereby social, as well as acquiring a measure of 'tradition and permanence' sedimenting into a recognizable if slowly changeable 'holistic pattern'. He also speculates that much everyday talk tends to be about material objects (artefacts from hoes to houses or trainers to televisions) which assume a 'symbolic' quality in the sense of having pinned to them a cluster of meanings with a 'capacity' for impelling reactions from people who encounter the said objects. In addition, as a geographer, he proposes that a given group of people in a given part of the world, interacting with each other and the material objects within their environs, cannot but formulate a shared 'image' of their place:

Th[is] image is not the underlying scheme talked about by students of mental maps [people's ability to represent, say, the layout of a local neighbourhood] as being antecedent to acts, but rather it is an explanation, an interpretation arising from our acts. The image may be explicit and detailed or barely formed in our conscious, but in either case we 'extended it out' or 'hang it on' the material setting so as to give a sense of place to that place. (1981: 286)

Our claim would be that, while not specifically following Richardson's lead, much that has since appeared in the guise of a 'new' cultural geography has effectively taken on board such a Geertzian semiotic line on culture. It has become assumed that the everyday production, circulation and consumption of meanings can, in principle at least, be 'read' by the careful human geography researcher.

Such a sense of 'reading' meanings is doubtless present in the work of the artisans, as discussed earlier, even if the ambition of reconstructing overall worldviews or 'local knowledges' has featured less than the recovery of meanings swirling around specific issues. None the less, some geographers – notably ones working on the boundaries between geography and anthropology such as Donald Moore (1997) or Tad Mutersbaugh (1998), and most likely researching peoples and places beyond the West – have sought to understand the human realm in a fashion akin, albeit not always explicitly referencing, Geertz's writing of 'thick descriptions' that embrace the rich totality of local cultural meanings. Moore's patient inquiries into the state-administered Kaerezi Resettlement Scheme in Zimbabwe's Eastern Highlands, fringing the border with Mozambique, might be an example. When he reflects upon the 'livelihood struggles' of Angela, a woman in her late 50s, and her neighbours, running up against the demands of patriarchal, chiefly and state power, a host of meanings come into play that evidence the extent to which '*place matters*; it has a politics and is produced through myriad and symbolic struggles' (Moore, 1997: 103, emphasis in original).

Another example might be Mutersbaugh's similarly patient inquiries into the Oaxacan (Mexican) indigenous community of Santa Cruz, where he lived for a year with his spouse undertaking intensive ethnographic work full of interviews, conversations and participant observation to establish the gendered dimensions of labour within the overall village order. 'Labour–organisation discourse encompasses both techniques of power (sanctions, surveillance) *qua* material practices and negotiation of meanings of labour', writes Mutersbaugh (1998: 440), all of which depend upon 'socially constructed sites – twilit front porches, smoky kitchens, the village assembly hall – where villagers converse and map out present and future projects'. It is true that both Moore and Mutersbaugh are interested in wider considerations to do with power and resistance or gender, labour and technology, and are thereby asking particular questions of their study areas, but in the process they are clearly 'reading' the 'texts' of local meanings. We would suggest, moreover, that in order to answer their questions they are indeed striving to 'read' the whole book of locally interwoven cultural meanings, not just a few scattered pages. As such, they are very much, in our terms here, Geertzian ethnographers in their mode of understanding human geographies.

The iconographer

The fourth figure we wish to consider is that of 'the iconographer', and here we move to considering a mode of understanding dealing not with words per se, whether written or spoken, but instead with images. As we stressed in Chapters 3 and 4, visual images in photographs, paintings, cartoons, films, television programmes and advertisements, to name but a few possibilities, can all be sources of geographical data. Once a researcher has decided that he or she will draw upon images, it becomes important to establish just how these images are to be interpreted so as to reveal matters of interest about the particular human situations under study. The researcher might elect to treat images in a formal manner, perhaps classifying and counting up particular sorts of images appearing in, say, a run of newspapers as a prelude to reaching conclusions about what meanings the images are supposed to convey. A very specific example might be finding lots of images in a newspaper showing supposedly homeless vendors of *The Big Issue* magazine apparently wearing expensive trainers, leading the researcher to suppose that the newspaper in question wishes to convey a message about such vendors not necessarily being homeless and poor but really 'scamming' it. As this indicates, however, it is hard to avoid the conclusion that most (if not all) attempts to understand images are ultimately about teasing out the meanings present within images, whether as purposefully 'coded' into them by their producers or as 'derived' from them by their consumers (and there may be many different constituencies of consumer for the same images, each of whom may derive different meanings from the images).

It would, therefore, not be inappropriate to think in terms of meanings being both 'written into' and 'read from' images, thereby extending the textual metaphor to the work the human geographer might do in relation to images. It is easy to find instances of geographers explicitly talking about 'reading' images as 'texts', one instance being Stuart Aitken's (1997) reference to 'media texts' and, more specifically, 'film texts' where there is normally a confluence of the image (moving visuals) and the written (the script) to generate an overall text. Tim Cresswell offers an example when he 'reads' the film *Falling Down*, extracting the possible meanings about power, domination and resistance wrapped up in the story of how the mainstream white defence industry worker, D-Fens, turns into

the hero/anti-hero who, gun in hand, 'questions urban constructions of space, time and money in turn as he wanders across the city' (Cresswell, 2000: 256). The meanings considered are both those of the film-makers, who wished to emphasize that alienation in the USA from the forces of both capital and the city is not solely a province of African-Americans, and different audiences, such as the liberal critics who saw it as a worrying expression of white male angst directed violently against women and people of colour. Cresswell does not lay out any explicit methodology guiding his account of *Falling Down* – it might actually be regarded as tending towards being deconstructionist (see below) in its focus upon faultlines of meaning in the film text – and it would have to be acknowledged that it is only recently that 'visual methodologies' (Rose, 2001) have shifted on to the agenda of human geography in a major way.

There is a significant exception to this rule, though, in that a handful of geographers concerned with landscape have for some time been influenced by an approach known as 'iconography'; and hence our naming of the figure of the iconographer to cover this mode of understanding meanings in images. We will return to landscape presently but, for the moment, let us stick with iconography, which began life (and here the link with the history of hermeneutics is instructive) as the 'interpretation of symbolic imagery' (Daniels and Cosgrove, 1988: 1) derived from the repertoire of the Classical world and adapted within the medieval Christian tradition. The project of iconography was revived in the early twentieth century by the school of art history under Aby Warburg, and a summary of what it came to entail runs as follows:

> In opposition to the purely formalistic tradition of art interpretation associated with Heinrich Wölfflin (which analysed pictures purely in terms of the surface patterns of colour, chiaroscuro, line and volume, relating

them principally to other works of art), iconographic study sought to probe meaning in a work of art by setting it in its historical context and, in particular, to analyse the ideas implicated in its imagery. While, by definition, all art history translates the visual into the verbal, the iconographic approach consciously sought to conceptualise pictures as encoded texts to be deciphered by those cognisant of the culture as a whole in which they were produced. The approach was systematically formulated by Warburg's pupil, Erwin Panofsky. (Daniels and Cosgrove, 1988: 2)

There are hints here of a distinction akin to that between the practices of sifting, sorting and enumerating, possibly together with the more formal 'coding' exercises of the artisan (see above), and the practice of understanding – and most specifically that as pursued by the Geertzian ethnographer – which calls for taking seriously the broad sweep of meanings constitutive of a given 'culture as a whole'. For Panofsky, iconography (he actually favoured the term 'iconology' in this respect) aimed to reveal the 'intrinsic meaning' of artworks as linked into the 'attitudes' of a given period, national, religious or even class consciousness. To some extent, the successful achievement of this task would depend upon the skills of the iconographer since, as he acknowledged:

> There were no established conventions or specific methods that would ascertain these principles; they were to be reconstructed by a kind of detective synthesis, searching out analogies between disparate forms like poetry, philosophy, social institutions and political life: 'To grasp these principles', wrote Panofsky, 'we need a mental faculty comparable to that of a diagnostician'. (Daniels and Cosgrove, 1988: 2)

The iconographer as the one who diagnoses the complex cross-cuttings of meanings within images, tracing from them to the play of poetic, philosophical, social and political forces in the time and place of their production:

these are hence the traits requisite of this mode of understanding. Once again, and as with both the man of letters and the ethnographer, it is possible to go only so far in itemizing the particular practices constitutive of successful work in this mode; it is possible to go only so far in learning the skills apparently needed by the iconographer since, ultimately, a measure of individual creative response cannot be denied.

There are other traditions of image analysis, including the Marxist-inspired approach of someone like John Berger (1972), which recognizes that different 'ways of seeing' (of composing and responding to images) are always bound up with different forms of economic, social and political organization. In particular, the likes of Berger suppose that modern capitalism has an 'interest' in deploying images both to 'mystify' the masses (to obscure from them the real conditions of their existence) and to 'sell' things to them (to prompt them into the consumption of commodities and experiences). This is not the place to debate such traditions, however, and instead we wish to underline, with Stephen Daniels and Denis Cosgrove (1988: 3), that 'Panofsky applied this approach of "reading what we see" to built as well as to painted forms'. In other words, he opened up the possibility for an iconography of buildings, townscapes and, by extension, landscapes of all kinds. A landscape can be the actual visual scene before the researcher's eye – the image here being what the eye latches on to when beholding the scene – or the *re*presentation of a given landscape in a photograph, drawing or painting. While entirely natural landscapes in the world – if any remain – cannot be taken as *deliberately* encoded with meanings, except perhaps by God, most worldly landscapes bearing human imprints and all represented landscapes are clearly open to an interpretation of their meaningful contents as both inscribed within and extractable from them.

In short, such landscapes are amenable to the interpretative practices of a Panofsky, a Berger, a feminist critic (Rose, 2001) or equivalent, leaving the possibility of a human geographical treatment of landscapes embedded within (and influencing) a more general human geographical treatment of images. In Chapter 4 we provided a familiar example of critiquing the meanings held within and prompted by a landscape image: that of Thomas Gainborough's late seventeenth-century painting of Mr and Mrs Andrews seated before the scene of their country estate.

Cosgrove and Daniels's *The Iconography of Landscape* (1988) collection paved the way in this regard, building upon past achievements by Cosgrove (1984) when relating 'symbolic landscapes' to 'social formations' (or, putting things more cryptically, connecting spaces of meaning to places of material life). This book also anticipated future accomplishments by both Cosgrove (1993), who 'reads' real and imagined 'Palladian' landscapes, and Daniels (1993), who 'reads' the cultivated, painted, cartooned and photographed landscapes integral to English national identity. A more recent collection edited by Cosgrove (1999) extends these principles to the 'reading' of maps from a host of different periods and places, showing how such maps, as very particular kinds of images, have often been as much vectors of cultural meaning as expressions of factual scientific (geographical) knowledge. One chapter here (by Paul Carter) focuses on both maps and other images, including an engraving by W. Hamilton called 'The View of the Island of Ouby from Freshwater Bay on Batchian' taken from Thomas Forrest's *A Voyage to New Guinea and the Moluccas* (1779) (see Figure 10.2). As Carter discusses:

> Th[is] remarkable engraving ... may pose as an ethnographically inflected picturesque view – in reality it assimilates the coast to the exhibition diorama. Type specimens of a multiplicity of exotic objects present themselves to the scientific gaze, washed-up, exposed, already detached from their environmental matrix, lost it seems unless they can be classified and correctly arranged. The native collectors, handling physically what the English savant examines mentally, pleasantly flatter the latter's sense of intellectual superiority. (1999: 133)

This image is hence subjected to a critical interrogation – again, it teeters on being a deconstruction (see below) – exposing the meanings that the white European (English) educated viewer might bring to bear, ones informed by an embryonic 'scientific racism' and as bound into a philosophical appreciation of how the products (or 'trophies') of exploration get lifted from distant shores ready for cataloguing at leisure in the scholar's museum. Carter as iconographer thereby signals the complexity of meanings that can indeed be 'read' from such an image, one that at first might seem quite straightforward in appearance. It might therefore be argued that what he provides (along with Cosgrove, Daniels and others) is akin to a Geertzian 'thick description', here of an image (or a selection of images) rather than of a particular people in a particular place.

The conversationalist

We now move to two rather different figures for whom the act of understanding draws away from the assumption that the human world comprises texts and situations somehow 'full of' meanings that are more or less *straightforwardly* available for interpretation by the researcher. Actually, in practice, all the previous four figures, perhaps excepting the critic, have run up against problems with this assumption. Indeed, whenever we have mentioned researchers being aware of their *own* role in deciding what are the meanings present in a text or situation, as when the artisan

Figure 10.2 'The View of the Island of Ouby from Freshwater Bay on Batchian' by W. Hamilton
Source: Carter (1999)

debates 'etic' coding, we have anticipated the figures that we will now call 'the conversationalist' and 'the therapist'.

For reasons to be explained shortly, the figure of the 'conversationalist' is meant to characterize researchers who are fully aware that the meanings of which they speak – the meanings that become the basis for their own studies in 'interpretative human geography' (Smith, D.M., 1988b) – cannot but be generated in the context of a *two*-way 'relationship' between researcher and research subjects. Revisiting the textual metaphor, the key recognition is that the reader cannot simply 'read' out the meanings intended by the 'author' since, in practice, what really occurs here is that meanings are produced precisely through the encounter *between* reader and author. Readers cannot avoid bringing to the act of reading what they find meaningful from their own vantage point, their own 'place' in the world, and, as such, there is no possibility of them ever being that neutral conduit of an

author's original meanings hoped for by the man of letters. Readers respond to an author's intentions, picking up on some but probably not all the meanings the latter has striven to imprint on the pages, and responses from readers may span the whole gamut of empathy, sympathy, indifference, hostility or even flying off on an wildly imaginative curve spurred by (yet in no sense contained by) the text before them. Seen in this light, meaning becomes a messy, fragile and multiple thing, devoid of the stability that a more scientific, explanatory take on research would ideally require. Yet we think it vital and appropriate to identify a mode of understanding that proceeds from a starting point wherein the relationship between the researcher and his or her research subjects – whether 'authors' of documents, utterers in interviews, actors in real events and places or whatever – is directly akin to that of the reader responding to an author.

We can now return to the 'conversationalist', a figure partially inspired by the participant in something like the staged *conversaziones* of 'polite' society, someone happily exchanging with others not just chit-chat but more developed opinions and thoughts on a variety of weighty matters. An important feature of such conversations is that, in principle at least, all participants are supposed to bring to the 'table' their own informed viewpoints – respect and 'space' duly being accorded to all the viewpoints present and expressed. This model has a number of significant theoretical off-shoots, whether in Richard Rorty's (e.g. 1979) pragmatist call for intellectuals to 'keep the conversation going' (not to cut off possible positions and arguments) or in Jürgen Habermas's (e.g. 1984) 'theory of communicative action' influenced by the image of a conversation-rich eighteenth-century 'public sphere'. For us, though, it simply introduces the sense of the researcher who 'converses' with his or her research subjects, the people at the heart of a study being conducted, whether literally (as an interviewer or a participant observer) or imaginatively (when reading a written document or perhaps a transcript). More than this, though, for what such a researcher also appreciates is the extent to which it is during 'conversations' that the very meanings integral to his or her research are actively produced. These meanings are hence neither solely the 'property' of the researcher, as it were, nor of the research subjects. Rather, both put their own viewpoints into the conversation so that the meanings which emerge reflect both parties and in effect are hybrid positions – summations of the 'discussion', resolutions of any 'debate' – irreducible to either or any of those initially taken by participants in the conversation.

The figure of the conversationalist is central to thinking within the realm of hermeneutics, as introduced above, and most particularly in the writing of Hans-Georg Gadamer, a German philosopher. One commentator summarizes Gadamer's insistence

that 'we' as scholars cannot avoid 'making sense' of whatever we study through the meaning-endowing lenses of *who we are, when and where we are living* and *what we have witnessed or learned about*:

> We never come upon situations, issues or facts without already placing them within some context, connecting them to some other situations, issues or facts and, in short, interpreting them in one way or another. The parameters of these interpretations, moreover, derive from our own circumstances and experience, and these circumstances and experience are always already informed by the history of the society and culture to which we belong. (Warnke, 1987: 169)

Gadamer usefully underlines that the meanings with which we make sense of the world are never entirely personal but rather have roots in, if not being determined by, many dimensions of the historically and geographically specific 'society and culture' where we live. He nevertheless believes that this should not render us closed to what others different from ourselves, perhaps from other times and places, can tell us about themselves and their worlds; and it is revealing that here he looks to 'the structure of dialogue' (Warnke, 1987: 169) for guidance on what might be possible. 'As he depicts genuine conversation,' continues Georgia Warnke in her exegesis, 'it is rather characteristic of it that all participants are led beyond their initial positions towards a consensus that is more differentiated and articulated than the separate views with which the conversation-partners began' (1987: 169). Warnke admits that the notion of 'consensus' is troublesome in Gadamer's work but concludes that his most fruitful account runs as follows:

> [C]onsensus refers simply to a 'fusion of horizons', an integration of differing perspectives in a deeper understanding of the matters in question. According to this second idea of dialogic consensus, one is required to take account of the positions of others in discussing an issue or subject-matter

with them. Here, even if one holds to one's initial point of view, one has nonetheless to deal with the objections, considerations and counter-examples that others introduce. In the end, whether one changes one's position or maintains it, the view that results is more developed than the one with which one began, and the same holds for the views of all participants. Whether they conclude by agreeing or disagreeing in a substantive sense, their positions are now informed by all the other positions. They are able to see the worth of different considerations, incorporate different examples and defend themselves against different criticisms. In this way their views acquire a greater warrant; they are less ... one-sided. (1987: 169–70)

It is worth quoting this lengthy passage, partly because it so neatly builds upon the figure of the conversationalist but also because it strikes us as a highly apt description of a mode of understanding that is now pivotal – if not quite expressed like this – to how a great many human geographers proceed when interpreting data from their encounter with research subjects.

For Gadamer, 'hermeneutics is a form of justification involving the dialogic adjudication of both beliefs and standards of rationality' (Warnke, 1987: 170). Such an interest in establishing the parameters of rationality is not that relevant to human geographers, but the model of conversation, dialogue and mediating between 'horizons' of meaning has clearly excited several. Anne Buttimer talks of 'dialogue' in her famous 1976 paper, for instance, proposing that each 'participant in a dialogue needs to become aware of his [sic] own stance, and the stance assumed by the other, so that language for dialogue could emerge, ie. be jointly created, or at least jointly accepted, by both participants' (p. 278). On one level she is asking for a dialogue between geography and phenomenology (a formal philosophy) but, on another, she wishes to effect a dialogue between the 'world horizons' (1976: 281) – recall the reference above

to achieving a 'fusion of horizons' – of the academic 'stranger' exploring the senses of 'social space' integral to a 'foreign culture':

> The knowledge acquired in one's own society is inadequate; one has to question the former 'givens' of social life and search for common denominators for dialogue with the other. To gain a foothold, or basis for dialogue, one needs to grasp the inner subjective meanings common to that other group, its sociocultural heritage, and its 'stream of consciousness'. (1976: 285–6)

Much the same construction, this time explicitly borrowing from the hermeneutics of Gadamer, shapes Derek Gregory's remark that 'interpretation is not a process of somehow *overcoming* the distance between one frame of reference and another' (1978a: 145, emphasis in original). He contemplates this issue chiefly in terms of the temporal distance that confronts historical geographers (see also Gregory, 1978b) but he accepts that the 'distance' between researcher and research subjects can arise for other reasons too, perhaps spatially (as 'us, here' research 'them, there') but often due to social, cultural, bodily and other differences between the 'positionalities' of researcher and researched (see also Chapter 1). Recognizing that there is merit in researchers 'immersing' themselves in the worlds of their research subjects, he adds that 'this does not mean and cannot involve the substitution of one frame of reference for another' (1978a: 145), quoting Gadamer in the process to the effect that the researcher's 'placing' of him- or herself in a situation does not require self-effacement but rather a full recognition of 'ourselves' – and our frame of reference – standing in a dynamic, even tense, encounter with the people whom we are studying.

We would argue that just such a sense of what has variously been termed the 'hermeneutic circle' or the 'double hermeneutic' has now become an article of faith for many human geographers. The result is a widespread recognition that human geographical research cannot but grow out of an exchange between two different bodies of meaning: that of the researcher, on the one hand, and that of the researched on the other. In-depth interviews where questioning and answering becomes co-equal discussion; participant observation full of shared practices and associated everyday talk; and researchers getting research subjects to comment on transcripts, field diaries or summary reports – all these common aspects of current human geography accept and actively promote a *shared* 'meaning production' in the construction of geographical data (see Chapters 5 and 6). Moreover, this sharing has almost become an unspoken assumption, the bridge between acknowledging one's 'positionality' (Jackson, 2000) in a research project – being self-aware and even self-critical of one's own attributes, attitudes, biography and social contexts (see Chapter 1) – and seeking to unearth all manner of meanings circulating 'out there' in a world of texts, images and situations available to us for interpretation. Very few human geographers would now take the stance adopted by Leonard Guelke (e.g. 1974; 1982) in his 'idealist human geography', wherein he stated that the researcher is an empty, neutral vessel simply finding and recording the ideas (the theories or meanings) present in the heads of research subjects (which for Guelke were usually people from the past, such as the Dutch colonizers of South Africa). Most human geographers today would hence be in no doubt that they are bringing their *own* meaningful baggage to a project – Harrison and Livingstone (1979; 1980) speak in their response to Guelke of the geographer's 'presuppositions' – and that somewhere in the heart of the research there *must* be an intermingling of meanings ('ours' and 'theirs'). As Trevor Barnes (2000c: 335–6, emphasis added) writes:

the explicit working out of the hermeneutical approach has become less significant ... Nonetheless, the spirit of hermeneutical inquiry, that is, the recognition of the importance of interpretation, open-mindedness, and a critical, *reflexive* sensibility, is as great as it has ever been, and certainly evident in the discipline's recent cultural turn and interest in the poetics and politics of representation.

Barnes admits that not all that much has been done recently by human geographers in terms of *explicitly* theorizing the hermeneutic balancing act between the 'frames of reference' held by both researcher and researched. Yet the balancing act *is* now a familiar one, so he continues, such that all manner of work now conducted by geographers within the 'cultural turn' (Cook et al., 2000b; Philo, 2000) or when discussing the 'crisis of representation' (see Chapter 11) is indeed alert to the reflexive moment in interpretation. In short, these geographers worry deeply about how the researcher's interpretations of a research subject's interpretations of the world are inextricably locked within a play of meanings across lifeworlds, a conversation to be heard and its points of (dis)connection registered, but more when debating the substance of a project's conclusions than as part of any grander epistemological hand-wringing.

The therapist

In order to introduce our sixth figure (that of 'the therapist'), it is worth pausing to underline – and to return to a theme of Chapter 5 – just how significant to contemporary human geography is the practice of conversation between researchers and those people under research. More than just a metaphor allowing us to consider the hermeneutics of meaning, as above, much human geography has come to rest on *real* conversations, real strips of dialogue, in grounded field settings. David Demeritt and Sarah Dyer (2002: 229–30) coin the phrase 'dialogical geography' to capture what is going on here, suggesting that such a term 'refers to qualitative research based on dialogue, or conversation, between the researcher and other human informants'. Moreover, and as a valuable cue for our arguments, they elaborate that

> Some forms of qualitative research involve literal dialogues with other people. The prevalence – indeed almost dominance – of such qualitative research methods in human geography today has given rise to a set of debates about the proper practice and the credibility or truth status of research based upon methods of literal dialogue. (2002: 229)

Something of these debates has already been rehearsed above when discussing the conversationalist, but what we want to stress now is one very recent strand of inquiry where the grounded conversation between researcher and researched has itself very much been foregrounded. This is the growing adoption of psychoanalytic methods within human geography (Rose, 2000), as buttressed by a more deep-seated turn to taking seriously the claims of psychoanalysis that we will mention shortly.

Building on contributions from Jacqueline Burgess (Burgess et al., 1988; Burgess and Pile, 1991), Steve Pile (1991) has introduced geographers to what he regards as an 'interpretative' approach that goes beyond the standard forms of 'qualitative' research to embrace a still more intensely 'intersubjective relationship' between researcher and researched. Such a relationship is seen as full of interpersonal dynamics, the recognition, enhancing and problematization of which are central to how psychoanalysis operates in the therapeutic encounter between therapist (or analyst) and client (or analysand):

> The patient works with the analyst because of the desire to relieve her or his symptoms [of mental distress], and because of the support provided by the analyst; this unspoken and provisional agreement of work

together is known as the 'therapeutic alliance' ... I suggest that there is an analogous, though different, 'research alliance', and we can use concepts developed in the course of therapy to excavate the multiple-layered interactions between researcher and researched. (Pile, 1991: 461)

Pile explores quite technical psychoanalytic notions of 'transference' and 'counter-transference', as played out in an exercise predicated on bringing to consciousness the 'repressed' experiences of the patient to be critically worked through (and hopefully resolved). This is an exercise where 'the therapist's unconscious and conscious reactions to the patient's transferences' (Parr, 1998b: 344) cannot be overlooked as part of the overall process, the basis of which is the 'therapeutic alliance' that Pile wishes to take as a possible model for interpersonal 'research alliances'. Objections have been raised to Pile's suggestions, with Linda McDowell (1992a) wondering if more needs to be known about power/social relations between therapist (researcher) and patient (research subject). Hester Parr (1998b: esp. 351) then worries that the therapeutic model commits geographers to an approach that is ultimately too rigid, too hung up on control and insufficiently open to 'messy methodologies which seek to "tune in" to research participants' different ways of telling'. A handful of human geographers have since considered further what methodological prompts can be derived from psychoanalysis, perhaps in terms of Donald Winnicot's therapeutic session as facilitative 'holding environment' or even the use of a Jungian 'sand play' for probing people's deepest interpretations of landscape.[4]

What we must now clarify, however, is that within the broader 'psychoanalytic turn' (Philo and Parr, 2003) affecting some corners of human geography there is also a trajectory of *criticizing* standard modes of understanding in the discipline. Such a critical stance arises because the psychoanalytic emphasis on the unconscious, positing that so much of a human being's mental life or 'libidinal economy' is occurring at a subterranean level unavailable to conscious inspection, challenges the conventional assumption that people can know, reflect upon and articulate what is meaningful to them. Even more so than for the humanistic geographers drawing upon phenomenological notions of the preconscious (Relph, 1970; Seamon, 1979), psychoanalytic geographers propose that very significant domains of the human condition – so many things, events, actions and processes within the human world, from both everyday incidents such as shunning a person unlike ourselves on the street to mass projects of 'ethnic cleansing' – are ultimately energized by unconscious drives about which we are dimly if at all aware. It is argued that such drives stem from buried passions, desires and fears whose contents remain 'repressed' from consciousness but which are, none the less, the foundation of an individual and even collective 'psychic economy' impelling all manner of goings-on central to worldly human geographies.[5]

Such drives could arguably be termed 'meanings', and it is possible to envisage psychoanalytic procedures (even the use of hypnotherapy) that might enable these subterranean strata of meaning to be exposed for critical inspection. Without a doubt, though, we are now working with a very different sense of what meaning entails from that favoured by the likes of the critic, the artisan, the ethnographer, the iconographer and even the conversationalist. The figure of the therapist (as someone who strives to retrieve threads of intelligibility from the debris of a person's unconscious) thus acquires a role very different from that of just another participant in a two-way conversation or dialogue. Whether geographers could or should attempt to deploy the skills of the trained therapist or counsellor is a vexed matter, but it is interesting that a small number *are* now

seeking to gain just such an expertise as an input to their future research practice.

It also needs to be stated that in the literature of psychoanalytic geography there is uncertainty about whether psychic materials can ever be truly 'knowable', given the suspicion that any conscious articulation of them will always amount to a mis-translation. One of Gillian Rose's (1997a;.1997b) key claims – rooted partially in psychoanalytic theory – is that much of the human world is 'inoperable' in the sense that it precisely does *not* entail human subjects communicating preformed meanings to one another through language but rather a flow of conducts, including speech acts, that in effect comprise a resistance 'to all the forms and all the violences of subjectivity' (Nancy, 1991: 35, cited in Rose, 1997b: 188). Much of the content of this flow is hence unknowable and uncertain, simply unavailable to the conscious researcher, and such a claim undercuts much that is assumed about (how to access) meanings by at least the first five of the figures described above. As Rose (1997b: 184) ponders, are there not some 'things beyond discourse, representation, interpretation, translation'; and she then asks 'what [are] the implications for my interpret[at]ive project, with its transcripts, codes, categories, records and papers?' The answer appears to be that such a project hits big problems, now being unable to interpret meanings as had been hoped, and Rose is left supposing that we cannot do much more than attend imaginatively to the 'silences' and 'absences' revealing that there is often *nothing beneath*, no coherent meanings, just 'emptiness'. Perhaps the truth about one's unconscious is not its capacity to generate meaning but, rather, its locus as the 'death' of meaning where things fall apart, become no-things, rather than becoming assembled – parcelled up as recognizable frames of reference – in the guise supposed by all who cleave to the hope of understanding human situations. Whatever the unconscious impels people to do thus cannot be taken as to do with meaning, at least not in anything like a conventional sense of what meanings entail, and so any equation of drives with meanings should be made only with the greatest of theoretical care.

The deconstructionist

Although they vary substantially in their approach to the interpretation of meaning, the figures we have discussed so far share a common starting point: namely, that the interpretation of meaning is just about possible. They all work from the assumption that texts, images, conversations and so on are sources of more or less stable and coherent meanings and that these meanings are in principle accessible to researchers with the appropriate skills and experience. In other words, they take the view that the role of the researcher is either to recover meanings that inhere in texts and other sources or, in the case of the 'conversationalist', to construct meaning in dialogue with the source. Our final figure takes a different view of meaning and the possibility of understanding it. The deconstructionist starts from the assumption that texts do *not* contain stable and coherent meanings. No matter how skilled the researcher, the deconstructionist argues, it is not possible to reconstruct the meaning of a source because the very idea of meaning on which such reconstructions are based is flawed. According to the deconstructionist, the meanings of any text are always unstable, partial and contradictory. Deconstructionists are particularly scathing about the idea that meanings are put into texts by their authors. Under the slogan 'death of the author' (proposed originally by the semiotician, Roland Barthes), they argue that authors never control their texts and that through the process of *writing* the text takes on an independent existence beyond authorial intention.

Deconstruction (see also Box 7.8) is most closely associated with the work of the French philosopher Jacques Derrida. According to Derrida, the truth claims of any text are undermined by the unspoken (or perhaps we should say unwritten) assumptions on which the explicit arguments of the text depend. Derrida argues that the Western philosophical tradition, and by extension Western thought in general, is *logocentric*. That is, it assumes that there is some ultimate foundation for knowledge (for example, 'logic' or 'rationality') and that this reflects the fundamentally ordered character of the universe. In other words, we usually assume that behind the words, concepts and theories through which we understand the world an underlying principle ('logic', 'rationality', etc.) is present that guarantees the validity of our accounts. This assumption is what Derrida calls a 'metaphysics of presence'. He uses the term 'metaphysics' here because we can have no independent verification of the presence of the underlying principle; we understand it only through the very words, concepts and theories that it supposedly validates. Because we cannot access the underlying principle directly (it is, in the end, just an assumption), all logocentric systems are without the secure external foundations they claim. What purports to be an ultimate foundation is actually part of the system itself, or may in fact contradict it. Deconstruction examines this disjuncture between the claims of a text and the basis for those claims. More specifically, Derrida argues that claims to truth typically rest on a series of binary oppositions. Examples of such binary pairs include inside/outside, self/other, male/female and rational/irrational. Moreover, these binary concepts are also hierarchical so that the self is more important than the other, male dominates female, rationality is view positively and irrationality negatively and so on. Derrida seeks to show is that it is never possible to exclude all traces of the subordinate element from its dominant pair so that any argument based on a sharp distinction between the binary elements can be shown to be internally incoherent. This process of uncovering the necessary instability of apparently logical and coherent texts is an example of deconstruction.

References to Derrida's work and the process of deconstruction have become widespread in contemporary human geography, although there are still relatively few examples of geographers working with the techniques of deconstruction in any sustained way. Trevor Barnes provides one of the clearest attempts to follow the spirit of Derrida's arguments in a study of geographical thought (1994), though it is perhaps worth noting in passing that valuing conventional 'clarity' in writing is an example of the very logocentrism that Derrida seeks to unsettle, as Barnes himself recognizes (1994: 1023). Barnes looks at the role of mathematics in geography and particularly at the value placed on mathematical procedures by proponents of the so-called quantitative revolution in geography in the 1960s and early 1970s. Supporters of quantification argued that mathematics would provide a solid foundation for geographers' accounts of the world that was *universal, logical, objective, simplifying* and *precise*. Through a deconstructive approach, Barnes shows that each of these claims about mathematics is untenable because each of them depends on its binary opposite. For example, drawing on philosophers such as Wittgenstein and Russell, Barnes argues that attempts to demonstrate the universality of mathematics (the argument that mathematics is true in all times and all places) always founder because any justification for universality must depend on conventions that are socially and temporally specific (1994: 1029–30). There are similar problems with the idea that mathematics offers greater precision than accounts written in ordinary language. This is because the

concept of 'precision' is not itself part of the mathematical system but is actually an ordinary language interpretation of the relationship of mathematical models to the things they are supposed to represent. One can only claim that mathematics is more precise than other languages by using those other, supposedly imprecise and thus fallible, languages to express the claim (Barnes, 1994: 1033–4). In his conclusion Barnes re-emphasizes the value of the deconstructive approach:

> I argued in this paper that mathematics is a prime example of logocentrism. For it is above all an attempt to impose order on the world. But ... that order neither inheres in mathematics nor the world as such, but only in the local institutional authorities that control mathematical practices. Of course, there have been attempts to justify those practices as something more than practices: for example, the claims that mathematics is universal, or that statistics is a foolproof method for making inferences. But such justifications always unravel ... They unravel because at bottom they assert that there is a realm of certainty that lies beyond language. It is here that the arguments of deconstructionist are so valuable. For they show that there is nothing outside the text; our truths are only those that we write to ourselves. (1994: 1037)

In addition to the analytic approach taken by Barnes, a number of other geographers have followed Derrida in a somewhat different way, but seeking to introduce something of his experimental approach to the use of language and writing into geography. Although not explicitly a deconstructionist, Gunnar Olsson was an early geographical proponent of playing with the style and form of written texts to try to destabilize their dominant meanings (1980). More recently the geographer Marcus Doel, among others, has adopted a self-consciously non-traditional writing style as a way of subverting the foundational knowledge claims of conventional geographical texts (1994; 1999).

Conclusion: between understanding and explanation

In this chapter we have sought to unpack the complex notions of 'meaning' and 'understanding' and suggested that geographers have approached both of these in strikingly different ways. Sometimes these approaches are complementary. For example, the 'artisan' and the 'ethnographer' are really close cousins, differing in their use of particular techniques perhaps but not fundamentally at odds over the role of meaning in social life and the tasks of the researcher. In other cases there are what appear to be irreconcilable differences. Most dramatically the 'deconstructionist' doubts the very existence of the authorial intentions that some 'critics' seek to reveal.

Whatever their differences, however, it seems to us that, taken together, our seven figures do speak to a more general issue inasmuch as that they confirm Derek Gregory's view that 'there can be no clearly defined distinction between understanding and explanation' (1978a: 145). At the start of this chapter we referred to the ideas of Dilthey and Weber. Gregory argues that there is an important difference between their approaches: 'whereas Dilthey had regarded *verstehen* (understanding) and *erklären* (explanation) as two quite separate procedures, Weber always connected them closely together … the ideal type was supposed to provide a means of moving between *verstehen* and *erklären*' (1978a: 133).

There are thus important connections to be drawn between the arguments of this chapter and those of Chapter 9. Explanations cannot afford to ignore meanings, particularly where the meanings with which social actions are imbued constitute an important part of the motivation of the actors involved. In this sense, meanings can be part of the explanation for the processes and patterns studied by human geographers. At the same time, people reflect on their circumstances and devise their own explanations for their own activities and those of others (which may or may not accord with the explanations advanced by geographers). In this way explanations can become part and parcel of the meaningful character of social life.

Notes

1 For a detailed discussion of Dilthey's ideas and their application in geography, see Rose (1981).
2 A development with many echoes in contemporary human geography – e.g. Burgess and Gold (1985), Aitken and Zonn (1994a) and Leyshon et al. (1998).
3 For more information on semiotics, go to www.aber.ac.uk/media/Documents/S4B/semiotic.html. This website has now been published as a book – see Chandler (2001).
4 See Bingley (2003) and Kingsbury (2003); see also Bondi (1999; 2003a; 2003b) for very recent ideas about the spaces of counselling.
5 Sibley (1995), Pile (1991; 1996), Wilton (1998), Nast (2000) and Callard (2003).

Representing Human Geographies

Introduction

The literal meaning of 'geography' is 'earth writing' – taken from the Greek *geo* ('the earth') and *graphien* ('to write') – and writing is indeed central to the practice of human geography. For while in principle geographers could be involved in the production of all sorts of texts – films, paintings, soundtracks and, of course, maps – it is writing that every sort of human geography and human geographer has in common. Academic libraries are full of books and journals containing the written output of various types of geographical practice, and you will present your own research results to your tutors largely in the form of written essays, projects and dissertations. It is not only the form of communication that is common. Whether you are writing a book, a journal article, a dissertation or an essay, many of the issues involved in the process of writing are similar. The 'terror' of the blank page and the ensuing trials and tribulations of authorship are faced by everyone sitting down to write: the authors of this book can all vouch for the fact that they are not restricted to students. This chapter, however, seeks to move beyond a consideration of the mechanics of writing to explore issues of representation. In other words, it concentrates on writing as a critical part of those 'practices by which meanings are constituted and communicated' (Duncan, 2000: 703).

Thus the central motif running through the chapter is that writing is a form of representation (or indeed re-presentation) which helps to create, rather than simply reflect, our geographical experiences. As part of this creative process, the author is inevitably interpreting his or her data through the very process of writing.

This in turn means that writing, and textual construction more generally, are not precursors ('writing things down') nor later add-ons ('writing up') to the practices of data construction and interpretation; they are a critical part of those constructions and interpretations. Writing and the construction of academic texts, whether whole books, single chapters, articles, essays, dissertations or reports, are not something separate that come at the end of the research process as later add-ons. Writing needs to be seen as something integral to research – better expressed perhaps as 'writing through' – and indeed it is only in the process of writing that our own ways of seeing and interpreting the world are conveyed to others. All practitioners of human geography, whether undergraduate dissertation students or senior professors, have to find productive ways of writing in order to tell others about the geographies they have found. In this sense academic writing concerns the construction of a narrative in order to communicate with, engage, and, ultimately, convince your particular audience. This audience could be other academics reading a journal paper; it could be students reading a textbook; it could be government policy-makers reading a report; or it could be a lecturer reading and marking undergraduate essays. But because language is never simply mimetic, and cannot mimic or imitate the world it seeks to convey, writing for any audience is always about the translation and interpretation of that world to others.

In this chapter we explore this role in terms of a move from issues of presentation (of writing as a craft, a skill, a tool) to issues of representation – writing as about the production of knowledge and ways of seeing/conceiving the world. Thus we look at the work of writing and at ways of presenting research before we move on to consider various moments of textual representation. Although the chapter is largely concerned with writing, it also presents a brief consideration of other ways of representing geographical practice, through film and notation. We should say at the outset that some geographers are very suspicious about undue attention being given to questions of writing and representation, fearing they deflect attention away from more pressing matters of substance and the politics of research to questions of poetics. Nigel Thrift, for instance, cautioned against 'over-wordy worlds' well over a decade ago, in 1991. We share many of his concerns over the worrying hegemony of culture within social research, for instance, and the move towards philosophical rather than empirical means of argumentation, and recognize the pitfalls of stressing poetics above politics (Barnes and Gregory, 1997) – not least because of the danger that this all becomes part of a cultural repositioning of geography as (second-rate) art or fiction, and geographers as nothing other than cultural intermediaries. Given this, we should perhaps stress that, in the argument presented

here, a concern with textual production is not separate from questions of empirical substance, matters concerning the relations of the researcher to the researched or issues around the point and politics of research; indeed, our key point is that it is heavily implicated in all these.

The work of writing

This section looks at the practice of textual construction as work. It argues that all practitioners of human geography research, from undergraduates sweating at their dissertations to lecturers labouring over their latest book, ultimately have to find fruitful and rewarding ways to 'write through' their data and ideas. While each individual will vary in how easy or difficult, how fun or painful he or she finds writing, all human geographers have to work at it. Box 11.1 builds on this idea of writing as work by setting out Northedge's ideas on the craft of writing, where he develops the notion that writing must be seen as taking place through a number of distinct, but interlinked, stages.

This section explores some of these stages and offers some advice on how to organize the craft of writing. In doing so it seeks to open up for debate some of the feelings people have about writing they often keep private. Although it acknowledges that 'writing through' one's research is far from easy, it does attempt to combine this with grounds for optimism and ways forward. It starts from the position that there is no neat and simple transition, or straightforward translation, from constructed data and interpretations into final texts. We have already seen in earlier chapters how the data that form the 'results' of research are subject to many manoeuvrings, from sifting and sorting onwards. At each stage the form these data take in any account of the

Box 11.1: The craft of writing

According to Northedge in his book *The Good Study Guide* (1994: 157–8, emphasis in original), writing can be conceived of as a craft. He first makes a case for this view by counterposing it to writing as inspiration:

> How do you picture really experienced writers setting about the job of writing? Do you imagine them sitting down with a blank sheet of paper, or a blank word processor screen and just spilling out words? Do you picture them being visited by *inspiration* and immediately starting to pour out beautiful and compelling sentences, which are ready to be sent off for publication? It is almost never like that.

He then makes the point that one reason why this never happens is because of the nature of the writing which we all do as geographers. As in most other academic disciplines, geographers are largely concerned with 'expository' writing – writing which seeks to analyse, understand, explain and argue – in contrast to 'expressive' writing which attempts to express feelings, and 'narrative' writing which tells stories. As we shall see later in the chapter, these three types of writing are not necessarily totally distinct, and the human geographer may well draw on all three. But for now we can stick with Northedge when he explains that the process of trying to develop a carefully constructed argument, and maintaining it over a number of pages drawing on empirical evidence and the ideas of others, is actually a very complex one. As he notes: 'Putting well-formed sentences on to paper ready for sending to your tutor is only the last of a series of stages in the process of putting together an essay. Before that a great deal of thinking and preparatory work has to be undertaken' (1994: 157–8).

Finally, he develops the notion of writing as a craft by drawing an analogy with furniture-making – in this case making a table. Like writing, this does not simply 'happen':

> First [the furniture-maker] has to conceive of a design,… then choose the wood, prepare it, measure it, mark it, cut it, shape it, make the joints and finally put it together. And even then it still has to be smoothed, waxed and polished. Writing essays may not be quite as elaborate … but it does have some of that quality – requiring you to work methodically through a whole series of closely linked activities. If you simply sit down when you have finished reading the course texts and try to write a whole essay in a single sweep you will get nowhere. The job is too big. You *have* to break it down into stages. Then you can take it stage by stage and work your way to a finished product. (pp. 157–8, emphasis in original)

When we write elsewhere in the chapter of 'writing as a craft', it is in this sense that we do so.

research is actively constructed through the process of writing; and in turn, this process will inevitably help to construct data and interpretations rather than simply reporting on them. Research writing, and the representation of geographical practice, is thus always about the work of writing through, and not just writing up.

Why writing through research can be hard

Two main reasons are usually advanced as to why writing through research can be difficult. First, it involves organizing and rationalizing from vast masses of data and ideas. As already noted, this process begins at the very earliest stages of sifting and sorting. This is intrinsically difficult for most people. It can also be quite scary. There can be worries about what one might lose as one organizes the data, accompanied by concerns about doing justice to the research you have done and to the data you and your research subjects have constructed. There will also be worries about how to package the interesting, but still somewhat vague, ideas you have had about how to interpret those data, and about the sheer complexity of the phenomena and lives you've found out about. Indeed, perhaps at the root of all this there may well be worries about whether any sort of order will ever emerge. As Becker (1986: 133, 134) puts it:

> That is the deepest cause of the anxiety that strikes writers when they begin. What if we cannot, just cannot, make order out of that chaos? I don't know about other people, but beginning a new paper gives me anxiety's classical physical symptoms: dizziness, a sinking feeling in the pit of the stomach, a chill, maybe even a cold sweat … The … world may be ordered, but not in any simple way that dictates which topics should be taken up first. That's why people stare at blank sheets of paper and rewrite first sentences a hundred times. They want those mystical exercises to flush out the One Right Way of organizing all that stuff.

The second reason writing can be difficult is because it involves presenting one's ideas, and indeed oneself, for scrutiny. Scrutiny by others – both real (actual readers you give materials to) and imagined (those whom, as you write, you imagine reading your text, often with the result you immediately delete whatever you have written as visions of their disappointed frowns pass through your head: your supervisors, examiners, peers, role models). Scrutiny too from your own critical and often unrealistically demanding eye. Such scrutiny means that writing can be considered a risky process – it can become a very emotional form of labour through which you are offering yourself for external judgement. Box 11.2 sets out

some thoughts from the sociologist Pamela Richards who developed her own notion of writing as a 'risk'.

As Richards concludes, the process of writing may be difficult but it can also be immensely rewarding and give you a great sense of achievement. However, there are certain strategies that can be followed in order to maximize the pleasurable aspects of writing and minimize the costs of self-doubt. Many books are now available which help students with different aspects of the writing process and, together, they give a huge array of advice.[1] As our emphasis in this chapter is on writing as representation and not writing as presentation, we will not seek to duplicate or summarize these but will pick out four relatively simple steps as a reminder that writing can be organized in productive ways, before we move on to look at some of the more detailed mechanics behind 'good' writing.

Organizing the work of writing

The first suggestion is to view writing as an ongoing constructive process rather than as a one-off act of research presentation. Practically, this means being prepared to write multiple drafts (even beginning with what de Souza, 1988, calls a 'spew draft'), starting writing as early as possible (including in memos and letters), and not expecting early writing to meet the same criteria one would apply to later drafts. As Becker (1986: 14, 17) puts it:

> writing need not be a one-shot, all-or-nothing venture. It [can] have stages, each with its own criteria of excellence … An insistence on clarity and polish appropriate to a late version was entirely inappropriate to earlier ones meant to get the ideas on paper … a mixed-up draft is no cause for shame … Knowing that you will write many more drafts, you know that you need not worry about this one's crudeness and lack of coherence. This one is for discovery, not for presentation.

Box 11.2: Writing as emotional risk

Richards explains how she came to view the process of writing as a risk, after having vivid dreams about people commenting on her work and being exposed as an academic fraud. In trying to understand why she had these dreams of self-doubt she came to the following conclusion:

> For me, sitting down to write is risky because it means that I open myself up to scrutiny [from colleagues and myself]. To do that requires that I trust myself, and it also means that I have to trust my colleagues. By far the more critical of these is the latter, because it is colleagues' responses that make it possible for me to trust myself. So I have dreams of self-doubt and personal attacks by one of my closest and most trusted friends ... If you give someone a working draft to read ... you're asking them to decide whether you are smart or not ... If there are no flashes of insight, no riveting ideas, what will the reader conclude? That you're stupid ... Hence the fear of letting anyone see working drafts. I cannot face the possibility of people thinking I'm stupid. (Richards, 1986: 113–15)

She then goes on to explain the classic Catch-22 situation that this gives rise to – she can receive the affirmation she seeks from others only through the process of writing, but it is only this affirmation that frees her to write:

> I trust myself (and can therefore risk writing down my ideas – things that I have made up) primarily because others I trust have told me that I am OK. But no one can tell me that until I actually do something, until I actually write something down. So there I am, faced with a blank page, confronting the risk of discovering that I cannot do what I set out to do, and therefore am not the person I pretend to be. I haven't yet written anything, so no one can help me affirm my commitment and underscore my sense of who I am. (1986: 117)

However, there is considerable light at the end of this particular tunnel. As Richards goes on to explain, this fear of risk does not paralyse her writing abilities completely. Instead, through the process of writing she comes to the conclusion that it is not an all-or-nothing proposition. What gets written is usually neither a literary gem nor pure rubbish. Most writing is somewhere in between and, by recognizing this, Richards was able to come to terms with writing as a process – and one which, like most things, is helped considerably by practice:

> In some ways, writing gets easier the more you do it, because the more you do it, the more you learn that it's not really as risky as you fear. You have a history on which to draw for self confidence, you have a believable reputation among a wider number of people whom you can call on the phone, and best of all, you have demonstrated to yourself that taking the risk can be worth it. You took the risk, you produced something, and voila! Proof that you are who you claim to be. Though I must also admit that it's not as easy as I'm making it sound. My writing history gives me some confidence, but I look at my past work with mixed emotions. It looks awkward and full of errors ... This means that every time I sit down to write I find myself wondering whether I can really do this stuff at all. So writing is still a risky activity. But what I seem to be learning as I spend more time writing is that the risks are worth taking. Yes, I produce an appalling amount of crap, but most of the time I can tell it's crap before anyone else gets a chance to look at it. And occasionally I produce something that fits ..., something that captures exactly what I want to say. Usually, it's just a sentence or two, but the number of those sentences grows if I just keep plugging away. This small hoard of good stuff also helps me take risks ... It reminds me there are two sides to risk. You can lose, but you can also win. (1986: 118–19)

Although the notion of seeing writing as a multistage process can be helpful, you should be aware that experience of writing for one purpose may not necessarily suit another. The preparation of writing many short essays, for instance, can actually be unhelpful when it comes to writing through one's own research in the form of a much longer dissertation or fieldwork report. Indeed Becker claims that:

> The student's situation rewards quick, competent preparation of short, passable papers, not the skills of rewriting and redoing ... Students find the skill of writing short papers quickly less useful as they advance ... eventually they have to write longer papers, making more complicated arguments based on more complicated data. Few people can write such papers in their head and get it right on the first try, though students may naively think that good writers routinely do ... So students flounder, 'fear getting things wrong', and don't get it done on time. (1986: 19)

Although Becker is writing from the standpoint of American sociology, the step up to doing a 10 or 12 000-word geography dissertation can be similar. The moral here is to adapt your writing strategy to the task in hand. What suits the production of a 1500-word essay linked to a tutorial or seminar discussion may not work for a dissertation ten times as long based on extensive fieldwork. But in each case multiple drafts and revisions will be far more beneficial than one-off bursts of activity as the deadline approaches.

Secondly, approaching writing as an ongoing part of the practice of human geography also means recognizing it needs to be organized: in the same way that any other craftsperson organizes his or her work. It should be viewed as an extended process rather than as something done in short, last-minute bursts, and as something one purposefully arranges in one's own diary, finding the times when and the places where one writes most productively. It can always be arranged as something one tries to make a pleasure,

through rewards and incentives (a few hours then a snack, short walk or some music). There are even tips on how to start and conclude each writing session carefully (so begin when fresh and not jaded; finish so that you have something easy to pick up on upon return).

Thirdly, this organization of the writing craft is not just individual and private. Indeed, one potential problem researchers may have with writing is the fact that by its very nature it tends to be organized and carried out in highly individualized and isolating ways. One means of overcoming such isolation commonly used by professional writers is to develop a circle of 'reading friends' who will look through early drafts and offer constructive advice – to be given, and taken, in the right spirit. According to Becker, these friends will always treat 'as preliminary what is preliminary' (1986: 21) and help the author sort out matters of style and presentation. It is surprising how often an external eye can pick up passages of writing which are unfocused or imprecise and point to sections where the ideas are unclear and muddled. This tends to be because the author gets too close to what he or she has written and finds it difficult to stand back and take a detached look. This is precisely what a writing friend can offer. These relationships work best if they are reciprocal – form a group and take it in turns to read through each other's work.

Fourthly, and finally, one needs to develop strategies for dealing with different stages of the writing process. Wolcott (1990) lists these as 'getting going', 'keeping going', 'linking up', 'tightening up' and 'finishing up', and devotes a chapter to each stage. Getting going covers the sorting, organizing and filing of materials, as well as deciding where, when and how to begin. We have already covered the preliminary stages of this in our discussion of sifting and sorting in Chapter 7, but this also involves the initial processes of editing and selecting appropriate data. As Wolcott puts it: 'The critical task … is not to accumulate all the data you can, but to "can" (i.e., get rid of) most of the data you accumulate' (1990: 35). So even in these very preliminary stages of the process we encounter the notion of writing being heavily implicated in the construction and interpretation of data rather than simply reflecting them. Once you have made a start, the trick is to keep going. Keeping going will involve more than writing in its literal sense – putting pen to paper or fingers to keyboard – and encompasses reading, organizing, refining and revising. Wolcott (1990: 51) even includes the process of 'just staring into space' as part of this phase. The key point here is to keep going; as Wolcott says, 'you have to have *something* written before you can begin to improve it' (1990: 55, emphasis in original). Try not to get stuck, either because of material you do not have or cannot understand. Make a note to yourself of what needs inserting or revising and go on to the next section. Once you have that crucial first draft, even with some spaces and unfinished sections, it can become a better second draft and a good third draft.

Linking up reminds us that all writing is embedded in broader academic and social contexts. It is done for particular reasons, to meet specific goals. These may necessitate linking your work to that of others – whether in terms of literature reviews or broader considerations of concepts, theories and methods. Tightening up takes place once these links have been drawn and everything is in place. It refers to another round of reviewing and editing but takes place once the material is more or less complete. This is the final stage in the construction of data, as the entire script is evaluated and appraised as a complete entity:

- Does it hang together?
- Are its points well made?
- Are the arguments illustrated with reference to other work or your own research?

- Does the empirical material relate to the more abstract sections?
- Is any material superfluous to the argument?
- Is each section introduced appropriately?
- Can the reader follow where the argument is going?

Once all these issues have been attended to, the piece of writing needs to be 'finished' and made fit for its purpose. Issues such as the title, references, word length, headings and subheadings, footnotes and endnotes, figures and appendices all have to be addressed. Once this stage is complete, you are ready for 'getting it out the door' (Becker, 1986: 121) where the tension between making it better and getting it done is finally resolved in favour of the latter. Usually for students this resolution is forced by deadlines, but if these do not come into play a sense of the productive and rewarding possibilities of participating in the collective intellectual endeavour that is the practice of human geography should help you to let go.

Presenting research

In this section we will keep the same instrumental, craft quality of writing to the fore but will do so in terms of accounts that advise not only on how to write but also how to write well. It deals with the presentation of research data, as part of its writing through, in three parts. In the first we will look at the ways in which some academics have sought to offer guidance on the textual presentation of research, and at how this guidance makes public what are usually more private commentaries from examiners, journal editors and referees. There are some recurrent themes in this which place an emphasis on clarity and direct expression (including the avoidance of passive sentence constructions) and highlight how to engage audiences through the use of particular academic conventions and the adoption of a certain 'author-itative' persona. The second section examines some of the positive aspects that stem from these kinds of prescriptions, emphasizing in particular the way they make us think about the process of writing as a skill that has to be, and indeed can be, worked at and developed. They also help us to focus on issues of audience engagement and remind us that ultimately writing should be viewed as relational – its purpose is to forge a relationship with the reader, however solitary a process it feels. Lastly, however, we go on to point out that, while there is much of value in these kinds of calls for 'good writing', they also tend to oversimplify and close down many of the issues they seek to confront. These issues will be picked up more fully as we move on to discuss representation and rhetoric later in this chapter.

Good writing: 'overcoming the academic pose'

There are now many books and guides available which set out tips and prescriptions for 'good writing' and most frame these tips in terms of clarity and brevity. Again, because our emphasis is on representation rather than presentation, we will simply outline the main points of these prescriptions. De Souza summarizes the advice as follows: 'Change passive to active constructions. Take long sentences apart. Eradicate redundancies. Eliminate pompous phrases … Get rid of qualifications … clarity is a virtue' (1988: 2–3). Time and time again one comes across the same injunctions to avoid unnecessary words or phrases, to eliminate the passive voice (especially any form of the verb 'to be'), and to make a virtue out of short and simple sentences whenever you can.[2]

One reason why most books on academic writing emphasize these points about brevity and clarity is that there tends to be a misplaced feeling that 'scholarly' writing is of necessity complex and convoluted. Indeed for some

people it seems that the more difficult a passage is to follow, the more intelligent it must be. This feeling is utterly misplaced and confuses the style of an argument with its substance. The complexity of the latter is not indicated by the complexity of the former – indeed, quite the opposite. As Hanson (1988: 6) puts it:

> Also necessary is the author's ability to communicate that argument in lucid, jargon-free prose. Ideas that are heavily draped with jargon are likely to set an editor musing about the moral of 'The Emperor's New Clothes'. Let us not convey to our students – however unintentionally – the idea that obscure, convoluted or jargon-laden writing is somehow 'scholarly'.

Many academics, however, seem unwilling to let go of this idea. According to de Souza:

> Articles published in scholarly geographical journals should clearly communicate ideas and results to those who read those journals. Yet many of us ignore the rules that govern good writing and turn out stilted and overblown prose. We use langauge that obscures what we want to say, drowning meaning in murkiness. (1988: 1)

According to Mills, this use of excessive jargon and obscure language actually has very little to do with the complexity of the subject-matter being written about. Instead he links the reason to the psychological state of the writer and to academic insecurity:

> lack of ready intelligibility, I believe, usually has … nothing at all to do with the profundity of thought. It has to do almost entirely with certain confusions of the academic writer about his [sic] own status … Desire for status is one reason why academic men [sic] slip so easily into unintelligibility … To overcome the academic prose you have to overcome the academic pose. (1959: 239–40)

For the student, overcoming the academic pose with good prose can help to raise marks considerably. As Steinberg says:

> better prose produces better examination results. When examiners say vaguely that so

and so wrote an 'economical' answer or use adjectives like 'neat', 'precise' or 'to the point', they have, without always noticing it, reacted to vigorous, energetic writing. A healthy verb cannot transform a plodding waffler into Maitland but it can make the waffle sound '2.lish' and not '2.2ish'... Strong verbs give the writer a sense of power. (1985: 90)

Identifying the positives in these kinds of prescriptions

The moral in these prescriptions is to avoid the 'academic pose' through active, clear and precise writing. There are four main benefits to this kind of advice. First, it does force you to think about writing as a craft skill, as something that you have consciously to develop abilities in rather than something you simply pick up through an imitation of what seems scholarly. Thus these prescriptions do remind us that writing does not simply spring on to the page ready formed through mental inspiration – it is something that, quite literally, has to be worked at and worked on. In this sense it can be understood as a kind of carving or sculpture where the rough edges are constantly honed and refined before we reach the finished product. These kinds of prescriptions tell you how to undertake this refinement. Secondly, this advice reminds you that the impact of your style of writing is never neutral – it is something that has a huge influence on what any reader makes of what you are saying. Thus the meaning of what you are saying is intimately connected to how you are saying it. As we noted at the beginning of the chapter, this is because writing both constructs and communicates meaning rather than simply reflecting or presenting an external reality.

Thirdly, most of these texts on how to write well do highlight the particular problem which exists in a temptation to use certain sorts of academic, exclusive vocabularies – both

as a means to make oneself sound scholarly and as a way of associating oneself with those whose intelligence and scholarliness one admires. Consider, for instance, Box 11.3, which cites part of a letter to Howard Becker from one of his students talking through how she related to academic language in this way. As we shall go on to examine later in the chapter, the final point that Rosanna makes about scholarly writing and being a scholar is not actually the case – as she herself realized. Often, the very best scholars are those who are able to explain complex ideas in the most accessible writing style.

This brings us on to the fourth positive aspect of these prescriptions. They centre issues about audience engagement and so make sure that writing is seen as a practice which is all about building a relationship to readers rather than a solitary, solipsistic activity. This helps to remind us that the point of 'writing through' research is to engage readers, not to convince yourself of your own worth. The audiences that read the results of geographical research are wide and varied: students, other academics, policy-makers, specialist researchers or even the public at large. What all these audiences have in common, though, is that they are at a distance and removed from the writer. In this sense, writing is a 'very special form of conversation' (Northedge, 1994: 146). The particular concerns this raises are set out in Box 11.4.

Problems with these sorts of prescriptions

However, these kinds of prescriptions have two sorts of problems. First, they oversimplify the criteria upon which the quality of writing can and should be judged. For instance, they define clarity in very instrumental ways – about the conveyance of a clear and strong message or argument – and ignore other roles

Box 11.3: On the temptation of academic jargon

In his book, *Writing for Social Scientists*, Howard Becker (1986: 26–8) recounts how he once spent a considerable time editing and shortening a chapter written by one of his PhD students – Rosanna Hertz. Becker polished the grammar, shortened overlength sentences, removed superfluous words and cut repetition. Instead of being pleased with his efforts, Rosanna was a bit put out. Although she accepted that his version was tighter and clearer, she maintained that her original version was 'classier' (1986: 28).

Becker, in turn, was somewhat taken aback by this and asked her to explain, in writing, what she meant by 'classier'. This is what she wrote by way of explanation:

> Somewhere along the line, probably in college, I picked up on the fact that articulate people used big words, which impressed me. I remember taking two classes from a philosophy professor simply because I figured he must be really smart since I didn't know the meanings of the words he used in class. My notes from these classes are almost non-existent. I spent class time writing down the words he used that I didn't know, going home and looking them up. He sounded so smart to me simply because I didn't understand him ... The way someone writes – the more difficult the writing style – the more intellectual they sound. (1986: 28)

Becker goes on to explain that these types of feelings partly arise from the very hierarchical nature of academic life, where professors and lecturers are seen just to 'know more' and are viewed as figures of intellectual authority – who should be imitated whether or not they make sense. As Rosanna explains:

> When I read something and I don't know immediately what it means, I always give the author the benefit of the doubt. I assume this is a smart person and the problem with my not understanding the ideas is that I'm not as smart. I don't assume ... that the author is not clear because of their own confusion about what they have to say. I always assume that it is my inability to understand or that there is something more going on than I'm capable of understanding. (1986: 29)

Rosanna is pointing to the classic feeling of inferiority which many students (and indeed some lecturers) feel when first confronted with academic writing which is complex and even unintelligible. She goes on in her letter to highlight the processes which help to maintain this situation:

> academic elitism is part of every graduate student's socialization. I mean that academic writing is not English, but written in a shorthand that only members of the profession can decipher ... Ideas are supposed to be written in such a fashion that they are difficult for untrained people to understand. This is scholarly writing. And if you want to be a scholar you need to reproduce this way of writing. (1986: 30)

that (differently clear) writing can play – to describe, evoke, give a feel for, to express ambiguities and a lack of certainty. This tends to presume particular sorts of audiences (a general geography readership, for example, or policy-makers) and may not be most effective with some others (for example, those whose lives are being written about).

Miller (1998), for instance, describes how one of her research subjects, having read her research report, comes back at her on the lack of humanity, feeling and relationship in the text. The forceful, assertive, active and vigorous language used by Miller in the report made her feel that this was the author's text and her own involvement had only been as a

Box 11.4: Writing as a conversation: speaking to your reader

Writing is a very particular kind of conversation in the sense that you cannot see the person you are talking to, and he or she never joins in the conversation. Nevertheless you are charged with the responsibility of convincing him or her of your argument via the written word. To make matters worse, you know he or she will be 'listening' hard, harder sometimes than in a real conversation, since he or she will usually be reading every word, in contrast to a conversation where the listener may get distracted or words are drowned out by background noise. Northedge (1994: 146–7, 166–7) sets out some pointers for taking part in such a conversation:

- You have to develop a sense of your audience because the style used will vary from audience to audience. When you read research reports written for policy makers, they will be in a different 'voice' from those written for an academic argument.
- The standard formula for students is to 'write for the intelligent person in the street': assume that your reader has not read the books and articles you have used but that he or she is interested in the topic and can pick up your arguments quickly, provided you spell them out clearly.
- You have to take the reader carefully through all the points of the conversation, illustrating any points made, constructing a logical route through the points and making sure they follow on from one another.
- The talking you do in this conversation takes a lot of effort. This is why the content should have been prepared and sorted out in advance. If you try to think up the content at the same time as addressing yourself to your reader, you will usually do one or the other badly.
- Since writing is a very formal activity compared to speaking, it sometimes takes a lot of practice to find your 'voice'. Don't be worried if you do not initially convey what you intended. This is what drafts are for.

resource: that she was just a life to be slotted into the researcher's argument (here we link into the question of research ethics discussed in the next chapter).

One might also challenge the (once and for all) equation of good writing and clarity with accessibility. In certain contexts, for certain purposes, there is nothing wrong with working through a specialized vocabulary (for example, of particular theoretical schools) if one's aim is to engage with other initiates or initiate others into that conceptual framework. Technical language is also necessary to think through aspects of human geographical realities that are not articulated in everyday life and language. It can also be used as a device to think otherwise, beyond current orthodoxies. Moreover, the reverse also holds – maths is a clear language but it isn't highly

accessible; and we don't always expect it to be so (even if sometimes we do expect mathematicians to be able to explain something of their work in lay terms). The same is true of medicine. Sometimes writing in complex, technical and sophisticated language is the best option. The problem with prescriptions is that they can be too prescriptive.

Secondly, these kinds of prescriptions for 'good writing' not only oversimplify evaluative criteria but they also fail to follow through on the implications of many of the important issues they raise. They close them down into disciplining templates and standard voices rather than using them to open up the possibilities (and responsibilities) of writing as textual production. They also don't situate the issues they raise in a wider consideration of the character of academic knowledge, or do so only implictly. For example, the question of authorial persona and voice is reduced into a binary opposition between imitative, jargon-ridden pretentiousness and plain speaking, confident 'neutrality'. This ignores broader issues about the social evaluations of certain kinds of voices. Rose (1993b), for instance, describes how a great deal of geographical writing conveys a male perspective on both nature and culture; indeed, she argues that the separation between the two is artificial, constructed through 'man-made language [which] divides what should be united' (1993b: 71). As a consequence, the subjectivity expressed through geographical thought and writing is one which is 'dependent on separation and domination, especially on the separation between masculine Culture and feminine Nature' (1993b: 71).

Moreover, the languages that such prescriptions applaud are bound up with certain framings of the world (for example, a separation of the known and knower, the latter untroubled and undisturbed by the former). Rose again picks up on this theme in her discussion of the way geographers have traditionally looked at landscape. As she notes: 'the pleasure of the masculine gaze at beautiful Nature is tempered by geography's scientism ... The gaze of the scientist has been described ... as part of masculinist rationality, and to admit an emotional response to Nature would destroy the anonymity on which that kind of scientific objectivity depends' (1993b: 88). Hence the injunction to clear, simple and precise writing can mask, and indeed stem from, a false separation and artificial distanciation between the observer and the observed. If one takes the view, presented in Chapter 1 and developed throughout Part I, of research as a social activity, where data are constructed through a relationship between the researcher and the researched, it becomes difficult, if not impossible, to maintain this separation.

This in turn means that concerns with good writing style can take us only so far in thinking about the nature and significance of textual production as a human geographical research practice. As Richardson (1988: 202) puts it, the 'solution to the writing problem is not the extermination of jargon, redundancies, passive voice, circumlocution and (alas) multisyllabic conceptualization and referential indicators'. Instead we have to involve ourselves with issues of representation and rhetoric and consider the numerous ways in which writing as a process extends far beyond the words and sentences on the page.

Representation and rhetoric

This section begins by setting out the notions of representation and rhetoric as ways to think about writing as a process that is implicated in the very creation of 'interpretations', 'data' and 'reality'. It then develops the argument that textual construction and writing should be viewed not so much as regrettably necessary processes of communication – and thus something to be undertaken in ways that draw as little attention to themselves as possible – but as stylistic processes of persuasion that are fundamental to research practice. These points are then illustrated across two contrasting fields that seem to sit very differently in relation to textual construction and its importance. The first is regional and landscape description, an area of geography where creative writing has been explicitly encouraged in order to convey to the reader a sense of how a particular landscape looks or even feels. The second draws on examples of supposedly more 'objective' types of representation in order to establish the point that even scientific objectivity does not lie outside rhetoric but, rather, is in part constructed through it. Thus, despite the differences between the textual styles and strategies adopted across these two fields, a number of common issues concerning the nature of writing materialize across both. These issues will then be explored in more detail later in this chapter when we examine a number of specific facets, or moments, of textual construction which all human geographers have to negotiate.

The device of rhetoric

For much of its history, human geography has rather unproblematically accepted the conventional view of language which sees words as mirrors of the worlds they represent. For Eagleton, the centrepiece of such a view is that words are 'felt to link up with their thoughts or objects in essentially right and incontrovertible ways' (cited in Barnes, 2000a: 588). Thus the trick of writing, of representing the world to others, becomes reduced to a mechanical one of lining up the right words in the right order (Barnes, 2000a: 588). Recently, however, this view has been challenged to emphasize that the relationship between words and the worlds they purport to represent is less straightforward and more complex. As Richardson puts it, 'writing is not simply a true representation of an objective reality, out there, waiting to be seen. Instead, through literary and rhetorical structures, writing creates a particular view of reality' (1990: 9). This sensibility towards the inherent complexity of representation foregrounds the notion of poetics within geography. By this we do not mean the appreciation or practice of poetry. Instead we are referring to poetics as a 'critical practice that involves taking into account the force, exactness and power of words' (Barnes and Gregory, 1997: 4). In other words poetics is about 'treating words with respect, recognizing their power, passion and potential' (Barnes, 2000a: 588). Said (1977; 1993), for instance, has shown how the power of words was used to construct very particular ideas of 'orientalism'. This notion in turn was

used to structure representations of Asia (and Asians) into a worldview that was seen from, and compatible with, projects of Western imperialism in the nineteenth and early twentieth centuries, where the Western cultural self was increasingly defined in contrast to non-Western 'Others'. This definition was then important in helping to underpin and validate the Western exploitation of these 'other' lands.

Once we accept that words are not neutral mirrors of the world they represent, we are led to examine just how they do gain their 'power, passion and potential'. Richardson has noted how 'all language has grammatical, narrative, and rhetorical structures that construct the subjects and objects of our research, bestow meaning, and create value' (1990: 12). Hence, part of the ability of words to be powerful and 'bestow meaning' comes from the 'rhetorical' devices used by their writer to establish authority and persuade readers. Rhetoric is defined here as the art of argument, and not used in its more pejorative everyday sense of pretence and display. As part of such an 'art', metaphors, imagery, citations, appeals to authority, anecdotes and even jokes are all used as 'instrumental devices' in the 'persuasive discourse of science' (Richardson, 1990: 15). In other words all texts seek to 'persuade' readers of the truth of their claims, and most will make use of rhetorical devices to do this. What is interesting about this is that diferent elements of science, different disciplines and subdisciplines all have their own set of literary devices which help this process of persuasion. We will explore some of the devices used in geography later in this chapter. For now we simply want to concentrate on rhetoric as a fundamental, and unavoidable, part of research practice: as a device which is constructive in the making of data and interpretations, as well as in the making of authors and readers.

Making these broad claims about the importance of rhetoric is not about privileging style (how something is written or represented) over substance (what is being said) but recognizing that style is a part of that very substance:

> The distinction between style and substance has burrowed like a worm deep into our culture ... yet it has few merits. It is all style and no substance. Consider. What is the distinction of style and substance in ice-skating or still-life painting or economic analysis?... By style we mean properly the details of substance ... Style is not a frosting added to a substantial cake ... The substance of a cake is not the list of basic ingredients. It is the style in which they are combined. (McCloskey, 1988: 286)

If one accepts that the style and the substance of an argument are bound up together, and that the device of rhetoric plays a key role in the binding, one can then think about the productive uses of a concern with rhetoric in both the writing and reading of human geographical research. These seem threefold. First, an appreciation of rhetoric allows you to explore how different understandings of 'what research is for' are put into practice. Secondly, they help you gain a critical sense of why or why not something 'grabs you' and holds your interest – of why a particular piece of writing might appeal to you more than another. Thirdly, this concern with rhetoric can be used in analysing how the stylistic substance of research texts is not only a matter of the free choices of authors but also stems from the social contexts and institutional structures of academic research practice, which privilege certain genres over others. We will explore these issues in more depth now, beginning by looking at the way geographers have tackled the 'art' of regional and landscape description.

Regional and landscape description

We discussed in Chapter 1 geography's long engagement with regional and landscape description. We return to it here because it is a field of geographical research within which issues of representation and rhetoric have been debated explicitly for some time. Indeed, it is an area of the discipline where it has long been acknowledged that geographers should write 'creatively' in order to represent their subject-matter to the reader. For many geographers, the representation of regions and landscapes placed geography firmly within the arts and humanities, and they sought to introduce large elements of creativity into their own work. One of the most influential of all geographers, H.C. Darby, for instance, drew a firm distinction between geography as a science and geography as an art:

> Geography is a science in the sense that what facts we perceive must be examined, and perhaps measured, with care and accuracy. It is an art in that any presentation (let alone perception) of those facts must be selective and so involve choice, and taste and judgment ... It would be interesting to organise ... a competition for geographers, and to invite an account of the regional geography of south Lancashire in the manner of say, Estyn Evans or Dudley Stamp or S.W. Wooldridge. I venture to say that the various accounts would not be identical, any more than a portrait of a man by Graham Sutherland would be identical with one of the same man by Henry Lamb. Each picture would have its own validity. (1962: 6)

Here Darby is applauding creativity through description and is well aware that different authors will inevitably produce different accounts of the same region – but, as he says, 'each picture would have its own validity'. Hart (1982: 2) drew a similar distiction between geography as a science and geography as an art, and argued that 'the highest form of the geographer's art is producing good regional geography' – which he defined as 'evocative descriptions that facilitate an understanding and an appreciation of places, areas and regions'.

Darby, though, also recognizes that such evocation and creativity are not particular strengths of geographical writing. As he puts it: 'good description of landscape or town-scape is not an outstanding feature of the writing of professional geographers … We look in vain for – to use an old-fashioned word – a "likeness" of, say, chalk downlands or clay vales or mountain uplands' (1962: 1). Another famous regional geographer, E.W. Gilbert, also drew a contrast between the ability of geographers to portray the 'person-ality' of a region and the skills of novelists. He wrote that

> English regional novelists display many mer-its that geographers can recognise and envy … The regional novelists have been able to produce a synthesis, 'a living picture of the unity of place and people', which so often eludes geographical writing. The geogra-pher often speaks of the 'personality' of the region and this is exactly what some novelists have brought out so strongly. (1972: 124–5)

By making explicit this contrast Gilbert is highlighting a particular geographical problem with how we might represent landscapes and regions textually. In many ways novelists have more licence to weave images of people and places together across the length of a piece of fiction. For geographers the difficulty remains of translating the geographies that we per-ceive into words. This difficulty partly arises because 'what one sees when one looks at geographies is stubbornly simultaneous' (Soja, 1989: 2) – picture an urban street scene of people driving to work, young children walk-ing to school, local residents going shopping, old men chatting, all set against a backdrop of varied architectures and styles. Even the most physical of landscapes will have several, if not many, different elements: topography, vegeta-tion, aspect, climate. According to Soja, the 'problem' with trying textually to represent such simultaneity is that 'what we write down is successive, because language is successive'

(1989: 247). In other words, the sequential, linear flow of textual narrative is not best suited to describing the complexity of any given space at any given moment in time. Ultimately Soja is left to conclude that 'the task of comprehensive, holistic regional description may therefore be impossible' (1989: 247).

However, he has sought his own way out of this dilemma by using a particular rhetorical device which runs through his powerful writ-ings on Los Angeles. Soja aims to describe that city in a 'free-wheeling' style (1989: 2), which makes its argument through the composition of that description. In this case Soja aims to 'creatively juxtapose … assertions and inser-tions of the spatial' (1989: 2). It is through the force of this juxtaposition that the description comes alive. Box 11.5 illustrates the manner in which he 'asserts and inserts' the spatial.

Soja, then, uses the rhetorical devices of insertion and juxtaposition to try to capture the multiple meanings of Los Angeles within his 'fulsome geographical text'. But this, of course, is not the only route to creative regional description. Figures 11.1 and 11.2 show Quoniam's annotated paintings of Arizona. The rhetorical device used here is one which is centred on the shape and appearance of the text rather than on its structure and content as with Soja. The figures represent Quoniam's notes from a trip through Arizona. As Cosgrove and Domosh point out:

> The notes are in fact illegible, except for the occasional stimulus word like 'Coronado', so that the text for the most part is meaning-less as language. Its meaning derives from its formal qualities as blocks of inscribed signs which we recognize as written notes and thus assume to have recorded and produced meaning. (1993: 32)

In other words, we assume that these notes hold geographical knowledge, even if the nature of such knowledge is not immediately

Box 11.5: Taking Los Angeles apart: the art of description through juxtaposition

All the following quotations come from Chapter 9 of Soja's (1989) *Postmodern Geographies*, where the reader is taken on a textual tour of Los Angeles.

What is this place? Even knowing where to focus, to find a starting point, is not easy ... (p. 222). What follows ... is a succession of fragmentary glimpses, a freed association of reflective and interpretive field notes (p. 223). However ... Los Angeles ... must be reduced to a more familiar and localized geometry to be seen. Appropriately enough, just such a reductionist mapping has popularly presented itself. It is defined by an embracing circle drawn sixty miles out from a central point located in the downtown core of the City of Los Angeles (p. 224). An imaginative cruise directly above the contemporary circumference of the Sixty-Mile Circle can be unusually revealing ... The circle cuts the south coast ... near one of the key check-points regularly set up to intercept the northward flow of undocumented migrants ... The first rampart to watch, however, is Camp Pendleton Marine Corps base, the largest military base in California in terms of personnel ... After cruising over the moors of Camp Pendleton ... we can land directly in Rampart no 2, March Air Force Base (p. 225). What in the world lies behind this Herculean Wall? What appears to need such formidable protection? In essence we return to the same question with which we began: What is this place? (p. 229). The Sixty-Mile Circle is ringed with a series of ... outer circles at varying stages of development, each a laboratory for exploring the contemporaneity of capitalist urbanization. At least two are combined in Orange County, seamlessly webbed together into the largest and probably fastest growing outer city complex in the country (world?) ... The Orange County complex has also been the focus for detailed research into the high technology industrial agglomerations that have been recentralizing the urban fabric of the Los Angeles region (p. 231). These new territorial complexes seem to be turning the city inside-out, recentering the urban to transform the metropolitan periphery into the core region of advanced industrial production (p. 233). The downtown core of the City of Los Angeles ... is the agglomerative and symbolic nucleus of the Sixty-Mile Circle (pp. 235–6). Looking down and out from City Hall the site is especially impressive ... Immediately below and around is the largest concentration of government offices and bureacracy in the country outside the federal capital district ... Included within this carceral wedge are the largest women's prison in the country and the seventh largest men's (p. 236). Looking westward now, toward the Pacific and the smog-hued sunsets which brilliantly paint the nightfalls of Los Angeles, is first the Criminal Courts Building, then the Hall of Records and Law Library ... Along the northern flank is the Hall of Justice, the US Federal Courthouse and the Federal Building, com-pleting the ring of local, city, state and federal government authority which com-prises this potent civic center (p. 237). From the center to the periphery ... the Sixty-Mile Circle today encloses a shattered metro-sea of fragmented yet homoge-nized communities, cultures and economies, confusingly arranged into a contin-gently ordered spatial division of labor and power (p. 244). I have been looking at Los Angeles from many different points of view and each way of seeing assists in sorting out the interjacent medley of the subject landscape. The perspectives explored are purposeful, eclectic, fragmentary, incomplete and frequently contra-dictory, but so too is Los Angeles and, indeed, the historical geography of every urban landscape. Totalizing visions ... can never capture all the meanings and signi-fications of the urban when the landscape is critically read and envisioned as a ful-some geographical text (p. 247).

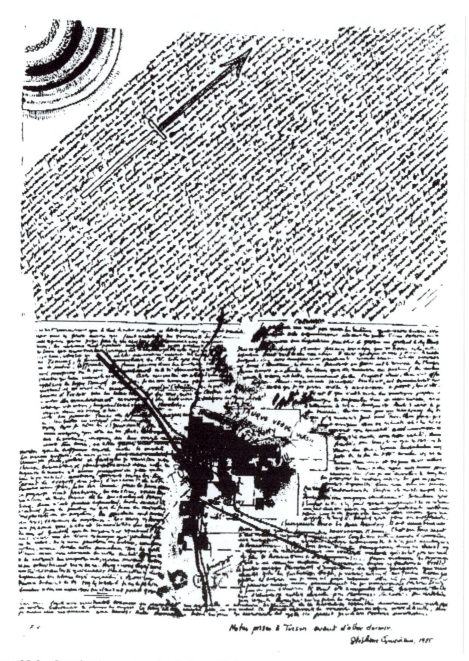

Figure 11.1 Quoniam's annotated painting of Arizona: Tucson

Source: Cosgrove and Domosh (1993: 33)

recognizable. The form of the notes is also important. Text is not always aligned horizontally but is sometimes inserted at an angle, resembling more the usual method of representing geological strata. Cosgrove and Domosh conclude that the 'structure of writing thus becomes the structure of language, its meanings layered like those of the earth itself'

Figure 11.2 Quoniam's annotated painting of Arizona: Coronado

Source: Cosgrove and Domosh (1993: 34)

(1993: 32). Here also the rhetorical device extends beyond words to encompass drawings, maps and sketches. However, in Quoniam's art it is these pictures which becomes the text. As he says, 'painting becomes the text of my ideas and the text becomes the descriptive picture of the landscape' (cited in Cosgrove and Domosh, 1993: 35). Cosgrove and Domosh conclude that 'by revealing the interchangeability of word and picture, he forces us to recognize words as simply another surface. But that surface, like a painting or an illustrated geological section, is not superficial: it contains and creates meanings' (1993: 35).

This section has sought to convey that the meanings created by our text are partly structured by the rhetorical devices we use. As we noted, the use of different rhetorical devices is expected, and even encouraged, in the 'art' of regional description. However, we do not see rhetoric and representation as optional extras in human geographical research that only some human geographers – the more arty types, those leaning to the humanities – engage in. The degree to which human geographers are explicit about rhetoric and representaion within their texts varies, but the degree to which they engage in them does not. Rhetoric is as implicated in constructions of scientific objectivity as it is in constructions of academic subjectivity. As Richardson (1990: 15) puts it 'science does not stand in opposition to rhetoric; it uses it'. We will now illustrate how science uses rhetoric by looking at some of the devices contained within the most 'objective' pieces of writing.

Rhetorics of objectivity

This section develops the notion that scientific objectivity does not stand outside rhetoric but, rather, is in part constructed through it. In so doing it helps to confirm the message that, ultimately, all research is textually constructed. It is, as we noted at the beginning of the chapter, written through rather than simply written up. Before we come on to the way research is presented in written accounts we will spend a short time examining what might be claimed as some of the most objective sort of geographical information we can find: railway timetables. We do this to illustrate how even these are highly rhetorical and implicated in wider connotations. As Bonsiepe puts it: 'Information without rhetoric is a pipe-dream ... "Pure" information exists for the designer only in arid

abstraction. As soon as he [*sic*] begins to give it concrete shape, to bring it within the range of experience, the process of rhetorical infiltration begins' (cited in Kinross, 1985: 30).

Figures 11.3 and 11.4 show two London North Eastern Railway (LNER) and timetables, the first from 1926 and the second from a redesign shortly after this date. The major difference between them is that of typeface, although dashes have been substituted for dots in alternate rows and the dots are further apart. The redesigned version shown in Figure 11.4 uses Gill Sans typeface. It was chosen to give all LNER printed matter a common identity. More than this, however, it was felt to possess some intrinsic qualities which made it especially suited to this purpose. As explained at the time, 'it is so "stripped for action" that as far as *glance* reading goes, it is the most efficient conveyor of thought' (Kinross, 1985: 23, emphasis in original). Such efficiency was necessary to help the 'passenger being jostled on a crowded platform on a winter evening, and trying with one eye on the station clock to verify the connections of a given train' (Kinross, 1985: 23). Kinross goes on to contrast these examples with the design of later timetables from both British Rail (Figure 11.5) and the Dutch railways (Figure 11.6). Each of these uses different rhetorical combinations of typeface, type style, symbols, spaces, dots and lines to impart their 'information'. When taken together these carefully designed combinations of textual features tell us something 'about the nature of the organization that publishes them' (Kinross, 1985: 21). The 'sense one has of ... their respective contexts ... is a consequence of the rhetorical devices they employ'; in short, these examples remind us that 'nothing is free of rhetoric' (1985: 29), and that even the most straightforward geographical information (the time and place of the train) is conveyed to the reader through deliberate rhetorical devices.

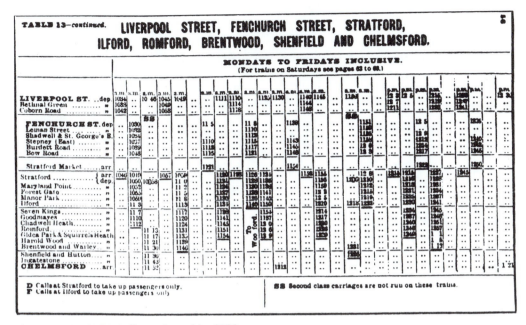

Figure 11.3 LNER Railway timetable, 1926

Source: Kinross (1985: 19)

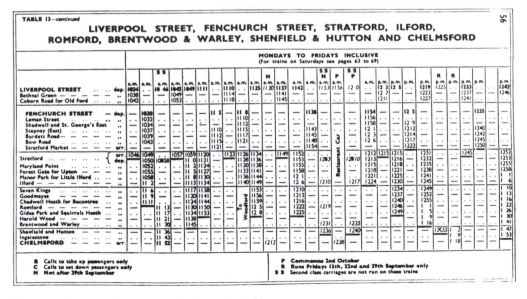

Figure 11.4 Redesigned LNER Railway timetable

Source: Kinross (1985: 20)

Table 8 **Mondays to Fridays**

London to Colchester, Walton-on-Naze and Clacton

Saturday service will apply Bank Holiday
Mondays, 27 May, 26 August and 31 March

								B					A						WTh0				
London Liverpool Street	d	18 22		18 40		18 42	19 00	19 04	19 30	19 34	20 04		20 05		20 30	21 04	21 30		22 04			23 00	23 06	23 55
Ilford	d	18b53				18b34		19 18		19b38	20 18		19 58			21 18			22 18				23 20	23 38
Romford	d	18b27				18b46		19 25		19b45	20 25		20 05	20 25		21 25			22 25				23 27	23 45
Shenfield	d	18 49				19 10		19 41			20 41		20 31	20 41		21 41	21 53		22 41				23 46	00 19
Ingatestone	d					19 15		19 46						20 46		21 46			22 46				23 51	
Chelmsford	d	19 00				19 23	19 31	19a54			20 14		20 40	20a54	21 04	21a54	22 03		22 55			23 44	23a59	00 33
Hatfield Peverel	d					19 31					20 22						22 09							
Witham	d	19 11				19 36	19 41				20a26		20 49		21 15		22 14		23 06			23 56		00 44
Kelvedon	d						19 46						20 54				22 19		23 11					
Marks Tey	10 d						19 51					20 54	21 00				22 25		23 18	23 21				
Colchester	10 a	19 25	19 31			19 52	19 58		20 21			21 01	21 08		21 29		22 35		23 26	23 29	00 13			08 58
St. Botolphs	d	19 26		19 36			20 00						21 10				22 37							
	a			19 42																				
Hythe	d			19 48																				
Wivenhoe	d			19 52								21 16				22 43								
Alresford	d			19 56								21 20				22 47								
Great Bentley	d	19 38		20 00								21 25				22 53								
Weeley	d			20 04																				
Thorpe-le-Soken	a	19 43		20 07		20 15						21 31				22 58								
Thorpe-le-Soken	d	19 44				20 18						21 34						23 03						
Kirby Cross	d	19 51				20 22						21 38						23 07						
Frinton	d	19 54				20 25						21 41						23 10						
Walton-on-Naze	a	19 57				20 28						21 44						23 13						
Thorpe-le-Soken	d		19 49	20 08		20 21						21 36				22 59								
Clacton	a		19 56	20 15		20 28						21 43				23 07								

Figure 11.5 British Rail timetable, 1974–5

Source: Kinross (1985: 21)

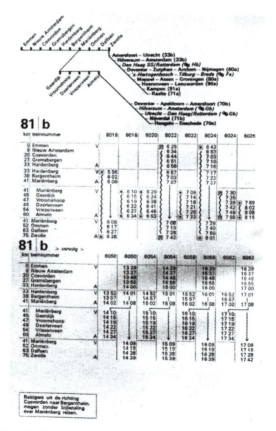

Figure 11.6 Dutch national railway timetable, 1970–1

Source: Kinross (1985: 22)

Our second example of the deliberate crafting of particular forms of expression in the context of varying social fields is concerned with the process of research and, more particularly, with the way this process is presented in objective scientific papers. It is drawn from the work of Gilbert and Mulkay (1984), which explores the contextually specific written and spoken discourses of biochemists in the field of 'oxidative phosphorylation'. Although not directly taken from geography this is a powerful example of the way that the same piece of research is accounted for differently at different times and to different audiences – including in journal papers, coffee-room chat and gossip, and formal explanations to sociologists in an interview. Their central message is not to pull down the façade of objectivity (which is presented through written journal articles) but to recognize it as a constructed voice, and as only one of the many voices constructed during research. When seen in this light 'scientific' writing loses its position as the voice of universal remit and is seen instead as a particular rhetorical construction set up to serve a specific purpose of persuasion.

Gilbert and Mulkay are particularly concerned to compare accounts of the same piece of experimental science as represented in a published paper and in conversations about the research. The opening paragraph of the introduction to the scientific paper reads as follows:

A long held assumption concerning oxidative phosphorylation has been that the energy available from oxidation-reduction reactions is used to drive the formation of the terminal covalent anhydride bond in ATP. Contrary to this view, recent results from several laboratories suggest that energy is used primarily to promote the binding of ADP and phosphate in a catalytically competent mode [ref. 1], and to facilitate the release of bound ATP [refs 2 & 3]. In this model, bound ATP forms at the catalytic site from bound ADP and phosphate with little change in free energy. (Gilbert and Mulkay, 1984: 41)

The formal account of the research presented here contains several rhetorical devices to persuade the reader of the validity of the paper's argument. The opening sentence portrays other scientists' work, but not as reasonably held beliefs associated with experimental evidence. Instead these existing accounts are labelled as mere 'assumptions'. This prepares the reader for the alternative view, which is duly delivered in the second sentence. This time, however, the reader is led to believe that this alternative opinion is based on hard data obtained from several laboratories – clearly a case of 'one group's results undermining the other group's assumptions' (Gilbert and Mulkay, 1984: 43). This point is then driven home in the third sentence which refers to a model. Gilbert and Mulkay conclude that 'in the course of three or four sentences, the text has conveyed a strong impression … that the rest of the paper is based upon a well established analytical position which constitutes a major advance on prior work' (1984: 43). However, no biochemical findings have been presented to support this impression: it has been achieved through rhetorical devices which convey particular views.

That these are also partial and highly selective views, one voice among many, becomes apparent when conversations with the paper's author are explored. In these he describes his reaction to the ideas embodied in the 'model' referred to in the paper when these were first suggested to him by the head of his laboratory:

He came running into the seminar, pulled me out along with one of his other post-docs and took us to the back of the room and explained this idea that he had … He was very excited. He was really high … It took him about 30 seconds … And so we sat down and designed some experiments to prove, test this…. It took him about 30 seconds to sell it to me. It really was like a bolt. I felt,

'Oh my God, this must be right! Look at all the things it explains'. (1984: 47)

Thus contrary to the formal paper, where we are told that experimental results suggested a model which seemed an improvement on earlier 'assumptions', the interview account suggests a dramatic revelation which was immediately seen to be right. Gilbert and Mulkay point out that 'Whereas in the [formal account] the model is presented as if it followed impersonally from experimental findings, in the [informal account] the sequence is reversed and the importance of intuitive insights is emphasised' (1984: 48).

Several other differences are revealed between the two accounts, concerning the status of experiments, the degree of commitment to the model and the fact that only two laboratories seem to have been involved rather than 'several'. Gilbert and Mulkay conclude that the participants' actions and beliefs were systematically different in the two settings, leading them to identify 'two major interpretative repertoires, or linguistic registers, which occur repeatedly in scientific discourse' (1984: 39). Our point here is that these competing repertoires are built up through different kinds of rhetoric. The formal scientific register, used in the paper, constructs the research process and those who take part in it in a particular way – as rational, logical and impartial. A certain style of writing is used which constructs the scientists' actions and beliefs as following unproblematically from the results of impersonal empirical findings. However, in conversation, the scientists constructed accounts of their actions which were much more contingent and dependent on personal and social circumstance. They also revealed their thoughts to be inspired and institutional, even quasi-religious. The differences in these two 'linguistic registers' reveal that scientific objectivity is also partly constructed through sets of rhetorical devices which are deliberately employed as part of a particular writing strategy.

Representation in practice

The previous section has shown how rhetoric is used to structure all written accounts of research, whether humanistic and cultural or scientific and objective. This general point about the importance of rhetoric can now be broken down into more detail by looking at some specific aspects of textual representation within human geography. All geographers have to negotiate these different aspects when they sit down to write, although they will do so to varying degrees depending on their object and method of study. However, they usually deal with them implicitly and, in many cases, almost automatically. We will consider them here in order to raise a set of explicit questions around key facets or moments of textual construction.

The textual construction of data: inscription, transcription and description

The interpretation of data through the process of writing begins with the notes, comments and observations which are made during the actual research process. Part I of the book drew attention to these from the perspective of data construction. In this section we will shift the emphasis to look at the ways in which what we write when we undertake research also begins to shape our interpretation of the data we produce. As we pointed out at the beginning of this chapter, writing is not something which is done only at the end of the research when we are 'writing up'. Writing is something which is carried out all the way through the research process – and which is integral to it – from the jotting down of the very first ideas, to the notes made in the field, to the finished article or dissertation.

The way in which elements of interpretation are present in such writing, long before any formal analysis is undertaken, can be illustrated through the example of ethnographic research which we explored in Chapter 6. There we referred to the many layers of description which we undertake in the field and through which ethnograhic data are textually constructed. Clifford (1990) notes that running across all these different layers of data collection are the three elements of inscription, transcription and description (see also Sanjek, 1990). For the ethnographer, inscription refers to the notes and jottings (sometimes just a mnemonic word or single phrase) made by the researcher so that he or she can fix an observation or later recall what someone has said. These are attempts to record social life 'in motion' as it happens around the researcher and are thus usually swift and somewhat partial. They are made contemporaneously with, or soon after, the events which are observed or words which are heard. Description is the name given to the element of writing which takes place at a slight distance. The account made is fuller than in inscription and may well be done in the evening when reflecting on the fieldwork undertaken during the day – often using the inscribed notes made in the field. The idea is to put together, to construct, a more or less coherent representation of what the ethnographic researcher has seen and heard during his or her research. Transcription refers to the careful recording of an informant's precise words as they are spoken during the fieldwork encounter.

The point we wish to make here is that all three elements are inevitably part of a creative process, one in which data are always as much textually and intertextually constructed as socially constructed. This is because the recording of data in the field is far from mechanical. As Clifford (1990: 57) puts it, 'the "facts" are selected, focused, initially interpreted, cleared up'. In many ways they have to be because what we are creating through

fieldwork is a 'recontextualised, portable account' (Clifford, 1990: 64). The 'present moment' of the actual fieldwork encounter is 'held at bay' and a 'reordering goes on'. Thus 'fieldnotes are written in a form that will make sense elsewhere, later on' (Clifford, 1990: 64). Hence the experiences that are recorded in the field become 'enmeshed in writing … that extends before, after and outside the experience of empirical research' (1990: 64). In this process of reordering, so that events can be made portable and revealed elsewhere (in a paper, report or dissertation), an initial round of interpretation has already taken place. Certain events and particular insights have been privileged and others given lesser prominence. In particular, in our desire to present a rounded coherent narrative, much of the complexity and messiness of the real world gets written out and smoothed over. Hence Clifford remarks that 'we seldom encounter in published work any cacophony or discursive contradiction of the sort found in actual cultural life and often reflected in fieldnotes. A dominant language has overridden, translated and orchestrated these complexities' (1990: 59).

However, the manner in which we textually fashion those whose behaviour and actions we are researching does not rely solely on these 'technical' issues of recording, selecting and reordering. There is always a moral dimension to how we describe and represent others. As Richard Rorty puts it, 'whatever terms are used to describe human beings become "evaluative" terms' (1987: 245). Even if the individual researcher uses particular terminology in a neutral 'descriptive' manner, 'there is no way to prevent anybody using any term "evaluatively"' (Rorty, 1987: 245). Thus, while many social scientists, including social geographers, have used the term 'underclass' to describe that proportion of the population who are denied access to work and as a result end up having to cope with poverty and deprivation, others have used the term in a more evaluative sense to depict a shiftless, lazy and dangerous substratum of society. We should be wary of the ways in which our initial categorizations and descriptions of those we are studying can carry a moral interpretation as well as an analytical interpretation. However, whether our interpretation is moral or analytical, the key point to make here is that even the most preliminary notes and observations made in the field are already casting a particular interpretive light on the eventual 'findings' of the research.

The textual construction of interpretations: metaphors

A second major way in which geographers have begun to cast an interpretive light on their data through the style of their writing is via the use of metaphors. Metaphors are powerful rhetorical devices which function to make something intelligible to an audience through equating it to something else. In this manner the characteristics of one object (usually familiar) are transferred to those of another object (usually less familiar) (Barnes and Gregory, 1997: 510). However, in this transference metaphors become central to the construction of interpretations and ideas. Barnes and Duncan (1992) distinguish between big and small metaphor use.[3] Small metaphors are those we use in an almost everyday sense, which 'pepper' (Barnes and Duncan, 1992: 11) our writings. An example is, of course, the peppering of our prose with metaphors. Many of these are unconscious and we use them all the time – we accuse others of glaring errors, we praise someone for tying up the loose ends of his or her argument, we rip the work of our competitors to shreds. Some of these everyday metaphors are geographical in their content and are built around spatial elements. Pratt (1992), for instance, writes of the way social and cultural theorists use

metaphors of mobility, travel, marginality and borderlands to help construct their arguments. Geographers, for instance, have been especially fond of the metaphor of the frontier to help convey all kinds of images: of diffusion, spread, development, economic change, social contestation (see Smith, N., 1996). Small metaphors such as these are now part of the language and cannot be avoided – we simply need to recognize they are there and look to mobilize (another metaphor) the most effective ones to help make our arguments as convincing as we can.

Big or large, metaphors are also important in academic writing. They help to structure whole schools of thought and 'shape the very explanatory framework' (Barnes, 2000b: 500) which these schools offer. They can also help to determine research agendas. Barnes (1992; 1996) uses the example of economic geography to discuss the manner in which these large metaphors have been employed. He shows how spatial scientists drew on metaphors of the gravity model and rational economic man to construct their models of spatial economic behaviour in the 1960s. Places became masses which interact according to a function of their distance apart, and urban land use is determined by the rational use of that space by individual actors. Of course neither situation corresponds to the social and economic world we study, but the metaphors drawn from physics were used to construct models which were in turn used to 'understand' urban and regional change (see also Chapter 9 on the type of explanation favoured by spatial science). However, Barnes shows that the use of metaphor has continued in economic geography and was not restricted to spatial science. Warde (1985), for instance, has criticized Massey's use of a geological metaphor where she conceives of distinct localities in terms of successive layers of economic activity, laid down on top of one another. Similarly, Amin and Robins (1990) note that Scott and Storper's work on economic agglomeration relies on the metaphor of flexibility. Once accepted, these big metaphors all helped to structure the way human geographers thought about and explained the uneven development of economic activity. Again, such metaphors are difficult to avoid. One simply has to be aware of them and employ them to maximum effect in one's own writing.

Questions of genre

This section will concentrate on how textual construction operates in social and institutional contexts and will argue that one of the main expressions of this is through conceptions of genre and generic conventions. These structure what authors (can) write, and seek to connect authors and readers into common frames of reference. Sometimes the genre used is defined by specific conventions. For example, most universities have rules and regulations which determine the length and style of undergraduate dissertations. Different academic journals also have different specifications in terms of word length and stylistic features, such as footnotes. Publishers will have particular expectations of textbooks and research monographs in terms of how these are laid out and constructed. However, we need to move beyond these basic stylistic features of different genres to appreciate that modes of interpretation are also predicated upon different textual conventions. Sayer (1989: 262), for instance, argues that one of the problems with much philosophical literature is that it assumes that 'the content of knowledge is indifferent to its form'. In contrast, Sayer highlights how the content of knowledge is intimately connected to the form. Thus enumeration, understanding and explanation are all predicated on different textual conventions. Using the example of regional geography, Sayer shows how different authors structure their texts in different ways

depending on whether they are presenting atheoretical descriptive narratives, positivist models, structural analysis or realist concrete accounts. This means that 'writing … is influenced by its location within ongoing debates and against rival academic and political groupings. So we not only write about things but for, with and against others. In this sense, the researcher's motives are not individual but relational' (Sayer, 1989: 269). Because writing is relational it is always partly defined and understood within, and in opposition to, particular genres. Broadly, it flows from this that genres and their institutional enforcement need to be understood and recognized by readers (see Agger, 1989a; 1989b), understood and used by authors for strategic purposes, and pushed and bent by authors when appropriate.

Composition and style

This final section on individual elements of rhetoric examines how the overall construction of the text should be considered in terms of interpretation and is an important part of what one actually says. Put simply, different sorts of research, based on different materials and with different audiences in mind, may suggest different compositional strategies. We can illustrate this with references to two pieces of geographical research on consumption – Alan Pred's book *Recognizing European Modernities* (1995) and Ben Malbon's book *Clubbing: Dancing, Ecstacy, Vitality* (1999). Pred's book is subtitled 'A montage of the present' and is very carefully composed in a montage style. At the outset Pred (1995: 11) quotes approvingly from Walter Benjamin: 'Method of this work: literary montage. I have nothing to say, only to show.' Pred uses visits to three spectacular spaces – the Stockholm Exhibition of 1897, the Stockholm Exhibition of 1930 and a contemporary arena, the Globe – to examine how new forms of consumption are entangled in changing discourses of power. He deliberately

361

uses a montage style, producing a new composite whole from fragments of words and pictures. In so doing Pred (1995: 28) notes that the avant-garde Soviet film-maker, Lev Kuleshov, referred to his own 'rapid intersplicing of differently situated people and urban landscapes as "creative geography"'. His book represents his own attempt at creative geography, the montage of text, photographs, poems, illustrations and citations being used to explore different elements of European modernity.

The idea is to use these different textual devices in a montage to 'make visible the multiple pasts layered within the present moment' (Pred, 1995: 264). Pred is attempting to use montage to reveal connections and interrelationships between elements of modernity usually held separate. The style and composition of the book are very much a part of the story being told. A new style of narrative is demanded in order to convey the complex intermeshings of political economy and cultural politics, at different – yet related – historical moments. Pred's ultimate message is that the multiple modernities that have existed within any one European country, and that continue to exist in the face of hyper- and postmodernity, can be revealed only through a textual strategy which invokes multiple meanings and representations. The book ends (1995: 264) with an echo of the Benjamin quotation which began it (along with a nod to one of Lenin's most important works):

What are the tension-riddled images to be evoked?
What is to be shown?
What is to be done?

In Pred's case the very deliberate composition of the book through a montage of fragments serves to highlight the complexity and interrelation of different elements of consumption within European modernity. Malbon (1999) uses a different compositional strategy to inform his understanding of consumption. In his case the textual motive is not montage but a night out. Malbon's object of study are the cultures and spaces of clubbing. He uses these to explore a key element of the geography of contemporary youth culture in the UK. As the book is about the experience of going clubbing, Malbon divides it up into three parts which mirror a night out. The opening part is labelled 'Beginnings' and is designed to contextualize the night out yet to come. In this opening part Malbon considers the key academic starting points for analysing clubbing – young people at play, consumption and identity formation. The second part comprises the main empirical section of the book, which investigates the various spaces and practices of clubbing. The reader is taken through four sections in this part, beginning with club entry and finishing at the end of the night. The final part mirrors those postclubbing times of reflection in seeking to understand the events and experiences of the previous night. Malbon deliberately mirrors the chronology of a night out in the narrative structure of the book. His self-avowed aim is to examine the practices, imaginations and emotions of the clubbers themselves, and he has carefully designed the composition of the book to reflect this. Perhaps Pred and Malbon are extreme examples of the way that composition can mirror and enhance content, but they both remind us that the structure of our writing can help inform the reader's interpretation of what we write.

Beyond the book

As we move into an age of multimedia presentation, digital photography, video streaming, text messaging and digital storage, exciting new possibilities are opening up for the presentation of geographical research. In short, the potential is now enormous for textualizing research through means other than books and journals. Geographers have only just begun to grasp this potential, so examples are limited – partly because of the disciplinary effect of accepted genres referred to earlier in this chapter. One example which we can refer readers to is the recent work of Michael Pryke on the redevelopment of Berlin following reunification. Potsdamer Square, at the economic and cultural centre of the former East Berlin, is being rebuilt with Western capital to form one of the largest redevelopment sites in Europe. Pryke has sought to capture the emerging economic, visual and audio rhythms of this new urban landscape with a visual and audio montage of Potsdamer Square. The resulting 'paper' (Pryke, 2000) can be found at www.open.ac.uk/socialsciences/geography/research/berlin/. This contains both audio and video clips of the area being redeveloped, and the 'reader' can click on different images and sounds which convey an impression of the new Berlin. These visual and audio montages are amenable to manipulation by the 'reader': you can zoom in or out of the video clips and stop and mix the audio sounds. The sounds available through the mini-disk recordings complement the visuality of the video clips. As a result, the 'reader' is introduced to a range of sensory experiences and can gain an appreciation of the visual and audio rhythms emerging in the new Berlin which would be impossible through written text alone. The different relations between 'author' (or photographer and sound recordist) and 'reader' (or watcher and listener) open up a realm of new interpretations and perceptions. The result is a fascinating glimpse of the possibilities which the new digital media are opening up for geographical communication (see also Pryke, 2002).

Conclusion

There are two main points to draw out of this chapter and to reiterate in this conclusion.

The first is that all human geography is fiction in the sense that it is textually made and fashioned. But unlike most novels it is largely about real people and real lives and real situations. The moral is that human geographers have to find ways to craft accounts that best suit the character of those lives and people and situations, while also attending to the trick of engaging their audience. This is not an easy process and partly explains why writing is such a difficult part of the research process. We have been at pains to stress how writing is indeed part of the whole research process and not just something which is tagged on at the end when the 'writing up' occurs. Secondly, this chapter has sought to present some helpful tips on what to consider when writing through research – but without undue prescription. It does not seek to use writing as a disciplining device, though it acknowledges it is that. Rather, it has tried to offer some routes through the process of textual construction that will let authors develop their own voices, their own styles, their (and their research subjects') own materials in ways which are appropriate for their research. It has also sought to provide some criteria upon which to judge and defend the choices they have made. But, ultimately, these choices will be made within a host of competing political, moral and ethical contexts. It is to these that we now turn in the final chapter.

Notes

1 See, for example, Becker (1986) and Wolcott (1990).
2 For those who wish to follow up such advice in detail, Northedge (1994) contains two excellent chapters on the craft and mechanics of writing.
3 See also Barnes and Curry (1992) who write of large and small metaphors.

12 The Politics of Practising Human Geography

We want to end the book with some reflections on the politics of geographical practice and on the power relations involved in both constructing and interpreting geographical data. One of the central messages running through this book has been the idea that all geographical research is a social activity. Far from operating in some kind of sealed environment or social vacuum (let alone an ivory tower), individual researchers are always working in particular social contexts. Indeed, the practice of research should not be viewed as a neutral, or technical, procedure but should instead be seen as a social process which is heavily laden with competing, and often conflicting, sets of demands and values. May (1997: 46) points out that the values of those involved enter the research process at all stages – in the interests leading to research (of both researcher and, where appropriate, funding agency); in the aims, objectives and design of individual projects; in data construction and interpretation; and in the use made of research findings. Through an investigation of each of these stages this chapter looks to situate the context within which research practice takes place, as well as to provide an appreciation of the demands and pressures which are placed on the researcher from a number of different sources. It is these demands and pressures, and the values which inform a response to them, which underpin what we term the 'politics' of research. The critical point is that all

research takes place within these multiple and overlapping sets of social relations and will inevitably be influenced by them.

We will indicate some of the key ways in which these social relations might affect the research process by focusing on different elements within the overall practice of research. We are not seeking here to be exhaustive, and there are complete books devoted to the politics of research.[1] Rather, we will pick out a range of indicative issues which will help us to show that human geographical research can never be viewed as value-free. We will begin by looking at those 'personal' politics which are implicated in the nature of the research we choose to do and in the ways we do it. We then move on to look at some of the broader politics which press in on the individual researcher and help to structure the context within which academic research takes place – both those which operate within the research process itself and those which influence research practice externally. We will finish the chapter with a look at how the twin issues of ethics and morality underlie each of these aspects of research.

The 'personal' politics of geographical practice

The first stage at which these various influences are felt occurs at the very outset of the research process when the research topic is devised and initial research questions are formulated. We might find the inspiration for these ideas in the books and articles that others have written or in discussions we have had with colleagues or fellow students. Field trips can be another source of inspiration, and encounters with new places and new peoples can trigger all kinds of research ideas. Ultimately, though, we all carry with us our own unique geographical imaginations which are fired in different ways through different means. As one of us has written elsewhere:

> An understanding of why we are interested in particular subjects/objects/modes of human geographical study will not solely be found in the quality and persuasiveness of the canon of study which already exists, important though that is. Such an understanding will also demand some scrutiny of the subjectivities, identities, positionalities and situated knowledges that we as individuals bring to the collectivity of that canon. (Cloke, 2004a: 1)

In this section we will talk through some of the ways in which these subjectivities, identities, positionalities and situated knowledges provide the prompts for our individual practice of human geography. We can combine these influences under the label of 'personal' politics – deliberately written with a small 'p' because they are not reducible to the more formal political world of elections, parties and pressure groups. This is not to deny the influence of formal politics (Harvey, 1996; 2000; Watts, 2001) but to acknowledge that for many of us the impulses which push our geographical sensibilities in a particular direction are much more diverse. Tracey Skelton (2001: 89), for instance, has written of how the fieldwork for her PhD research 'was initially about my wishes, dreams and desires'. These in turn were originally prompted by a school visit from Miss Peacock (a missionary who had been in India) which left a childhood dream of spending time abroad and learning about other ways of life. When combined with a desire to conduct research from a feminist standpoint, which emerged through her degree, this led to a decision by Tracey to pursue research into gender relations in the Caribbean. Here we see an example of the way that the 'persuasiveness of the canon of study' – in this case feminist geography – combined with Tracey's own subjectivity and positionality to forge her particular research practice.

In the opening chapter of the book we introduced you to Carl and Linda and told you something of the way they practised their particular human geographies. We also introduced you to some of the sensibilities and values which influenced such practice. We now want to introduce you to a third geographer, whom we will call Rachel. Like Carl and Linda, Rachel is a real individual. She could well be you, for her story begins when she was an undergraduate student. Recounting Rachel's story allows us to illustrate the manner in which personal identities, emotions and values help to shape the nature of the research we do, and also how the very practice of that research may well encourage us to reflect back on, and even change, our own values and identities. We have chosen Rachel to illustrate these points because she has written so powerfully, eloquently and honestly about the manner in which her own geography has been practised. Her route into geographical research, and the influences on her subsequent journey through it, are outlined in Box 12.1.

Rachel's journey is instructive for several reasons. It confirms how the impetus to do research often comes from deep within us, out of our personal engagement and desires. It also shows how the influence of others – in this case, academics and fellow students – can help to shape our geographical imagination (see Cloke, 1994, for a similar account of such influences). And it shows how our positionality

Box 12.1: Rachel's journey

In 1992 Rachel left a small Yorkshire mining town to begin a geography degree. Her initial experiences at university were far from positive and the social and cultural environment she found jarred very strongly with the close-knit working-class community she had just left. As she has written, 'I arrived at university and suffered massive culture shock. I felt like an alien from another planet…my first year at university was a nightmare, and all I wanted was to wake up' (Saltmarsh, 2001: 142). But out of this nightmare came anger – at academia and academics – and this began to inform her writing and research. Gradually she replaced the opinion of others with her own, and her work became semi-autobiographical. As she explained:

> I was being forced to think about the social and cultural implications of my being [at university] due to the very nature of my degree subject. Lectures about class, urban and social geography, cultural geography and the way in which we were introduced early on to the politics of knowledge construction and research techniques, really forced me to think about how people and places were being described and analysed by mainly middle-class geographers and how this was tied up with my own experiences of being caught between two cultures. Crucially it was also encouraging me to think about who was speaking for whom and why. It was about power and I felt at the time that I had none. (2002: 4)

In an effort give voice to those who were powerless, Rachel decided to undertake her undergraduate dissertation on pit closures and the subsequent loss of mining cultures. That it was written from an autobiographical standpoint helped her to deal with her past and her present; as she put it, 'It was at [university] that I discovered who I was, who I wasn't and who I would never be' (2002: 2).

Rachel's dissertation topic and style had come directly from her need to explore her own roots and her desire to let working-class people speak for themselves. Her geographical imagination had been fired by her experience of university and her loss of community. She explained how writing this dissertation in turn 'kick-started my journey into further PhD research based around my class-culture, situated knowledge and autobiographical writing' (2002: 1).

But when she returned to her community to undertake her PhD research and continue with her particular form of geographical practice, she found that something had changed:

> The most startling issue I found during my PhD was the fact that my own class cultural identity was being re-created…In contrast to the experiences of my dissertation research, I often felt out of place in the town where I was born…The irony is that my class identity has kind of reversed itself on my journey through my education. I was held up by some of my interviewees as being living proof that it was possible to transform yourself from being working class into someone who is middle class through the education system. (2002: 7)

But it wasn't just Rachel's interviewees who saw her in a new light and were placing a new identity on her. In the course of her PhD research she also experienced what she has described as a 'loss of faith'. As she puts it:

> I can't explain what this loss of faith is exactly, but it has something to do with my changing identity. Going back to the places and people where I came from has made me realize that I have been gradually changing over time, without even

---

Box 12.1 (Continued)

realizing it...I am a different person to the one who left the working class, council housing estate to go to university. I'm a different person to the one who came to begin my PhD...I am re-negotiating my identity constantly. (2001: 146–7).

Rachel's account of her journey is evocative and moving. In the course of it she opens up all kinds of issues about class, gender and community, and the painful process (for some) of 'gaining' an education. We can only deal here with the elements which concern research practice, but her writing reveals much much more about the wider cultural and social politics of higher education.

---

within the research process, discussed in Chapter 1, is constantly changing and always having to be renegotiated. In the course of two or three years, between undertaking fieldwork for her undergraduate dissertation and fieldwork for her PhD – in the same place – Rachel had changed in the eyes of some of those she was interviewing. They no longer saw her as one of them but, instead, positioned her as part of an educated middle class. This repositioning by the research subjects will have inevitably affected the practice of research itself – for what people tell you, what they reveal and what they let you know are conditioned by how they see you and how they respond to you. As Claire Madge puts it, the practice of fieldwork involves

> playing out a multiplicity of changing roles during the course of the research. These roles, which are sometimes complementary, sometimes clashing, and which are contingent on our positionality, will affect the data given/gained and our subsequent interpretations. In other words they will influence what we produce as knowledge. Personal relationships with people will influence the ethical decisions we make regarding what we create as knowledge. Power, ethics and knowledge are interconnected. (cited in WGSG, 1997: 94–5)

We will now explore this interconnection a little further by talking about the politics of the research process itself. In doing so we will shift attention from the subjectivities of the researcher, which help to shape what and how

he or she chooses to research, to look at the wider politics which impinge upon the practice of research once it is under way.

The politics of research practice

Whatever the motivations for the shape and content of the research questions, there is an increasing recognition that the practices of research are themselves highly politicized. This issue can be explored on a number of levels but all involve viewing the research process itself as an exercise of power – whether power over the type of knowledge produced, power over those who are being 'researched' or power within the research team itself. We have been at pains throughout the book to stress that research is not simply a neutral activity for unveiling information but, instead, is intimately involved in the actual construction of that information. What is under consideration here is the manner in which this is done.

One of the reasons why it could be maintained for so long that the purpose of research was rather unproblematically to 'collect facts' is the distance that has traditionally been drawn, and indeed encouraged, between the researcher and the researched. It is a long-held notion that the more 'objective' the researcher is, the better and more reliable are the research results. But such 'objectivity' is impossible in social research and, as Acker et al. put it, 'neither the subjectivity of the researcher nor the subjectivity of the researched can be eliminated in the [research] process' (cited in May, 1997: 52). Indeed, as we have already noted in Chapter 5, the practice of research can never be a neutral exercise since 'the very act of entering the worlds of other people means that the research and the researcher become part co-constituents of those worlds. Therefore we cannot *but* have impact on those with whom we come into contact' (Cloke et al., 2000a: 151, emphasis in original). Thus, rather than being viewed as two separate parts of a purely technical research process, the researcher and the researched are co-constituted during the practice of research. Each influences the other and is affected by the other.

This situation is made more complex, and even more difficult to negotiate, by the fact that neither researcher nor researched are stable and singular categories. Each will shift as differing identities come to the fore during the process of research. The identity of both researcher and researched will be multiple – embracing and crossing, for instance, categories of gender, class, race, sexuality, age and disability. This helps to problematize the somewhat straightforward categories of researcher and researched. We have already given an example of how this might happen in Chapter 5 where we quoted a researcher reflecting on her position vis-à-vis those she was interviewing for a project on rural homelessness (see also Box 5.7):

> The stronger part of my self-identity is that of a single parent, and ...I have, on occasion, had difficulty (as many people do) with trying to maintain my rent payments and provide for my family. These experiences allowed for a certain degree of identification with my interviewees but, in certain situations, this proved problematic – at times it felt as if I was at once the researcher and the

> researched, perhaps too much of an 'insider' in some ways. (Cloke et al., 2000a: 143)

Such experiences are worth repeating here not only because they alert us to the sensitivity of the relation between researcher and researched but also because they indicate the micro-politics involved in the very practice of research, as the researcher strives to balance distance and closeness. Just how these issues are balanced, and the position taken on the continuum between insider and outsider, will affect the construction of research data. We can explore these micro-politics of research a little further by introducing you to a fourth geographer, called Robina. Her account of her research with Muslim women is set out in Box 12.2.

The complexity of the identities negotiated by Robina and her respondents during the act of research, and the impact these had on the data constructed, remind us that research is an embodied process and that people's positionalities can enable as well as constrain. Hence the outcome of research is always dependent on the micro-politics of the research process. As Robina puts it, 'the research "findings" can be seen to be "produced" out of a set of interactions and there will be differences in what even different researchers of the same colour, gender and/or class may find or give significance to' (Mohammad, 2001: 108).

Thus, in addition to considering the multiple roles and positions of any individual researcher, it is also the case that different researchers will approach the same research situation differently, and thereby construct different data from it. Cloke et al. draw our attention to these differences, and to the politics involved in negotiating them, when they recount the experience of a particular research situation:

> Although I was equipped with a mobile phone, there were some situations in which I felt I was putting myself in danger and this restricted my ability to do some of the research. There were some places I just wasn't

Robina Mohammad was involved in a series of research projects with Muslim women between 1994 and 1996, using participant observation, questionnaires, focus groups and semi-structured interviews. The way that she has written about these experiences reveals important insights into the positionality of the researcher and the politics of the research process (Mohammad, 2001). Robina is a Pakistani Muslim by birth but was brought up in Britain. Despite cultural and social differences forged through her upbringing and education, the Pakistani women Robina was researching saw her as one of 'them'. In order to play out this role she was expected to attend Friday prayers at the Muslim Women's Association (2001: 107). However, this status as insider was both beneficial and detrimental to the research. It was beneficial because it allowed access to material that other researchers would have been denied. As Robina puts it:

> I would often hear remarks such as 'we shouldn't really say this since it is disloyal [to the group] but since you are one of us it is all right', or 'I can tell you because you'll know what I'm talking about'. These comments were followed by the sharing of gossip about the group or personal experience that otherwise, it was suggested, would have been withheld or presented differently. (2001: 108)

As well as opening up closed knowledge, her status as 'insider' also hampered the construction of data. In these cases, Robina almost appeared 'too close for comfort, making people wary of sharing information' (2001: 109). Ultimately, Robina had to negotiate her positionality with each of her respondents, illustrating that the position of researcher is not fixed and immutable but highly fluid and somewhat vulnerable. But her position also affected the conduct of the research practice. Robina tended to dress in a Westernized style, which for some interviewees placed her on the '"insider"/"outsider" border [and] mediated both mine and the researched's performances in the situation of the interview' (2001: 110). She was seen variously as 'same', 'other' and simultaneously 'same' and 'other' – some women spoke to her only because she wore Western dress, while for others this created a barrier. The implications for the construction of data were manifold. As Robina concludes: 'I had a rough research agenda to follow, but what was created out of the process of fieldwork has never been entirely what could have been predicted...the knowledges produced are always versions. They represent one out of other possible truths' (2001: 113).

able to visit [and] missed opportunities which resulted from my position as a lone female researcher who lacked experience in handling certain situations. (2000a: 145)

The research process is always a complex social situation and the data constructed stem from an interactive set of activities which are constantly negotiated and renegotiated between the researcher and the researched. The politics surrounding these negotiations – in terms of positionality, identity and power – will of necessity impact upon the data constructed. As Robina suggested (see Box 12.2),

the knowledges which are constructed through these negotiations and renegotiations are always 'versions'. On a different day, with other participants, other possible (and competing) narratives would emerge.

Another dimension to the identity of the researcher is added by their potential position in a research team. Both undergraduate and postgraduate research tends to be carried out on an individual basis, but academic staff will often work in research teams. Large-scale funded projects are often undertaken by more than one person and may involve a team of several staff. The politics surrounding the operation of such teams are at once external and internal. Internally, the research team will aim to function as a unit, like all teams. Project directors, research staff, administrative and secretarial staff will all attempt to work together and will have various tasks within the overall framework of the research project. Inevitably different sets of social relations will be involved between the different members of the team, and the quality of the research data will be partly dependent on the characteristics of these relations. Externally, the politics involved will operate around the fact that different staff members perform different tasks and are involved in different elements of the research process. Cloke et al. (2000a) note how hierarchical positioning within the project team may lead to particular responsibilities being discharged within the research process. As one of the research staff put it:

> I can't help but wonder whether my increased sense of ethical dilemma was not also partly due to the fact that, as an employee and part of a team, I felt somehow more beholden to my employers and colleagues to do a particularly good job of work and come up with the stories I was collecting…the 'status' and 'role' of any researcher in a team are bound to affect the different ethical dilemmas and political 'conflicts of loyalty' they are likely to experience. (2000a: 144)

The politics of the academy

The task of research teams is to begin to shift our attention towards those external pressures which push down on the researcher in certain situations. In many ways the intellectual history of the discipline of geography can be read through the lens of these external pressures. In his book tracing this history, David Livingstone wrote that 'The heart of my argument is simply that geography changes as society changes, and that the best way to understand the tradition to which geographers belong is to get a handle on the different social and intellectual environments within which geography has been practised' (1992: 347). Hence geography's formation as a university subject in the late nineteenth century was linked to its utility as a science of exploration and mapping to be used in the service of colonial expansion, and the famous explorer H.M. Stanley noted how geography 'has been and is intimately connected with the growth of the British Empire' (cited in Livingstone, 1992: 219). The shift to quantification in the 1950s and 1960s has been interpreted by Harvey (1984) as 'a strategic manoeuvre to escape the political suspicion falling on social science' (cited in Livingstone, 1992: 324) during the McCarthy era of witch-hunts and repression in the USA, and even if one accepts the alternative argument that quantification stemmed from a search for scientific credibility, there is still a broader politics here in terms of what counts as science and what is seen as credible. Likewise, the backlash against quantitative spatial science that swept through the discipline from the late 1960s onwards was fuelled in part by a political desire to provide a critique of social and economic inequality. The opening editorial of the newly founded journal *Antipode* in 1969 (Vol. 1, no. 1) advocated the 'replacement of institutions and institutional arrangements in our society that can no longer respond to changing societal needs, that stifle attempts to provide us with a more viable pattern for living, and often serve no other purpose than perpetuating themselves'. And, of course, the more recent rise of new concerns within geography (such as those operating around issues of gender, race and sustainability) has mirrored the growth of these as political movements within society more generally.

Those seeking a handle on the social and intellectual environments which are helping to shape the current practice of geography are forced to confront significant change. Over the past two or three decades the rather congenial relationship between well-funded universities, government and society has begun to breakdown. Under this system, public trust and government funds combined to allow research to be largely driven by individual intellectual curiosity. This is no longer the case and, as Demerrit notes, the change has had implications for research as 'older narratives about scientific knowledge as a public good or as an end in itself no longer hold sway' (2000: 309). Instead, university funding is increasingly subject to tight budgetary control and external accountability. This is not necessarily a bad thing but, as Demerrit notes, 'the problem with the new social contract [sustaining academic research] is not so much with the idea of making academic research publicly relevant and accountable, but with the narrowness with which public relevance, accountability and value have been defined' (2000: 324).

The narrowness of the definition of relevance has had a significant impact on the content of geographical research, especially in the USA. Here, the US National Research Council published a report entitled *Rediscovering Geography: New Relevance for Science and Society* (1997). In it, a case was made for geography's utility as 'good science and societally relevant science'. But such utility was interpreted in a very particular and partial manner. As Johnston points out in his review of the report:

> it has identified a market for geographers with GIS and related analytical skills and has decided that this is the market that the discipline should be orientated towards... Radical geography, despite its major contributions – theoretical, empirical and practical – to the discipline's development over the last three decades of the 20th century, was implicitly characterised as 'nonscientific' and excluded from...the priorities for...teaching and research. (2000c: 988)

There are thus all kinds of politics surrounding the practice of geographical research which, together, help to define and delimit particular spheres of research activity as 'relevant' and therefore more legitimate than others. Some of these are internal: the US National Research Council report, for instance, was drafted by a committee of 16 practising geographers from within the discipline.[2] In other instances the pressure to define legitimate and acceptable fields of research inquiry comes from outside the discipline – especially from those who actually provide funds for research. In the UK those public sector agencies and charities that fund large-scale social science – such as the Economic and Social Research Council (ESRC), the Joseph Rowntree Foundation, the Leverhulme Trust and the Nuffield Foundation – all distinguish a set of particular themes as priorities for research funding (their current priorities can be found on their respective websites). It remains easier to obtain funding for research which is inside these themes rather than outside them. Individual researchers, or research teams, are able to shape their research proposals to some extent but, in order to win research monies, they must remain within the contours set by the funders. In this sense most research is reactive rather than proactive in that it responds to calls from funding agencies who will all have a particular politics to their research themes and funding streams. In other words, the model of the lone scholar producing research on items of his or her choosing is now fading and most large-scale geography research projects now take place within a system which is shaped by the demands of the research funders.

The private sector also has a role in determining which aspects of research are deemed as 'relevant' or 'legitimate' and therefore worthy of funding. These pressures are perhaps most clearly felt in those areas of academia which have a directly applicable economic benefit (such as biotechnology, genetics, software engineering or pharmaceuticals) but they do impinge upon particular areas of geography. Curry (1998), for instance, shows how industry sponsorship of GIS research and teaching laboratories has tended to undermine academic autonomy (see Demeritt, 2000: 324). Another example of the encroachment of private sector concerns into academic geography is provided by what became known as the 'Shell debate' within the Royal Geographical Society–Institute of British Geographers (RGS-IBG). In 1995 the RGS-IBG became the professional body representing academic geographers in the UK following a merger of its two constituent organizations – although some have argued that, given the unequal power relations which existed between the parties involved, the word merger is perhaps not entirely appropriate (Tickell, 1999). The oil company Shell was a corporate sponsor of the RGS, giving around £45 000 a year to its Expedition Advisory Centre for young people (Gilbert, 1999: 224; Woods, 1999: 229). There were anxieties about such sponsorship in discussions over the merger, with many academic geographers expressing concern about the environmental impacts of Shell's mining and drilling activities.

These concerns reached a peak in November 1995 when the Nigerian writer and political activist, Ken Saro-Wiwa, and eight other activists from the Ogoni region of the Niger Delta, were murdered by the Nigerian military regime (Gilbert, 1999). Saro-Wiwa was a critic of Shell's oil operations in the Niger Delta, claiming they devastated the environment and did little to benefit the Ogoni people. Following the executions, a campaign was mounted by some academic geographers to end the RGS-IBG's relationship with Shell. This was unsuccessful and Shell is still a sponsor of the merged society. Many of the geographers active in the campaign against Shell's sponsorship have resigned

from the RGS-IBG. Others have remained in the merged organization. This episode raises a whole host of issues concerning the relationship between academic research and private sponsorship (Kitajima, 1999; Watts, 1999b). As Gilbert notes:

> the debates over the future of the Shell connection have forced British geographers to think about environmental and political ethics, about the public role of their professional body, about the relationship between external funding and academic independence, and about the tensions between the academy and activism…It is hard to imagine an issue more calculated to test understandings of the nature of the discipline, and the political and ethical consequences of 'being a geographer'. (1999: 220)

What is interesting is that it was the actions of a private company on another continent which stimulated such a fundamental questioning of the consequences of 'being a geographer'.

For undergraduates, the consequences of 'being a geographer' may well be a temporary concern, one which is over in three years. But for many academics the nature, if not the consequences, of 'being a geographer' is something which has to be confronted and negotiated on a fairly frequent basis. A large part of being a geographer in a contemporary academic setting is bound up with research, and there are innumerable choices that have to be faced over both the content and direction of this research activity. Some of these choices are closely linked to the kinds of external pressures we have just sketched out. Most academic research, for instance, requires funding of some kind or another – to pay for research staff; to fund visits to specialist libraries and archives; to finance fieldwork; to buy computers and software; to pay for secretarial support; to purchase research reports written by others; and so on. Yet as we noted above, those who fund research are becoming increasingly prescriptive. Particular types of research on particular subjects stand more chance of finance than do others.

In addition to determining the nature of the research agenda, the funding agencies will also play a key role in shaping research output. Hence what is written within the discipline is also shaped by external pressures. Just as students have to respond to a set of rules and regulations concerning what they produce for their degrees, academics have to meet external requirements which shape what they write. Each research project will almost certainly have its own output requirements. The ESRC has traditionally expected research dissemination to concentrate on academic outputs in the form of books, chapters or journal articles but, in its desire to make itself more 'relevant', it now expects the projects it funds to produce outputs which are also aimed at the 'users' of research, ranging from the general public to the policy community.[3] The Joseph Rowntree Foundation requires a final report to be presented to it rather than any academic output, but it also expects user groups to be informed of the research results. The Nuffield Foundation also requires a research report. These requirements will inevitably shape the interpretation and dissemination of the research and help to determine how, where and in what format the research is presented. Again, this is not a matter purely for the researcher to determine, and we need to move away from the notion of the individual academic huddled over a desk writing up his or her research according to his or her own wishes. While the desires of the individual researcher to make a particular intervention in a certain set of academic debates will play a part in determining what gets written, a larger factor will be the needs and requirements of the funding agencies.

In the UK, research output is also driven by the Research Assessment Exercise (RAE). Those who fund higher education now use this exercise to determine the level of baseline research funding to each academic department. In total, around £5 billion of research funds were distributed in response to the results of the 2001 exercise. These exercises, which have been running since 1986, have considerably heightened the culture of performance and evaluation within UK universities (see Demeritt, 2000; Smith, D., 1995b; 2000). Each academic is expected to produce a certain number of publications over the lifetime of each research cycle, which are then rated for quality by a panel of expert reviewers. This in itself has been enough to alter the kind of research undertaken and the nature of the output produced. David Demeritt has written of his own experiences of these demands, as a geographer at the beginnings of an academic career. As he puts it:

> I feel such pressures impinging on my career path. While academic employment depends on a generalized 'excellence' in research, some research clearly pays better than does other research. Having trained as a historical geographer, I have found success in the academic labour market only by emphasizing my environmental expertise, which promises the kind of relevance valued by funding agencies. (2000: 321)

In addition to impinging upon the careers of individual researchers and academics, the pressure to deliver certain types of research in particular formats to satisfy performance review criteria has also had an impact on research dissemination. Publishers and journal editors have reported a large rise in the types of submissions most favoured by the RAE – journal articles and research monographs – and a fall in the type of work not considered RAE relevant, most noticeably textbooks and unpublished research reports. This may well be a major reason for the fact that, despite a growing sensitivity within human geography to exploring the world through a 'polyphony of voices', this has not yet been matched by a full recognition of multiple audiences, other than those which are demanded by research funders (Milbourne, 2000). As Cloke et al. conclude, 'New words and new worlds have not really travelled to new ears and eyes' (2000a: 147). In this sense research is still primarily carried out 'on' others rather than being done 'for' others.

To raise these issues is to begin to appreciate the various 'politics' which are involved in geographical research. As we noted earlier, in the words of Claire Madge, power, ethics and knowledge are indeed interconnected. And as we have seen, these politics and ethical considerations operate at every level throughout the research process – from the macro-scale politics involved in national-level research funding, to the micro-scale politics involved in the relation between researcher and researched. When wrapped up and looked at as a whole, what they tell us is that the design and implementation of research should not be seen as a narrowly technical issue, somehow uncontaminated by political and ethical questions. Instead, all stages of the research process are affected by these kinds of issues. Questions of how to negotiate and resolve them are complex and multidimensional, involving research funders and users as well as the researchers and the researched themselves. One way to plot a route through such complexity is to seek guidance from broader sets of moral and ethical considerations. This is what we will turn to next in the final section of the chapter.

Ethics, morality and geographical research

By embracing the larger concerns of ethics and morality we can start to open a window on a number of other questions which are implicated in the process of geographical research. The questions involve whom and what we choose to research; how we conduct that research and how we relate to our research 'subjects'; how we use interpretative and authorial power in constituting our research findings; and what use we make, and allow others to make, of these findings (Cloke, 2002; 2004b). To a certain extent these questions have been implicit in much of what we have written throughout this book. Engaging with issues of ethics and morality makes them explicit.

Philo (1991b) identified three main ways in which geographers might engage with the concerns of moral philosophy. He referred to the first as the geography of everyday moralities which are established through the 'different moral assumptions and supporting arguments that particular peoples in particular places make about "good" and "bad"/"right" and "wrong"/"just" and "unjust"/"worthy" and "unworthy"' (1991b: 16). As Philo points out, there can be little argument that these assumptions do vary considerably – from one nation to another, one community to another and one street to another. A second form of engagement might come from the way in we we investigate the geography in everyday moralities, which refers to the ways in which 'moral assumptions and arguments often have built into their very heart thinking about space, place, environment, landscape, and (in short) geography' (Philo, 1991b: 16). Again, there is little doubt about the ways in which particular behaviours and practices are viewed as suitable for particular spaces and places.

It is, however, the third form of engagement between geography and morality which we want to concentrate on in the remainder of the chapter, for this refers to the morality of geographical practices. This engagement arises through geographers reflecting 'upon their own morality, their own sense of "good" and "bad"/"right" and "wrong" and so on that permeates and maybe informs their own practice as academic geographers' (Philo, 1991b: 17) – including the practice of research. Such reflection may often be uncomfortable but, in many senses, geographers are ideally placed to undertake it for our whole discipline is founded on investigations of social and spatial inequality. If people and places were equal, geography would lose its rationale: it is built around uncovering and exploring division, difference and uneven development.

Yet the notion of a 'moral human geography' would surely have to do more than uncover and explore these differences. As Harvey remarked some 30 years ago, our task should not entail 'mapping even more evidence of man's patent inhumanity to man' (1973: 144). It would instead have to involve a commitment to helping to secure a more equal and just society. In the words of David Smith, 'it is in seeking the new institutions and social arrangements required to equalise human life prospects, whoever and wherever people are, that the creation of a moral geography can be envisaged' (2004: 207). There are many places where geographers can contribute; their work, after all, embraces a wide range of relevant issues, including the distribution of employment and wealth; the location and provision of welfare services; the exploitation and conservation of natural resources; the emergence of new cultural and political identities; and the development of inequality by gender, race, sexuality, disability or age. We could go on to produce a score of relevant examples where geographical work is central to issues of equality and justice. All will offer the human geographer the chance of explicitly engaging with the 'articulation of the moral and the spatial' (Philo, 1991b: 26).

However, as David Smith (2004) goes on to point out, a project with such huge implications is conceivable only if driven by the most powerful moral argument. As individuals, geographers will bring their own moral understandings to such a task – and these will be variously informed by a range of personal commitments. One possible way to unite these various understandings through the practice of research is to draw on the work of the French anthropologist Marc Augé. Augé argues that two strands of sensitivity are required in relation to others (to those we do research 'on'): a sense of the other and a sense for the other (Cloke, 2002; 2004b). We would argue that in recent years geographers have increasing shown a sensitivity towards other voices as their research has broadened to gain a sense of what has meaning for others. However, Augé argues that our research should show a sense for the other, and for otherness, in the same way that we develop a sense of direction, of rhythm, of belonging. We would argue that this has been much less evident in geography, making it much more difficult to engage with others in an emotional, committed and connected manner. If we can unite a sense of the other with a sense for the other it becomes possible to do research with people and for people rather than just on people. And if we can make these connections it becomes more likely that we will be able to redirect our future research in order to engage with a truly moral human geography which seeks to contribute to a more just and equal society.

Notes

1 Bell and Roberts (1984), Punch (1986) and Hammersley (1993; 1995).
2 See Johnston (2000c: 982) for a list of members.
3 On the link between geographical research and public policy, see Peck (1999), Massey (2000) and Martin (2001).

References

Abler, R., Adams, J.S. and Gould, P. (1971) *Spatial Organization: The Geographer's View of the World*. Englewood Cliffs, NJ: Prentice Hall.

Ackerman, E.A. (1965) *The Science of Geography*. Washington, DC: National Academy of Sciences, National Research Council.

Agger, B. (1989a) 'Do books write authors? A study of disciplinary hegemony', *Teaching Sociology*, 17: 365–9.

Agger, B. (1989b) *Reading Science: A Literary, Political and Sociological Analysis*. Dix Hills, NY: General Hall.

Agnew, J. (1995) 'Classics in human geography revisited: Thrift, N. "On the determination of social action in space and time": commentary 1', *Progress in Human Geography*, 19: 525–6.

Aitken, S.C. (1994) 'I'd rather watch the movie than read the book', *Journal of Geography in Higher Education*, 18: 291–307.

Aitken, S.C. (1997) 'Analysis of texts: armchair theory and couch-potato geography', in R. Flowderdew and D. Martin (eds) *Methods in Human Geography: A Guide for Students Doing a Research Project*. Harlow: Longman, 197–212.

Aitken, S.C. and Zonn, L.E. (1993) 'Weir(d) sex: representation of gender–environment relations in Peter Weir's "Picnic at Hanging Rock" and "Gallipoli"', *Environment and Planning D: Society and Space*, 11: 191–212.

Aitken, S.C. and Zonn, L.E. (eds) (1994a) *Place, Power, Situation and Spectacle: A Geography of Film*. Lanham, MD: Rowman & Littlefield.

Aitken, S.C. and Zonn, L.E. (1994b) '*Re*-presenting the place pastiche', in S.C. Aitken and L.E. Zonn (eds) *Place, Power, Situation and Spectacle: A Geography of Film*. Lanham, MD: Rowman & Littlefield, 3–25.

Alaasutari, P. (1995) *Researching Culture: Qualitative Methods and Cultural Studies*. London: Sage.

Allen, J., Massey, D. and Cochrane, A. (1998) *Rethinking the Region*. London: Routledge.

Amaral, D.J. and Wisner, W.B. (1970) 'Participant observation, phenomenology and the rules for judging sciences: a comment', *Antipode*, 2: 42–51.

Amedeo, D. and Golledge, R. (1975) *An Introduction to Scientific Reasoning in Geography*. Chichester: Wiley.

Amin, A. and Robins, K. (1990) 'The re-emergence of regional economies? The mythical geography of flexible accumulation', *Environment and Planning D: Society and Space*, 8: 7–34.

Anderson, J. (2002) 'Researching environmental resistance: working through secondspace and thirdspace approaches', *Qualitative Research*, 2: 301–21.

Anderson, J.R. (1961) 'Towards more effective methods of obtaining land use data in geographic research', *The Professional Geographer*, 13: 15–18.

Anderson, K. (1999) 'Introduction', in K. Anderson and F. Gale (eds) *Cultural Geographies*. Harlow: Longman, 1–21.

Anderson, K. and Gale, F. (eds) (1992) *Inventing Places: Studies in Cultural Geography*. Melbourne: Longman Cheshire.

Anderson, K. and Smith, S. (2001) 'Editorial: emotional geographies', *Transactions, Institute of British Geographers*, 26: 7–10.

Angus, T., Cook, I., Evans, J. et al. (2001) 'A manifesto for cyborg pedagogy?', *International Research in Geographical and Environmental Education*, 10: 195–201.

Anon (2001) 'Kinaesthesia', in *Collins Concise Dictionary*. Glasgow: HarperCollins, 809.

Archer, J.E. and Dalton, T.H. (1968) *Fieldwork in Geography*. London: Batsford.

Asad, T. (ed.) (1973) *Anthropology and the Colonial Encounter*. London: Ithaca Press.

Bailey, C., White, C. and Pain, R. (1999) 'Evaluating qualitative research: dealing with the tension between "science" and "creativity"', *Area*, 31: 169–83.

Bailey, T. and Gatrell, A. (1995) *Interactive Spatial Data Analysis*. Harlow: Longman.

Baker, A.R.H., Hamshere, J. and Langton, J. (eds) (1970) *Geographical Interpretations of Historical Sources*. Newton Abbot: David & Charles.

Bakhtin, M. (1986) *The Dialogical Imagination*. Austin, TX: University of Texas Press.

Barnes, T. (1992) 'Reading the texts of theoretical economic geography', in T. Barnes and J. Duncan (eds) *Writing Worlds*. London: Routledge, 118–35.

Barnes, T. (1994) 'Probable writing: Derrida, deconstruction and the quantitative revolution in human geography', *Environment and Planning A*, 26: 1021–40.

Barnes, T. (1996) *Logics of Dislocation: Metaphors, Models and Meanings of Economic Space*. New York, NY: Guilford.

Barnes, T. (1998) 'A history of regression: actors, networks, machines and numbers', *Environment and Planning A*, 30: 225–34.

Barnes, T. (2000a) 'Poetics of geography', in R.J. Johnston et al. (eds) *The Dictionary of Human Geography* (4th edn). Oxford: Blackwell, 588–9.

Barnes, T. (2000b) 'Metaphor', in R.J. Johnston et al. (eds) *The Dictionary of Human Geography* (4th edn). Oxford: Blackwell, 499–501.

Barnes, T. (2000c) 'Hermeneutics', in R.J. Johnston et al. (eds) *The Dictionary of Human Geography* (4th edn). Oxford: Blackwell, 334–6.

Barnes, T. (2001a) 'Lives lived and lives told: biographies of geography's quantitative revolution', *Environment and Planning D: Society and Space*, 19: 409–30.

Barnes, T. (2001b) 'Retheorising economic geography: from the quantitative revolution to the "cultural turn"', *Annals of the Association of American Geographers*, 91: 546–65.

Barnes, T. and Curry, M. (1992) 'Postmodernism in economic geography: metaphor and the construction of alterity', *Environment and Planning D: Society and Space*, 10: 57–68.

Barnes, T. and Duncan, J. (eds) (1992) *Writing Worlds*. London: Routledge.

Barnes, T. and Gregory, D. (eds) (1997) *Reading Human Geography*. London: Arnold.

Barnett, C. (1997) '"Sing along with the common people": politics, postcolonialism and other figures', *Environment and Planning D: Society and Space*, 15: 137–54.

Barrell, J. (1982) 'Geographies of Hardy's Wessex', *Journal of Historical Geography*, 8: 347–61.

Batterbury, S. (1994) 'Alternative affiliations and the personal politics of overseas research: some reflections', in E. Robson and K. Willis (eds) *Postgraduate Fieldwork in Developing Areas: A Rough Guide*. IBG Developing Areas Research Group Monograph 8. London: IBG, 60–89.

Batty, M., Dodge, M., Doyle, S. and Smith, A. (1998) 'Modelling virtual environments', in P. Longley et al. (eds) *Geocomputation: A Primer*. Chichester: Wiley, 139–61.

Batty, M. and Longley, P. (1994) *Fractal Cities*. London: Academic Press.

Batty, M., Longley, P. and Fotheringham, S. (1989) 'Urban growth and form: scaling fractal geometry and diffusion-limited aggregation', *Environment and Planning A*, 21: 1447–72.

Baxter, J. and Eyles, J. (1997) 'Evaluating qualitative research in social geography: establishing rigour in interview analysis', *Transactions, Institute of British Geographers*, 22: 505–25.

Bayliss-Smith, T. and Owens, S. (1990) 'Introduction: aerial photography, the hidden agenda', in T. Bayliss-Smith and S. Owens (eds) *Britain's Changing Environment from the Air*. Cambridge: Cambridge University Press, 7–14.

Beauregard, R. (1993) *Voices of Decline*. Blackwell: Oxford.

Beaver, S.H. (1962) 'The Le Play Society and field work', *Geography*, XLVII: 225–40.

Becker, H. (1986) *Writing for Social Scientists*. Chicago, IL: University of Chicago Press.

Behar, R. (1995) 'Introduction: out of exile', in R. Behar and D. Gordon (eds) *Women Writing Culture*. Berkeley, CA: University of California Press, 1–29.

Behar, R. and Gordon, D. (eds) (1995) *Women Writing Culture*. Berkeley, CA: University of California Press.

Bell, C. and Roberts, H. (1984) *Social Researching: Politics, Problems, Practice*. London: Routledge & Kegan Paul.

Bell, J. (1987) *Doing your Research Project*. Milton Keynes: Open University Press.

Bell, M. (1993) '"The pestilence that walketh in darkness." Imperial health, gender and images of South Africa *c*. 1880–1910', *Transactions, Institute of British Geographers*, 18: 327–41.

Benedikt, M. (ed.) (1991) *Cyberspace: First Steps*. Cambridge, MA: MIT Press.

Benko, G. and Strohmayer, U. (eds) (1997) *Space and Social Theory: Interpreting Modernity and Postmodernity*. Oxford: Blackwell.

Bennett, K. (2002) 'Participant observation', in P. Shurmer-Smith (ed.) *Doing Cultural Geography*. London: Sage, 139–49.

Bennett, K., Ekinsmyth, C. and Shurmer-Smith, P. (2002) 'Selecting topics for study', in P. Shurmer-Smith (ed.) *Doing Cultural Geography*. London: Sage, 81–93.

Bennett, K. and Shurmer-Smith, P. (2001) 'Writing conversation', in M. Limb and C. Dwyer (eds) *Qualitative Methodologies for Geographers*. London: Arnold, 251–63.

Bennett, K. and Shurmer-Smith, P. (2002) 'Handling case studies', in P. Shurmer-Smith (ed.) *Doing Cultural Geography*. London: Sage, 199–209.

Bennett, R. (1979) *Spatial Time Series*. London: Pion.

Bennett, R. and Graham, D.J. (1998) 'Explaining size differentiation of business service centres', *Urban Studies*, 35: 1457–80.

Bennett, R.J. and Chorley, R.J. (1978) *Environmental Systems: Philosophy, Analysis and Control*. London: Methuen.

Benton, L. (1995) 'Will the real/reel Los Angeles please stand up?', *Urban Geography*, 16: 144–64.

Berg, B. (1989) *Qualitative Research Methods for the Social Sciences* (3rd edn). London: Allyn & Bacon.

Berger, J. (1972) *Ways of Seeing*. London: Penguin Books.

Berleant-Schiller, R. (1991) 'Hidden places and Creole forms: naming the Barbundan landscape', *The Professional Geographer*, 43: 92–101.

Berry, B.J.L. (1964) 'Approaches to regional analysis: a synthesis', *Annals of the Association of American Geographers*, 54: 2–11.

Bertalanffy, L. van (1968) *General Systems Theory: Foundation, Development, Applications*. London: Allen Lane.

Billinge, M.D. (1983) 'The Mandarin dialect: an essay on style in contemporary geographical writing', *Transactions, Institute of British Geographers*, 8: 400–20.

Billinge, M.D., Gregory, D. and Martin, R. (1984) *Recollections of a Revolution: Geography as Spatial Science*. New York, NY: St Martin's Press.

Bingham, N. and Thrift, N. (1999) 'Some new instructions for travellers: the geography of Bruno Latour and Michel Serres', in M. Crang and N. Thrift (eds) *Thinking Space*. London: Routledge, 281–301.

Bingley, A. (2003) 'In here and out there: sensations between self and landscape', *Social and Cultural Geography*, 4: 329–46.

Blake, L.A. (1999) 'Pastoral power, governmentality and cultures of order in nineteenth-century British Columbia', *Transactions, Institute of British Geographers*, 24: 79–93.

Blakemore, M.J. (1994) 'Cartography', in R.J. Johnston et al. (eds) *The Dictionary of Human Geography* (3rd edn). Oxford: Blackwell, 44–8.

Blaut, J. (1961) 'Space and process', *The Professional Geographer*, 13: 1–7.

Blomley, N. (1994) 'Activism and the academy', *Environment and Planning D: Society and Space*, 12: 383–5.

Blunt, A. (1994a) 'Mapping authorship and authority: reading Mary Kingsley's landscape descriptions', in A. Blunt and G. Rose (eds) *Writing Women and Space: Colonial and Post-colonial Geographies*. New York, NY: Guilford, 51–72.

Blunt, A. (1994b) *Travel, Gender and Imperialism: Mary Kingsley and West Africa*. New York, NY: Guilford.

Blunt, A., Gruffudd, P., May, J., Ogborn, M. and Pinder, D. (eds) (2003) *Cultural Geography in Practice*. London: Arnold.

Bondi, L. (1999) 'Stages on a journey: some remarks about human geography and psychotherapeutic practice', *The Professional Geographer*, 51: 11–24.

Bondi, L. (2003a) 'A situated practice for (re)situating selves: trainee counsellors and the promise of counselling', *Environment and Planning A*, 35: 853–70.

Bondi, L. (2003b) '"Unlocking the cage door": counselling, care and space', *Journal of Social and Cultural Geography*, forthcoming.

Bonnett, A. (1992) 'Art, ideology, and everyday space: subversive tendencies from Dada to postmodernism', *Environment and Planning D: Society and Space*, 10: 69–86.

Boots, B. and Getis, A. (1988) *Point Pattern Analysis*. Harlow: Sage.

Bowen, M. (1981) *Empiricism and Geographical Thought: From Francis Bacon to Alexander von Humboldt*. Cambridge: Cambridge University Press.

Brandth, B. (1995) 'Rural masculinity in transition: gender image in tractor advertisements', *Journal of Rural Studies*, 11: 123–34.

Brewer, J. (1989) *The Sinews of Power: War, Money and the English State, 1688–1783*. London: Unwin Hyman.

Brice, A.M.H. and Bain, J.A. (n.d., *c*. 1918) *Two Great Explorers: Sir Henry M. Stanley and Fridtjof Nansen*. London: Partridge & Co.

Bridge, G. (1992) 'Questionnaire surveys', in A. Rogers et al. (eds) *The Student's Companion to Geography*. Blackwell: Oxford, 196–206.

Bristow, J. (1991) 'Life stories: Carolyn Steedman's history writing', *New Formations*, 13: 113–31.

Brown, L., Flavin, C. and Kane, H. (1992) *Vital Signs 1992: The Trends that are Shaping our Future*. London: Earthscan.

Bunge, W. (1962) *Theoretical Geography*. Lund: Gleerup.

Bunge, W. (1971) *Fitzgerald: The Geography of a Revolution*. Cambridge, MA: Schenkman.

Bunge, W. (1975) 'The point of production: a second front', *Antipode*, 9: 60–76.

Bunge, W. (1988) *Nuclear War Atlas*. Blackwell: Oxford.

Burgess, J. (1985) 'News from nowhere: the press, the riots and the myth of the inner city', in J. Burgess and J. Gold (eds) *Geography, the Media and Popular Culture*. London: Croom Helm, 192–228.

Burgess, J. (1987) 'Landscapes in the living-room: television and landscape research', *Landscape Research*, 12: 1–7.

Burgess, J. (1990) 'The production and consumption of environmental meanings in the mass media: a research agenda for the 1990s', *Transactions, Institute of British Geographers*, 15: 139–61.

Burgess, J. (1996) 'Focusing on fear: the use of focus groups in a project for the Community Forest Unit, Countryside Commission', *Area*, 28: 130–55.

Burgess, J. and Gold, J. (eds) (1986) *Geography, the Media and Popular Culture*. London: Croom Helm.

Burgess, J., Harrison, C. and Maiteny, P. (1991) 'Contested meanings: the consumption of news about nature conservation', *Media, Culture and Society*, 13: 499–520.

Burgess, J., Limb, M. and Harrison, C. (1988) 'Exploring environmental values through the medium of small groups. 2. Illustrations of a group at work', *Environment and Planning A*, 20: 457–76.

Burgess, J. and Pile, S. (1991) 'Constructions of the self: some thoughts on the need for a psychoanalytic perspective', in C. Philo (compiler) *New Words, New Worlds: Reconceptualising Social and Cultural Geography*. Lampeter: Social and Cultural Geography Study Group of the Institute of British Geographers, 31–2.

Burgess, R. (1984) *In the Field: An Introduction to Field Research*. London: Allen & Unwin.

Burgess, R. (1986) 'Being a participant observer', in J. Eyles (ed.) *Qualitative Approaches in Social and Geographical Research. Occasional Paper 26*. London: Department of Geography and Earth Science, Queen Mary College, University of London, 47–66.

Burke, P. (1988) 'Verstehen', in A. Bullock et al. (eds) *The Fontana Dictionary of Modern Thought*. London: Fontana, 894.

Burrough, P. (1998) 'Dynamic modelling and geocomputation', in P. Longley et al. (eds) *Geocomputation: A Primer*. Chichester: Wiley, 165–92.

Burton, F. and Carlen, P. (1979) *Official Discourse*. London: Routledge.

Burton, I. (1963) 'The quantitative revolution and theoretical geography', *The Canadian Geographer*, 7: 151–62.

Butler, R. (2001) 'From where I write: the place of positionality in qualitative writing', in M. Limb and C. Dwyer (eds) *Qualitative Methodologies for Geographers: Issues and Debates*. London: Edward Arnold, 264–76.

Buttimer, A. (1974) *Values in Geography. Commission on College Geography Resource Paper* 24. Washington, DC: Association of American Geographers.

Buttimer, A. (1976) 'Grasping the dynamism of lifeworld', *Annals of the Association of American Geographers*, 66: 277–92.

Buttimer, A. (1978) 'Charisma and context: the challenge of *la géographie humaine*', in D. Ley and M. Samuels (eds) *Humanistic Geography: Prospects and Problems*. London: Croom Helm, 58–76.

Buttimer, A. (2001) 'Home-reach-journey', in P. Moss (ed.) *Placing Autobiography in Geography*. Syracuse, NY: Syracuse University Press, 22–40.

Buttimer, A. and Seamon, D. (eds) (1980) *The Human Experience of Space and Place*. London: Croom Helm.

Bygott, J. (1934) *Mapwork and Practical Geography*. London: University Tutorial Press.

Bynner, J. and Stribley, K. (1978) *Social Research: Principles, Procedures*. London: Longman.

Byron, M. (1993) 'Using audiovisual aids in geography research: questions of access and responsibility', *Area*, 25: 379–85.

Callard, F. (2003) 'The taming of psychoanalysis in geography', *Social and Cultural Geography*, 4: 295–312.

Callaway, H. (1992) 'Ethnography and experience: gender implications in fieldwork and texts', in J. Okely and H. Callaway (eds) *Anthropology and Autobiography*. London: Routledge, 29–49.

Campbell, D. (1974) 'Role relationships in advocacy geography', *Antipode*, 6: 102–5.

Capel, H. (1981) 'Institutionalisation of geography and strategies of change', in D.R. Stoddart (ed.) *Geography, Ideology and Social Concern*. Oxford: Blackwell, 37–69.

Capps, P. (2001) 'The retail workplace of gender, space and power dynamics: a case study of a camping business in south-east England', in M. Limb and C. Dwyer (eds) *Qualitative Methodologies for Geographers: Issues and Debates*. London: Arnold, 291–3.

Carr, M. (1987) *Patterns: Process and Change in Human Geography*. London: Macmillan.

Carter, H. (1990) 'Towns and urban systems 1730–1914', in R. Dodgshon and R. Butlin (eds) *An Historical Geography of England and Wales*. London: Academic Press, 401–51.

Carter, P. (1999) 'Dark with excess of bright: charting the coastlines of knowledge', in D. Cosgrove (ed.) *Mappings*. London: Reaktion Books, 125–47.

Cassidy, F.G. (1947) *The Place-names of Dane County, Wisconsin. Publications of the American Dialect Society* 7. Greensboro, NC: American Dialect Society.

Castells, M. (1989) *The Informational City*. Oxford: Oxford University Press.

Castree, N. (1996) 'Birds, mice and geography: Marxisms and dialectics', *Transactions, Institute of British Geographers*, 21: 342–62.

Chamberlain, P.G. (1995) 'Metaphorical vision in the literary landscape of William Shakespeare', *The Canadian Geographer/Le Géographie canadien*, 39: 306–22.

Chandler, D. (2001) *Semiotics: The Basics*. London: Routledge.

Chapman, G.P. (1977) *Human and Environmental Systems*. London: Academic Press.

Chisholm, M. (1975) *Human Geography: Evolution or Revolution*. Harmondsworth: Penguin Books.

Chorley, R.J. (1964) 'Geography and analogue theory', *Annals of the Association of American Geographers*, 54: 127–37.

Chouinard, V. (1994) 'Radical geography: is all that's Left right?', *Environment and Planning D: Society and Space*, 12: 2–5.

Chrisman, N. (1997) *Exploring Geographic Information Systems*. New York, NY: Wiley.

Clanchy, M.T. (1979) *From Memory to Written Record: England 1066–1307*. London: Edward Arnold.

Clark, M.J. (1998) 'GIS – democracy or delusion?', *Environment and Planning A*, 30: 303–16.

Clarke, D. (1984) 'Film/space', *Environment and Planning A*, 26: 1821–3.

Clarke, D. (ed.) (1997) *The Cinematic City*. London: Routledge.

Clarke, G. (1999) 'Geodemographics, marketing and retail location', in M. Pacione (ed.) *Applied Geography: Principles and Practice*. London: Routledge, 577–92.

Clarke, G. and Cooke, D. (1983) *A Basic Course in Statistics* (2nd edn). London: Edward Arnold.

Clarke, K. (1998) 'Visualising different geofutures', in P. Longley et al. (eds) *Geocomputation: A Primer*. Chichester: Wiley, 119–37.

Cliff, A. and Ord, J. (1973) *Spatial Autocorrelation*. London: Pion.

Clifford, J. (1986) 'Introduction: partial truths', in J. Clifford et al. (eds) *Writing Culture: The Poetics and Politics of Ethnography*. Berkeley, CA: University of California Press, 1–26.

Clifford, J. (1990) 'Notes on (field)notes', in R. Sanjek (ed.) *Fieldnotes*. Ithaca, NY: Cornell University Press, 47–70.

Clifford, J. (1992) 'Travelling cultures', in L. Grossberg et al. (eds) *Cultural Studies*. London: Routledge, 96–112.

Clifford, J. (1997) *Routes: Travel and Translation in the Late Twentieth Century*. London: Harvard University Press.

Clifford, J. and Marcus, G. (eds) (1986) *Writing Culture: The Poetics and Politics of Ethnography*. Berkeley, CA: University of California Press.

Cloke, P. (1994) '(En)culturing political economy: a day in the life of a "rural geographer"', in P. Cloke et al. (eds) *Writing the Rural: Five Cultural Geographies*. London: Paul Chapman, 149–90.

Cloke, P. (1996) 'Rural lifestyles: material opportunity, cultural experience, and how theory can undermine policy', *Economic Geography*, 72: 433–49.

Cloke, P. (1997) 'Country backwater to virtual village? Rural studies and "the cultural turn"', *Journal of Rural Studies*, 13: 367–75.

Cloke, P. (2002) 'Deliver us from evil: prospects for living ethically and acting politically in human geography', *Progress in Human Geography*, 26: 587–604.

Cloke, P. (2004a) 'On vision and envisioning…', in P. Cloke et al. (eds) *Envisioning Human Geographies*. London: Arnold, 1–10.

Cloke, P. (2004b) 'Spiritual geographies', in P. Cloke et al. (eds) *Envisioning Human Geographies*. London: Arnold, 228.

Cloke, P., Cooke, P., Cursons, J., Milbourne, P. and Widdowfield, R. (2000a) 'Ethics, reflexivity and research: encounters with homeless people', *Ethics, Place and Environment*, 3: 133–54.

Cloke, P., Crang, P. and Goodwin, M. (1999a) *Introducing Human Geography*. London: Arnold.

Cloke, P., Goodwin, M., Milbourne, P. and Thomas, C. (1995b) 'Deprivation, poverty and marginalisation in rural lifestyles in England and Wales', *Journal of Rural Studies*, 11: 351–65.

Cloke, P., Milbourne, P. and Thomas, C. (1994) *Lifestyles in Rural England*. London: Rural Development Commission.

Cloke, P., Milbourne, P. and Thomas, C. (1995a) 'Poverty in the countryside: out of sight and out of mind', in C. Philo (ed.) *Off the Map: A Social Geography of Poverty*. London: Child Poverty Action Group, 83–102.

Cloke, P., Milbourne, P. and Widdowfield, R. (1999b) 'Homelessness in rural areas: an invisible issue', in P. Kennett and A. Marsh (eds) *Homelessness: Exploring the New Terrain*. Bristol: Policy Press, 61–80.

Cloke, P., Milbourne, P. and Widdowfield, R. (2000b) 'Homelessness and rurality: out of place in purified space', *Environment and Planning D: Society and Space*, 18: 715–35.

Cloke, P. and Perkins, H. (1998) '"Cracking the Canyon with the Awesome Foursome": representations of adventure tourism in New Zealand', *Environment and Planning D: Society and Space*, 16: 185–218.

Cloke, P., Philo, C. and Sadler, D. (1991) *Approaching Human Geography: An Introduction to Contemporary Theoretical Debates*. London: Paul Chapman.

Close, C.F. (1911) 'The purpose and position of geography', *Bulletin of the American Geographical Society*, 43: 740–53.

Coan, T.M. (1899) 'Hawaiian ethnography', *Journal of the American Geographical Society of New York*, 31: 24–30.

Coffey, W.J. (1981) *Geography: Towards a General Spatial Systems Approach*. London: Methuen.

Cohen, S. (1995) 'Sounding out the city: music and the sensuous reproduction of place', *Transactions, Institute of British Geographers*, 20: 434–46.

Cole, J. and King, C. (1968) *Quantitative Geography*. London: Wiley.

Coleman, A. (1954) 'Fieldwork in schools: a sample traverse in east Kent', *Geography*, XXXIV: 262–71.

Coleman, A. and Lukehurst, C.T. (eds) (1974) *Field Excursions in South East England. Field Studies for Schools* 11. London: Rivingtons.

Colenutt, R. (1971) 'Postscript on the Detroit Geographical Expedition', *Antipode*, 6: 85.

Cook, I. (1997) 'Participant observation', in R. Flowerdew and D. Martin (eds) *Methods in Human Geography: A Guide for Students Doing Research Projects*. Harlow: Longman, 127–49.

Cook, I. (2001) '"You want to be careful you don't end up like Ian, he's all over the place": autobiography in/of an expanded field', in P. Moss (ed.) *Placing Autobiography in Geography*. Syracuse, NY: Syracuse University Press, 99–120.

Cook, I. et al. (2000a) 'Social sculpture and connective aesthetics: Shelley Sacks' "exchange values"', *Ecumene*, 7: 338–44.

Cook, I. et al. (2002) *Commodities: The DNA of Capitalism* (http://apm.brookes.ac.uk/exchange/texts/Ian Cook2.pdf) accessed 23 June 2003.

Cook, I. et al. (forthcoming) 'Follow the thing: papaya', *Antipode*.

Cook, I. et al. (in press) 'Positionality/situated knowledge', in D. Atkinson et al. (eds) *Cultural Geography: A Critical Dictionary of Key Ideas*. London: I.B. Tauris.

Cook, I. and Crang, M. (1995) *Doing Ethnographies*. Norwich: Environmental Publications.

Cook, I. and Crang, P. (1996) 'Commodity systems, documentary filmmaking and new geographies of food: Amos Gitai's *Ananas*.' Paper presented at the 1996 annual conference of the Institute of British Geographers/Royal Geographical Society conference, Glasgow, January (http://www.ges.bham.ac.uk/draftpapers/iancook-pineapple.htm) accessed 23 June 2003.

Cook, I., Crang, P. and Thorpe, M. (1999) 'Eating into Britishness: multicultural imaginaries and the identity politics of food', in S. Roseneil and J. Seymour (eds) *Practising Identities: Power and Resistance*. London: Macmillan, 223–48.

Cook, I., Crouch, D., Naylor, S. and Ryan, J. (eds) (2000b) *Cultural Turns/Geographical Turns: Perspectives on Cultural Geography*. Harlow: Longman.

Cook, I. and Harrison, M. (2003) 'Cross over food: re-materialising postcolonial geographies', *Transactions, Institute of British Geographers*, 28: 296–317.

Cope, M. (2003) 'Coding transcripts and diaries', in N.J. Clifford and G. Valentine (eds) *Key Methods in Geography*. London: Sage, 445–59.

Corrigan, P. and Sayer, D. (1985) *The Great Arch: English State Formation as Cultural Revolution*. Oxford: Blackwell.

Cosgrove, D. (1984) *Social Formation and Symbolic Landscape*. London: Croom Helm.

Cosgrove, D. (1993) *The Palladian Landscape*. Leicester and London: Leicester University Press.

Cosgrove, D. (ed.) (1999) *Mappings*. London: Reaktion Books.

Cosgrove, D. and Daniels, S. (eds) (1988) *The Iconography of Landscape: Essays on the Symbolic Representation, Design and Use of Past Environments*. Cambridge: Cambridge University Press.

Cosgrove, D. and Domosh, M. (1993) 'Author and authority: writing the new cultural geography', in J. Duncan and D. Ley (eds) *Place/Culture/Representation*. London: Routledge, 25–38.

Cosgrove, D. and Jackson, P. (1987) 'New directions in cultural geography', *Area*, 19: 95–101.

Couclelis, H. (1998) 'Geocomputation in context', in P. Longley et al. (eds) *Geocomputation: A Primer*. Chichester: Wiley, 17–29.

Crang, M. (1995) *Doing Ethnographies (CATMOG 58)*. Norwich: Environmental Publications.

Crang, M. (1996) 'Watching the city: video, surveillance, and resistance', *Environment and Planning A*, 28: 2099–104.

Crang, M. (1997) 'Analysing qualitative materials', in R. Flowerdew and D. Martin (eds) *Methods in Human Geography: A Guide for Students Doing a Research Project*. Harlow: Longman, 183–96.

Crang, M. (2000) 'Playing nymphs and swains in pastoral idyll', in A. Hughes et al. (eds) *Ethnography and Rural Research*. Cheltenham: Countryside and Community Press, 158–78.

Crang, M. (2001) 'Filed work: making sense of group interviews', in M. Limb and C. Dwyer (eds) *Qualitative Methodologies for Geographers: Issues and Debates*. London: Edward Arnold, 215–33.

Crang, M. (2002) 'Qualitative methods: the new orthodoxy?', *Progress in Human Geography*, 26: 647–55.

Crang, M., Crang, P. and May, J. (eds) (1999) *Virtual Geographies*. London: Routledge.

Crang, M., Hudson, A., Reimer, S. and Hinchliffe, S. (1997) 'Software for qualitative research. 1. Prospectus and overview', *Environment and Planning A*, 29: 771–87.

Crang, P. (1992) 'The politics of polyphony: reconfigurations of geographical authority', *Environment and Planning D: Society and Space*, 10: 527–49.

Crang, P. (1994) 'It's showtime: on the workplace geographies of display in a restaurant in southeast England', *Environment and Planning D: Society and Space*, 12: 675–704.

Crang, P. (2000) 'Cultural turn', in R.J. Johnston et al. (eds) *The Dictionary of Human Geography* (4th edn). Oxford: Blackwell, 141–3.

Crapanzano, V. (1977) 'On the writing of ethnography', *Dialectical Anthropology*, 2: 69–73.

Crapanzano, V. (1986) 'Hermes' dilemma: the masking of subversion in ethnographic description', in J. Clifford and G. Marcus (eds) *Writing Culture: The Poetics and Politics of Ethnography*. Berkeley, CA: University of California Press, 51–76.

Cresswell, T. (1993) 'Mobility as resistance: a geographical reading of Kerouac's *On the Road*', *Transactions, Institute of British Geographers*, 18: 249–62.

Cresswell, T. (1996) *In Place/Out of Place: Geography, Ideology and Transgression*. Minneapolis, MN: University of Minnesota Press.

Cresswell, T. (2000) 'Falling down: resistance as diagnostic', in J.P. Sharp et al. (eds) *Entanglements of Power: Geographies of Resistance/Domination*. London: Routledge, 256–68.

Cresswell, T. (2001) *The Tramp in America*. London: Reaktion Books.

Crewe, L. and Goodrum, A. (2000) 'Fashioning new forms of consumption: the case of Paul Smith', in S. Bruzzi and P. Church Gibson (eds) *Fashion Cultures: Theories, Explorations and Analysis*. London: Routledge, 25–48.

Crush, J. (1994) 'Post-colonialism, de-colonisation and geography', in A. Godlewska and N. Smith (eds) *Geography and Empire*. Oxford: Blackwell, 333–50.

Cullen, M.J. (1975) *The Statistical Movement in Early Victorian Britain*. Hassocks: Harvester.

Curry, M. (1995) 'GIS and the inevitability of ethical inconsistency', in J. Pickles (ed.) *Ground Truth: The Social Implications of Geographic Information Systems*. New York, NY: Guilford, 68–87.

Curry, M.R. (1998) *Digital Places: Living with Geographic Information Technologies*. London: Routledge.

Czaja, R. and Blair, J. (1996) *Designing Surveys: A Guide to Decisions and Procedures*. Thousand Oaks, CA: Pine Forge Press.

Dandeker, C. (1990) *Surveillance, Power and Modernity: Bureaucracy and Discipline from 1700 to the Present Day*. Cambridge: Polity Press.

Daniels, S. (1985) 'Arguments for a humanistic geography', in R.J. Johnston (ed.) *The Future of Geography*. London: Methuen, 143–58.

Daniels, S. (1989) 'Marxism, culture and the duplicity of landscape', in R. Peet and N. Thrift (eds) *New Models in Geography*. London: Unwin Hyman, 196–220.

Daniels, S. (1993) *Fields of Vision*. Cambridge: Cambridge University Press.

Daniels, S. and Cosgrove, D. (1988) 'Introduction: iconography and landscape', in D. Cosgrove and S. Daniels (eds) *The Iconography of Landscape: Essays on the Symbolic Representation, Design and Use of Past Environments*. Cambridge: Cambridge University Press, 1–10.

Daniels, S. and Rycroft, S. (1993) 'Mapping the modern city: Alan Sillitoe's Nottingham novels', *Transactions, Institute of British Geographers*, 18: 460–80.

Dant, T. (1999) *Material Culture in the Social World: Values, Activities, Lifestyles*. Philadelphia, PA: Open University Press.

Darby, H.C. (1950–1) 'Domesday woodland', *Economic History Review*, 3: 21–43.

Darby, H.C. (1951) 'The economic geography of England A.D. 1000–1250', in H. Darby (ed.) *An Historical Geography of England before A.D. 1800*. Cambridge: Cambridge University Press, 165–229.

Darby, H.C. (1962) 'The problem of geographical description', *Transactions, Institute of British Geographers*, 30: 1–14.

Darby, H.C. (1977) *Domesday England*. Cambridge: Cambridge University Press.

Darden, J. (1976) 'Review of Ley, D. (1974) *The Black Inner City as Frontier Outpost*', *The Geographical Review*, 66: 116–17.

Davies, G. (1998) 'Networks of nature: stories of natural history film-making from the BBC.' Unpublished PhD thesis, University College London.

Davies, G. (1999) 'Exploiting the archive: and the animals came in two by two, 16 mm, CD-ROM and BetaSp', *Area*, 31: 49–58.

Davies, G. (2000) 'Narrating the Natural History Unit: institutional orderings and spatial practices', *Geoforum*, 31: 539–51.

Davies, G. (forthcoming) 'Researching the networks of natural history television', in M. Ogborn et al. (eds) *Cultural Geography in Practice*. London: Arnold.

Davis, C.M. (1954) 'Field techniques', in P.E. James and C.F. Jones (eds) *American Geography: Inventory and Prospect*. Syracuse, NY: Syracuse University Press for the AAG, 496–529.

Davis, D.H. (1926) 'Objectives in a geographic field study of a community', *Annals of the Association of American Geographers*, 16: 102–9.

Davis, M. (1990) *City of Quartz: Excavating the Future of Los Angeles*. London: Verso.

Dean, M. (1999) *Governmentality: Power and Rule in Modern Society*. London: Sage.

Dear, M. (1986) 'Postmodernism and planning', *Environment and Planning D: Society and Space*, 4: 367–84.

Dear, M. (1988) 'The postmodern challenge: reconstructing human geography', *Transactions, Institute of British Geographers*, 13: 262–74.

de Blij, H. (1982) 'Review of *Outcast Cape Town*', *The Geographical Review*, 72: 493–4.

Deleuze, G. and Guattari, F. (1984) *Anti-Oedipus: Capitalism and Schizophrenia. Volume 1*. London: Athlone.

Deleuze, G. and Guattari, F. (1988) *A Thousand Plateaus: Capitalism and Schizophrenia. Volume 2*. London: Athlone.

DeLyser, D. (1999) 'Authenticity on the ground: engaging the past in a California ghost town', *Annals of the Association of American Geographers*, 89: 602–32.

DeLyser, D. and Starrs, P. (eds) (2001) Special issue on 'Doing Fieldwork', *The Geographical Review*, 91: ix–viii, 1–508.

Demeritt, D. (1996) 'Social theory and the reconstruction of science and geography', *Transactions, Institute of British Geographers*, 14: 635–58.

Demeritt, D. (2000) 'The new social contract for science: accountability, relevance and value in US and UK science and research policy', *Antipode*, 32: 308–29.

Demeritt, D. and Dyer, S. (2002) 'Dialogue, metaphors of dialogue and understandings of geography', *Area*, 34: 229–41.

DeMers, M. (1997) *Fundamentals of Geographic Information Systems*. Oxford: Oxford University Press.

Denzin, N. (1978) *The Research Act* (2nd edn). New York, NY: McGraw-Hill.

Derrida, J. (1967) *De la Grammatologie*. Paris: Les Editions de Minuit. Translated as *Of Grammatology*. Baltimore, MD: Johns Hopkins University Press.

Derrida, J. (1991) *The Derrida Reader: Between the Blinds*. Hemel Hempstead: Harvester.

Desforges, L. (2001) 'Tourism consumption and the imagination of money', *Transactions, Institute of British Geographers*, 26: 353–64.

de Souza, A. (1988) 'Writing matters', *The Professional Geographer*, 40: 1–3.

Deutsche, R. (1991) 'Boy's town', *Environment and Planning D: Society and Space*, 9: 5–30.

De Vaus, D. (1991) *Surveys in Social Research* (3rd edn). London: UCL Press.

Dewar, N. (1999) 'Classics in human geography revisited: Western, J. (1981) *Outcast Cape Town*: commentary 1', *Progress in Human Geography*, 23: 421–2.

Dewsbury, J.D. and Naylor, S. (2002) 'Practising geographical knowledge: fields, bodies and dissemination', *Area*, 34: 253–60.

Dey, I. (1993) *Qualitative Data Analysis*. London: Routledge.

Dickens, P., Duncan, S., Goodwin, M. and Gray, F. (1985) *Housing, States and Localities*. London: Methuen.

Dickinson, G.C. (1963) *Statistical Mapping and the Presentation of Statistics*. London: Edward Arnold.

Dilke, M.S. (ed.) (1965) *Field Excursions in North West England. Field Studies for Schools 2*. London: Rivingtons.

Dillman, D. (1978) *Mail and Telephone Surveys: The Total Design Method*. New York, NY: Wiley.

Dixon, C. and Leach, B. (1978) *Questionnaires and Interviews in Geographic Research. Concepts and Techniques in Modern Geography 18*. Norwich: Geobooks.

Dixon, D.P. and Jones, J.-P. III (1998a) 'My dinner with Derrida, or spatial analysis and poststructuralism do lunch', *Environment and Planning A*, 30: 247–60.

Dixon, D.P. and Jones, J.-P. III (1998b) 'For a *supercalifragilisticexpialidocious* scientific geography', *Annals of the Association of American Geographers*, 86: 767–79.

Dodds, K. (1993) 'War stories: British elite narratives of the 1992 Falklands/Malvinas war', *Environment and Planning D: Society and Space*, 12: 619–40.

Dodgshon, R. (1998) *Society in Time and Space*. Cambridge: Cambridge University Press.

Doel, M.A. (1994) 'Deconstruction on the move: from libidinal economy to liminal materialism', *Environment and Planning A*, 26: 1041–59.

Doel, M.A. (1999) *Poststructuralist Geographies: The Diabolical Art of Spatial Science*. Edinburgh: Edinburgh University Press.

Doel, M.A. (2001) '1a. Qualified quantitative geography', *Environment and Planning D: Society and Space*, 19: 555–72.

Domosh, M. (1991a) 'Toward a feminist historiography of geography', *Transactions, Institute of British Geographers*, 16: 95–104.

Domosh, M. (1991b) 'Beyond the frontiers of geographical knowledge', *Transactions, Institute of British Geographers*, 16: 488–90.

Dorling, D. (1995a) 'The visualisation of local urban change across Britain', *Environment and Planning B: Planning and Design*, 22: 269–90.

Dorling, D. (1995b) *A New Social Atlas of Britain*. Chichester: Wiley.

Dorling, D. (1998) 'Human geography: when it is good to map', *Environment and Planning A*, 30: 277–88.

Dorling, D. and Fairbairn, D. (1997) *Mapping: Ways of Representing the World*. London: Longman.

Dorling, D. and Simpson, S. (eds) (1999a) *Statistics in Society*. London: Arnold.

Dorling, D. and Simpson, S. (1999b) 'Introduction to statistics in society', in D. Dorling and S. Simpson (eds) *Statistics in Society*. London: Arnold, 1–6.

Dowler, L. (2001) 'Fieldwork in the trenches: participant observation in a conflict area', in M. Limb and C. Dwyer (eds) *Qualitative Methodologies for Geographers: Issues and Debates*. London: Arnold, 153–64.

Downs, R. (1970) 'Geographic space perception: past approaches and future prospects', *Progress in Geography*, 2: 65–108.

Driver, F. (1988) 'Moral geographies: social science and the urban environment in mid-nineteenth century England', *Transactions, Institute of British Geographers*, 13: 275–87.

Driver, F. (1989) 'The historical geography of the workhouse system, 1834–1883', *Journal of Historical Geography*, 15: 269–86.

Driver, F. (1991) 'Henry Morton Stanley and his critics: geography, exploration and empire', *Past and Present*, 133: 134–66.

Driver, F. (1992) 'Geography's empire: histories of geographical knowledge', *Environment and Planning D: Society and Space*, 10: 23–40.

Driver, F. (1993) *Power and Pauperism: The Workhouse System, 1834–1884*. Cambridge: Cambridge University Press.

Driver, F. (1998) 'Hints to travellers: the Royal Geographical Society and the culture of exploration.' Typescript.

Driver, F. (2000a) 'Field-work in geography', *Transactions, Institute of British Geographers*, 25: 267–8.

Driver, F. (2000b) *Geography Militant: Cultures of Exploration and Empire*. Oxford: Blackwell.

du Gay, P. (ed.) (1997) *Production of Culture/Cultures of Production*. London: Sage.

Duncan, J.S. (1980) 'The superorganic in American cultural geography', *Annals of the Association of American Geographers*, 70: 181–98.

Duncan, J.S. (1998) 'Classics in human geography revisited: "The superorganic in American cultural geography": author's response', *Progress in Human Geography*, 22: 571–3.

Duncan, J.S. (2000) 'Representation', in R.J. Johnston et al. (eds) *The Dictionary of Human Geography* (4th edn). Oxford: Blackwell, 703–5.

Dwyer, C. and Limb, M. (2001) 'Introduction: doing qualitative research in geography', in M. Limb and C. Dwyer (eds) *Qualitative Methodologies for Geographers: Issues and Debates*. London: Arnold, 1–20.

Dwyer, K. (1977) 'The dialogic of ethnology', *Dialectical Anthropology*, 4: 205–24.

Ebdon, D. (1985) *Statistics in Geography* (2nd edn). Oxford: Blackwell.

Edwards, K.C. (1970) 'Organisation for fieldwork', *The Geographical Magazine*, 42: 314.

Ekinsmyth, C. (2002) 'Feminist methodology', P. Shurmer-Smith (ed.) *Doing Cultural Geography*. London: Sage, 177–85.

Ekinsmyth, C. and Shurmer-Smith, P. (2002) 'Humanistic and behavioural geography', in P. Shurmer-Smith (ed.) *Doing Cultural Geography*. London: Sage, 19–27.

Elton, G.R. (1972) *Policy and Police: The Enforcement of the Reformation in the Age of Thomas Cromwell*. Cambridge: Cambridge University Press.

E.M.Y. (1967) 'Review of Dilke's *The Purpose and Organisation of Field Studies and Field Excursions in North-west England*', *Geography*, LII: 228–9.

England, K. (1994) 'Getting personal: reflexivity, positionality and feminist research', *The Professional Geographer*, 46: 80–9.

Ennion, E.A.R. (ed.) (1949–52) *Field Study Books* (a series of books published under the auspices of the Council for the Promotion of Field Studies). London: Methuen.

Entrikin, J.N. (1981) 'Philosophical issues in the scientific study of regions', in D. Herbert and R.J. Johnston (eds) *Geography and the Urban Environment. Volume 4*. Chichester: Wiley, 1–27.

Entrikin, N. (1976) 'Contemporary humanism in geography', *Annals of the Association of American Geographers*, 66: 615–32.

Entrikin, N. (1979) 'Review of *Humanistic Geography: Prospects and Problems*, ed. Ley and Samuels', *Economic Geography*, 55: 253–7.

Evans, M. (1988) 'Participant observation: the researcher as research tool', in J. Eyles and D. Smith (eds) *Qualitative Methods in Human Geography*. Cambridge: Polity Press, 197–218.

Eyles, J. (1985) *Senses of Place*. Warrington: Silverbrook Press.

Eyles, J. (1986) 'Qualitative methods: a new revolution?', in J. Eyles (ed.) *Qualitative Approaches in Social and Geographical Research. Occasional Paper* 26. London: Department of Geography and Earth Science, Queen Mary College, University of London, 1–22.

Eyles, J. (ed.) (1988a) *Research in Human Geography: Introductions and Investigations*. Oxford: Blackwell.

Eyles, J. (1988b) 'Interpreting the geographical world: qualitative approaches in geographical research', in J. Eyles and D.M. Smith (eds) *Qualitative Methods in Human Geography*. Cambridge: Polity Press, 1–16.

Eyles, J. and Smith, D.M. (eds) (1988) *Qualitative Methods in Human Geography*. Cambridge: Polity Press.

Farrow, H., Moss, P. and Shaw, B. (eds) (1997) 'Symposium on feminist participatory research' (a series of papers), *Antipode*, 27: 71–101.

Fielding, N. and Fielding, J. (1986) *Linking Data*. London: Sage.

Fielding, N. and Lee, R. (eds) (1991) *Using Computers in Qualitative Research*. London: Sage.

Finch, J. (1993) 'It's great to have someone to talk to: ethics and the politics of interviewing women', in M. Hammersley (ed.) *Social Research: Philosophy, Politics and Practice*. London: Sage, 166–80 (originally published in Bell, C. and Roberts, H. (eds) (1984) *Social Researching: Politics, Problems, Practice*. London: Routledge, 70–87).

Fink, A. and Kosecoff, J. (1985) *How to Conduct Surveys: A Step by Step Guide*. London: Sage.

Fishman, P. (1990) 'Interaction: the work women do', in J. McCarl Nielson (ed.) *Feminist Research Methods: Exploring Readings in the Social Sciences*. London: Westview Press, 224–37.

Flowerdew, R. (1998) 'Reacting to *Ground Truth*', *Environment and Planning A*, 30: 289–302.

Flowerdew, R. and Martin, D. (eds) (1997) *Methods in Human Geography: A Guide for Students Doing Research Projects*. Harlow: Longman.

Floyd, B. (1963) 'Quantification: a geographic deviation?', *The Professional Geographer*, 15: 15–17.

Foddy, W. (1993) *Constructing Questions for Interviews and Questionnaires*. Cambridge: Cambridge University Press.

Foord, J. and Gregson, N. (1986) 'Patriarchy: towards a reconceptualisation', *Antipode*, 18: 186–211.

Foot, S., Rigby, D. and Webber, M. (1989) 'Theory and measurement in historical materialism', in A. Kobayashi and S. Mackenzie (eds) *Remaking Human Geography*. London: Unwin Hyman, 116–33.

Forbes, D. (1988) 'Getting by in Indonesia: research in a foreign land', in J. Eyles (ed.) *Research in Human Geography*. Oxford: Blackwell, 100–20.

Ford, L. (1994) 'Sunshine and shadow: lighting and colour in the depiction of cities on film', in S.C. Aitken and L.E. Zonn (eds) *Place, Power, Situation and Spectacle: A Geography of Film*. Lanham, MD: Rowman & Littlefield, 119–36.

Fotheringham, S. (1997) 'Trends in quantitative methods. 1. Stressing the local', *Progress in Human Geography*, 21: 88–96.

Fotheringham, S. (1998) 'Trends in quantitative methods. 2. Stressing the computational', *Progress in Human Geography*, 22: 283–92.

Fotheringham, S. (1999) 'Trends in quantitative methods III. Stressing the visual', *Progress in Human Geography*, 23: 597–606.

Fotheringham, S., Brunsdon, C. and Charlton, M. (2000) *Quantitative Geography: Perspectives on Spatial Data Analysis*. London: Sage.

Fotheringham, S. and O'Kelly, M. (1989) *Spatial Interaction Models: Formulations and Applications*. London: Kluwer.

Foucault, M. (1986) *The Care of the Self: The History of Sexuality. Volume 3*. New York, NY: Pantheon Books.

Foucault, M. (1991) 'Governmentality', in G. Burchell et al. (eds) *The Foucault Effect: Studies in Governmentality*. Hemel Hempstead: Harvester Wheatsheaf, 87–104.

Fowler, F. (1988) *Survey Research Methods* (revised edn). Beverly Hills, CA: Sage.

Fowler, F. and Mangione, T. (1990) *Standardized Survey Interviewing*. London: Sage.

Fowler, R. (1991) *Language in the News*. London: Routledge.

Franklin, B. and Murphy, D. (1990) *What News?* London: Routledge.

French, W.W. (1940) 'Local survey at Tiverton', *Geography*, XXV: 76–84.

Friedman, J. (1990) 'Being in the world: globalization and localization', in M. Featherstone (ed.) *Global Culture: Nationalism, Globalization and Modernity*. Newbury Park, CA: Sage, 311–28.

Fuller, D. (1999) 'Part of the action, or "going native"? Learning to cope with the "politics of integration"', *Area*, 31: 221–7.

Galbraith, V.H. (1961) *The Making of the Domesday Book*. Oxford: Oxford University Press.

Gale, S. (1972) 'Inexactness, fuzzy sets and the foundations of behavioural geography', *Geographical Analysis*, 4: 285–322.

Gamson, W., Croteau, D., Hoynes, W. and Sasson, T. (1992) 'Media images and the social construction of reality', *Annual Review of Sociology*, 18: 373–93.

Gandy, M. (1996) 'Visions of darkness: the representation of nature in the films of Werner Herzog', *Ecumene*, 3: 1–21.

Geertz, C. (1973) *The Interpretation of Cultures*. New York, NY: Basic Books.

Geertz, C. (1983) *Local Knowledge: Further Essays in Interpretive Anthropology*. New York, NY: Basic Books.

Genthe, M.K. (1912) 'Comment on Colonel Close's address to the purpose and position of geography', *Bulletin of the American Geographical Society*, 44: 27–38.

Gerard, R.W. (1969) 'Hierarchy, entitation and levels', in L.L. Whyte et al. (eds) *Hierarchical Structures*. New York, NY: American Elsevier, 215–30.

Gibbons, M., Limoges, C., Nowotny, H., Schwartzman, S., Scott, P. and Trow, S. (1994) *The New Production of Knowledge*. London: Sage.

Giddens, A. (1981) *A Contemporary Critique of Historical Materialism. Volume 1*. Basingstoke: Macmillan.

Giddens, A. (1984) *The Constitution of Society*. Cambridge: Polity Press.

Giddens, A. (1985) *The Nation-state and Violence*. Cambridge: Polity Press.

Gilbert, D. (1995) 'Between two cultures: geography, computing and the humanities', *Ecumene*, 2: 1–13.

Gilbert, D. (1999) 'Sponsorship, academic independence and critical engagement: a forum on Shell, the Ogoni dispute, and the Royal Geographical Society (with the Institute of British Geographers)', *Ethics, Place and Environment*, 2: 219–28.

Gilbert, E. (1998) 'Ornamenting the facade of hell: iconographies of nineteenth-century Canadian paper money', *Environment and Planning D: Society and Space*, 16: 57–80.

Gilbert, E.W. (1972) *British Pioneers in Geography*. Newton Abbot: David & Charles.

Gilbert, N. and Mulkay, M. (1984) *Opening Pandora's Box*. Cambridge: Cambridge University Press.

Glennie, P. (1990) *'Distinguishing Men's Trades': Occupational Sources and Debates for Pre-census England. Historical Geography Research Series 25*. Bristol: Institute of British Geographers.

Gober, P., Glasmeier, A., Goodman, J., Plane, D., Stafford, H. and Wood, S. (1995) 'Employment trends in geography', *The Professional Geographer*, 47: 336–46.

Goddard, S. (ed.) (1983) *A Guide to Information Sources in the Geographical Sciences*. London: Croom Helm.

Godkin, M.A. (1980) 'Identity and place: clinical applications based on notions of rootedness and uprootedness', in A. Buttimer and D. Seamon (eds) *The Human Experience of Space and Place*. London: Croom Helm, 73–85.

Godlewska, A. and Smith, N. (eds) (1994) *Geography and Empire*. Oxford: Blackwell.

Goffman, E. (1968) *Asylums: Essays on the Social Situation of Mental Patients and other Inmates.* Harmondsworth: Penguin Books.

Gold, J. (1985) 'From *Metropolis* to *The City*: film visions of the future city, 1919–39', in J. Burgess and J. Gold (eds) *Geography, the Media and Popular Culture.* London: Croom Helm, 123–43.

Gold, J. (1994) 'Locating the message: place promotion as image communication', in J. Gold and S. Ward (eds) *Place Promotion.* Chichester: Wiley, 19–37.

Goldman, A. and McDonald, S. (1987) *The Group Depth Interview: Principles and Practices.* Englewood Cliffs, NJ: Prentice Hall.

Goodchild, M. (1984) 'Geocoding and geosampling', in G.L. Gaile and C.J. Willmott (eds) *Spatial Statistics and Models.* Dordrecht: Reidel, 33–54.

Goodchild, M. (1994) 'Geographical information systems (GIS)', in R.J. Johnston et al. (eds) *The Dictionary of Human Geography* (3rd edn). Oxford: Blackwell, 219–20.

Goodchild, M. and Longley, P. (1998) 'The future of GIS and spatial analysis', in P. Longley et al. (eds) *Geographical Information Systems: Principles, Techniques, Management and Applications* (2 vols). Cambridge and New York, NY: Geo Information International, 567–80.

Gorden, R. (1987) *Interviewing.* Homewood, IL: Dorsey Press.

Gormley, N.B. (1994) 'Letter from the field: reflections half way through', in E. Robson and K. Willis (eds) *Postgraduate Fieldwork in Developing Areas: A Rough Guide. IBG Developing Areas Research Group Monograph 8.* London: IBG, 103–10.

Goss, J. (1993) 'Placing the market and marketing place: tourist advertising of the Hawaiian Islands, 1972–92', *Environment and Planning D: Society and Space*, 11: 663–88.

Goss, J. (1995a) 'Marketing the new marketing: the strategic discourse of geodemographic information systems', in J. Pickles (ed.) *Ground Truth: The Social Implications of Geographic Information Systems.* New York, NY: Guilford, 130–70.

Goss, J. (1995b) 'We know who you are and we know where you live: the instrumental rationality of geomarketing information systems', *Economic Geography*, 71: 171–98.

Goss, J. (1996) 'Introduction to focus groups', *Area*, 28: 113–14.

Goss, J. (1999) 'Once-upon-a-time in the commodity world: an unofficial guide to mall of America', *Annals of the Association of American Geographers*, 89: 45–75.

Gould, P. (1979) 'Geography 1957–1977: the Augean period', *Annals of the Association of American Geographers*, 69: 139–51.

Gould, P. (1981) 'Letting the data speak for themselves', *Annals of the Association of American Geographers*, 71: 166–76.

Gouldner, A. (1967) *Enter Plato.* London: Routledge & Kegan Paul.

Graham, E. (1997) 'Philosophies underlying human geography research', in R. Flowerdew and D. Martin (eds) *Methods in Human Geography.* Harlow: Longman, 6–30.

Greenbaum, T. (1988) *The Practical Handbook and Guide to Focus Group Research.* Lexington, MA: Lexington Books.

Gregory, D. (1978a) *Ideology, Science and Human Geography.* London: Hutchinson.

Gregory, D. (1978b) 'The discourse of the past: phenomenology, structuralism and historical geography', *Journal of Historical Geography*, 4: 161–73.

Gregory, D. (1980) 'The ideology of control: systems theory and geography', *Tidschrift voor Economische en Social Geografie*, 71: 327–42.

Gregory, D. (1981) 'Human agency and human geography', *Transactions, Institute of British Geographers*, 6: 1–18.

Gregory, D. (1982) *Regional Transformation and Industrial Revolution: A Geography of the Yorkshire Woollen Industry.* London: Macmillan.

Gregory, D. (1985) 'Suspended animation: the stasis of diffusion theory', in D. Gregory and J. Urry (eds) *Social Relations and Spatial Structures.* London: Macmillan, 296–336.

Gregory, D. (1986) 'Entitation', in R.J. Johnston et al. (eds) *The Dictionary of Human Geography* (2nd edn). Oxford: Blackwell, 129–30.

Gregory, D. (1989a) 'Presences and absences: time–space relations and structuration theory', in D. Held and J.B. Thompson (eds) *Social Theory of Modern Societies: Anthony Giddens and his Critics.* Cambridge: Cambridge University Press, 185–214.

Gregory, D. (1989b) 'Areal differentiation and post-modern human geography', in D. Gregory and R. Walford (eds) *Horizons in Human Geography*. London: Macmillan, 67–96.

Gregory, D. (1989c) 'The crisis of modernity', in R. Peet and N. Thrift (eds) *New Models in Geography*. London: Unwin Hyman, 348–85.

Gregory, D. (1994) *Geographical Imaginations*. Oxford: Blackwell.

Gregory, D. (1995) 'Between the book and the lamp: imaginative geographies of Egypt, 1849–50', *Transactions, Institute of British Geographers*, 20: 29–57.

Gregory, D. (1996) 'Areal differentiation and post-modern human geography', in J. Agnew et al. (eds) *Human Geography: An Essential Anthology*. Oxford: Blackwell, 211–32.

Gregory, D. (2000a) 'Structuration theory', in R.J. Johnston et al. (eds) *The Dictionary of Human Geography* (4th edn). Oxford: Blackwell, 798–800.

Gregory, D. (2000b) 'Contextual approach', in R.J. Johnston et al. (eds) *The Dictionary of Human Geography* (4th edn). Oxford: Blackwell, 110–12.

Gregory, D. (2000c) 'Compositional theory', in R.J. Johnston et al. (eds) *The Dictionary of Human Geography* (4th edn). Oxford: Blackwell, 103.

Gregory, D. (2000d) 'Model', in R.J. Johnston et al. (eds) *The Dictionary of Human Geography* (4th edn). Oxford: Blackwell, 508–10.

Gregory, D. (2000e) 'Empiricism', in R.J. Johnston et al. (eds) *The Dictionary of Human Geography* (4th edn). Oxford: Blackwell, 205–6.

Gregory, D. (2000f) 'Positivism', in R.J. Johnston et al. (eds) *The Dictionary of Human Geography* (4th edn). Oxford: Blackwell, 606–8.

Gregory, D. and Ley, D. (1988) 'Culture's geographies', *Environment and Planning D: Society and Space*, 6: 115–16.

Gregory, D. and Urry, J. (eds) (1985) *Social Relations and Spatial Structures*. London: Macmillan.

Gregory, S. (1978) *Statistical Methods and the Geographer* (4th edn). London: Longman.

Gregory, S. (1992) 'Thinking statistically', in A. Rogers et al. (eds) *The Student's Companion to Geography*. Oxford: Blackwell, 135–40.

Gregson, N. (1995) 'Classics in human geography revisited: Thrift, N. "On the determination of social action in space and time". Commentary 2', *Progress in Human Geography*, 19: 527–8.

Gregson, N., Brooks, K. and Crewe, L. (2000) 'Narratives of consumption and the body in the space of the charity shop', in P. Jackson et al. (eds) *Commercial Cultures: Economies, Practices, Spaces*. Oxford: Berg, 101–21.

Grigg, D. (1965) 'The logic of regional systems', *Annals of the Association of American Geographers*, 55: 465–91.

Grigg, D. (1967) 'Regions, models and classes', in R.J. Chorley and P. Haggett (eds) *Models in Geography*. London: Methuen, 461–509.

Guelke, L. (1974) 'An idealist alternative in human geography', *Annals of the Association of American Geographers*, 64: 193–202.

Guelke, L. (1978) 'Geography and logical positivism', in D. Herbert and R.J. Johnston (eds) *Geography and the Urban Environment*. Chichester: Wiley, 35–61.

Guelke, L. (1982) *Historical Understanding in Geography*. Cambridge: Cambridge University Press.

Habermas, J. (1972) *Knowledge and Human Interests*. London: Heinemann.

Habermas, J. (1984) *The Theory of Communicative Action. Volume 1. Reason and the Rationalisation of Society*. London: Heinemann.

Haggett, P. (1965) *Locational Analysis in Human Geography*. London: Edward Arnold.

Haggett, P., Cliff, A. and Frey, A. (1977) *Locational Models (Volume I); Locational Methods (Volume II)*. London: Arnold.

Haining, R. (1990) *Spatial Data Analysis in the Social and Environmental Sciences*. Cambridge: Cambridge University Press.

Hakim, C. (1980) 'Census reports as documentary evidence: the census commentaries 1801–1951', *Sociological Review*, 28: 551–80.

Halfacree, K. (1993) 'Locality and social space, discourse and alternative definitions of the rural', *Journal of Rural Studies*, 9: 1–15.

Hall, P. and Markusen, A. (eds) (1985) *Silicon Landscapes*. Boston, MA: Allen & Unwin.

Hall, P. and Preston, P. (1988) *The Carrier Wave: New Information Technology and the Geography of Innovation.* London: Unwin Hyman.

Hall, S. (ed.) (1997) *Representation: Cultural Representations and Signifying Practices.* London: Sage.

Hall, S., Hobson, D., Lowe, A. and Willis, P. (eds) (1980) *Culture, Media, Language.* London: Methuen.

Hall, T., Healey, M. and Harrison, M. (2002) 'Fieldwork and disabled students: discourses of exclusion and inclusion', *Transactions, Institute of British Geographers,* 27: 213–31.

Hammersley, M. (ed.) (1993) *Social Research: Philosophy, Politics and Practice.* 2nd edn. London: Sage.

Hammersley, M. (1995) *The Politics of Social Research.* London: Sage.

Hammersley, M. and Atkinson, P. (1983) *Ethnography: Principles in Practice.* London: Tavistock.

Hammersley, M. and Atkinson, P. (1995) *Ethnography: Principles in Practice.* 2nd edn. London: Routledge.

Hammond, P. and McCullagh, P. (1974) *Quantitative Techniques in Geography.* Oxford: Clarendon Press.

Hamnett, C. (1998) 'The owner-occupied housing market in Britain: a north–south divide?', in J. Lewis and A. Townsend (eds) *The North-south Divide.* London: Paul Chapman, 97–113.

Hannah, M.G. (1997) 'Space and the structuring of disciplinary power: an interpretive review', *Geografiska Annaler B,* 79: 171–80.

Hanson, S. (1988) 'Soaring', *The Professional Geographer,* 40: 4–7.

Haraway, D. (1988) 'Situated knowledges: the science question in feminism and the privilege of partial perspective', *Feminist Studies,* 14: 575–99.

Haraway, D. (1989) *Primate Visions: Gender, Race and Nature in the World of Modern Science.* New York, NY: Routledge.

Haraway, D. (1991) *Simians, Cyborgs and Women: The Reinvention of Nature.* New York, NY: Routledge.

Haraway, D. (1996) 'Situated knowledges: the science question in feminism and the privilege of partial perspective', in J. Agnew et al. (eds) *Human Geography: An Essential Anthology.* Oxford: Blackwell, 108–28.

Harley, J.B. (1988) 'Maps, knowledge and power', in D. Cosgrove and S. Daniels (eds) *The Iconography of Landscape: Essays on the Symbolic Representation, Design and Use of Past Environments.* Cambridge: Cambridge University Press, 277–312.

Harley, J.B. (1989) 'Deconstructing the map', *Cartographica,* 27: 1–23.

Harré, R. (1979) *Social Being.* Oxford: Blackwell.

Harris, T., Weiner, D., Warner, T. and Levin, R. (1995) 'Pursuing social goals through participatory GIS: redressing South Africa's historical political ecology', in J. Pickles (ed.) *Ground Truth: The Social Implications of Geographic Information Systems.* New York, NY: Guilford, 196–222.

Harrison, R.T. and Livingstone, D.N. (1979) 'There and back again: towards a critique of idealist human geography', *Area,* 11: 75–9.

Harrison, R.T. and Livingstone, D.N. (1980) 'Philosophy and problems in human geography: a presuppositional approach', *Area,* 12: 25–31.

Hart, J.F. (1982) 'The highest form of the geographer's art', *Annals of the Association of American Geographers,* 72: 1–29.

Hartshorne, R. (1939) *The Nature of Geography: A Critical Survey of Current Thought in the Light of the Past.* Lancaster, PA: Association of American Geographers.

Harvey, D. (1969) *Explanation in Geography.* London: Edward Arnold.

Harvey, D. (1973) *Social Justice and the City.* London: Edward Arnold.

Harvey, D. (1982) *The Limits to Capital.* London: Blackwell.

Harvey, D. (1985) *Consciousness and the Urban Experience.* Oxford: Blackwell.

Harvey, D. (1989a) *The Condition of Postmodernity: An Enquiry into the Origins of Cultural Change.* Oxford: Blackwell.

Harvey, D. (1989b) *The Urban Experience.* Oxford: Blackwell.

Harvey, D. (1990) 'Between space and time: reflections on the geographical imagination', *Annals of the Association of American Geographers,* 80: 418–34.

Harvey, D. (1995) 'A geographer's guide to dialectical thinking', in A.D. Cliff et al. (eds) *Diffusing Geography: Essays for Peter Haggett.* Oxford: Blackwell, 3–21.

Harvey, D. (1996) *Justice, Nature and the Geography of Difference.* Oxford: Blackwell.

Harvey, D. (2000) *Spaces of Hope.* Edinburgh: Edinburgh University Press.

Hay, A.M.H. (1994) 'Systems analysis', in R.J. Johnston et al. (eds) *The Dictionary of Human Geography* (3rd edn). Oxford: Blackwell, 615.

Hay, I. (1998) 'Making moral imaginations. Research ethics, pedagogy and professional human geography', *Ethics, Place and Environment*, 1: 55–76.

Hay, I. (2000) *Qualitative Research Methods in Geography (Meridian Series)*. Melbourne: Oxford University Press.

Hepple, L. (1998) 'Context, social construction, and statistics: regression, social science and human geography', *Environment and Planning A*, 30: 225–34.

Hepple, L. (2000) 'Spatial autocorrelation', in R.J. Johnston et al. (eds) *The Dictionary of Human Geography* (4th edn). Oxford: Blackwell, 775.

Herbert, S. (2000) 'For ethnography', *Progress in Human Geography*, 24: 550–68.

Hetherington, K. and Law, J. (2000) 'After networks', *Environment and Planning D: Society and Space*, 18: 127–32.

Hill, M.R. (1981) 'Positivism: a "hidden" philosophy in geography', in M.E. Harvey and B.P. Holly (eds) *Themes in Geographic Thought*. London: Croom Helm, 38–60.

Hillis, K. (1994) 'The virtue of becoming a no-body', *Ecumene*, 1: 177–96.

Hinchliffe, S. (2000) 'Performance and experimental knowledge: outdoor management training and the end of epistemology', *Environment and Planning D: Society and Space*, 18: 575–95.

Hodge, D.C. (ed.) (1995) 'Should women count? The role of quantitative methodology in feminist geographical research' (a series of papers), *The Professional Geographer*, 47: 426–66.

Hoggart, K., Lees, L. and Davies, A. (2002) *Researching Human Geography*. London: Arnold.

Holbrook, B. and Jackson, P. (1996) 'Shopping around: focus group research in north London', *Area*, 28: 136–42.

Holloway, S.L., Rice, S.P. and Valentine, G. (eds) (2003) *Key Concepts in Geography*. London: Sage.

Holstein, J. and Gubrium, J. (1997) 'Active interviewing', in D. Silverman (ed.) *Qualitative Research: Theory, Method and Practice*. London: Sage, 113–29.

hooks, b. (1997) *Real to Reel*. Chicago, IL: University of Chicago Press.

Horvarth, R.J. (1971) 'The "Detroit Geographical Expedition and Institute" experience', *Antipode*, 3: 73–84.

Howard, S. (1994) 'Methodological issues in overseas fieldwork: experiences from Nicaragua's northern Atlantic coast', in E. Robson and K. Willis (eds) *Postgraduate Fieldwork in Developing Areas: A Rough Guide*. IBG Developing Areas Research Group Monograph 8. London: IBG, 19–35.

Hubbard, P.J., Kitchin, R., Bartley, B. and Fuller, D. (2002) *Thinking Geographically: Space, Theory and Contemporary Human Geography*. London: Continuum.

Huck, S. and Sandler, H. (1979) *Rival Hypotheses: Alternative Interpretations of Data Based Conclusions*. New York, NY: HarperCollins.

Hudson, R. (1995) 'Making music work? Alternative regeneration strategies in a deindustrialized locality: the case of Derwentside', *Transactions, Institute of British Geographers*, 20: 461–73.

Huff, D. (1973) *How to Lie with Statistics*. Harmondsworth: Penguin Books.

Huggett, R. (1980) *Systems Analysis in Geography*. Oxford: Clarendon Press.

Hughes, A. and Cormode, L. (eds) (1998) 'Theme issue: researching elites and elite spaces', *Environment and Planning A*, 30: 2095–180.

Hughes, A., Morris, C. and Seymour, S. (eds) (2000) *Ethnography and Rural Research*. Gloucester: Countryside and Community Press.

Hutchings, G.E. (1949) 'The geographer as field naturalist', *School Nature Study*, 44: 33–8.

Hutchings, G.E. (1960) *Landscape Drawing*. London: Methuen.

Hutchings, G.E. (1962) 'Geographical field teaching', *Geography*, XLVII: 1–14.

Jackson, P. (1983) 'Principles and problems of participant observation', *Geografiska Annaler Series B: Human Geography*, 65: 39–46.

Jackson, P. (1985) 'Urban ethnography', *Progress in Human Geography*, 9: 157–76.

Jackson, P. (1988a) 'Definitions of the situation: neighbourhood change and local politics in Chicago', in J. Eyles and D.M. Smith (eds) *Qualitative Methods in Human Geography*. Cambridge: Polity Press, 49–74.

Jackson, P. (1988b) 'Street life: the politics of carnival', *Environment and Planning D: Society and Space*, 6: 213–27.

Jackson, P. (1989) *Maps of Meaning: An Introduction to Cultural Geography*. London: Unwin Hyman.

Jackson, P. (1993) 'Changing ourselves: a geography of position', in R.J. Johnston (ed.) *The Challenge for Geography: A Changing World, a Changing Discipline*. Oxford: Blackwell, 198–214.

Jackson, P. (1998) 'Classics in human geography revisited: Ley 1974 *The Black Inner City as Frontier Outpost*, commentary 1', *Progress in Human Geography*, 22: 75–6.

Jackson, P. (1999) 'Postmodern urbanism and the ethnographic void', *Urban Geography*, 20: 400–2.

Jackson, P. (2000) 'Positionality', in R.J. Johnston et al. (eds) *The Dictionary of Human Geography* (4th edn). Oxford: Blackwell, 604–5.

Jackson, P. (2001) 'Making sense of qualitative data', in M. Limb and C. Dwyer (eds) *Qualitative Methodologies for the Geographer: Issues and Debates*. London: Arnold, 119–214.

Jackson, P. and Smith, S. (1984) *Exploring Social Geography*. Hemel Hempstead: Allen & Unwin.

Jackson, P., Stevenson, N. and Brooks, K. (1999) 'Making sense of men's lifestyle magazines', *Environment and Planning D: Society and Space*, 17: 353–68.

Jacobs, J.M. (1996) *Edge of Empire: Postcolonialism and the City*. London: Routledge.

James, P.E. and Mather, C. (1977) 'The role of periodic field conferences in the development of geographical ideas in the United States', *The Geographical Review*, 67: 446–61.

Jameson, A.H. and Ormsby, M.T.M. (1934) *Mathematical Geography*. London: Sir Isaac Pitman & Sons.

Jayaratne, T. (1993) 'The value of quantitative methodology for feminist research', in M. Hammersley (ed.) *Social Research*. London: Sage, 109–23.

Jensen, M.A.P. (1946) 'The promotion of field studies', *Geography*, XXXI: 153–6.

Johnson, R. (1986) 'The story so far: and further transformations?', in D. Punter (ed.) *Introduction to Contemporary Cultural Studies*. London: Longman, 277–313.

Johnston, R.J. (1978) *Multivariate Statistical Analysis in Geography*. London: Longman.

Johnston, R.J. (1979) *Political, Electoral and Spatial Systems*. Oxford: Clarendon Press.

Johnston, R.J. (1997a) 'W(h)ither spatial science and spatial analysis', *Futures*, 29: 323–36.

Johnston, R.J. (1997b) *Geography and Geographers: Anglo-American Human Geography since 1945*. London: Arnold.

Johnston, R.J. (2000a) 'On disciplinary history and textbooks: or where has spatial analysis gone?', *Australian Geographical Studies*, 38: 125–37.

Johnston, R.J. (2000b) 'Questionnaire', in R.J. Johnston et al. (eds) *The Dictionary of Human Geography* (4th edn). Oxford: Blackwell, 668.

Johnston, R.J. (2000c) 'Intellectual respectability and disciplinary transformation? Radical geography and the institutionalisation of geography in the USA since 1945', *Environment and Planning A*, 27: 971–90.

Johnston, R.J., Hepple, R., Hoare, A., Jones, K. and Plummer, P. (2003) 'Contemporary fiddling in human geography while Rome burns: has quantitative analysis been largely abandoned – and should it be?', *Geoforum*, 34: 157–61.

Johnston, R.J. and Pattie, C. (2000) 'New Labour, new electoral system, new electoral geographies? A review of proposed constitutional changes in the United Kingdom', *Political Geography*, 19: 495–515.

Johnston, R.J., Pattie, C. and Allsopp, J.G. (1988) *A Nation Dividing? The Electoral Map of Great Britain 1979–1987*. London: Longman.

Jones, E. (1980) 'Review of *Ideology, Science and Human Geography* by Gregory, and *Humanistic Geography: Prospects and Problems*, ed. Ley & Samuels', *The Geographical Journal*, 146: 113–15.

Jones, J.P. III and Casetti, E. (eds) (1992) *Applications of the Expansion Methods*. London: Routledge.

Jones, K. (1991) 'Specifying and estimating multilevel models for geographical research', *Transactions, Institute of British Geographers*, 16: 148–59.

Jones, K. and Duncan, C. (1996) 'People and places: the multilevel model as a general framework for the quantitative analysis of geographical data', in P. Longley and M. Batty (eds) *Spatial Analysis: Modelling in a GIS Environment*. Cambridge: GeoInformation International, 79–104.

Jones, W.D. (1931) 'Field mapping of residential areas in Metropolitan Chicago', *Annals of the Association of American Geographers*, 21: 207–14.

Jones, W.D. (1934) 'Procedures in investigating human occupance of a region', *Annals of the Association of American Geographers*, 24: 93–107.

Jones, W.D. and Sauer, C.O. (1915) 'Outline for fieldwork in geography', *Bulletin of the American Geographical Society*, 47: 520–6.

Katz, C. (1992) 'All the world is staged: intellectuals and the projects of ethnography', *Environment and Planning D: Society and Space*, 10: 495–510.

Katz, C. (1994) 'Playing the field: questions of fieldwork in geography', *The Professional Geographer*, 46: 67–72.

Kearns, R. (2000) 'Being there: research through observing and participating', in I. Hay (ed.) *Qualitative Research Methods in Human Geography*. South Melbourne: Oxford University Press, 103–21.

Keat, R. and Urry, J. (1975) *Social Theory as Science*. London: Routledge & Kegan Paul.

Keith, M. (1991) 'Knowing your place: the imagined geographies of racial subordination', in C. Philo (ed.) *New Words, New Worlds: Reconceptualising Social and Cultural Geography*. Aberystwyth: Cambrian Printers, 178–92.

Keith, M. (1992) 'Angry writing: (re)presenting the unethical world of the ethnographer', *Environment and Planning D: Society and Space*, 10: 551–68.

Kennedy, B.A. (1979) 'A naughty world', *Transactions, Institute of British Geographers*, 4: 550–8.

Kennedy, C. and Lukinbeal, C. (1997) 'Towards a holistic approach to geographical research on film', *Progress in Human Geography*, 21: 33–50.

Kimble, G.H.T. (1938) *Geography in the Middle Ages*. London: Methuen.

Kindon, S. (2003) 'Participatory video in geographic research: a feminist practice of looking?', *Area*, 35: 142–53.

Kingman, M.M. (1969) 'Field study for the non-sighted', *The Professional Geographer*, 21: 199.

Kingsbury, P. (2003) 'Psychoanalysis, a gay spatial science?', *Social and Cultural Geography*, 4: 347–68.

Kinross, R. (1985) 'The rhetoric of neutrality', *Design Issues*, 2: 18–30.

Kinsman, P. (1995) 'Landscape, race and national identity: the photography of Ingrid Pollard', *Area*, 27: 300–10.

Kitajima, S. (1999) 'Changing universities and the question of academic practice', *Ethics, Place and Environment*, 2: 254–6.

Kitchin, R.M. and Hubbard, P.J. (eds) (1999) 'Research, action and "critical geographies"' (a series of papers), *Area*, 31: 195–246.

Kitchin, R.M. and Tate, N. (2000) *Conducting Research in Human Geography: Theory, Methodology and Practice*. Harlow: Prentice Hall.

Kneafsey, M. (2000) 'Changing roles and constructing identities: ethnography in the Celtic periphery', in A. Hughes et al. (eds) *Ethnography and Rural Research*. Cheltenham: Countryside Commission, 52–65.

Kobayashi, A. (2001) 'Negotiating the personal and the political in critical qualitative research', in M. Limb and C. Dwyer (eds) *Qualitative Methodologies for Geographers*. London: Arnold, 55–70.

Kong, L. (1993) 'Ideological hegemony and the political symbolism of religious buildings in Singapore', *Environment and Planning D: Society and Space*, 11: 23–45.

Kong, L. (1995a) 'Music and cultural politics: ideology and resistance in Singapore', *Transactions, Institute of British Geographers*, 20: 447–59.

Kong, L. (1995b) 'Popular music in geographical analyses', *Progress in Human Geography*, 19: 193–8.

Kong, L. (1996) 'Popular music in Singapore: exploring local cultures, global resources, and regional identities', *Environment and Planning D: Society and Space*, 14: 273–92.

Kong, L. and Goh, E. (1995) 'Folktales and reality: the social construction of race in Chinese tales', *Area*, 27: 261–7.

Kreuger, R. (1988) *Focus Groups: A Practical Guide for Applied Research*. Beverly Hills, CA: Sage.

Krippendorff, K. (1980) *Content Analysis*. London: Sage.

Kunzru, H. (1997) 'You are cyborg: for Donna Haraway, we are already assimilated', *Wired*, 5 (www.wired.com/wired/archives/5.02/ffharaway.html).

Langton, J. (1972) 'Potentialities and problems in adopting a systems approach to the study of change in human geography', *Progress in Human Geography*, 4: 125–79.

Latour, B. (1991) 'Technology is society made durable', in J. Law (ed.) *A Sociology of Monsters: Essays on Power, Technology and Domination*. London: Routledge, 103–31.

Latour, B. (1993) *We Have Never Been Modern*. Brighton: Harvester Wheatsheaf.

Latour, B. (1996) *Aramis, or the Love of Technology*. Cambridge, MA: Harvard University Press.

Latour, B. (1999) *Pandora's Hope: Essays on the Reality of Science Studies*. Cambridge, MA: Harvard University Press.

Laurier, E. (2001) 'Why people say where they are during mobile phone calls', *Environment and Planning D: Society and Space*, 19: 485–504.

Laurier, E. and Philo, C. (1999) 'X-morphising: review essay of Bruno Latour's *Aramis, or the Love of Technology*', *Environment and Planning A*, 31: 1047–71.

Laurier, E. and Philo, C. (2003) 'The region in the boot: mobilising lone subjects and multiple objects', *Environment and Planning D: Society and Space,* 21: 85–106.

Law, J. (1986) 'On the methods of long distance control: vessels, navigation and the Portuguese route to India', in J. Law (ed.) *Power, Action and Belief.* London: Routledge & Kegan Paul, 234–63.

Law, J. (1991) 'Introduction: monsters, machines and sociotechnical relations', in J. Law (ed.) *A Sociology of Monsters.* London: Routledge, 1–23.

Law, J. (1994) *Organising Modernity.* Oxford: Blackwell.

Lee, R. (ed.) (1992) 'Teaching qualitative geography: a JGHE written symposium' (a series of papers), *Journal of Geography in Higher Education,* 16: 123–84.

Leighley, J. (ed.) (1963) *Land and Life: A Selection from the Writings of Carl Ortwin Sauer.* Berkeley and Los Angeles, CA: University of California Press.

Leslie, D. and Reimer, S. (2003) 'Gender, modern design, and home consumption', *Environment and Planning D: Society and Space,* 21: 293–316.

Levitas, R. (1996) 'Fiddling while Britain burns? The "measurement" of unemployment', in R. Levitas and W. Guy (eds) *Interpreting Official Statistics.* London: Routledge, 45–65.

Lewis, C. and Pile, S. (1996) 'Woman, body, space: Rio Carnival and the politics of performance', *Gender, Place and Culture,* 3: 23–42.

Lewis, P. (1985) 'Beyond description', *Annals of the Association of American Geographers,* 75: 465–77.

Ley, D. (1974) *The Black Inner City as Frontier Outpost: Images and Behavior of a Philadelphia Neighborhood.* Washington, DC: Association of American Geographers.

Ley, D. (1977) 'Social geography and the taken-for-granted world', *Transactions, Institute of British Geographers,* 2: 498–512.

Ley, D. (1978) 'Review of *Prisoners of Space?* by Graham Rowles', *Economic Geography,* 54: 355–6.

Ley, D. (1980) *Geography without Man: A Humanistic Critique. School of Geography Research Paper* 24. Oxford: University of Oxford.

Ley, D. (1981a) 'Behavioural geography and the philosophies of meaning', in K.R. Cox and R.G. Golledge (eds) *Behavioural Problems in Geography Revisited.* London: Methuen, 209–30.

Ley, D. (1981b) 'Cultural/humanistic geography', *Progress in Human Geography,* 5: 249–57.

Ley, D. (1982) 'Rediscovering man's place', *Transactions, Institute of British Geographers,* 7: 248–53.

Ley, D. (1988) 'Interpretative social research in the inner city', in J. Eyles (ed.) *Research in Human Geography: Introductions and Investigations.* Oxford: Blackwell, 121–38.

Ley, D. (1998) 'Classics in human geography revisited: Ley, D. (1974) *The Black Inner City as Frontier Outpost:* author's response', *Progress in Human Geography,* 22: 78–80.

Ley, D. and Cybriwsky, R. (1974) 'Urban graffiti as territorial markers', *Annals of the Association of American Geographers,* 64: 491–505.

Ley, D. and Mountz, A. (2001) 'Interpretation, representation, positionality: issues in field research in human geography', in M. Limb and C. Dwyer (eds) *Qualitative Methodologies for Geographers: Issues and Debates.* London: Edward Arnold, 234–47.

Ley, D. and Samuels, M.S. (eds) (1978) *Humanistic Geography: Prospects and Problems.* London: Croom Helm.

Leyshon, A., Matless, D. and Revill, G. (1995) 'The place of music', *Transactions, Institute of British Geographers,* 20: 423–33.

Leyshon, A., Matless, D. and Revill, G. (eds) (1998) *The Place of Music.* New York: Guilford.

Lie, M. (1992) 'Teknologi og kjonnsidentitet: Har datamaskinen kjonn?', in A. Andenaes et al. (eds) *Epler Fra Var Egen Hage.* Trondheim: Centre for Women's Studies.

Limb, M. and Dwyer, C. (eds) (2001) *Qualitative Methodologies for Geographers: Issues and Debates.* London: Edward Arnold.

Lindsay, J.M. (1997) *Techniques in Human Geography.* London: Routledge.

Livingstone, D.N. (1992) *The Geographical Tradition: Episodes in the History of a Contested Enterprise.* Oxford: Blackwell.

Longley, P. (1998) 'Foundations', in P. Longley et al. (eds) *Geocomputation: A Primer.* Chichester: Wiley, 17–29.

Longley, P., Brooks, S., McDonnell, R. and Macmillan, B. (eds) (1998b) *Geocomputation: A Primer.* Chichester: Wiley.

Longley, P. and Clarke, G. (eds) (1995) *GIS for Business and Service Planning*. Cambridge: GeoInformation International.

Longley, P., Goodchild, M., Maguire, D. and Rhind, D. (eds) (1998a) *Geographical Information Systems: Principles, Techniques, Management and Application* (2 vols). New York, NY: Wiley.

Lorimer, H. and Spedding, N. (2002) 'Editorial: putting philosophies of geography into practice', *Area*, 34: 227–8.

Lowenthal, D. (1961) 'Geography, experience and imagination: towards a geographical epistemology', *Annals of the Association of American Geographers*, 51: 241–60.

Lutwack, L. (1984) *The Role of Place in Literature*. Syracuse, NY: Syracuse University Press.

MacKinnon, D. (2000) 'Managerialism, governmentality and the state: a neo-Foucauldian approach to local economic governance', *Political Geography*, 19: 293–314.

Macmillan, W. (1998) *Computing and the Science of Geography*. University of Oxford Economic Research Group Working Paper. Oxford: University of Oxford Economic Research Group.

Madge, C. (1993) 'Boundary disputes: comments on Sidaway (1992)', *Area*, 25: 294–9.

Madge, C. (1994) 'The ethics of research in the "Third World"', in E. Robson and K. Willis (eds) *Postgraduate Fieldwork in Developing Areas: A Rough Guide. Developing Areas Research Group Monograph* 9. London: RGS and the IBG, 91–102.

Madge, C., Raghuram, P., Skelton, T., Willis, K. and Williams, J. (1997) 'Methods and methodologies in feminist research', in WGSG (ed.) *Feminist Geographies: Diversity and Difference*. Harlow: Longman, 86–111.

Maitland, F.W. (1897) *Domesday Book and Beyond: Three Essays in the Early History of England*. Cambridge: Cambridge University Press.

Malbon, B. (1999) *Clubbing: Dancing, Ecstacy, Vitality*. London: Routledge.

Mallory, W.E. and Simpson-Houseley, P. (1987) *Geography and Literature: A Meeting of the Disciplines*. Syracuse, NY: Syracuse University Press.

Mann, M. (1986) *The Sources of Social Power. Volume I. A History of Power from the Beginning to AD 1760*. Cambridge: Cambridge University Press.

Marcus, G. (1986) 'Contemporary problems of ethnography in the modern world system', in J. Clifford and G. Marcus (eds) *Writing Culture: The Poetics and Politics of Ethnography*. Berkeley, CA: University of California Press, 165–93.

Marcus, G. (1995) 'Ethnography in/of the world system: the emergence of multi-sited ethnography', *Annual Review of Anthropology*, 24: 95–117.

Marcus, G. (1998) *Ethnography through Thick and Thin*. Princeton, NJ: Princeton University Press.

Marcus, G. (2000) 'The twistings and turnings of geography and anthropology in the winds of millennial transition', in I. Cook et al. (eds) *Cultural Turns/Geographical Turns: Perspectives on Cultural Geography*. Harlow: Longman, 13–25.

Marcus, G. and Fischer, M. (1986) *Anthropology as Cultural Critique: An Experimental Moment in the Human Sciences* (1st edn). Chicago, IL: University of Chicago Press.

Marcus, G. and Fischer, M. (1999) *Anthropology as Cultural Critique: An Experimental Moment in the Human Sciences* (2nd edn). Chicago, IL: University of Chicago Press.

Marsh, C. (1988) *Exploring Data*. Cambridge: Polity Press.

Martin, D. (1996) *Geographic Information Systems: Socio-economic Applications* (2nd edn). London: Routledge.

Martin, D. (1997) 'Geographic information systems and spatial analysis', in R. Flowerdew and D. Martin (eds) *Methods in Human Geography*. Harlow: Addison Wesley Longman, 213–29.

Martin, D. (2000) 'Towards the geographies of the 2001 UK Census of Population', *Transactions, Institute of British Geographers*, 25: 321–32.

Martin, R. (2001) 'Geography and public policy: the case of the missing agenda', *Progress in Human Geography*, 25: 189–210.

Martin, R., Sunley, P. and Willis, J. (1993) 'The geography of trade union decline: spatial dispersal or regional resilience?', *Transactions, Institute of British Geographers*, 18: 36–62.

Marx, K. (1973) *Grundrisse*. Harmondsworth: Penguin Books.

Mason, J. (1996) *Qualitative Researching*. London: Sage.

Massey, D. (1984) *Spatial Divisions of Labour*. London: Macmillan.

Massey, D. (1991a) 'Flexible sexism', *Environment and Planning D: Society and Space*, 9: 31–57.

Massey, D. (1991b) 'A progressive sense of place', *Marxism Today*, June: 24–9.

Massey, D. (1992) 'Politics and space/time', *New Left Review*, 196: 65–84.

Massey, D. (1993) 'Power-geometry and a progressive sense of place', in J. Bird et al. (eds) *Mapping the Futures: Local Cultures, Global Change*. London: Routledge, 59–69.

Massey, D. (1994) *Space, Place and Gender*. Cambridge: Polity Press.

Massey, D. (1999a) 'Space-time, "science" and the relationship between physical geography and human geography', *Transactions, Institute of British Geographers*, 24: 261–76.

Massey, D. (1999b) 'Spaces of politics', in D. Massey et al. (eds) *Human Geography Today*. Cambridge: Polity Press, 279–94.

Massey, D. (2000) 'Practising political relevance', *Transactions, Institute of British Geographers*, 25: 131–3.

Matless, D. and Revill, G. (1995) 'A solo ecology: the erratic art of Andy Goldsworthy', *Ecumene*, 2: 423–48.

Matthews, J. (1981) *Quantitative and Statistical Approaches to Geography: A Practical Manual*. Oxford: Pergamon.

Matthewson, K. (1998) 'Classics in human geography revisted, Duncan, J.S. "The superorganic in American cultural geography": commentary 2', *Progress in Human Geography*, 22: 569–71.

May, J. (1991) 'Putting some space back into time: the spatio-temporal self and the turn to the postmodern', in C. Philo (ed.) *New Words, New Worlds: Reconceptualising Social and Cultural Geography*. Aberystwyth: Cambrian Printers, 168–71.

May, T. (1997) *Social Research: Issues, Methods and Process*. Buckingham: Open University Press.

McCloskey, D. (1988) 'The consequences of rhetoric', in A. Klamer et al. (eds) *The Consequences of Economic Rhetoric*. Cambridge: Cambridge University Press, 280–93.

McDowell, L. (1983) 'Towards an understanding of the gender division of urban space', *Environment and Planning D: Society and Space*, 1: 15–30.

McDowell, L. (1988) 'Coming in from the dark: feminist research in geography', in J. Eyles (ed.) *Research in Human Geography: Introductions and Investigations*. Oxford: Blackwell, 155–73.

McDowell, L. (1989) 'Women, gender and the organisation of space', in D. Gregory and R. Walford (eds) *Horizons in Human Geography*. London: Macmillan, 136–51.

McDowell, L. (1992a) 'Doing gender: feminism, feminists and research methods in human geography', *Transactions, Institute of British Geographers*, 17: 399–416.

McDowell, L. (1992b) 'Multiple voices: speaking from inside and outside "the project"', *Antipode*, 24: 56–72.

McDowell, L. (1995) 'Understanding diversity: the problem of/for theory', in R.J. Johnston et al. (eds) *Geographies of Global Change*. Oxford: Blackwell, 280–94.

McDowell, L. (1997) *Capital Culture: Gender at Work in the City*. Oxford: Blackwell.

McDowell, L. (1999) *Gender, Identity and Place: Understanding Feminist Geographies*. Cambridge: Polity Press.

McDowell, L. and Court, G. (1994) 'Performing work: bodily representations in merchant banks', *Environment and Planning D: Society and Space*, 12: 727–50.

McDowell, L. and Massey, D. (1984) 'A woman's place?', in D. Massey and J. Allen (eds) *Geography Matters! A Reader*. Cambridge: Cambridge University Press, 128–47.

McDowell, L. and Sharp, J.P. (eds) (1997) *Space, Gender, Knowledge: Feminist Readings*. London: Arnold.

McEwan, C. (2000) *Gender, Geography and Empire: Victorian Women Travellers in West Africa*. Aldershot: Ashgate.

McKibben, B. (1989) *The End of Nature*. New York, NY: Anchor Books.

Mead, W. (1981) 'Review of *The Human Experience of Space and Place*, ed. Buttimer & Seamon', *The Geographical Journal*, 147: 94–5.

Medick, H. (1976) 'The proto-industrial family economy: the structural function of household and family during the transition from peasant society to industrial capitalism', *Social History*, 3: 291–315.

Meinig, D. (1983) 'Geography as an art', *Transactions, Institute of British Geographers*, 8: 314–28.

Mercer, D. and Powell, J. (1972) 'Phenomenology and related non-positivistic viewpoints in the social sciences', *Monash Publications in Geography*, 1972–6. Melbourne: Monash University.

Merrifield, A. (1994) 'Situated knowledge through exploration: reflections on Bunge's "Geographical Expeditions"', *Antipode*, 27: 49–70.

Merton, R., Fiske, M. and Kendall, P. (1990) *The Focused Interview* (2nd edn). New York, NY: Free Press.

Michael, J. (1982) *The Politics of Secrecy*. London: Penguin Books.

Michael, M. (2000) *Reconnecting Culture, Technology and Nature: From Society to Heterogeneity*. London: Routledge.

Milbourne, P. (2000) 'Exporting "other" rurals: new audiences for qualitative research', in A. Hughes et al. (eds) *Ethnography and Rural Research*. Cheltenham: Countryside and Community Press, 179–97.

Miller, E.J.W. (1969) 'The naming of the land in the Arkansas Ozarks', *Annals of the Association of American Geographers*, 59: 240–51.

Miller, M. (1998) '(Re)presenting voices in dramatically scripted research', in A. Banks and S. Banks (eds) *Fiction and Social Research: By Ice or Fire*. Walnut Creek, CA: AltaMira Press, 67–78.

Mills, C.W. (1970) *The Sociological Imagination*. Harmondsworth: Penguin Books.

Mills, I. (1987) 'Developments in census-taking since 1841', *Population Trends*, 48: 37–44.

Milner, A. (2002) *Re-imagining Cultural Studies: The Promise of Cultural Materialism*. London: Sage.

Mitchell, B. and Draper, D. (1983a) *Relevance and Ethics in Geography*. London: Longman.

Mitchell, B. and Draper, D. (1983b) 'Ethics in geographical research', *The Professional Geographer*, 35: 9–17.

Mitchell, D. (1995) 'There's no such thing as culture: towards a reconceptualisation of the idea of culture in geography', *Transactions, Institute of British Geographers*, 20: 102–16.

Mitchell, D. (2000) *Cultural Geography: A Critical Introduction*. Oxford: Blackwell.

Mitchell, T. (1989) 'The world-as-exhibition', *Comparative Studies in Society and History*, 31: 217–36.

Mitchell, W. (1986) *Iconography: Image, Text, Ideology*. Chicago, IL: University of Chicago Press.

Miyares, I. and McGlade, M. (1994) 'Specialisation in "jobs in geography"', *The Professional Geographer*, 46: 170–7.

Mohammad, R. (2001) '"Insiders" and/or "outsiders": positionality, theory and praxis', in M. Limb and C. Dwyer (eds) *Qualitative Methodologies for Geographers: Issues and Debates*. London: Edward Arnold, 101–17.

Molotch, H. (1996) 'LA as design product', in A. Scott and E. Soja (eds) *The City*. Berkeley, CA: University of California Press, 225–77.

Moon, G. and Brown, T. (2000) 'Governmentality and the spatialised discourse of policy: the consolidation of the post-1989 NHS reforms', *Transactions, Institute of British Geographers*, 25: 65–76.

Moore, D. (1997) 'Remapping resistance: "Ground for struggle" and the politics of place', in S. Pile (ed.) *Geographies of Resistance*. London: Routledge, 87–105.

Morgan, D. (1993) *Successful Focus Groups: Advancing the State of the Art*. Beverly Hills, CA: Sage.

Morgan, D. (1997) *Focus Groups as Qualitative Research* (2nd edn). Beverly Hills, CA: Sage.

Morgan, K. and Sayer, A. (1988) *Microcircuits of Capital: 'Sunrise' Industries and Uneven Development*. Cambridge: Polity Press.

Morgan, R. (1967) 'Grass roots of geography', *The Geographical Magazine*, 40: 145–6.

Morley, D. and Robins, K. (1995) *Spaces of Identity: Global Media, Electronic Landscapes and Cultural Boundaries*. London: Routledge.

Moser, C. and Kalton, G. (1971) *Survey Methods in Social Investigation* (2nd edn 1985). London: Heinemann.

Murdoch, J. (1995) 'Actor networks and the evolution of economic forms: combining description and explanation in theories of regulation, flexible specialization and networks', *Environment and Planning A*, 27: 731–57.

Murdoch, J. (1997) 'Inhuman/nonhuman/human: actor network theory and the prospects for a non-dualistic and symmetrical perspective on nature and society', *Environment and Planning D: Society and Space*, 18: 731–56.

Murdoch, J. (1998) 'The spaces of actor-network theory', *Geoforum*, 29: 357–74.

Murdoch, J. and Marsden, T. (1995) 'The spatialization of politics: local and national actor-spaces in environmental conflict', *Transactions, Institute of British Geographers*, 20: 368–80.

Murdoch, J. and Ward, N. (1997) 'Governmentality and territoriality – the statistical manufacture of Britain's "national farm"', *Political Geography*, 16: 307–24.

Mutersbaugh, T. (1998) 'Women's work, men's work: technology acquisition in a Oaxacan village (Mexico)', *Environment and Planning D: Society and Space*, 16: 439–58.

Myers, G. (1996) 'Naming and placing the other: power and the urban landscape in Zanzibar', *Tijdschrift voor Economische en Sociale Geografie*, 87: 237–46.

Myers-Jones, H. and Brooker-Gross, S. (1994) 'Newspapers as promotional strategists for regional definition', in J. Gold and S. Ward (eds) *Place Promotion: The Use of Publicity and Marketing to Sell Towns and Regions*. Chichester: Wiley, 195–212.

397

Nagar, R. (1997a) 'Exploring methodological borderlands through oral narratives', in J.P. Jones III et al. (eds) *Thresholds in Feminist Geography*. Oxford: Rowman & Littlefield, 203–24.

Nagar, R. (1997b) 'The making of Hindu communal organizations, places and identities in postcolonial Dar Es Salaam', *Environment and Planning D: Society and Space*, 15: 707–30.

Nancy, J.-L. (1991) *The Inoperative Community*. Minneapolis, MN: University of Minnesota Press.

Nash, C. (2000) 'Performativity in practice: some recent work in cultural geography', *Progress in Human Geography*, 24: 653–64.

Nash, C. (2003) 'Cultural geography: anti-racist geographies', *Progress in Human Geography*, 27: 637–48.

Nast, H. (ed.) (1994) 'Women in the field: critical feminist methodologies and theoretical perspectives' (a series of papers), *The Professional Geographer*, 46: 54–102.

Nast, H. (1998) 'The body as "place": reflexivity and fieldwork in Kano, Nigeria', in H.J. Nast and S. Pile (eds) *Places through the Body*. London: Routledge, 93–116.

Nast, H. (2000) 'Mapping the unconscious: racism and the Oedipal family', *Annals of Association of American Geographers*, 90: 215–55.

Natter, W. and Jones, J.P. III (1993) 'Signposts towards a poststructuralist geography', in J.P. Jones III et al. (eds) *Postmodern Contentions: Epochs, Politics, Space*. New York, NY: Guilford, 165–203.

Natter, W. and Jones, J.P. III (1997) 'Identity, space and other uncertainties', in G. Benko and U. Strohmayer (eds) *Space and Social Theory: Interpreting Modernity and Postmodernity*. Oxford: Blackwell, 141–61.

Norcliffe, G. (1977) *Inferential Statistics for Geographers*. London: Hutchinson.

Northedge, A. (1994) *The Good Study Guide*. Milton Keynes: Open University Press.

Oakley, A. (1979) *From Here to Maternity: Becoming a Mother*. Harmondsworth: Penguin Books.

Oakley, A. (1984) *Taking it Like a Woman*. London: Fontana.

Oakley, A. (1990) 'Interviewing women', in H. Roberts (ed.) *Doing Feminist Research*. London: Routledge & Kegan Paul, 30–61.

O'Brien, L. (1992) *Introducing Quantitative Geography*. London: Routledge.

Ogborn, M., Blunt, A., Gruffudd, P., May, J. and Pinder, D. (eds) (forthcoming) *Cultural Geography in Practice*. London: Arnold.

Olsson, G. (1980) *Birds in Egg/Eggs in Bird*. London: Pion.

Olsson, G. (1988) 'The eye and the index finger: bodily means to cultural meaning', in R. Golledge et al. (eds) *A Ground for Common Search*. Santa Barbara, CA: Santa Barbara Geographical Press, 126–37.

Olsson, G. (1991a) *Lines of Power/Limits of Language*. Minneapolis, MN: University of Minnesota Press.

Olsson, G. (1991b) 'Invisible maps', *Geografiska Annaler*, 73B: 85–91.

Olsson, G. (1994) 'Heretic cartography', *Ecumene*, 1: 213–34.

Olsson, G. (1995) 'Signs of persuasion', in W. Natter et al. (eds) *Objectivity and its Other*. New York, NY: Guilford, 21–32.

Openshaw, S. (1984) *The Modifiable Areal Unit Problem (CATMOG 38)*. Norwich: GeoAbstracts.

Openshaw, S. (1991) 'A view of the GIS crisis in geography, or, using GIS to put Humpty-Dumpty back together again', *Environment and Planning A*, 23: 621–8.

Openshaw, S. (1996) 'Fuzzy logic as a new paradigm for doing geography', *Environment and Planning A*, 28: 761–8.

Openshaw, S. (1998) 'Towards a more computationally minded scientific human geography', *Environment and Planning A*, 30: 317–32.

Openshaw, S. and Openshaw, C. (1997) *Artificial Intelligence in Geography*. Chichester: Wiley.

Oppenheim, A. (1992) *Questionnaire Design, Interviewing and Attitude Measurement*. London: Pinter.

Ord, J. and Getis, A. (1995) 'Local spatial autocorrelation statistics: distributional issues and an application', *Geographical Analysis*, 27: 286–306.

Paassen, C.V. (1976) 'Human geography in terms of existential anthropology', *Tijdschrift voor Economische en Sociale Geografie*, 67: 324–41.

Page, N. and Preston, P. (eds) (1993) *The Literature of Place*. London: Macmillan.

Painter, J. (2002) 'Governmentality and regional economic strategies', in J. Hillier and E. Rooksby (eds) *Habitus: A Sense of Place*. Aldershot: Ashgate, 115–39.

Palm, R. (1975) 'Review of Ley, D. (1974) *The Black Inner City as Frontier Outpost*', *Annals of the Association of American Geographers*, 65: 572–3.

Palm, R. (1998) 'Classics in human geography revisited: Ley 1974 *The Black Inner City as Frontier Outpost*, commentary 2', *Progress in Human Geography*, 22: 77–8.

Panofsky, E. (1970) *Meaning in the Visual Arts*. Harmondsworth: Penguin Books.

Parfitt, J. (1997) 'Questionnaire design and sampling', in R. Flowerdew and D. Martin (eds) *Methods in Human Geography*. Harlow: Longman, 76–109.

Parr, H. (1998a) 'Mental health, ethnography and the body', *Area*, 30: 28–37.

Parr, H. (1998b) 'The politics of methodology in "post-medical geography": mental health research and the interview', *Health and Place*, 4: 341–53.

Parr, H. (2000) 'Interpreting the "hidden social geographies" of mental health: ethnographies of inclusion and exclusion in semi-institutional spaces', *Health and Place*, 6: 225–37.

Parr, H. (2001) 'Negotiating different ethnographic contexts and building geographical knowledges: empirical examples from mental health research', in M. Limb and C. Dwyer (eds) *Qualitative Methodologies for Geographers: Issues and Debates*. London: Arnold, 181–95.

Parr, H. and Philo, C. (2003) 'Introducing psychoanalytic geographies', *Social and Cultural Geography*, 4: 283–93.

Pattie, C. and Johnston, R.J. (1993) 'Surface change but underlying stability? The geography of the flow of the vote in Great Britain, 1979–1992', *Area*, 25: 257–66.

Peck, J. (1999) 'Grey geography?', *Transactions, Institute of British Geographers*, 24: 131–5.

Peet, R. (1977) 'The development of radical geography in the United States', *Progress in Human Geography*, 1: 240–63.

Peet, R. (ed.) (1978) *Radical Geography: Alternative Viewpoints on Contemporary Social Issues*. London: Methuen.

Peet, R. (1998) *Modern Geographical Thought*. Oxford: Blackwell.

Penslar, R. (ed.) (1995) *Research, Ethics, Cases and Materials*. Bloomington, IN: Indiana University Press.

Philbrick, A.K. (1957) 'Principles of areal functional organisation in regional human geography', *Economic Geography*, 33: 299–336.

Philip, L.J. (1998) 'Combining quantitative and qualitative approaches to social research in human geography – an impossible mixture?', *Environment and Planning A*, 30: 261–76.

Philo, C. (1984) 'Reflections on Gunnar Olsson's contribution to the discourse of contemporary human geography', *Environment and Planning D: Society and Space*, 2: 217–40.

Philo, C. (ed.) (1991a) *New Words, New Worlds*. Aberystwyth: Cambrian Printers.

Philo, C. (1991b) 'Delimiting human geography: new social and cultural perspectives', in C. Philo (ed.) *New Words, New Worlds*. Aberystwyth: Cambrian Printers, 14–27.

Philo, C. (1992) 'Neglected rural geographies: a review', *Journal of Rural Studies*, 8: 193–207.

Philo, C. (1994) 'Escaping Flatland: a book review essay inspired by Gunnar Olsson's *Lines of Power/Limits of Language*', *Environment and Planning D: Society and Space*, 12: 229–52.

Philo, C. (ed.) (1995) *Off the Map: The Social Geography of Poverty in the UK*. London: Child Poverty Action Group.

Philo, C. (ed.) (1998) 'Reconsidering quantitative geography' (a series of papers), *Environment and Planning A*, 30: 191–332.

Philo, C. (2000) 'More words, more worlds: reflections on the "cultural turn" and human geography', in I. Cook et al. (eds) *Cultural Turns/Geographical Turns: Perspectives on Cultural Geography*. London: Prentice Hall, 26–53.

Philo, C., Mitchell, R. and More, A. (1998) 'Reconsidering quantitative geography: the things that count', *Environment and Planning A*, 30: 191–202.

Pickles, J. (ed.) (1995) *Ground Truth: The Social Implications of Geographic Information Systems*. New York, NY: Guilford.

Pile, S. (1991) 'Practising interpretative geography', *Transactions, Institute of British Geographers*, 16: 458–69.

Pile, S. (1996) *The Body and the City: Psychoanalysis, Space and Subjectivity*. London: Routledge.

Pile, S. and Thrift, N. (eds) (1995) *Mapping the Subject*. London: Routledge.

Pirie, G. (1984) 'Review of Western, J. (1981) *Outcast Cape Town*', *Economic Geography*, 60: 257–9.

Platt, R.S. (1935) 'Field approach to regions', *Annals of the Association of American Geographers*, 25: 153–72.

Plummer, P. and Sheppard, E. (2001) 'Must emancipatory economic geography be qualitative?' *Antipode*, 33: 194–9.

Pocock, D. (1979) 'The novelist's image of the north', *Transactions, Institute of British Geographers*, 4: 62–76.

Pocock, D. (ed.) (1981a) *Humanistic Geography and Literature*. Beckenham: Croom Helm.

Pocock, D. (1981b) 'Imaginative literature and the geographer', in D. Pocock (ed.) *Humanistic Geography and Literature*. Beckenham: Croom Helm, 9–19.

POD (1969) 'Research', in *Pocket Oxford Dictionary (POD)*. Oxford: Clarendon Press, 703.

Pool, I. (1957) 'A critique of the twentieth anniversary issue', *Public Opinion Quarterly*, 21: 190–8.

Poole, R. (1988) 'Deconstruction', in A. Bullock and O. Stallybrass (eds) *The Fontana Dictionary of Modern Thought*. London: Fontana, 205–6.

Popkin, K. (1998) 'Corporate imaginations of rural idyll and community.' Unpublished PhD thesis, School of Geographical Sciences, University of Bristol.

Powell, R.C. (2002) 'The Sirens' voices? Field practices and dialogue in geography', *Area*, 34: 261–72.

Pratt, G. (1989) 'Quantitative techniques and humanistic-historical materialist perspectives', in A. Kobayashi and S. Mackenzie (eds) *Remaking Human Geography*. London: Unwin Hyman, 101–5.

Pratt, G. (1992) 'Spatial metaphors and speaking positions', *Environment and Planning D: Society and Space*, 10: 241–4.

Pratt, G. (1994) 'Poststructuralism (including deconstruction)', in R.J. Johnston et al. (eds) *The Dictionary of Human Geography* (3rd edn). Oxford: Blackwell, 468–9.

Pratt, G. (2000a) 'Focus group', in R.J. Johnston et al. (eds) *The Dictionary of Human Geography* (4th edn). Oxford: Blackwell, 272.

Pratt, G. (2000b) 'Post-structuralism', in R.J. Johnston et al. (eds) *The Dictionary of Human Geography* (4th edn). Oxford: Blackwell, 625–6.

Pred, A. (1995) *Recognizing European Modernities*. London: Routledge.

Preston, P. and Simpson-Houseley, P. (eds) (1994) *Writing the City: Eden, Babylon and the New Jerusalem*. London: Routledge.

Prigogine, I. (1984) *Order out of Chaos: Man's New Dialogue with Nature*. London: Heinemann.

Prince, H. (1980) 'Review of *Humanistic Geography: Prospects and Problems*, ed. Ley & Samuels', *Annals of the Association of American Geographers*, 70: 294–6.

Proctor, J.D. (1998) 'Ethics in geography: giving moral form to the geographical imagination', *Area*, 30: 8–18.

Pryke, M. (2000) 'Tracing economic rhythms through visual and audio montage' (available at http://www.open.ac.uk/socialsciences/geography/research/bellin/).

Pryke, M. (2002) 'The white noise of capitalism: audio and visual montage and sensing economic change', *Cultural Geographies*, 9: 472–7.

Punch, M. (1986) *The Politics and Ethics of Fieldwork*. Beverly Hills, CA: Sage.

Punch, S. (2001) 'Multiple methods and research relations with children in rural Bolivia', in M. Limb and C. Dwyer (eds) *Qualitative Methodologies for Geographers: Issues and Debates*. London: Arnold, 165–80.

Quiller-Couch, A.T. (1916) *On the Art of Writing: Lectures Delivered in the University of Cambridge, 1913–1914*. Cambridge: Cambridge University Press.

Quinton, A. (1988) 'Dialectic', in A. Bullock and O. Stallybrass (eds) *The Fontana Dictionary of Modern Thought*. London: Fontana, 225.

Rabinow, P. (1989) *French Modern: Norms and Forms of the Social Environment*. Cambridge, MA: MIT Press.

Raco, M. (2003) 'Governmentality, subject-building, and the discourses and practices of devolution in the UK', *Transactions, Institute of British Geographers*, 28: 75–95.

Radcliffe, S. (1994) '(Re)presenting post-colonial women: authority, difference and feminisms', *Area*, 26: 25–32.

Reah, D. (1998) *The Language of Newspapers*. London: Routledge.

Reichert, D. (1992) 'On boundaries', *Environment and Planning D: Society and Space*, 10: 87–98.

Relph, E. (1970) 'An inquiry into the relations between phenomenology and geography', *The Canadian Geographer*, 14: 193–201.

Relph, E. (1976) *Place and Placelessness*. London: Pion.

Relph, E. (1980) 'Review of *Humanistic Geography: Prospects and Problems*, ed. Ley & Samuels', *The Geographical Review*, 70: 115–17.

Revill, G. (1995) 'Hiawatha and pan-Africanism: Samuel Coleridge-Taylor (1875–1912), a black composer in suburban London', *Ecumene*, 2: 247–66.

Richards, I. (1929) *Practical Criticism: A Study of Literary Judgement*. London: Kegan Paul, Trench, Trübner & Co.

Richards, P. (1986) 'Risk', in H. Becker (ed.) *Writing for Social Scientists*. Chicago, IL: University of Chicago Press, 108–20.

Richardson, L. (1988) 'The collective story: postmodernism and the writing of sociology', *Sociological Focus*, 21: 199–208.

Richardson, L. (1990) *Writing Strategies: Reaching Diverse Audiences*. Newbury Park, CA: Sage.

Richardson, M. (1981) 'Commentary on the "superorganic" in American cultural geography', *Annals of the Association of American Geographers*, 71: 284–7.

Roberts, H. (ed.) (1990) *Doing Feminist Research*. London: Routledge & Kegan Paul.

Robinson, G.M. (1998) *Methods and Techniques in Human Geography*. Chichester: Wiley.

Robson, E. (1994) 'From teacher to taxi driver: reflections on research roles in developing areas', in E. Robson and K. Willis (eds) *Postgraduate Fieldwork in Developing Areas: A Rough Guide. Developing Areas Research Group Monograph 9*. London: RGS with the IBG, 36–59.

Robson, E. and Willis, K. (eds) (1994a) *Postgraduate Fieldwork in Developing Areas: A Rough Guide. Developing Areas Research Group Monograph 9*. London: RGS with the IBG.

Robson, E. and Willis, K. (1994b) 'Introduction', in E. Robson and K. Willis (eds) *Postgraduate Fieldwork in Developing Areas: A Rough Guide. Developing Areas Research Group Monograph 9*. London: RGS with the IBG, 1–4.

Rodaway, P. (1994) *Sensuous Geographies: Body, Sense and Place*. London: Routledge.

Rogers, A., Viles, H. and Goudie, A. (eds) (1992) *The Student's Companion to Geography*. Oxford: Blackwell.

Rorty, R. (1979) *Philosophy and the Mirror of Nature*. Oxford: Blackwell.

Rorty, R. (1987) 'Method, social science and social hope', in M. Gibbons (ed.) *Interpreting Politics*. Oxford: Blackwell, 241–59.

Rosaldo, R. (1986) *Culture and Truth: The Remaking of Social Analysis*. Boston, MA: Beacon.

Rosaldo, R. (1993) *Culture and Truth: The Remaking of Social Analysis*. London: Routledge.

Rose, C. (1981) 'Wilhelm Dilthey's philosophy of historical understanding: a neglected heritage of contemporary humanistic geography', in D. Stoddart (ed.) *Geography, Ideology and Social Concern*. Oxford: Blackwell, 99–113.

Rose, D. and Sullivan, O. (1996) *Introducing Data Analysis for Social Scientists*. Buckingham: Open University Press.

Rose, G. (1990) 'Imagining Poplar in the 1920s: contested concepts of community', *Journal of Historical Geography*, 16: 425–37.

Rose, G. (1991) 'On being ambivalent: women and feminisms in geography', in C. Philo (comp.) *New Words, New Worlds: Reconceptualising Social and Cultural Geography*. Lampeter: Social and Cultural Geography Study Group, 156–63.

Rose, G. (1992) 'Geography as a science of observation: the landscape, the gaze and masculinity', in F. Driver and G. Rose (eds) *Nature and Science: Essays in the History of Geographical Knowledge*. London: Historical Geography Research Group, 8–18.

Rose, G. (1993a) 'Some notes towards thinking about the spaces of the future', in J. Bird et al. (eds) *Mapping the Futures: Local Cultures, Global Change*. London: Routledge, 70–83.

Rose, G. (1993b) *Feminism and Geography: The Limits of Geographical Knowledge*. Cambridge: Polity Press.

Rose, G. (1994) 'The cultural politics of place: local representation and oppositional discourse in films', *Transactions, Institute of British Geographers*, 19: 46–60.

Rose, G. (1995) 'Distance, surface, elsewhere: a feminist critique of the space of phallocentric knowledge', *Environment and Planning D: Society and Space*, 13: 761–81.

Rose, G. (1997a) 'Situating knowledges: positionality, reflexivities and other tactics', *Progress in Human Geography*, 21: 305–20.

Rose, G. (1997b) 'Performing inoperative community: the space and the resistances of some community arts projects', in S. Pile (ed.) *Geographies of Resistance*. London: Routledge, 184–202.

Rose, G. (2000) 'Psychoanalytic theory, geography and', in R.J. Johnston et al. (eds) *The Dictionary of Human Geography* (4th edn). Oxford: Blackwell, 653–5.

Rose, G. (2001) *Visual Methodologies: An Introduction to the Interpretation of Visual Materials*. London: Sage.

Rossi, P., Wright, J. and Anderson, A. (1983) *Handbook of Survey Research*. San Diego, CA: Academic Press.

Roszak, T. (1994) *The Cult of Information*. Berkeley, CA: University of California Press.

Roth, J. (1966) 'Hired hand research', *The American Sociologist*, 1: 190–6.

Round, J.H. (1895) *Feudal England: Historical Studies on the XIth and XIIth Centuries*. London: Swan Sonnenschein.

Routledge, P. (1996) 'The third space as critical engagement', *Antipode*, 28: 398–419.

Rowles, G.D. (1978a) *Prisoners of Space? Exploring the Geographical Experience of Older People*. Boulder, CO: Westview Press.

Rowles, G.D. (1978b) 'Reflections on experiential fieldwork', in D. Ley and M.S. Samuels (eds) *Humanistic Geography: Prospects and Problems*. London: Croom Helm, 173–93.

Rowles, G.D. (1980) 'Towards a geography of growing old', in A. Buttimer and D. Seamon (eds) *The Human Experience of Space and Place*. London: Croom Helm, 55–72.

Rowles, G.D. (1981a) 'Geographical perspectives on human development', *Journal of Human Development*, 24: 67–76.

Rowles, G.D. (1981b) 'The surveillance zone as a meaningful space for the aged', *The Gerontologist*, 21: 304–11.

Rowles, G.D. (1983) 'Place and personal identity: observations from Appalachia', *Journal of Environmental Psychology*, 3: 299–313.

Rowles, G.D. (1987) 'A place called home', in L. Carstensen and B. Edelsten (eds) *The Handbook of Clinical Gerontology*. New York, NY: Pergamon, 335–53.

Rundstrom, R.A. and Kenzer, M.S. (1989) 'The decline of fieldwork in human geography', *The Professional Geographer*, 41: 294–303.

Ryan, J.R. (1997) *Picturing Empire: Photography and the Visualisation of the British Empire*. London: Reaktion Books.

Saarinen, T. (1966) *Perception of Drought Hazard on the Great Plains*. Department of Geography Research Paper 106. Chicago, IL: University of Chicago, Department of Geography.

Sack, R.D. (1980) *Conceptions of Space in Social Thought*. London: Macmillan.

Sadler, D. (1997) 'The role of supply chain management strategies in the "Europeanization" of the auto-mobile production system', in R. Lee and J. Wills (eds) *Geographies of Economies*. London: Arnold, 311–20.

Said, E. (1978) *Orientalism*. London: Penguin Books.

Said, E. (1993) *Culture and Imperialism*. New York, NY: Alfred A. Knopf.

Salter, C.L. (1969) 'The bicycle as a field aid', *The Professional Geographer*, 21: 360–2.

Salter, C.L. and Meserve, P. (1991) 'Life lists and the education of the geographer', *The Professional Geographer*, 43: 520–5.

Saltmarsh, R. (2001) 'A journey into autobiography: a coal miner's daughter', in P. Moss (ed.) *Placing Autobiography in Geography*. Syracuse, NY: Syracuse University Press, 138–48.

Saltmarsh, R. (2002) 'From mining town to university: a class-cultural experience of an education in Wales.' Paper presented to the Association of American Geographers' annual conference, Los Angeles.

Sandberg, L.A. and Marsh, J.S. (1988) 'Focus: literary landscapes – geography and literature', *The Canadian Geographer/Le Géographe canadien*, 32: 266–76.

Sanjek, R. (1990) 'A vocabulary for fieldnotes', in R. Sanjek (ed.) *Fieldnotes*. Ithaca, NY: Cornell University Press, 92–121.

Sauer, C.O. (1924) 'The survey method in geography and its objectives', *Annals of the Association of American Geographers*, 14: 17–33.

Sauer, C.O. (1956) 'The education of a geographer', *Annals of the Association of American Geographers*, 46: 287–99.

Sayer, A. (1984) *Method in Social Science: A Realist Approach* (1st edn). London: Hutchinson.

Sayer, A. (1989) 'The "new" regional geography and problems of narrative', *Environment and Planning D: Society and Space*, 7: 253–76.

Sayer, A. (1992) *Method in Social Science: A Realist Approach* (2nd edn). London: Routledge.

Sayer, A. (1997) 'Realism and geography', in T. Barnes and D. Gregory (eds) *Reading Human Geography*. London: Arnold, 112–24.

Sayer, A. and Storper, M. (1997) 'Ethics unbound: for a normative turn in social theory', *Environment and Planning D: Society and Space*, 15: 1–17.

Schaeffer, F.K. (1953) 'Exceptionalism in geography: a methodological examination', *Annals of the Association of American Geographers*, 43: 226–49.

Scott, A.J. (1984) 'Territorial reproduction and transformation in a local labour market: the animated film workers of Los Angeles', *Environment and Planning D: Society and Space*, 2: 277–307.

Scott, A.J. and Angel, D.P. (1987) 'The US semiconductor industry: a locational analysis', *Environment and Planning A*, 20: 1047–67.

Scott, J. (1990) *A Matter of Record*. Cambridge: Polity Press.

Seamon, D. (1976) 'Phenomenological examination of imaginative literature: a commentary', in G.T. Moore and R.G. Golledge (eds) *Environmental Knowing*. Stroudsburg, PA: Dowden, Hutchinson & Ross, 286–90.

Seamon, D. (1979) *A Geography of the Lifeworld: Movement, Rest and Encounter*. London: Croom Helm.

Selkirk, K.E. (1982) *Pattern and Place: An Introduction to Mathematical Geography*. Cambridge: Cambridge University Press.

Selwyn, T. (ed.) (1996) *The Tourist Image: Myths and Myth-making in Tourism*. Chichester: Wiley.

Semple, R.K. and Green, M.B. (1984) 'Classification in human geography', in G.L. Gaile and C.J. Willmott (eds) *Spatial Statistics and Models*. Dordrecht: Reidel, 55–79.

Serres, M. and Latour, B. (1995) *Conversations on Science, Culture, Time*. Ann Arbor, MI: University of Michigan Press.

Sharp, J.P., Routledge, P., Philo, C. and Paddison, R. (eds) (2000) *Entanglements of Power: Geographies of Domination/Resistance*. London: Routledge.

Shelley, F.M., Archer, J.C., Davidson, F.M. and Brunn, S.D. (1996) *Political Geography of the United States*. New York, NY: Guilford.

Sheppard, E. and Barnes, T. (1990) *The Capitalist Space Economy: Geographical Analysis after Ricardo, Marx and Sraffa*. London: Unwin Hyman.

Shields, R. (1996) *Cultures of the Internet*. London: Sage.

Shortridge, J.R. (1991) 'The concept of the place-defining novel in American popular culture', *The Professional Geographer*, 43: 280–91.

Shurmer-Smith, P. (1998a) 'Classics in human geography revisited: Duncan, J.S. "The superorganic in American cultural geography", commentary 1', *Progress in Human Geography*, 22: 567–9.

Shurmer-Smith, P. (1998b) 'Becoming a memsahib: working with the Indian administrative service', *Environment and Planning A*, 30: 2163–79.

Shurmer-Smith, P. (2000) 'Review of Cook et al. (eds) (2000) *Cultural Turns/Geographical Turns: Perspectives on Cultural Geography*', *Transactions, Institute of British Geographers*, 25: 524–5.

Shurmer-Smith, P. (2002a) 'Introduction', in P. Shurmer-Smith (ed.) *Doing Cultural Geography*. London: Sage, 1–7.

Shurmer-Smith, P. (2002b) 'Methods and methodology', in P. Shurmer-Smith (ed.) *Doing Cultural Geography*. London: Sage, 95–8.

Shurmer-Smith, P. (2002c) 'Postcolonial geographies', in P. Shurmer-Smith (ed.) *Doing Cultural Geographies*. London: Sage, 67–77.

Sibley, D. (1981a) *Outsiders in Urban Societies*. Oxford: Blackwell.

Sibley, D. (1981b) 'The notion of order in spatial analysis', *The Professional Geographer*, 33: 1–5.

Sibley, D. (1985) 'Travelling people in England: a regional comparison', *Regional Studies*, 19: 139–47.

Sibley, D. (1995) *Geographies of Exclusion: Society and Difference in the West*. London: Routledge.

Sibley, D. (1998) 'Sensations and spatial science: gratification and anxiety in the production of ordered landscapes', *Environment and Planning A*, 30: 235–46.

Sibley, D. (n.d.) *Spatial Applications of Exploratory Data Analysis (CATMOG 49)*. Norwich: Geo Books.

Sidaway, J.D. (1992) 'In other worlds: on the politics of research by "First World" geographers in the "Third World"', *Area*, 24: 403–8.

Sidaway, J.D. (1997) 'The production of British geography', *Transactions, Institute of British Geographers*, 22: 488–504.

Silk, J. (1979) *Statistical Concepts in Geography*. London: George Allen & Unwin.

Silk, J. (1984) 'Beyond geography and literature', *Environment and Planning D: Society and Space*, 2: 151–78.

Silverman, D. (1993) *Interpreting Qualitative Data: Methods for Analysing Talk, Text and Interaction*. London: Sage.

Simon, D. (1999) 'Classics in human geography revisited: Western, J. (1981) *Outcast Cape Town*', *Progress in Human Geography*, 23: 423–5.

Simpson, C.A. (1945) 'A venture in field geography', *Geography*, XXX: 35–44.

Simpson, S. and Dorling, D. (1999) 'Conclusion: statistics and "the truth"', in D. Dorling and S. Simpson (eds) *Statistics in Society*. London: Arnold, 414–20.

Skelton, T. (2001) 'Cross-cultural research: issues of power, positionality and "race"', in M. Limb and C. Dwyer (eds) *Qualitative Methodologies for Geographers: Issues and Debates*. London: Arnold, 87–100.

Sloman, A. (1988) 'Methodology', in A. Bullock and O. Stallybrass (eds) *The Fontana Dictionary of Modern Thought*. London: Fontana, 525–6.

Smailes, A.E. (1964) 'Urban survey', in R. Clayton (ed.) *The Geography of Greater London: A Source Book for Teacher and Student*. London: George Philip & Son, 202–21.

Smith, C. (1982) 'Review of *The Human Experience of Space and Place*, ed. Buttimer & Seamon', *The Geographical Review*, 72: 107–8.

Smith, D.M. (1975) *Patterns in Human Geography*. Harmondsworth: Penguin Books.

Smith, D.M. (1977) *Human Geography: A Welfare Approach*. London: Edward Arnold.

Smith, D.M. (1979) *Where the Grass is Greener: Living in an Uneven World*. Harmondsworth: Penguin Books.

Smith, D.M. (1988a) 'A welfare approach to human geography', in J. Eyles (ed.) *Research in Human Geography: Introductions and Investigations*. Oxford: Blackwell, 139–54.

Smith, D.M. (1988b) 'Towards an interpretative human geography', in J. Eyles and D.M. Smith (eds) *Qualitative Methods in Human Geography*. Cambridge: Polity Press, 255–67.

Smith, D.M. (1994) *Geography and Social Justice*. Oxford: Blackwell.

Smith, D.M. (1995a) 'Moral teaching in geography', *Journal of Geography in Higher Education*, 19: 271–83.

Smith, D.M. (1995b) 'Against differential research funding', *Area*, 27: 79–83.

Smith, D.M. (1997) 'Geography and ethics: a moral turn?', *Progress in Human Geography*, 21: 583–90.

Smith, D.M. (1998) 'Geography and moral philosophy: some common ground', *Ethics, Place and Environment*, 1: 7–34.

Smith, D.M. (2000) 'Moral progress in human geography: transcending the place of good fortune', *Progress in Human Geography*, 24: 1–18.

Smith, D.M. (2004) 'Moral geographies', in P. Cloke et al. (eds) *Envisioning Human Geographies*. London: Arnold, 195–209.

Smith, N. (1992) 'History and philosophy of geography: real wars, theory wars', *Progress in Human Geography*, 16: 257–71.

Smith, N. (1993) 'Homeless/global: scaling places', in J. Bird et al. (eds) *Mapping the Futures: Local Cultures, Global Change*. London: Routledge, 87–119.

Smith, N. (1996) *The New Urban Frontier*. London: Routledge.

Smith, S.J. (1981) 'Humanistic method in contemporary social geography', *Area*, 13.

Smith, S.J. (1984) 'Practising humanistic geography', *Annals of the Association of American Geographers*, 74: 353–74.

Smith, S.J. (1988) 'Constructing local knowledge: the analysis of the self in everyday life', in J. Eyles and D.M. Smith (eds) *Qualitative Methods in Human Geography*. Cambridge: Polity Press, 17–38.

Smith, S.J. (1994) 'Soundscape', *Area*, 26: 232–40.

Smith, S.J. (2001) 'Doing qualitative research: from interpretation to action', in M. Limb and C. Dwyer (eds) *Qualitative Methodologies for Geographers*. London: Arnold, 23–40.

Soja, E. (1989) *Postmodern Geographies*. London: Verso.

Sparke, M. (1994) 'Escaping the herbarium: a critique of Gunnar Olsson's "Chiasm of thought-and-action"', *Environment and Planning D: Society and Space*, 12: 207–20.

Sparke, M. (1996) 'Displacing the field in fieldwork: masculinity, metaphor and space', in N. Duncan (ed.) *BodySpace: Destabilising Geographies of Gender and Sexuality*. London: Routledge, 212–33.

Stacey, J. (1988) 'Can there be a feminist ethnography?', *Women's Studies International Forum*, 11: 21–7.

Stacey, J. (1991) 'Can there be a feminist ethnography?', in S. Gluck and D. Patai (eds) *Women's Words: The Feminist Practice of Oral History*. London: Routledge, 111–18.

Stacey, J. (1997) 'Can there be a feminist ethnography?', in L. McDowell and J. Sharp (eds) *Space, Gender, Knowledge: Feminist Readings*. London: Arnold, 115–23.

Steinberg, J. (1985) 'Verbal anaemia', *New Society*, 19 July: 90.

Stewart, D. and Shamdasani, P.N. (1990) *Focus Groups: Theory and Practice*. Beverly Hills, CA: Sage.

Stoddart, D.R. (1986) *On Geography and its History*. Oxford: Blackwell.

Stoddart, D.R. (1987) 'To claim the high ground: geography for the end of the century', *Transactions, Institute of British Geographers*, 12: 327–36.

Stoddart, D.R. (1991) 'Do we need a feminist historiography of geography – and if we do, what should it be?', *Transactions, Institute of British Geographers*, 16: 484–7.

Stoll, C. (1995) *Silicon Snake Oil: Second Thoughts on the Information Highway*. London: Doubleday.

Storper, M. (1994) 'The transition to flexible specialisation in the US film industry: external economies, the division of labour and the crossing of industrial divides', in A. Amin (ed.) *Post-Fordism: A Reader*. Oxford: Blackwell, 195–226.

Storper, M. and Christopherson, S. (1985) *The Changing Location and Organization of the Motion Picture Industry*. Los Angeles, CA: UCLA Graduate School of Architecture and Urban Planning.

Strohmayer, U. and Hannah, M. (1992) 'Domesticating postmodernism', *Antipode*, 24: 29–55.

Stutz, F. (1979) 'Review of Rowles, G. (1978) *Prisoners of Space*?', *The Geographical Review*, 69: 473.

Sudman, S. and Bradburn, R. (1982) *Asking Questions*. San Francisco, CA: Jossey-Bass.

Taylor, P.J. (1990) 'Editorial comment: GKS', *Political Geography Quarterly*, 9: 211–12.

Taylor, P.J. and Johnston, R.J. (1979) *Geography of Elections*. London: Penguin Books.

Taylor, P.J. and Johnston, R.J. (1995) 'Geographic information systems and geography', in J. Pickles (ed.) *Ground Truth*. London: Guilford, 51–7.

Thompson. K. (ed.) (1997) *Media and Cultural Representation*. London: Sage.

Thrall, G.I. (1985) 'Scientific geography: report on "scientific geography" conference, Georgia, USA, March 1995', *Area*, 17: 254.

Thrift, N. (1983a) 'Literature, the production of culture and the politics of place', *Antipode*, 15: 12–24.

Thrift, N. (1983b) 'On the determination of social action in space and time', *Environment and Planning D: Society and Space*, 1: 23–57.

Thrift, N. (1985) 'Flies and germs: a geography of knowledge', in D. Gregory and J. Urry (eds) *Social Relations and Spatial Structures*. Basingstoke: Macmillan, 366–403.

Thrift, N. (1991) 'Over-wordy worlds? Thoughts and worries', in C. Philo (ed.) *New Words, New Worlds*. Aberystwyth: Cambrian Printers, 144–8.

Thrift, N. (1995) 'Classics in human geography revisited: Thrift, N. "On the determination of social action in space and time": author's response', *Progress in Human Geography*, 19: 528–30.

Thrift, N. (1996) *Spatial Formations*. London: Sage.

Thrift, N. (1997) 'Cities without modernity, cities with magic', *Scottish Geographical Magazine*, 113: 138–49.

Thrift, N. (1999) 'Steps towards an ecology of place', in D. Massey et al. (eds) *Human Geography Today*. Cambridge: Polity Press, 295–322.

Thrift, N. (2000a) 'Dead or alive?', in I. Cook et al. (eds) *Cultural Turns/Geographical Turns: Perspectives on Cultural Geography*. London: Prentice Hall, 1–6.

Thrift, N. (2000b) 'Non-representational theory', in R.J. Johnston et al. (eds) *The Dictionary of Human Geography* (4th edn). Oxford: Blackwell, 556.

Thrift, N. (2000c) 'Entanglements of power: shadows?', in J. Sharp et al. (eds) *Entanglements of Power*. London: Routledge, 262–78.

Thrift, N. (2000d) 'Afterwords', *Environment and Planning D: Society and Space*, 18: 213–55.

Thrift, N. and Dewsbury, J.-D. (2000) 'Dead geographies and how to make them live', *Environment and Planning D: Society and Space*, 18: 411–32.

Tickell, A. (1995) 'Reflections on activism and the academy', *Environment and Planning D: Society and Space*, 13: 235–7.

Tickell, A. (1999) 'On getting inside the project', *Ethics, Place and Environment*, 2: 234–9.

Tillyard, E.M.W. (1972) *The Elizabethan World Picture*. London: Penguin Books.

Tivers, J. (1978) 'How the other half lives: the geographical study of women', *Area*, 10: 302–6.

Tomaselli, K.G. (1988) 'The geography of popular memory in post-colonial South Africa: a study of Afrikaans cinema', in J. Eyles and D.M. Smith (eds) *Qualitative Methods in Human Geography*. Cambridge: Polity Press, 136–55.

Tozer, H.F. (1897) *A History of Ancient Geography*. New York, NY: Biblo & Tanner.

Tuan, Y.-F. (1972) 'Structuralism, existentialism and environmental perception', *Environment and Behaviour*, 3: 319–31.

Tuan, Y.-F. (1976a) 'Humanistic geography', *Annals of the Association of American Geographers*, 66: 266–76.

Tuan, Y.-F. (1976b) 'Literature, experience, and environmental knowing', in G.T. Moore and R.G. Golledge (eds) *Environmental Knowing*. Stroudsburg, PA: Dowden, Hutchinson & Ross, 260–72.

Turner, T. (1991) 'Social dynamics of video media in an indigenous society: the cultural meaning and personal politics of video-making', *Visual Anthropology Review*, 7: 68–76.

Turner, T. (1992) 'Defiant images: the Kapayo appropriation of video', *Anthropology Today*, 8: 5–16.

Twyman, C., Morrison, J. and Sporton, D. (1999) 'The final fifth: autobiography, reflexivity and interpretation in cross-cultural research', *Area*, 31: 313–25.

Unwin, D. (1981) *Introductory Spatial Analysis*. London: Methuen.

Upton, G. (1999) 'Illuminating social statistics', in D. Dorling and S. Simpson (eds) *Statistics in Society*. London: Arnold, 400–13.

Urry, J. (1990) *The Tourist Gaze: Leisure and Travel in Contemporary Societies*. London: Sage.

US National Research Council (1997) *Rediscovering Geography: New Relevance for Science and Society*. Washington, DC: National Research Council.

Valentine, G. (1989) 'The geography of women's fear', *Area*, 21: 385–90.

Valentine, G. (1995) 'Creating transgressive space: the music of kd lang', *Transactions, Institute of British Geographers*, 20: 474–86.

Van Hoven, B. (2003) 'Using CAQDAS in qualitative research', in N.J. Clifford and G. Valentine (eds) *Key Methods in Geography*. London: Sage, 461–76.

Walford, G. (ed.) (1994) *Researching the Powerful*. London: UCL Press.

Walford, N. (1995) *Geographical Data Analysis*. Chichester: Wiley.

Wallis, R. and Malm, K. (1987) 'The international music industry and transcultural communication', in J. Lull (ed.) *Popular Music and Communication*. Newbury Park, CA: Sage.

Ward, D. (1978) *Cities and Immigrants: A Geography of Change in Nineteenth Century America*. New York, NY: Oxford University Press.

Warde, A. (1985) 'Spatial change, politics and the division of labour', in D. Gregory and J. Urry (eds) *Social Relations and Spatial Structures*. London: Macmillan, 190–212.

Warnke, G. (1987) *Gadamer: Hermeneutics, Tradition and Reason*. Cambridge: Polity Press.

Warren, C. et al. (2000) 'Writing the other, inscribing the self', *Qualitative Sociology*, 23: 183–99.

Watts, M. (1999a) 'Commodities', in P. Cloke et al. (eds) *Introducing Human Geographies*. London: Arnold, 305–15.

Watts, M. (1999b) 'Privatization and governance', *Ethics, Place and Environment*, 2: 256–7.

Watts, M. (2001) '1968 and all that…', *Progress in Human Geography*, 25: 157–88.

Webb, S. and Webb, B. (1932) *Methods of Social Study*. London: Longmans Green & Co.

Weber, M. (1962) *Basic Concepts in Sociology*. London: Peter Owen.

Weber, M. (1978) *Economy and Society: An Outline of Interpretative Sociology*. Berkeley, CA: University of California Press.

Weitzman, E.A. (2000) 'Software and qualitative research', in N.K. Denzin and Y.S. Lincoln (eds) *Handbook of Qualitative Research* (2nd edn). London: Sage, 803–20.

Weitzman, E.A. and Miles, M.B. (1995) *Computer Programs for Qualitative Data Analysis: A Software Sourcebook*. Thousand Oaks, CA: Sage.

Wells, L. (ed.) (1997) *Photography: A Critical Introduction*. London: Routledge.

West, R.C. (1954) 'The term bayou in the United States', *Annals of the Association of American Geographers*, 44: 219–21.

Western, J. (1978) 'Knowing one's place: "the coloured people" and the Group Areas Act', in D. Ley and M. Samuels (eds) *Humanistic Geography: Prospects and Problems*. London: Croom Helm, 297–318.

Western, J. (1981a) *Outcast Cape Town*. Minneapolis, MN: University of Minnesota Press.

Western, J. (1981b) 'Review of *The Human Experience of Space and Place*, ed. Buttimer and Seamon', *Economic Geography*, 57: 275–8.

Western, J. (1999) 'Classics in human geography revisited: Western, J. (1981) *Outcast Cape Town*', *Progress in Human Geography*, 23: 425–7.

Whatmore, S.J. (2002) *Hybrid Geographies: Natures, Cultures and Spaces*. London: Sage.

Whatmore, S.J. and Thorne, L. (1997) 'Nourishing networks: alternative geographies of food', in D. Goodman and M. Watts (eds) *Globalising Food: Agrarian Questions and Global Restructuring*. London: Routledge, 287–304.

Wheeler, K.S. and Harding, M. (eds) (1967) *Geographical Fieldwork*. London: Bland Education.

Wheeler, P.T. (1967) 'The development and role of the Geographical Field Group', *East Midland Geographer*, 4: 185–91.

Whitehead, F. (1987) 'The GRO use of social surveys', *Population Trends*, 48: 45–54.

Whittlesey, D.S. (1927) 'Devices for accumulating geographic data in the field', *Annals of the Association of American Geographers*, 17: 72–8.

Widdowfield, R. (2000) 'The place of emotions in academic research', *Area*, 32: 199–208.

Williams, R. (1979) *Politics and Letters: Interviews with New Left Review*. London: New Left Books.

Williams, R. (1981) *Culture*. London: Fontana.

Williams, R. (1986) *Intermediate Statistics for Geographers and Earth Scientists*. London: Macmillan.

Willis, P. (1977) *Learning to Labour: How Working Class Kids Get Working Class Jobs*. London: Gower.

Willis, P. and Trondman, M. (2000) 'A manifesto for *Ethnography*', *Ethnography*, 1: 5–16.

Wilson, A.G. (1974) *Urban and Regional Models in Geography and Planning*. London: Wiley.

Wilson, A.G. (1981a) *Catastrophe Theory and Bifurcation: Applications to Urban and Regional Systems*. London: Croom Helm.

Wilson, A.G. (1981b) *Geography and the Environment: Systems Analytical Methods*. Chichester: Wiley.

Wilson, A.G. (1986) 'System', in R.J. Johnston et al. (eds) *The Dictionary of Human Geography* (2nd edn). Oxford: Blackwell, 476–7.

Wilson, A.G. (1992) *The Culture of Nature*. Oxford: Blackwell.

Wilson, A.G. and Bennett, R. (1985) *Mathematical Methods in Human Geography and Planning*. Chichester: Wiley.

Wilton, R. (1998) 'The constitution of difference: space and psyche in landscapes of exclusion', *Geoforum*, 29: 173–85.

Winchester, H. (2000) 'Qualitative research and its place in human geography', in I. Hay (ed.) *Qualitative Research Methods in Human Geography*. South Melbourne: Oxford University Press, 1–22.

Winchester, H. and Costello, L. (1995) 'Living in the street: social organisation and gender relations of Australian street kids', *Environment and Planning D: Society and Space*, 13: 329–48.

Wolcott, H. (1990) *Writing up Qualitative Research*. London: Sage.

Wolfinger, N. (2002) 'On writing fieldnotes: collection strategies and background expectancies', *Qualitative Research*, 2: 85–95.

Women and Geography Study Group (WGSG) (1984) *Geography and Gender: An Introduction to Feminist Geography*. London: Hutchinson.

Women and Geography Study Group (WGSG) (1997) *Feminist Geographies: Explorations in Diversity and Difference*. London: Longman.

Wood, D. (1993) *The Power of Maps*. London: Routledge.

Woods, M. (1998) 'Researching rural conflicts: hunting, local politics and actor-networks', *Journal of Rural Studies*, 14: 321–40.

Woods, M. (1999) 'Power, professors and protest: reflections on the RGS–IBG debate and activism in 1990s academe', *Ethics, Place and Environment*, 2: 228–33.

Woodward, R. (1996) '"Deprivation" and "the rural": an investigation into contradictory discourses', *Journal of Rural Studies*, 12: 55–67.

Wooldridge, S.W. (1955) 'The status of geography and the role of fieldwork', *Geography*, XL: 73–83.

Woolf, P. (1988) 'Symbol of the Second Empire: cultural politics and the Paris Opera House', in D. Cosgrove and S. Daniels (eds) *The Iconography of Landscape*. Cambridge: Cambridge University Press, 214–35.

Wright, G. (1987) *Gadamer: Hermeneutics, Tradition and Reason*. Cambridge: Polity Press.

Wright, J.K. (1947) 'Terrae incognitae: the place of the imagination in geography', *Annals of the Association of American Geographers*, 37: 1–15.

Wright, M. (1997) 'Crossing the factory frontier: place and power in the Mexican maquiladora', *Antipode*, 29: 278–302.

Wrigley, N. (1976) *Introduction to the Use of LOGIT Models in Geography (CATMOG 10)*. Norwich: GeoAbstracts.

Wrigley, N. (1985) *Categorical Data Analysis for Geographers and Environmental Scientists*. London: Longman.

Wrigley, N. and Bennett, R. (eds) (1981) *Quantitative Geography: A British View*. London: Routledge & Kegan Paul.

Yates, E.M. and Robertson, M.F. (1968) 'Geographical field studies', *Geography*, LIII: 53–67.

Youngman, M. (1986) *Analysing Questionnaires*. Nottingham: School of Education, University of Nottingham.

Index